Biogeography of Australasia

A Molecular Analysis

Over the last decade, molecular studies carried out on the Australasian biota have revealed a new world of organic structure that exists from submicroscopic to continental scale. Furthermore, in studies of global biogeography and evolution, DNA sequencing has shown that many large groups, such as flowering plants, passerine birds and squamates, have their basal components in this area.

Using examples ranging from kangaroos and platypuses to kiwis and birds of paradise, the book examines the patterns of distribution and evolution of Australasian biodiversity and explains them with reference to tectonic and climatic change in the region. The surprising results from molecular biogeography demonstrate that an understanding of evolution in Australasia is essential for understanding the development of modern life on Earth.

A milestone in the literature on this subject, this book will be a valuable source of reference for students and researchers in biogeography, biodiversity, ecology and conservation.

Michael Heads is a Research Associate at the Buffalo Museum of Science, Buffalo, New York, USA. He is also an independent scholar living in New Zealand. He has carried out most of his field work in rainforest and in alpine areas and authored over 70 publications in the area of biogeography and taxonomy, including his most recent book, *Molecular Panbiogeography of the Tropics* (2012; University of California Press).

Biogeography of Australasia

A Molecular Analysis

MICHAEL HEADS

Buffalo Museum of Science, Buffalo, NY, USA

CAMBRIDGE
UNIVERSITY PRESS

University Printing House, Cambridge CB2 8BS, United Kingdom

Published in the United States of America by Cambridge University Press, New York

Cambridge University Press is part of the University of Cambridge.

It furthers the University's mission by disseminating knowledge in the pursuit of education, learning and research at the highest international levels of excellence.

www.cambridge.org
Information on this title: www.cambridge.org/9781107041028

© Cambridge University Press 2014

First published 2014

Printed in the United Kingdom by TJ International Ltd. Padstow Cornwall

A catalogue record for this publication is available from the British Library

Library of Congress Cataloguing in Publication data
Heads, Michael J.
Biogeography of Australasia : a molecular analysis / Michael Heads, Buffalo Museum of Science, Buffalo, NY, USA.
  pages   cm
Includes bibliographical references and index.
ISBN 978-1-107-04102-8 (hard back)
1. Biogeography – Australasia.   2. Biology – Australasia – Classification – Molecular aspects.
3. Species – Australasia.   I. Title.
QH196.8.H43   2014
577.2′2099 – dc23   2013016825

ISBN 978-1-107-04102-8 Hardback

Contents

Preface

The theme of this book is the distribution of plants and animals in Australasia, the region made up of Australia and the larger islands that fringe its east coast – New Zealand, New Caledonia and New Guinea. Geographical information on particular clades (taxonomic groups) has been collated for this book from molecular studies carried out over the past decade or so. The molecular revolution has revealed a whole new world of beautiful, intricate structure in nature, just as the microscope and the telescope did four centuries ago. Sequencing studies are providing a fabulous new wealth of data on geographical distribution in all kinds of organisms and from an intercontinental scale down to a local level.

Ever since the region was discovered by the outside world, biologists have had a special interest in the flora and fauna of Australasia. The habitats range from subantarctic islands to alpine peaks, deserts, hot, humid forests and coral reefs, and Australasia has some of the most unusual plants and animals in the world. These give a different perspective on the 'normal' groups found in most places. For example, the New Caledonian plant *Amborella* is the sister-group of all other flowering plants, the New Zealand tuatara is the sister of all the lizards and snakes, and the New Zealand wrens (Acanthisittidae) are the sisters of all other perching birds. Many other groups show similar patterns and this book examines possible explanations for this.

Most of the studies discussed here are based on molecular evidence, as it allows such great phylogenetic and biogeographic resolution. On the other hand, traditional taxonomic revisions based on morphology often examine larger samples and suggest relationships for even the rarest and most inaccessible populations. Morphological studies can shed light on detailed aspects of distribution and are cited below for some areas or groups in which molecular surveys have not yet been carried out.

The work discussed here has been published in exemplary accounts based on well-conceived surveys, using the latest techniques. Nevertheless, there are often geographical or phylogenetic gaps in the sampling, the resolution of the phylogenies is sometimes incomplete and statistical support for the clades varies. Other important aspects of the studies include the parts of the genome sequenced, the methods used to establish sequences and the methods used to construct a phylogeny from the sequence data. Information on all of these is available in the original papers and the details are not referred to here. Instead, the focus is on the results – the geographical patterns – that have been obtained so far.

The phylogenies represent interim hypotheses and will change, to some extent, with further work. But although the situation is still fluid, in many groups the phylogenies are stabilizing in a remarkable way and well-supported clades with coherent distributions are emerging. How can this intriguing new information best be put to use? Can it help answer the many unresolved questions about evolution and ecology?

This book describes and illustrates some of the main distribution patterns involving Australasia and also explores ideas on their historical development. Although fascinating new phylogenies are now appearing almost daily, the interpretation of the distributions is lagging far behind the descriptive studies; authors still seem more concerned with producing new results than explaining them. Most of the interpretations that are being offered are based on old concepts inherited from the Modern Synthesis or on minor variations of these.

Instead of relying on these approaches, the method of interpretation adopted here is that of panbiogeography, a synthesis of plant geography, animal geography and geology (Craw *et al.*, 1999; Heads, 2012a). This method dissects the geographic patterns of molecular groupings, compares them with patterns in other groups, and synthesizes the results with current ideas on Earth history. Mapping is fundamental in this process. In geology, maps have been an integral part of the discipline since its origin and a mapping project is a standard component of first-year university courses. It is impossible to imagine a regional geological study without maps. Yet distribution maps did not become the norm in biological studies of groups, floras and faunas until the 1960s; even now, many taxonomic monographs are published without any distribution maps at all. The situation is changing, though, and biologists are taking a much more active interest in distribution. This is because molecular studies have found so much impressive, unexpected geographic structure in most groups. In many cases the molecular groupings show a much closer relationship with geography than with morphology, or at least with traditional interpretations of morphology. Increasing numbers of papers in molecular phylogenetics are including distribution maps, as the clear-cut patterns are among their most interesting results.

The distribution patterns are compared here with the underlying tectonic developments in Earth's history and with the fossil record. The fossil record is useful for dating, but most groups have no fossils and even in groups with many fossils, these can only give minimum ages for clades – the actual age can be much older. Panbiogeography estimates the actual ages of groups by relating distributions with spatially associated tectonic and climatic events, such as the last, great rises in sea level in the Cretaceous period. Many authors are now abandoning the fossil record as a source of maximum clade ages and are instead calibrating phylogenies with single tectonic events. Panbiogeography extends this approach by employing multiple tectonic correlations to calibrate many nodes on a phylogeny. This approach involves a broad engagement of biogeography and tectonics.

The results in the proposed model are unexpected, as the groups are inferred to be much older and long-distance dispersal much less significant than has been thought. Yet the new model offers distinct advantages, as it does not rely on chance events to explain distribution patterns and instead provides a coherent synthesis of phylogeny and Earth history. If distributions were the result of chance dispersal, each group would have a different, idiosyncratic pattern and there would be little overall structure within or among groups. Instead, most distributions are shared by many unrelated plants and animals that often differ in their ecology and means of dispersal. This suggests that the repeated phenomena are probably the result of general causes, such as geological or climatic change. As the geographic patterns continue to be investigated with molecular work they are becoming clearer and, in addition, the same distributions and breaks are recurring in different groups. As these results accumulate, chance dispersal appears less and less likely to be a general explanation for biogeography.

The relationship between biogeography and ecology has often been problematic, but there are hopeful signs of a new integration of the two fields. Ricklefs and Jenkins (2011) wrote: 'The schism between ecology and biogeography possibly peaked during the 1970s, soon after Robert MacArthur (1965, 1972) explicitly excluded history from the purview of ecology . . . one could argue that ecologists further weakened the study of biogeography through the development of the equilibrium theory of island biogeography, which was essentially nonhistorical . . . Only after the general acceptance of plate tectonics in the 1960s and the development of increasingly analytical approaches to studying geographical distributions, such as panbiogeography (Croizat, 1958), vicariance biogeography (Wiley, 1988), analytical biogeography (Myers and Giller, 1988) and areography (Rapoport, 1982), did biogeography experience a resurgence that eventually commanded the attention of ecologists . . .'.

If a group's distribution represents inherited information, this can shed light on the group's ecology and its evolution. This book discusses many of the interactions among tectonics, biogeography and ecology; two of the main processes are passive uplift of populations during orogeny and stranding of coastal groups inland following retreat of marine incursions.

No two groups have identical distributions, but many share similar, distinctive features, including their main phylogenetic and geographic breaks or nodes. The location of these nodes in Australasia, their relationships with other nodes and their development in space and time are the main topics explored in this book.

Acknowledgements

I am very grateful for the help and encouragement that I have received from friends and colleagues, especially Lynne Parenti (Washington, DC), John Grehan (Buffalo), Isolda Luna-Vega and Juan Morrone (Mexico City), Jürg de Marmels (Maracay), Amparo Echeverry and Mauro Cavalcanti (Rio de Janeiro), Guilherme Ribeiro (São Paulo), Jorge Crisci (Buenos Aires), Andres Moreira-Muñoz (Santiago), Pierre Jolivet (Paris), Alan Myers (Cork), Robin Bruce (London), Pauline Ladiges and Gareth Nelson (Melbourne), Malte Ebach, Tony Gill and David Mabberley (Sydney), Rhys Gardner (Auckland), Karin Mahlfeld and Frank Climo (Wellington), Robin Craw and Bastow Wilson (Dunedin), and Brian Patrick (Alexandra). I also thank the production team – Dominic Lewis, Jo Breeze, Renee Duncan-Mestel, Lynette Talbot, Ilaria Tassistro and Sarah Beanland – for their friendly and efficient collaboration.

1   The spatial component of evolution

Molecular studies have documented high levels of geographic structure in most plant and animal groups and this finding has fundamental implications for the science of biogeography. The fascinating depth and detail of the geographic structure, together with the repetition of patterns in unrelated groups, all threaten to undermine the main theory of twentieth century biogeography – speciation by chance dispersal. Ever since the influential study by Matthew (1915), dispersal theory has dominated the field. The idea that dispersal is a mode of speciation and not just the movement of organisms is still often accepted as a fact, even in molecular work. Biologists producing molecular phylogenies sometimes suggest that their work has 'revealed' chance dispersal, but in a scientific study it is the facts that everyone agrees on – not inferences and interpretations – that are revealed. The molecular studies have indeed revealed spectacular distribution patterns that can be accepted as 'facts', at least for the purposes of discussion, but the underlying causes of these facts are much less obvious.

The distribution patterns shown in the new molecular work are interpreted here with reference not to chance, but to a small number of geological and climatic revolutions. These include pre-breakup tectonics in Gondwana (extension and magmatism), the rifting of Gondwana, and the last, great marine flooding in the mid-Cretaceous, probably caused by a combination of tectonic and climatic events. This approach does not deny that chance dispersal exists, but the focus here is on repeated patterns, not on idiosyncratic distributions found in only one or a few groups.

In the Modern Synthesis, chance dispersal is the main mode of speciation. Following its origin by dispersal across a barrier, a group then attains its distribution by a second dispersal, a range expansion outwards. In contrast, during a vicariance event different clades develop from a common ancestor that was already widespread, and each new clade originates at the same time as its distribution, more or less *in situ*. The debate concerning the relative importance of dispersal and vicariance had begun by the end of the Enlightenment, around the same time that evolutionary theory was introduced (White, 1789 [1977: 65]; Willdenow, 1798: 430 [as quoted in Weimarck, 1934]). Since then, ideas on evolution have changed in many ways, but the vicariance versus dispersal debate has continued, unresolved, to the present day.

Most writers over the millennia have assumed that distribution patterns are caused by physical movement – 'dispersal'. All organisms move, and, with a few exceptions, every individual organism has moved to its current location. In addition, while the distribution of a clade is represented on a map with dots or lines, it is in fact mobile,

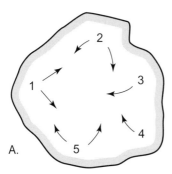

Fig. 1.1 The distribution of a hypothetical group, A. Its centre of origin might occur in the area with a highest diversity in the group (1), in the region of the oldest fossil (2), in the area of the most 'advanced' form (3), in the area of the most 'primitive' form (4), or in the area of the basal group (5). (From Michael Heads, *Molecular Panbiogeography of the Tropics*. ©2012 by the Regents of the University of California. Republished by permission of the University of California Press.)

or at least at a dynamic equilibrium. Likewise, if evolution involved only vicariance, every small area on the Earth's surface would have only one type of organism and it would be endemic there. This is far from the case and so, again, there must have been movement. Yet although physical movement is universal in all (or almost all) individuals and has occurred in many clades, it does not account for all aspects of distribution. Many phylogenetic/geographic breaks have instead resulted from cessation of movement or at least a decrease in the rate of interchange among populations.

The question is not so much whether dispersal or vicariance has occurred, but whether dispersal and speciation are the result of chance movement of individual organisms or general, underlying causes such as geological change. 'Dispersal biogeography' is a research programme that explains the geographical distribution of organisms based on processes of dispersal, while 'vicariance biogeography' explains distribution with reference to geological events (Gillespie *et al.*, 2012). In vicariance theory, range expansion and speciation (vicariance) are both caused by geological change (for example, marine incursion or mountain building). In dispersal theory, neither is – range expansion and speciation with long-distance dispersal occur in different groups at different times and are both due to chance.

Models of spatial evolution in biogeography

The centre of origin model

If a hypothetical group of plants or animals, A, has a geographical distribution as shown in Figure 1.1, how did the distribution develop? Most authors have accepted that the group originated in a restricted area and migrated outwards from there. Researchers locate this 'centre of origin' or 'ancestral area' using different criteria (Fig. 1.1). For a particular group, different methods locate the centre of origin in the area of: maximum

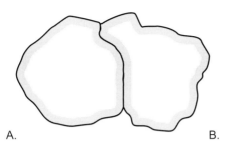

A.   B.

Fig. 1.2 Distribution of group A and its sister-group, B, two allopatric clades. (From Michael Heads, *Molecular Panbiogeography of the Tropics*. ©2012 by the Regents of the University of California. Republished by permission of the University of California Press.)

diversity of forms, the oldest fossil, the most 'advanced' form (Darwin, 1859; Briggs, 2005), the most 'primitive' form (Mayr, 1942; Hennig, 1966), or, in most modern studies, the 'basal' member of the group.

Bremer (1992) justified the search for a restricted centre of origin, reasoning that most modern species of Asteraceae, for example, are endemic to a single continent, and so the ancestor of the group would have been the same. Yet this denies the possibility of evolution in aspects of the group and there is no need to assume that ancestral taxa had the same ecology, range size, variability, or evolvability as their modern descendants. A centre of origin approach has been continued by modern authors, and programmes designed to find the centre of origin are popular. These include DIVA ('dispersal–vicariance'; Ronquist, 1997) and DEC (dispersal–extinction–cladogenesis analysis) implemented in LAGRANGE (Ree and Smith, 2008). In certain cases DIVA will indicate widespread ancestors, contradicting the core assumption of the centre of origin approach and instead supporting vicariance. DEC, in contrast, is guaranteed to locate a centre of origin (Clark *et al.*, 2008).

The vicariance model: differentiation of a widespread ancestor

In an alternative to the centre of origin model, the group A can be considered not on its own, but together with its closest relative, or sister-group, B (Fig. 1.2). In this pattern, analysed samples from different localities in the respective areas of A and B show a phylogeny: (A^1, A^2, A^3 . . .) (B^1, B^2, B^3 . . .). In other words, groups A and B show reciprocal monophyly; they are sister-groups and neither is related to any particular population of the other. In Figure 1.2 the distributions of the two sister-groups show precise allopatry. The pattern is compatible with each of the two groups having arisen not by spreading out from a point, but by geographic (allopatric) differentiation *in situ*. In this model, each group has evolved in its respective area, from a widespread ancestor that occurred throughout the areas of both A and B. This process, vicariance, is dominated by differentiation between the two populations; any physical movement of individuals between them is of less importance. In vicariance analysis, the aim is to find the originary break *between* the groups, not a point centre of origin *within* a group. The focus is not

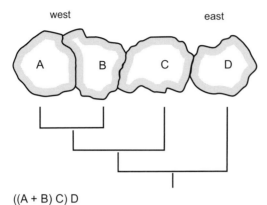

((A + B) C) D

Fig. 1.3 Four groups A–D with allopatric distributions in areas A–D. The phylogeny of the four groups is also shown. (From Michael Heads, *Molecular Panbiogeography of the Tropics*. ©2012 by the Regents of the University of California. Republished by permission of the University of California Press.)

on the group itself or on details of its internal geographic and phylogenetic structure, but on its geographic and ecological relationship with its sister. In a vicariance model, many younger groups have a distribution that is more or less the same as the 'centre of origin'.

In a phylogenetic tree diagram, two sister clades are shown joined at a node. A node is also expressed on the Earth's surface as a geographic boundary between two or more distributions. In a vicariance interpretation, a node does not represent a centre of origin or an ancestor, but a phylogenetic and biogeographic break or division, where two or more related groups have diverged from a common ancestor.

Origin of the ancestor

If the two descendant groups, A and B in Figure 1.2, originated by vicariance, what about their widespread ancestor? Was its wide range achieved by dispersal? That would be one possibility, but the ancestral group itself could have originated as an allopatric member of a widespread complex (Fig. 1.3). In this case, again, the group did not attain its distribution by spreading out from a localized centre of origin. The same pattern of allopatry can extend to additional relatives in other areas, and widespread or even global series of allopatric forms occur in many plants and animals. In contrast, in the centre of origin theory, each of the four allopatric groups in Figure 1.3 has a separate centre of origin within the group's area and the distributions have no direct relationship with the groups' origins – the groups form first and their distributions develop later. In this approach, the mutual boundaries of the four groups in the figure are secondary; the ranges only meet after the four individual groups each spread out from their respective centres of origin. Instead, in vicariance theory, the mutual boundaries are attributed to phylogeny. These breaks or nodes tend to recur at the same localities in many different

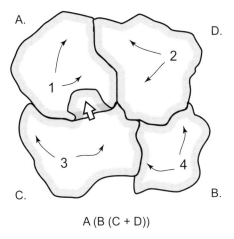

A (B (C + D))

Fig. 1.4 Four groups A–D with allopatric distributions in areas A–D. The phylogeny of the four groups is also shown. Numbers and black arrows indicate traditional centres of origin and dispersal routes. White arrow at zone of overlap indicates local range expansion in clade C. (From Michael Heads, *Molecular Panbiogeography of the Tropics*. ©2012 by the Regents of the University of California. Republished by permission of the University of California Press.)

groups, often with different ecology and means of dispersal, and so a chance explanation is unlikely.

If the groups in Figure 1.3 have a phylogeny: D (C (B + A)), as indicated in the figure, this represents a simple geographic sequence of differentiation in a widespread ancestor that has proceeded from east to west: D versus the rest, then C versus A + B, then A versus B. Often there is no simple geographic progression in the differentiation sequence, and this is shown in Figure 1.4. The groups have a phylogeny: A (B (C + D)) and so the sequence of differentiation 'jumps' from the first node (between A and the rest) to the second (between B and C + D) and then back to the middle (between C and D). The allopatry has developed by normal vicariance in a widespread ancestor. Localized centres of origin for each of the four clades (1–4 in Figure 1.4) and dispersal from these (black arrows) are not necessary.

Subsequent range expansion leading to geographic overlap

The overlap between groups C and A in Figure 1.4 (white arrow) is the result of range expansion of C after its origin. This has developed as a secondary process following the origin of the four groups by vicariance and allopatry. Simple range expansion occurs by the group's normal, observed means of dispersal and does not involve any phylogenetic differentiation. More extensive overlap is discussed further below.

Allopatric groups that show reciprocal monophyly

One common distribution pattern involves a group that occurs on a continent or mainland, A, and its sister-group that occurs on a smaller, offshore island, B (Fig. 1.5). The island

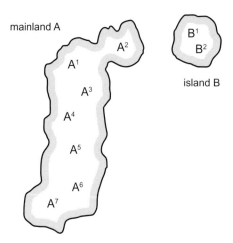

Fig. 1.5 Distribution of two sister-groups, A and B. The phylogeny is: $(A^1 + A^2 + \ldots A^7) (B^1 + B^2)$. (From Michael Heads, *Molecular Panbiogeography of the Tropics*. © 2012 by the Regents of the University of California. Republished by permission of the University of California Press.)

group is often assumed to have been derived from the mainland group by dispersal. This hypothesis predicts that island forms will be related to particular populations in their large sister-group, representing source regions. Yet well-sampled molecular studies show that in many of these cases the phylogeny instead has the pattern: $(A^1, A^2, A^3 \ldots)$ $(B^1, B^2, B^3 \ldots)$, with the superscripts indicating the different sample sites within areas A and B. The two groups in the two areas show reciprocal monophyly and the group in B is not related to any one population in A. In this type of pattern, dispersal can still be proposed as an explanation, but only if it occurred just once, and prior to any other differentiation in the group, and this is often unlikely. On the other hand, reciprocal monophyly is the standard signature of simple vicariance of a widespread ancestor at a break between A and B. Even if groups in the two areas A and B do not show reciprocal monophyly, a vicariance origin is still possible and this is discussed next.

'Basal' groups and centres of origin

A clade often has its basal phylogenetic break between a small group, with just a few species, and its more diverse sister-group (cf. Fig. 1.5). The smaller group is termed 'basal', although in fact only the nodes are basal; no group is more or less basal than its sister. 'Basal group' is a useful term for a smaller sister-group, but there is no reason to assume that a basal group is more primitive than its sister, or is the ancestor of its sister. The basal clade in a group is often interpreted as occupying the centre of origin for the group, but, again, this assumption is not warranted. There is no reason why the basal group, rather than its sister, should represent the centre of origin, the time of origin, the ancestral ecology or the ancestral morphology of the group as a whole. The focus instead should be on the phylogenetic and geographic details of the break between the two sister-groups.

Interpreting phylogenies as sequences of vicariance events, rather than sequences of dispersal events

In the Modern Synthesis, the sequence of nodes in a phylogeny is read as a sequence of dispersal events. The centre of origin is occupied by the basal group, with subsequent taxa invading a new region, differentiating and adapting, and then invading the next region. Instead, as indicated above, the sequence of nodes can represent a series of differentiation events in an ancestral complex that was already widespread.

Groups with a basal grade in one region

In many groups the subgroups show simple allopatry (Fig. 1.3) or minor overlap (Fig. 1.4). Other groups show more extensive overlap. In one common pattern, groups have a *basal grade*, a paraphyletic group, in one area, A, and a disjunct group in another area, B (Fig. 1.6). In centre of origin theory, the phylogeny of this pattern, for example, A (A (A (A+B (A)))), is explained as the result of dispersal within the clade, from A to B. Nevertheless, this cannot be assumed, as the ancestor may have already been widespread in both A and B (Fig. 1.6A). Allopatric differentiation around a node in A (Fig. 1.6B, C), and extinction of populations between A and B leads to a basal grade in area A (Fig. 1.6D). Subsequent local overlap can occur in area A, but even if the clades develop substantial overlap they often show slight but significant differences in their distribution (Fig. 1.6D) and these may represent traces of the original allopatry. Many biogeographic studies code all the species in areas equivalent to A in the same way, lumping them together and overlooking minor but critical differences.

The origin of the whole clade shown in Figure 1.6 is a separate question from the origin of the differentiation within it, and is not discussed here. The origin of a group cannot be deduced from the phylogeny or biogeography of the group itself; it requires comparison with the group's relatives.

A vicariance analysis of the pattern with a basal grade in one region, shown in Figure 1.6, infers local overlap within A by normal means of dispersal. Instead, a dispersal analysis suggests long-distance dispersal from A to B, often by unusual or extraordinary means. An actual example of the hypothetical pattern shown in Figure 1.6 is the subtribe Arctotidinae in the plant family Asteraceae. The group comprises a basal grade of three southern African clades, a clade including *Cymbonotus* of southern Australia, and its sister-group, a diverse southern African clade (Fig. 1.7; Funk *et al.*, 2007; McKenzie and Barker, 2008). Note that *Cymbonotus* is embedded in an Australian–southern African clade, not in a southern African clade, and the Australian representation is not just a secondary, derivative outlier. The five main clades are all present in Africa but it is likely that there are significant differences in the distributions.

Figure 1.6 suggests that in a simple, common case of evolution, the location of a basal grade will not indicate a centre of origin; this argument also applies to ecology and the idea of ancestral habitat. For example, an ecological phylogeny for a group: (freshwater (freshwater (freshwater (saline lakes)))), will not always mean that the saline lake clade was derived from freshwater habitat. The ancestor of the whole group may have already

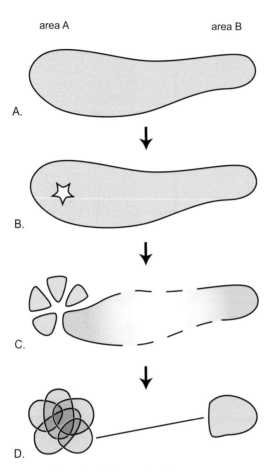

Fig. 1.6 A hypothetical example of evolution in a taxon found in two areas. (A) A widespread ancestor. (B) The ancestor begins to differentiate around a node (star) associated with the formation of a mountain range or inland sea, for example. (C) The ancestor has differentiated into five allopatric clades, four with a narrow range and one widespread. Their ranges begin to overlap while some of the populations of the widespread clade suffer extinction (broken line). (D) The clades now overlap but the ranges still show traces of their original allopatry. Following extinction of populations between areas A and B, the outlier in area B may appear to be a secondary feature and the result of long-distance dispersal from area A. (Reprinted from Heads, 2009b, by permission of John Wiley & Sons.)

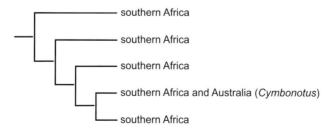

Fig. 1.7 The phylogeny and distribution of clades in the Arctotidinae (Funk *et al.*, 2007; McKenzie and Barker, 2008). (Reprinted from Heads, 2009b, by permission of John Wiley & Sons.)

occupied both freshwater and saline habitat before the differentiation of the groups, and this is even more likely if the ancestor was widespread.

A critique of the popular program DIVA admitted that it 'reconstructs histories accurately when evolution has been simple; that is, where speciation is driven mainly by vicariance' (Kodandaramaiah, 2010). But for a simple area phylogeny: (A (A (A, B))), programs such as DIVA and DEC (in LAGRANGE) will always find a centre of origin in A and dispersal from A to B, even if the ancestor in fact occurred in both A and B, and evolution proceeded entirely by vicariance (in area A). In this case the only dispersal was local range expansion within area A (Fig. 1.6).

The DEC model (Ree and Smith, 2008) assumes that if an ancestor is widespread across two or more areas, lineage divergence can only occur between a single area and the rest of the range, or between a single area and the entire range. It does not include a mechanism for the subdivision, in a single step, of a widespread ancestral range (comprising more than three unit areas) into two allopatric daughter ranges each comprising more than one unit area. The model requires secondary dispersal and extinction events to explain this case of normal vicariance. Many DEC analyses assume that ancestral range of a group should be similar in spatial extent to those of the living species, and this rules out a simple vicariance origin for allopatric species (see Heads, 2012a, Chapter 6).

In a valuable critique, Arias *et al.* (2011) observed that programs such as DIVA and DEC are based on discrete, pre-defined areas and that the definition of the areas, carried out prior to analysis, is far from trivial. Arias *et al.* instead suggested switching the focus away from relationships among pre-defined areas, onto the actual geographic breaks among the clades. This approach is adopted here.

Groups with a basal grade in one region and widespread apical clades

Asteraceae

Asteraceae, with ~24 000 species, is the largest plant family. The five basal branches in the family are all small groups with all or most of their members restricted to South America, while one of the apical clades is widespread (Fig. 1.8; Panero and Funk, 2008). Instead of a centre of origin in South America, the pattern suggests differentiation of a worldwide ancestor mainly at breaks in what is now South America. Following the differentiation, the diverse worldwide clade has expanded its range to include most of South America, overlapping with the basal clades there.

While the family is very large, the number of global clades in the distal group is quite small (several of the larger tribes and subtribes, some large genera). This means that through the history of the family, only a small number of widespread ancestors have existed; large groups in the family, such as the tribes Senecioneae and Astereae, each require their own global ancestor. These few ancestors have each undergone a phase of mobilism and range expansion during which they each occupied much of Earth's land surface, before settling down into a subsequent phase of immobilism and differentiation.

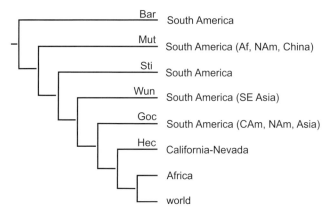

Fig. 1.8 Phylogeny of Asteraceae (Funk *et al.*, 2005; Panero and Funk, 2008). Areas listed in brackets have much lower levels of diversity (Af = Africa, CAm = Central America, NAm = North America). Subfamily abbreviations: Bar = Barnadesioideae, Mut = Mutisioideae, Sti = Stifftioideae, Wun = Wunderlichioideae, Goc = Gochnatioideae, Hec = Hecastocleidoideae. (Reprinted from Heads, 2009b, by permission of John Wiley & Sons.)

A worldwide bee: *Hylaeus*

The colletid bee genus *Hylaeus* has about 600 species, with 170 in Australia. Kayaalp *et al.* (2013) sampled 74 species and found the following phylogeny (Aus = Australia):

(Aus and New Zealand (Aus (Aus (Aus (Aus (Aus (world except Aus)))))))

The overall distribution conforms to the pattern shown in Figure 1.6 and can be explained by vicariance in a worldwide group, followed by overlap (dispersal) within Australia. The non-Australasian samples make up a single, allopatric clade and Kayaalp *et al.* (2013) described this as 'particularly striking'. In a vicariance model it is a case of simple vicariance but the authors adopted a dispersal/radiation model with a centre of origin in Australia and they discussed some of the questions this raises. For example, why has there been no back-migration of the worldwide clade into Australia, despite the convergence of Laurasia and Australia through the Cenozoic? Kayaalp *et al.* (2013) cited home advantage, which seems reasonable and also applies to a vicariance model. The authors also discussed another conundrum: 'what kinds of biogeographical and ecological factors could simultaneously drive global dispersal [in the single widespread clade], yet strongly constrain further successful migrations out of Australia when geographical barriers appear to be weak?' Kayaalp *et al.* argued that initial dispersal into new niches and enemy-free spaces would have been possible, but that subsequent dispersal by other lineages into the same areas would have been prevented because no further niche-space was available. Nevertheless, the other clades of *Hylaeus* show extensive overlap within Australia (along with some interesting differences; *Atlas of Living Australia*, 2012), and this contradicts the argument.

Kayaalp *et al.* (2013) concluded that: 'Biotic composition of a region may be determined more by the *early appearance of dispersal opportunities* than by later relaxation of geographic dispersal barriers.' [Italics added.] This seems reasonable and if the dispersal in *Hylaeus* or pre-*Hylaeus* took place early enough – before the seven modern clades began to differentiate – the differentiation itself could have been by vicariance. This model does not require speciation by chance dispersal. Instead, it proposes dispersal before the divergence of the seven clades, and after (within Australia), but not during divergence.

Overlap in distribution

The geographic allopatry between close relatives is often not perfect and there can be interdigitation or marginal overlap. More extensive overlap can develop over time, with most traces of allopatry obscured by range expansion. In most cases overlap is due to dispersal – secondary range expansion that has occurred after the group's origin. Widespread overlap between sister-groups is more common at higher taxonomic levels; for example, most orders and many families are worldwide and overlap more or less everywhere. This sort of overlap means that vicariance cannot be the only biogeographic process, as mentioned above. If it were, each area on Earth would have only one, endemic life form. In fact, most places have many kinds of plants and animals, and the overlap indicates phases of range expansion.

Phases of population mobilism leading to range expansion and colonization of new habitat are likely to have occurred, for example, along the extensive coastlines that formed during the vast marine transgressions of the Mesozoic. These floods, caused by tectonics and global warming, occurred on all the continents around the time of Gondwana breakup. At the peak of the Cretaceous transgression, one-third of Earth's present land area was submerged (Dixon *et al.*, 2001). Along with the extinction of many groups, weedy, pioneer taxa would have thrived in the new, marginal habitats. During this phase of mobilism, whole communities – not just individual clades – would have occupied new maritime habitat and become widespread. With the retreat of the seas, many of these groups would have been left, stranded inland, in the centres of North and South America, Africa, Europe, Asia and Australia.

In the model adopted here, mobilism (leading to range expansion and overlap) and immobilism (a prerequisite for vicariance) are both caused by geology, not by chance movement of individual organisms in each individual clade. Some groups never undergo range expansion; in others, range expansion only takes place once or a few times in tens of millions of years. Any range expansion is achieved by the group's everyday means of dispersal, not the unusual means that are often proposed for speciation dispersal. All groups have effective means of dispersal or they would not survive, but most clades, especially lower rank groups, tend to form allopatric components of vast, intercontinental mosaics and show little sign of active range expansion. A group's normal, ecological means of dispersal – ordinary means of survival – are only active in range expansion during particular times in history, and these phases of mobilism are caused by large-scale tectonic, climatic or anthropogenic changes.

Range expansion ('dispersal' in one sense) is not due to chance – very unusual or unique events that occur in a single, individual organism – but to events in Earth history. This normal process of dispersal involves many members of a community during a *general* phase of mobilism or 'geodispersal' (Lieberman, 2000). This last term was coined to contrast the process with 'chance dispersal', which proposes that single lineages have single founders (either an individual or small population). During a phase of mobilism, groups that were localized and more or less immobile can act as aggressive weeds, crossing 'barriers' and invading large areas of the world in a brief moment of geological time. In contrast, during a phase of immobilism, clades of even the most mobile creatures, such as sharks, sea birds, or whales, can maintain the same distribution, at least in their breeding sites, and so the population structure can be disrupted even by small-scale physiographic changes.

The four processes proposed in biogeography and the two that are accepted here

Differentiation (e.g. speciation) can be explained as the result of either *vicariance* of a widespread ancestor (dichopatric speciation) or *founder dispersal* (= chance dispersal, = peripatric speciation) from a centre of origin. In addition, overlap can be explained as the result of range expansion (by *normal ecological dispersal*, simple physical movement) or by *sympatric differentiation*.

Of the four processes just cited, **vicariance** and **normal ecological dispersal** are accepted as important by all authors. They are the two processes that are accepted here in explaining distributions. Normal ecological dispersal can involve movement within the distribution area or outside it, and in some cases the latter will lead to range expansion.

Sympatric differentiation was controversial but is now accepted in some groups. It is probably quite rare though. Low-level clades are often allopatric with their sisters, whereas higher level clades show more overlap. This general pattern of increasing sympatry with relative node age indicates that overlap of sister-groups is a secondary process that has developed over time, obscuring the original allopatry, rather than by sympatric evolution (Heads, 2012a). Many supposed cases of sympatry between sister-groups only involve partial geographic overlap and the groups show significant allopatry. Likewise, 'sympatric' sister species on individual islands (and in 'habitat islands' such as lakes) have been attributed to sympatric speciation, but they probably represent secondary overlap of different biogeographic elements. The sympatric speciation proposed for some plants on Lord Howe Island, between Australia and New Zealand, is discussed in Chapter 2.

The fourth process, *speciation by founder dispersal*, is controversial. It explains geographic distribution by chance – extraordinary events that are proposed to happen only once in tens of millions of years. For example, based on a phylogeny, a small, non-flighted forest arthropod that usually moves just metres in its life might be inferred to have 'jumped' – just once in its 90-million-year history – 10 000 km across the Pacific from Australasia to South America. Chance dispersal events do not correlate with any other physical or biological factor and can explain any distribution, but they are also difficult to investigate or test. Chance dispersal may occur, but it has not been necessary to invoke it for any of the Australasian distributions examined in this book.

Ecological speciation

Many authors, including Darwin (1859), have stressed that regions with similar environments often have very different biotas. This indicates that factors of the present environment – ecology – cannot explain these phylogenetic and structural differences, which instead are caused by biogeographic factors operating at larger scales of space and time. Nevertheless, some modern authors attribute great significance to ecology and have rejected allopatric differentiation with subsequent overlap, in favour of sympatric, ecological speciation. Of course, the plants and animals in a region sort themselves out into the different ecological habitats that are available, but this does not necessarily mean they have evolved by ecological speciation (Heads, 2012a, Chapter 1). Ecological, non-geographic speciation would lead to sister-species with different ecology and with geographic distributions that not only overlap, but are more or less the same. Within a biogeographic region, for example, an area of endemism in hill forest 50 kilometres across, one endemic species may occur on ridges, while its sister could be endemic to the same region but live in valleys. Pairs like this exist, but are very uncommon. In most cases there is a significant degree of geographic allopatry between two overlapping sister-species and this can be assumed to represent the original, allopatric break (whether this is caused by chance dispersal or vicariance).

Case-studies in and around Australasia

Ampelopsis and allies in the grape family (Vitaceae)

Ampelopsis and its allies are vines in the grape family and are widespread globally. The group has the following phylogeny (Nie *et al.*, 2012):

Northern hemisphere (Asia and SE United States). (*Ampelopsis* in part.)

Northern hemisphere (Europe, Asia and SE United States). (*Ampelopsis* in part.)

Southern hemisphere: Africa (*Rhoicissus*), sister to west and east Australia (*Clematicissus*) + South America (central Chile to Colombia and Brazil; a clade of three '*Cissus*' species).

The first two clades form a paraphyletic basal grade in the northern hemisphere, and so a dispersal model predicts a northern centre of origin, dispersal (once) to the southern hemisphere, then, once, across each of the Indian, Pacific and Atlantic Oceans (Nie *et al.*, 2012). Yet this does not explain the simple allopatry among the three southern hemisphere clades or the allopatry between the second northern hemisphere clade and the southern hemisphere clade. This allopatry at two levels in the hierarchy suggests that the overlap of the first two clades in the northern hemisphere is secondary and, at the time of their origin, the two occupied different parts of the northern hemisphere. The remarkable simplicity of the global pattern indicates that there has not been significant range expansion outside Asia or North America, as this would obscure the pattern. In particular, there has been no successful dispersal in any of the three clades between hemispheres or among continents in the southern hemisphere.

The break between the northern and southern clades (between North America and South America, between Europe and Africa, and between Asia and Australia) corresponds with the global belt of Mesozoic tectonism along the complex of Tethyan basins.

Evolution around the Indian Ocean: differentiation in Poaceae subfamily Danthonioideae

One subfamily of grasses, the Danthonioideae, has always interested biogeographers because of the disjunct, southern hemisphere distribution of its basal members. Humphreys *et al.* (2009) and Linder *et al.* (2010) presented a new phylogeny:

Eastern South Africa (Drakensberg, along the eastern Lesotho border) to Madagascar and Ethiopia (*Merxmuellera*).

Western South Africa ('Cape Floristic Region') (*Capeochloa* and *Geochloa*).

Africa (from the Cape – most species – north to the mountains of Cameroon and Ethiopia) (*Pentameris*).

Tasman basin (New Zealand, Lord Howe I., SE Australia) (*Chionochloa*).

Southern Africa (*Chaetobromus* and *Pseudopentameris*).

World (*Danthonia* and nine other genera).

This can be summarized: Africa (Africa (Africa (Tasman (Africa (world))))). A literal reading of the phylogeny would suggest a centre of origin in Africa, where the three basal clades are endemic, followed by dispersal to the Tasman region (10 000 km), followed by dispersal back again to Africa. This sequence of events seems unlikely, but the type of pattern itself, with alternating centres – in this case, in the south-west Indian and south-west Pacific basins – is common (cf. Fig. 1.7). In a vicariance model, the pattern can be explained by an ancestor that was already global undergoing repeated differentiation around southern Africa, with one differentiation event around Australasia. There is no need to invoke any dispersal across the Indian Ocean in the four basal members of the Danthonioideae, only within Africa. Even within Africa, the allopatry of the first two clades provides good evidence of the original vicariance. The only overlap involves *Pentameris* and so the only range expansion required is in this genus.

In Danthonioideae, the paraphyletic basal grade in the southern hemisphere suggests that the last, worldwide clade listed above in the subfamily began as a widespread northern group and underwent range expansion into the south. Distributions *within* the worldwide clade (which comprises an Indo–Pacific group and its Pacific–Atlantic sister) indicate that this overlap occurred early in the subfamily's history.

In the model proposed here, the worldwide distribution of subfamily Danthonioideae developed before any of the main breaks within it. How did the subfamily itself originate and attain its wide range? Biogeographic analysis of old events that involve extensive global overlap is more difficult than analysis of younger, allopatric groups and is not attempted here. As usual, dissecting out the biogeographic origin of the subfamily would require detailed phylogenetic and geographic comparison with its sister-group.

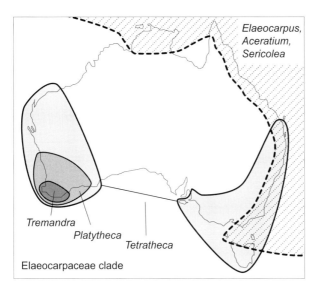

Fig. 1.9 Distribution of Elaeocarpaceae in Australia: *Elaeocarpus*, *Aceratium* and *Sericolea*, and their sister-group, the 'tremands' (Crayn *et al.*, 2006).

The tremands (Elaeocarpaceae): is the centre of diversity and the location of the basal grade a centre of origin?

Two important biogeographic centres occur in south-western and south-eastern Australia. The areas have mesic ecology and are separated by a desert, the Nullarbor Plain. There are many disjunct affinities between the two mesic regions, for example in a clade of Elaeocarpaceae known as the 'tremands' (formerly the Tremandraceae) (Fig. 1.9). The group has a phylogeny (Crayn *et al.*, 2006):

SW Australia (*Tremandra*).

SW Australia (*Platytheca*).

SW Australia and SE Australia (*Tetratheca*).

A western centre of origin and eastward dispersal of *Tetratheca* has been inferred (McPherson, 2006), but the basal paraphyletic grade in the west cannot be assumed to represent a centre of origin (Ladiges, 2006). If the sister-group of the complex as a whole is allopatric with it, an origin by vicariance would be just as likely as origin at a local centre. In fact, the 'tremands' as a whole are sister to a diverse clade (*Elaeocarpus*, *Aceratium* and *Sericolea*) that is widespread along the eastern coast of Australia and the Old World tropics, but is not present in south-western Australia. This suggests that initial vicariance between the two groups occurred somewhere around the region of overlap, in south-eastern Australia, and around a line through central–north-western Australia. This was followed by the vicariance in the tremands, overlap in the south-west and south-east, and extinction in the central areas with increased aridity in the Neogene. There is no need for eastward dispersal in tremands – they were already in the east when they originated, as indicated by the allopatry with the sister-group.

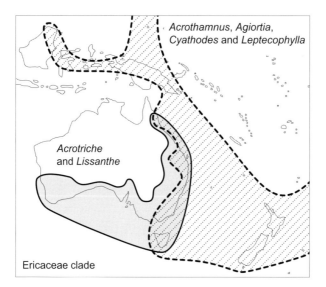

Fig. 1.10 Distribution of a clade of Ericaceae in Australasia: *Acrotriche + Lissanthe*, and the *Acrothamnus* group (Puente-Lelièvre *et al.*, 2013). The *Acrothamnus* group extends north of New Guinea to the Mariana Islands, and east of the Chatham Islands to Rapa Island and the Society, Marquesas and Hawaiian Islands.

Tremands are small shrubs with reduced 'leaves' that are interpreted here as sterilized inflorescence bracts. Comparative morphology suggests that the whole plant shoot system of tremands is equivalent to the inflorescence of the large, rainforest trees in *Elaeocarpus*. Inflorescence-type architecture in the plant is also present in the unusual 'juvenile' stage of *Elaeocarpus hookerianus* from New Zealand. This species, along with many other divaricate shrubs in New Zealand, inhabits rainforest on hills or in swampy, alluvial sites. This ecology suggests that, while the 'bract-plant' morphology is pre-adapted to survive in more arid conditions (as in parts of Australia), it probably did not develop as a direct result of these, through selection. The juvenile stage of *E. hookerianus* represents a trace of the tremands' inflorescence-type architecture in their sister-group. Before seafloor spreading in the Tasman, *E. hookerianus* or its ancestors would have been located next to the tremands of south-eastern Australia, suggesting that they inherited some of the 'bract-plant' genes in the region.

A similar distribution pattern to that of the tremands and their sister-group is seen in Ericaceae (Fig. 1.10). A clade of south-western and eastern Australia (*Acrotriche* and *Lissanthe*) is sister to a group of eastern Australia and the Pacific islands (*Cyathodes* s.str., *Agiortia*, *Acrothamnus* and *Leptecophylla*) (Puente-Lelièvre *et al.*, 2013). *Leptecophylla* has endemics in New Guinea, Mariana Islands, south-eastern Australia, New Zealand, also in the central Pacific in the Chatham Islands, Rapa Island and the Society, Hawaiian and Marquesas Islands. As with the break between the tremands and their sister-group, the boundary between *Acrotriche* and the *Cyathodes* group is located in eastern Australia.

Another group with a similar distributional break occurs in the parrots (Psittacidae). The subfamily Platycercinae has two clades (Schweizer *et al.*, 2013; IUCN, 2013):

Australia (widespread) and Tasmania (*Platycercus* etc.).

Tasmania (breeding), south-eastern Australia (somewhat inland of the main divide), Queensland islands (Curtis, Northumberland and Whitsunday Islands) (*Lathamus*), New Zealand, subantarctic islands, New Caledonia, Fiji and the Society Islands (*Cyanoramphus* etc.).

The overlap between the clades is restricted to a small part of the overall range, in Tasmania, south-eastern Australia, and the Queensland islands. The simplest explanation for the pattern would be a break in a widespread ancestor at the region of overlap, followed by local range expansion. A dispersal analysis (using DEC, implemented in LAGRANGE) instead located a centre of origin for the Platycercinae in southern Australia. From there, dispersal proceeded to central Australia in one lineage and to New Zealand and the Pacific islands in the other (Schweizer *et al.*, 2013). The authors concluded: 'our data clearly show that the Melanesian islands Fiji, Tonga and New Caledonia provided stepping stones for parrots in their colonization of the "mini continent" New Zealand', but it was the centre of origin program, not the data, that revealed a centre of origin. In any case, the authors did not mention the main break in the group, in eastern Australia/Tasmania, or provide a mechanism for it. One possibility is that the break developed at the east Australian Great Dividing Range or Bass Strait Basin. These coincide spatially with the phylogenetic break and formed soon before seafloor spreading began in the Tasman Sea. Both features are discussed further in Chapter 4.

Repeated breaks at the same nodes: pygmy perches (Percichthyidae) and east/west differentiation in Australia

Repeated breaks in a widespread ancestor around the same node can lead to paraphyletic basal grades located around the node. This will give area phylogenies of the type: A (A (A B)) (Fig. 1.6). In many groups, repeated differentiation has occurred over the same time period at more than one geographic locality. Often just two areas are involved, giving phylogenies of the type: A (B (A (B (A + B)))). The simplicity of the phylogenies indicates that two breaks have been active in a widespread ancestor, with local overlap within A and B. Evolution in global groups, for example, has often involved repeated breaks in the south-west Pacific and south-west Indian Oceans. In a dispersal interpretation, a phylogeny: SW Pacific (SW Indian (SW Pacific (SW Indian (SW Pacific)))), would require repeated long-distance dispersal events backwards and forwards between the two centres. A vicariance interpretation of the same pattern instead proposes repeated differentiation at the same nodes in a widespread ancestor, with local overlap in the regions. The breaks could have been caused by a series of tectonic events in a region or by reactivation of the same tectonic features.

As illustrated in the tremands, there is a biogeographic break between south-eastern and south-western Australia. Morgan *et al.* (2007) and Unmack *et al.* (2011) discussed the two hypotheses that have been used to explain the pattern. The 'Multiple Invasion' hypothesis proposes that multiple east–west movements have occurred, resulting in eastern and western faunas that do not show reciprocal monophyly. In contrast, the

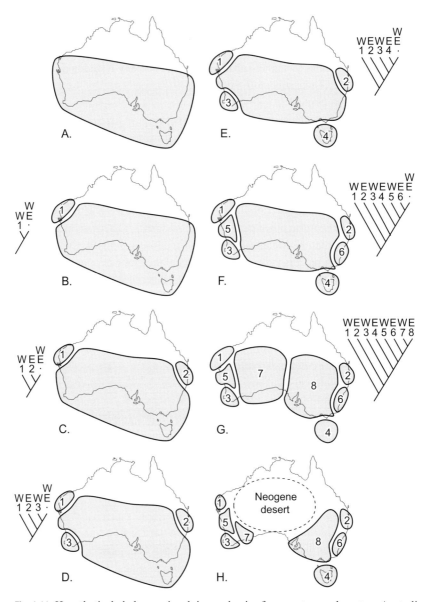

Fig. 1.11 Hypothetical phylogeny involving endemics from eastern and western Australia.

'Endemic Speciation' hypothesis proposes that groups in southern Australia were subject to a single east–west split, resulting in monophyletic eastern and western lineages.

These two scenarios are not the only possibilities though. Starting with an ancestor that was widespread in southern Australia, local allopatric speciation in the west and in the east over the same time period would produce a pattern that does not show reciprocal monophyly, without requiring any east–west movement. One hypothetical example is shown in Figure 1.11. Any subsequent local dispersal would lead to overlap within each region (not shown in the figure). Localities appear to 'jump around' in the phylogeny, but this is a natural result of the widespread ancestor gradually breaking up at different boundaries, not because of movement between regions.

In Australian pygmy perches (*Nannatherina* and *Nannoperca*: Percichthyidae), eastern and western species did not show reciprocal monophyly and so Unmack *et al.* (2011) supported the hypothesis of east–west dispersal. They wrote: 'This study appears to be the first example of an animal group displaying *clear multiple east–west movements* in southern Australia, as all other aquatic and terrestrial fauna previously examined [in molecular studies] displayed a single east–west split'. [Italics added.] The latter pattern has been shown in groups such as *Heleioporus* frogs (Morgan *et al.*, 2007), but whether a break between east and west occurs early in the phylogeny, as in *Heleioporus*, or later, as in the pygmy perches, east–west dispersal is not required to explain the pattern (cf. Ladiges *et al.*, 2012).

A diverse beetle genus: *Bembidion* (Carabidae)

This beetle group again shows the repetition of an area – this time the northern hemisphere – in a phylogeny. Again, this does not necessarily mean there has been any dispersal to or from the area. *Bembidion* is a global genus with over 1200 described species. It is most diverse in the Holarctic region, but there are also centres of diversity in temperate South America and New Zealand. New Zealand alone contains 36 described species, all endemic. Maddison (2012) sampled 256 species and found that: 'Among the more striking results from the molecular data are the biogeographic patterns and consistency of groups. Most of the species from Australia and all of the species from New Zealand form an endemic clade . . .'. This finding, repeated in other groups, is important because large, monophyletic southern groups contradict the traditional model of the Modern Synthesis. This predicted that southern groups were accumulated waifs and strays from the north; the biota was a chance assemblage of unrelated offshoots.

One of the two large clades in *Bembidion*, the *Bembidion* series, includes all species from South America, Australia, New Zealand, most species from southern Africa and Madagascar, and many from the northern hemisphere. It has the following phylogeny (NH = northern hemisphere):

South America plus NH.

South America plus Costa Rica.

NH.

Australia and New Zealand.

NH.

Madagascar.

Australia, South Africa, NH.

Further work is needed, but so far the evidence suggests a simple geographic structure for such a diverse group, with hundreds of species. The basal nodes involve breaks around

South America. The Australian members belong to just two clades, one shared with New Zealand, the other with South Africa, Madagascar, and the northern hemisphere. The overlap in the northern hemisphere may be apparent (the groups there may have different areas within the region) or it may be real, reflecting the great range expansion that has occurred in many northern groups. In any case, there is no need to postulate dispersal of the clades between northern and southern hemispheres or among the southern continents. Vicariance of a global ancestor among these larger regions, followed by some local overlap by range expansion within the regions, is sufficient to explain the pattern.

Malurus: is Australia a centre of origin or a centre of differentiation?

In the passerine bird *Malurus* (Maluridae), the most diverse clade has three subgroups (Driskell *et al.*, 2011; Lee *et al.*, 2012) with the following distributions:

Widespread in mainland Australia.
Widespread in mainland Australia (not in the far north), one subspecies in Tasmania.
Widespread in mainland Australia (not in the far south), one species in New Guinea.

The three clades show extensive overlap in mainland Australia and so this area could be interpreted as a centre of origin, with dispersal to New Guinea and Tasmania. An alternative model is possible, though, if the process began with an ancestor that was already widespread in Australia, Tasmania and New Guinea. The pattern suggests that the three clades, at the time of their origin, were allopatric with distributions such as:

Central Australia.
Southern Australia and Tasmania.
Northern Australia and New Guinea.

Methods to find centres of origin have analysed relationships among assumed areas of endemism, such as 'Australia' and 'New Guinea', but have often left these geographic areas themselves unanalysed. This has led to the idea that the breaks have occurred between the supposed areas, but the areas themselves could be misconstrued. If the common ancestor of the *Malurus* clade was widespread and the clades evolved in allopatry, this would account for the 'outliers' in Tasmania and New Guinea, the scarcity of the second clade in northern Australia, and the scarcity of the third clade in southern Australia. There is no need for any dispersal between Australia and Tasmania, or between Australia and New Guinea. Dispersal is only required within mainland Australia, in order to explain the range overlap there. In this model, 'mainland Australia' is not a centre of origin or an original area of endemism and the outliers are not secondary.

The family Maluridae (Meliphagoidea) as a whole has the following phylogeny (Driskell *et al.*, 2011; Lee *et al.*, 2012):

Australia (*Amytornis*).

    Australia (*Stipiturus*).

      New Guinea (*Sipodotus*, *Clytomias*, *Chenoramphus*).

      New Guinea (*Malurus cyanocephalus*).

      Australia, Tasmania and New Guinea (main clade of *Malurus* – see above).

A dispersal analysis of this pattern might propose a centre of origin in Australia, as the basal, paraphyletic grade occurs there. This would be followed by three separate invasions of New Guinea, or one, and then once back to Australia. Yet the phylogeny is also compatible with differentiation of a widespread ancestor at breaks in Australia and New Guinea, with phases of overlap within the two areas, but with no migration between them. The details of overlap and allopatry among the five clades can be worked out and differences in the clade distributions within Australia and within New Guinea would be of special interest. For example, *Amytornis* is absent along the eastern seaboard of Australia and in Tasmania, while *Stipiturus* is absent in the far north, and these suggest traces of the original allopatry.

Pittosporaceae: south-eastern Australia as a centre of origin or a centre of differentiation

Pittosporaceae, a family of trees and shrubs, is widespread in the Old World. Chandler *et al.* (2003) found three main clades (their mutual relationships were not resolved):

Eastern Australia and New Guinea (*Hymenosporum*).
SW and SE Australia (*Marianthus* etc.).
Widespread in the Indo–Pacific region, from Africa to Hawaii (*Pittosporum* etc.).

Instead of spreading out from a centre of origin in south-eastern Australia, where all three clades now overlap, the family can be derived from an ancestral complex that was widespread through the Indo–Pacific. In the simplest model, it divided into the three main clades around a break in south-eastern Australia (cf. Elaeocarpaceae, Fig. 1.9). The global division was the primary event, the local overlap in Australia was secondary.

Biogeography and dispersal

Means of dispersal

Many authors have commented on the paradoxical nature of means of dispersal; large numbers of bird species are local endemics, while many small invertebrates are widespread across continents. Even within clades, the means of dispersal do not seem to correlate with actual distribution patterns. These counter-intuitive observations have important implications. For example, the cosmopolitan butterfly genus *Vanessa* (Nymphalidae) 'contains some of the most cosmopolitan and vagile species of butterflies on the planet, as well as some highly restricted taxa. . . . This is a clear paradox: how can a clade of butterflies containing some of the world's most mobile and widespread insects also have closely related species that are such narrow endemics?' (Wahlberg and Rubinoff, 2011).

In New Zealand, studies of two stream insects, one with a higher dispersal potential (the caddisfly *Orthopsyche fimbriata*) and one with lower dispersal potential (the short-lived mayfly *Acanthophlebia cruentata*), gave unexpected results: higher levels of genetic differentiation in the caddisfly than in the mayfly (Smith and Collier, 2001). The authors concluded that 'factors other than dispersal ability can have an over-riding influence on the genetic structure of stream invertebrate populations, and . . . an understanding of past geographical events can be important for the interpretation of results from genetic studies'. New Zealand straddles a plate boundary and the authors emphasized that 'land mass changes have occurred over relatively small spatial scales'.

In Western Australian angiosperms, Gove *et al.* (2009) found that geographic range sizes were related to location (the biogeographic region in which species were found), not to the mode of seed dispersal (wind-, ant-, or vertebrate-dispersed). Location is a major factor in evolution, but it is often overlooked in ecological studies that concentrate on aspects of the present environment.

Changing ideas on dispersal in marine groups

In marine ecosystems, many species have pelagic larvae that drift in the water column. These groups were thought to be panmictic and uniform through their range, because of larval movement with the currents. Molecular studies have now disproved this idea and instead indicate high levels of geographic structure in marine species (Heads, 2005a). Coastal marine clades that depend on shallow water often have distribution patterns that also occur in terrestrial groups, and so the general patterns are likely to reflect general causes, such as tectonic history. For example, the marine fish *Micromesistius australis* (Gadidae) comprises disjunct populations around the continental shelves of New Zealand and the Falkland Islands, and there is significant genetic differentiation between the two (Ryan *et al.*, 2002). This South Pacific disjunction has long been recognized in land groups. The sister species of *Micromesistius australis* is *M. poutassou* of the Black Sea, Mediterranean and North Atlantic. This is also a well-known distribution pattern in terrestrial groups and coincides with the former Tethys Seas. Differentiation between the South Pacific and Tethys species, and also within *M. australis* across the Pacific, are all compatible with simple vicariance around these ocean basins.

In the Hawaiian Islands, Toonen *et al.* (2011) sequenced 27 marine vertebrates and invertebrates. They found major breaks between Hawaii and Maui, and between Oahu and Kauai (as in many terrestrial organisms), and they concluded (Toonen *et al.*, 2011: 5):

These data are striking in that more than half of the species surveyed show significant concordant barriers to gene flow concentrated in [four regions of the archipelago] . . . Given the broad differences in taxonomy, life history, and ecology of the species surveyed, including limpets, sea cucumbers, vermetid tube snails, reef fishes, monk seals, and spinner dolphins . . . there is no *a priori* reason to expect that patterns of connectivity would be shared among the majority of the species. However, the four shared barriers to dispersal highlighted here indicate that these species are responding to common factors that limit dispersal and delineate independent units . . . *We hypothesize that the dominant factors are likely abiotic as opposed to biotic, given the diversity of species with radically divergent life histories that share the pattern of isolation. . . .* For most species *there are enigmatic restrictions to dispersal that appear to have little to do with geographic distance. . . . the patterns of genetic differentiation do not generally match predictions for larval dispersal based on water movement information . . . the basis for these shared genetic restrictions is poorly understood . . .* [Italics added.]

The basis for 'shared genetic restrictions' in Australasia is one of the main themes of this book. As in Hawaii, geographic distance, life history, ecology and means of dispersal often do not appear to be relevant to the history or location of the breaks. Instead, the same breaks are shared between groups with different ecology, including marine, freshwater and terrestrial clades.

Hidas *et al.* (2007) studied intertidal invertebrates across a biogeographic barrier in south-eastern Australia and found a similar result to that of Toonen *et al.* (2011): the distributions of the species were not related to their dispersal potential. Groups with planktonic larvae, assumed to be good dispersers, were sometimes restricted to one side of the barrier, while groups with direct developing juveniles and assumed to be 'poor' dispersers were present on both sides of the barrier.

Many molecular studies have reached similar conclusions and have brought about a revolution in marine biology (Heads, 2005a). Cryptic, geographic clades are often recognized and there are obvious implications for systematics, management and conservation. Before this molecular revolution, marine biologists were strong advocates of dispersal theory, but now they sometimes attach more importance to vicariance than terrestrial biologists do. For example, in galatheid deep-sea lobsters, Macpherson *et al.* (2010) found several centres of diversity in the South Pacific, including the Malay archipelago, the Coral Sea, New Zealand and French Polynesia. They argued that the high levels of endemism in each centre suggest that the groups and the centres have evolved by vicariance.

Changing ideas on dispersal in freshwater microorganisms

The molecular work has caused a similar paradigm shift in freshwater microbiology. For many years the microbiologist's credo was 'everything is everywhere and the environment selects'. Vyverman *et al.* (2007) wrote that 'There is a long-standing belief that microbial organisms have unlimited dispersal capabilities, [and] are therefore ubiquitous'. Yet new molecular evidence presents a different picture (Logares, 2006). The regional and local differentiation that is indicated – above, at and below species level – implies allopatric divergence along with restrictions to panmixis in groups that were thought to lack these.

In diatoms, studies using molecular markers and fine-grained morphological variation have shown that the geographic distribution of clades ranges from global to locally endemic (Vanormelingen *et al.*, 2007). Evans *et al.* (2009) reported high levels of differentiation among populations of the freshwater diatom *Sellaphora capitata* that suggest limited dispersal and enhanced opportunities for allopatric speciation. For freshwater diatoms in general, Vyverman *et al.* (2007) wrote: 'At regional to global scales, historical factors explain significantly more of the observed geographic patterns in genus richness than do contemporary environmental conditions'. The diatom genus *Eunophora*, for example, is endemic to Tasmania and New Zealand; these two areas make up an important centre of endemism for life in general and are discussed in more detail below for multicellular groups.

'Dispersal': one word, several concepts

Molecular phylogenies have recovered many clades that suggest either vicariance of a widespread ancestor or, in a dispersal model, repeated trans-oceanic jumps of thousands of kilometres. These jumps are proposed even for organisms that do not appear to have adequate means of dispersal and the process has been explained as the result of 'chance'. Li *et al.* (2011), for example, considered that for plants 'a lot of long-distance dispersal is caused by random incidents . . .'. As Li *et al.* also emphasized, 'the dispersal mechanism of the plant itself sometimes is irrelevant for long-distance dispersal'. This process of chance dispersal that does not use the normal means of dispersal is discussed next. To begin with, several distinct processes have been termed 'dispersal' and these can be contrasted as follows.

1. Normal ecological dispersal

This is the normal, physical movement seen in plants and animals within their distribution range. It is observed every day and does not involve differentiation of distinct clades. 'Normal dispersal' takes place over long, intercontinental distances, as in sea birds, or over much shorter distances, as in soil invertebrates, depending on the organism and its normal ecology. Following their origin by reproduction, all individual organisms have dispersed to where they are by normal dispersal. It is observed in the weeds that colonize a garden that has just been dug, an area of burnt vegetation, a landslide or a new volcano. This process of simple movement does not always explain the distribution area occupied by a taxon (in particular, any allopatry with related taxa), as it does not account for evolutionary differentiation. The distribution of a group can be caused or modified by evolution, as well as by physical movement.

2. Range expansion

Range expansion is seen in historical times in the anthropogenic spread of weeds. At particular times in geological and evolutionary history, conditions have led to general phases of range expansion involving many taxa. Range expansion, when it does occur, can be very rapid. Given suitable conditions, a more or less local plant or animal can become worldwide in hundreds, rather than millions, of years. This takes place by normal ecological dispersal of many individuals, using the normal means for the group, not the 'extraordinary means' often invoked for the unique events of founder speciation. Following the massive, long-lasting ice ages of the Permian–Carboniferous, for example, many groups would have expanded into the new deglaciated habitat before settling down into a round of relative immobilism and *in-situ* evolution through the Mesozoic. Later in the Jurassic–Cretaceous, during pre-breakup volcanism, rifting and marine flooding, taxa would again have expanded their range in a renewed phase of mobilism, spreading along the vast lengths of new coastline. Once the rifting was complete and the inland seas had receded, there was a new period of *in-situ* evolution. In this way, phases of mobilism in which normal means of dispersal are deployed alternate with phases of immobilism in which allopatric evolution (vicariance) takes place.

Range expansion is not due to chance, but to tectonic and climatic events. Range expansion is not due to means of dispersal per se, as locally endemic organisms with restricted modes of dispersal, including plants, worms and snails, can, given suitable changes in the environment, expand their range or even become aggressive, invasive 'weeds'. Modern humans can perhaps be considered as a 'geological' factor, as many aspects of current biogeography stem from human activities. These include moving plants and animals around the globe, as well as changing the environment in different ways and at different scales.

3. 'Chance dispersal' ('long-distance dispersal', 'speciation by founder dispersal')

The defining feature of speciation by chance dispersal is not so much the long distance involved, but the idea of a unique, extraordinary, dispersal event by a founder across a 'barrier'. This leads to speciation. This is a different process from individual movement or range expansion, processes that are seen every day. Long-distance dispersal with founder effect speciation is a theoretical inference. In contrast, normal ecological movement and range expansion, along with other kinds of 'dispersal' such as diurnal and annual migrations, are observable processes.

Dispersal theory proposes that 'When lineages arrive in new habitats they will usually diverge and sometimes speciate' (Renner, 2005). Again, the process differs from the dispersal observed every day. Any patch of cleared garden will soon be colonized by the local 'weedy' flora and fauna, and later by less weedy taxa, but none of these will diverge or speciate there. This is 'normal' or 'ecological' dispersal, simple physical movement.

One main problem with the process of founder dispersal is explaining why movement between populations would be occurring at one time, but then at some point stop or at least decrease, allowing differentiation to proceed. What is the reason for the change? In dispersal theory the factor is 'chance'. Reduced dispersal rates are sometimes attributed to changing behavioural patterns in animals or morphological means of dispersal in plants, which at least provides a concrete cause, but it does not explain repeated patterns in unrelated animals and plants. Geological or climatic change that is fundamental enough to cause phylogenetic breaks through entire communities is one possible mechanism for reduced dispersal and this is the basis of vicariance as an explanation for allopatry.

Dispersal theory accepts that normal ecological dispersal and also founder dispersal both occur in nature, and downplays the significance of vicariance. For example, Mayr (1965) wrote: 'Quite obviously, except for a few extreme [i.e. local] endemics, every species is a colonizer because it would not have the range it has, if it had not spread there by range expansion, by "colonization", from some original place of origin'. In contrast with this approach, many recent studies have accepted, or at least discussed, vicariance (the term has been cited in 5610 papers from 2010 until 2013; Google). The model used here accepts both vicariance and normal ecological dispersal, but does not require speciation by chance dispersal. Again, it is important to distinguish between normal dispersal (simple range expansion) and founder effect dispersal. Kodandaramaiah (2010) saw the failure to distinguish between the two as a serious drawback of centre of origin programs such as DIVA. This conflation of the two processes – founder dispersal speciation and normal ecological movement – is a key aspect of dispersal biogeography. It is also the basis of the confusing criticism that work in panbiogeography somehow denies dispersal, when it is only founder speciation/chance dispersal, not normal dispersal, that is being rejected.

Dispersal and two different concepts of chance

Although a long distance between populations is often cited as an important factor in chance dispersal, the process is also used to account for differentiation over very short distances, for example, between taxa on two sides of a river. The defining feature of 'jump dispersal' is not the distance but the fact that, following the single founder event (leading to a monophyletic clade), dispersal then stops. But why? The process is 'chance', not in the sense of being analysable with normal statistics, but in the sense of not being analysable. The most unlikely events are proposed to have occurred because 'anything can happen given enough geological time' but, apart from being so unlikely, chance dispersal is unfalsifiable and explains all distributions and none at the same time.

The new concept of chance as a calculated probability – not as luck or fortune – began with Pascal and Fermat in the seventeenth century and became a founding principle of modern science. In this sense, a probability-cloud diagram showing distance of seed dispersal away from a parent tree, for example, depicts the 'chance' aspect of normal ecological dispersal. On the other hand, 'chance' in the sense of 'factors beyond our understanding' (as in 'chance dispersal') is not a real explanation for anything.

Biogeography and genetics

Critique of founder effect speciation in population genetics studies

In traditional founder dispersal, the founder becomes isolated from its parent population by 'dispersing over a barrier' (an apparent contradiction) and then differentiates as a new species. This is achieved by a second process, the 'genetic revolution' produced by the founder effect. The founder effect itself is well-established in genetics, but many geneticists do not accept that it involves a genetic revolution or that it can lead to speciation (Rundle et al., 1998; Mooers et al., 1999; Tokeshi, 1999; Florin, 2001; Nei, 2002; McKinnon and Rundle, 2002; Rundle, 2003; Orr, 2005; Crow, 2008). An advocate of speciation by rare dispersal events even found it necessary to publish an article stressing 'The reality and importance of founder speciation in evolution' (Templeton, 2008). This was a reply to Coyne and Orr's (2004: 401) conclusion that 'there is little evidence for founder effect speciation'.

Just because geneticists such as Coyne and Orr (2004) have disputed speciation by founder effect does not mean they have rejected founder dispersal and founder events. But founder effect speciation was one of the pillars of dispersal theory and rejecting it has important consequences. If it is rejected, but founder dispersal is retained, other modes of differentiation must be invoked and ecological differentiation is often inferred. This is unlikely to be the initial cause of geographic variants, though (see p. 13), and this was the problem that Mayr (1954: 168) recognized in proposing his ideas on 'genetic revolution'. Mayr (1999: xix) emphasized that: 'The crucial process in speciation is not selection, which is always present in evolution even when there is no speciation, but isolation'.

Incongruence between patterns of variation in different genes and characters

Centre of origin/dispersal biogeography has two main conceptual bases. One is the idea that all clades except local endemics have attained their distribution by spreading out from a particular centre of origin. The other is the idea that the immediate common ancestor of the clade was monomorphic and homogeneous, either a single parent pair or at most a uniform species. The new clade is separated from the ancestor by a chance, one-off event. Each character in the ancestor had only one state and the new clade is defined by at least one new feature.

In contrast, the vicariance model accepts that ancestors are often polymorphic before the descendant groups began to differentiate, and that this prior polymorphism was inherited in the descendants. This process means that a new taxon is often no more than a recombination of ancestral polymorphism and has no new features. Rather than originating at point centre of origin from a single parent pair, the process suggests that a new group can emerge over a broad region within the ancestral range. In simple vicariance between, say, a northern hemisphere group and its southern hemisphere sister the groups have not had a traditional centre of origin, but have each emerged *in situ*, over a broad region, from a global ancestor.

What are the implications of a polymorphic ancestor? The first is that different characters can show different, incongruent distributions among the groups. In the palm tribe Areceae, Norup *et al.* (2006) stressed that several genera have no unique characters but are distinguished instead by different 'combinations of widespread character states'. This phenomenon is common in all groups – Reinert *et al.* (2006) described it in mosquito genera, to cite just one group of animals – but it is seldom acknowledged as a general principle. It is the antithesis of a neat, hierarchical arrangement of characters and taxa, and contradicts the idea that clades are defined by unique, derived characters. Recombination of characters would not be expected if every new mutation, form or taxon arose at a centre of origin and spread from there by physical movement. But it would be, if vicariant groups evolved as recombinations of widespread ancestral characters that were already distributed over the ancestral range before the evolution of the modern groups began. Apart from ancestral polymorphism and hybridism, recombination of characters can also occur through parallel evolution. Molecular phylogenies indicate

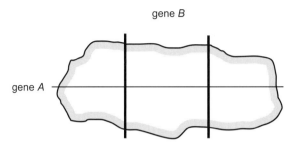

Fig. 1.12 Distribution of a hypothetical clade (grey) in which phylogenetic/geographic breaks in two parts of the genome lie at right angles to each other. (From Michael Heads, *Molecular Panbiogeography of the Tropics*. ©2012 by the Regents of the University of California. Republished by permission of the University of California Press.)

that parallel evolution is ubiquitous in morphology, and it is probably also widespread in molecular evolution.

In many groups the geographical variation of one gene is incongruent with that of another gene and this complexity cannot be depicted in a single tree. The incongruence is sometimes due to hybridism and this is often associated with community disturbance and mobilism. Following this phase, hybrid swarms can persist, more or less frozen in place, for millions of years until they are disturbed again. For some groups, a tree diagram is no longer regarded as an accurate representation of phylogeny. In prokaryotes, horizontal gene transfer is pervasive and 'very few gene trees are fully consistent, making the original tree of life concept obsolete' (Puigbó *et al.*, 2009). Hybridism is well-known in plants and there is an increasing appreciation of its role in animal evolution. Homoploid hybrid speciation has now been documented in many animals (Mallet, 2007), including birds (Brelsford *et al.*, 2011), and many more cases will probably be shown in future surveys.

Incongruence between gene trees can be caused by hybridism, but also by polymorphism already present in the ancestor before the descendant taxa began to differentiate. This variation can be inherited by the descendants if there is incomplete lineage sorting in the ancestor. The ancestral polymorphism can be incongruent, in both phylogeny and geography, with the variation that develops in the descendants. Figure 1.12 shows the distribution of a hypothetical clade in which patterns of variation in two parts of the genome are incongruent and the geographic trends lie at right angles to each other. Both patterns result from different phases of allopatry (these may have been almost simultaneous) and at least one of the patterns could represent variation present in the ancestor before the modern clades diverged. Figure 1.13 shows an example from the marine fish *Nemadactylus* (Cheilodactylidae) in which the variation in cytochrome *b* shows geographic incongruence with groupings shown in D-loop sequences (from Burridge, 1999, slightly simplified). Burridge discussed whether the pattern was due to hybridism or to incomplete lineage sorting and concluded it was the latter. In any case, patterns of 'incongruent' variation have long been known in morphological studies and are now well-documented in molecular work. The widespread incongruence suggests

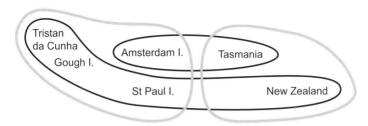

Fig. 1.13 Patterns of variation in the fish *Nemadactylus* (incl. *Acantholatris*) (Cheilodactylidae) (Burridge, 1999). Grey line = clades defined by cytochrome *b* sequences; black line = clades defined by D-loop sequences. (From Michael Heads, *Molecular Panbiogeography of the Tropics.* ©2012 by the Regents of the University of California. Republished by permission of the University of California Press.)

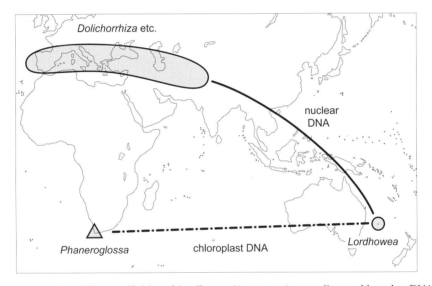

Fig. 1.14 The different affinities of *Lordhowea* (Asteraceae), according to chloroplast DNA (*ndhF, trnL, psbA – trnH*, 5′ and 3′ *trnK* and *trnL-F*) and nuclear DNA (ITA, ETS) (Pelser *et al.*, 2010).

that lineages do not evolve in a linear or hierarchical way. Instead, whether by hybridism or incomplete lineage sorting, phylogeny can develop through the recombination of ancestral characters and distributions that are often much older than the clades in which they are now present.

Examples of incongruence and incomplete lineage sorting

Lordhowea (Asteraceae: Senecioneae) is endemic on Lord Howe Island in the Tasman Sea. Data from the plastid genome indicate that its sister-group is *Phaneroglossa* of the western Cape region, South Africa (Fig. 1.14; Pelser *et al.*, 2010). This is a typical Indian Ocean pattern. Nuclear sequences instead show *Lordhowea* as sister to a group with a

Tethyan distribution (*Dolichorrhiza*: Causasus Mountains, Afghanistan, etc.; *Iranecio*: Iran etc.; *Caucasalia*: Caucasus; *Adenostyles*: Mediterranean and the Alps). The affinities here illustrate two of the four main connections of the Tasman basin biota, one spanning the Indian Ocean and the other the Tethyan belt. The double pattern in *Lordhowea* can be explained if the common ancestor of all the groups was already widespread and diverse before the origin of the modern groups. Some of the genetic variation, with its geographic distribution, dates back to a time before the modern clades began to differentiate.

In their study of Senecioneae, Pelser *et al.* (2010) found substantial incongruence between the nuclear and plastid phylogenies. They suggested that in some cases incomplete lineage sorting probably did not cause the pattern, as population sizes would have needed to be large for this to occur (for example, never falling below 20 600 in the *Lamprocephalus–Oresbia* clade). The authors argued that 'This is extremely improbable considering the present day population sizes of Senecioneae species', but, if the group has evolved by vicariance, earlier groups would have had much larger ranges and population sizes.

In early morphological studies, the moss *Echinodium* was recognized from Macaronesia and Australasia (south-eastern Australia, New Caledonia and New Zealand). This is a normal variant of the Lord Howe–Mediterranean track shown in Figure 1.14. Molecular studies found a different arrangement, in which the Macaronesian and Australasian clades did not form a monophyletic group (Stech *et al.*, 2008). Yet the authors wrote that: 'While the molecular data suggest that the peculiar distribution pattern of "*Echinodium*" is an artefact, the striking morphological similarity observed in Macaronesian and Australasian species cannot be dismissed'. Again, the different features show different distributions and this incongruence can reflect the inheritance of ancestral polymorphism.

Biogeography and ecology

Some authors have argued that biogeography is determined by ecology. Yet, as Darwin (1859: 346) wrote, 'the first great fact' in biogeography is 'that neither the similarity nor the dissimilarity of the inhabitants of various regions can be accounted for by their climatal and other physical conditions'. In a typical biogeographic region, such as the area of endemism in hill forest cited above, the taxa are sorted out into the standard ecological zones and habitats. For example, some taxa will occur on the valley floors and along streams, some on the ridges and summits, and some on the intermediate slopes. This ecological differentiation is spatial, but occurs on a smaller or much smaller scale than the biogeographic differentiation. Outside the biogeographic region the species pool changes and so even in equivalent habitats the biota can differ. Clades define habitat types within a biogeographic region, but at the scale of the region and above, historical and phylogenetic effects start to apply. Very few taxa define areas of similar ecology throughout the world. For example, groups that are restricted to rainforest are not found in all rainforests, while groups found in all rainforests also occur in other habitats.

Ecological centres of origin: ancestral habitats

Following the rise of the Modern Synthesis, most studies of Australasian biogeography have adopted a centre of origin approach. As discussed in the following chapters, many authors have proposed centres of origin for Australasian groups in particular *areas* outside the region. Studies examining the evolutionary ecology of a clade have adopted the same paradigm and assumed a restricted, ancestral *habitat type* – an ecological centre of origin. The agamids (dragon lizards) are a good example. They are diverse (>300 species) and widespread through mainland Africa and Asia, and occur throughout Australia. The family (with chameleons) is an Old World vicariant of iguanids, found in the New World, the Pacific islands and Madagascar (the two overlap only on Madagascar). As well as the allopatry at family level, there is also vicariance within the agamids. Collar *et al.* (2010) found strong support for three clades in the family, one in Africa and south-west Asia, one in Southeast Asia, and one mainly in Australia and New Guinea but including a single Southeast Asian species. The phylogeny indicates a gradual breaking down of a global group into allopatric units at different levels of the hierarchy (family, subfamily), and the simplest way of generating this is by vicariance. The agamids are so diverse and widespread that it is not surprising that they also occupy several habitat types; Collar *et al.* (2010) distinguished species as rock-dwelling, arboreal, semi-arboreal and terrestrial. Arboreal groups include *Draco* of Indonesia and Southeast Asia, in rainforest, while terrestrial groups include *Moloch* of central Australia, in desert. Collar *et al.* (2010) proposed that just one of the four habitats – terrestrial – was the ancestral habitat for the family. Yet the biogeography suggests that the ancestor was widespread through the Old World and this would imply that it already occupied more than one habitat type before the modern clades evolved.

The CODA paradigm: centre of origin–dispersal–adaptation

How does modern evolutionary theory interpret differentiation at biogeographic and ecological scales? Modern concepts of the processes leading to distribution are still often based on centres of origin, dispersal from there and adaptation to new conditions; Lomolino and Brown (2009) referred to this as the CODA model. The model might imply, for example, that a group inhabits the New Guinea highlands because it invaded the mountains from a lowland centre of origin and then adapted to the new conditions. In contrast, a group may be endemic in the New Guinea highlands because its ancestors were already in the region before the uplift of the mountains in the Cenozoic. The modern group has inherited its area rather than invading it. If an uplifted lowland group had suitable *pre*-adaptations permitting montane life it will survive the uplift; if not, it goes extinct. This non-CODA model is compatible with recent work on evolution at the genomic level (Lynch, 2007; Cutter *et al.*, 2009; Stoltzfus and Yampolsky, 2009). This suggests that adaptation caused by natural selection has only secondary significance and that most evolution is the result of broad, intrinsic trends in genome evolution, such as biased gene conversion (Duret and Galtier, 2009; review in Heads, 2012a, Chapter 10).

2   Evolution in time

Avise (2000: 9) argued that 'Vicariance and dispersal are historical phenomena whose relative roles in particular instances can be weighed on the scales of phylogeographic analysis.' Wallis and Trewick (2001) agreed: 'Molecular data offer an obvious solution' to problems interpreting vicariance and dispersal. Many authors now support the use of molecular clock dates to solve biogeographic problems. Still, some have admitted that 'molecular clocks are notoriously difficult to apply because of many well-recognized problems . . . the most difficult of which is that different lineages evolve at different rates' (Buckley *et al.*, 2001). Another fundamental problem is calibration.

Apart from cross-calibrating from another node with a known age, there are only three ways to date evolutionary development on geological time scales:

1. Use the oldest fossil of a group. However, this only gives a minimum age for the group.
2. Use the age of the island or the strata to which a group is endemic. However, young islands and strata often have older endemics in their biota.
3. Correlate the geographic distribution of a group with associated tectonic events. However, tectonic features may be re-activated at different times.

All three methods are flawed, but the first two have serious, inherent limitations, while the third suggests possible lines of research.

Equating the age of a node with the age of the oldest known fossil

Based on the age of the oldest fossil in one group, dates for other groups can be derived by cross-calibrating to other parts of a phylogeny. But a fossil can only give a *minimum* age for a group, and so, at best, extrapolating to other groups can only give minimum ages for these. This means that only younger events can be ruled out as irrelevant to the group's origin, not older events. Fossils are invaluable as they falsify certain clade ages (younger ones), not because they verify any. Many authors accept this in theory, but, in practice, treat oldest fossil dates and dates derived from these as *maximum* or absolute

This chapter incorporates material previously published in *Cladistics* (Heads, 2005b), reprinted by permission of John Wiley & Sons, and in *Systematic Biology* (Heads, 2011), reprinted by permission of Oxford University Press and the Society of Systematic Biologists.

dates for clades, despite the logical fallacy. Based on this flawed argument, *earlier* events can be ruled out as a causative factor in a group's origin; many fossil-calibrated studies claim that a group is *too young* to have been affected by one or other geological event, but this is not valid.

Hystricognath rodents are a classic biogeographic problem, as the New World caviomorphs are sister to the African phiomorphs. The oldest fossils of caviomorphs and of phiomorphs are dated at ~37 Ma. Honeycutt (2009) concluded: 'these dates are congruent with the hypothesis that the invasion of South America by African hystricognath ancestors involved overwater waif dispersal across a 1700-km expanse of the Atlantic subsequent to the separation of these two continents at a much earlier date'. If the fossil dates are treated not as maximum dates but as estimates of minimum age, they are just as congruent with vicariance as with dispersal. Honeycutt (2009) did not accept vicariance as a possibility because the dates were being treated as maximum clade ages, implying that the group evolved subsequent to the opening of the Atlantic. If fossil-based dates are treated as minimum estimates of age, it is not possible to argue that a group evolved 'subsequent' to anything, only that it evolved *before* such and such a date or event.

Overall, the fossil record does indicate broad evolutionary trends, with Paleozoic, Mesozoic and Cenozoic members of a group often showing great differences. Yet deducing the chronological details of individual lineages through a literal reading of the fossils' stratigraphic distribution can be very misleading; in A.B. Smith's (2007: 731) careful wording: 'the fossil record provides direct evidence . . . but it cannot be taken at face value'.

Age of fossil versus age of clade

Biologists have often dated the time course of evolution based on interpretations of the fossil record. Since Matthew (1915), studies of phylogeny and biogeography have attributed special significance to the oldest known fossil of a group. Although Darwin (1859) emphasized the fragmentary nature of the fossil record, Matthew (1915) based his influential model of global biogeography on a literal reading of the fossil data. A taxon was taken to be the same age as its oldest known fossil or, sometimes, the oldest fossil of a closely related group. In this method the absence of earlier fossils of larger groups is attributed to a genuine absence of the group at that time. This approach was adopted by the Modern Synthesis and the influential New York School of Zoogeography that followed Matthew's lead (Nelson and Ladiges, 2001). The ideas provide the conceptual basis for modern studies in evolutionary chronology. As Forest (2009: 790) wrote: 'There is general consensus that the fossil record provides by far the best information with which to transform relative time estimates into absolute ages'.

Nevertheless, doubts are often raised about the literal interpretation of the fossil record. The method relies on absence of earlier fossils, in other words, negative evidence, described by one paleontologist (Gould, 1989) as 'the most treacherous kind of argument that a scientist can ever use'. Many authors have suspected that groups are older than their fossils alone indicate. The oldest bat fossil is dated at 51 Ma (Gunnell and Simmons, 2005). However, based on the fossils of Messel, Germany, with their stunning preservation, G. Storch (quoted in Hoffmann, 2000: 48) proposed that 'Bats were already advanced 49 million years ago [Eocene]. I'm convinced they originated much earlier than you read in textbooks' (cf. Hooker, 2001). Estimates of the completeness of the bat fossil record have been attempted (Eiting and Gunnell, 2009), but these have been based on the fossil record itself rather than an independent set of evidence.

A literal reading of the fossil record accepts that taxa cannot be tens of millions of years older than their fossil record. But why not? Ages for clades much older than their fossil record are often proposed in molecular clock studies. (Knowing the branch lengths of a phylogeny – the degrees of difference between different clades – a fossil from one part of a phylogeny can be used to calculate the ages of other clades.) In addition, new discoveries of fossils much older than the previous oldest fossils of a group appear on a regular basis. For example, proscopiid grasshoppers were unknown as fossils until one was described from Early Cretaceous material (110 Ma; Heads, 2008). The oldest known bee fossils were dated as 55–65 Ma until Poinar and Danforth (2006) discovered fossils in Burmese amber dated at Early Cretaceous, ~100 Ma. Crown-group salamanders are now known from Middle Jurassic rocks, predating the previous record by ~100 m.y. (Gao and Shubin, 2003). Hummingbird fossils were known back to 1 Ma, until one dated at 30 Ma was found (Mayr, 2004). Metatheria (marsupials and their relatives) were known back to 75 Ma, until Luo et al. (2003) described a fossil dated at 125 Ma.

These new 'oldest fossils' are often regarded as very significant and their location is assumed to represent a new centre of origin. In the approach supported by Matthew (1915), the metatherian fossil represents 'a rich source of new information about the time and place of origin' of the group (Cifelli and Davis, 2003: 1899) and it corroborates Matthew's account of the origin of marsupials in the northern continents (Luo et al., 2003).

In the plant family Asteraceae, fossil pollen found in South Africa and dated at ~60 Ma almost doubled earlier estimates for the age of the family. Keeley et al. (2007: 99) discussed biogeography in members of Asteraceae and assumed that vicariance was impossible, 'given the relatively young age of the family with respect to continental separations (90–120 Ma) . . . ' – but what is the age of the family? The authors assumed that the actual age is the same as the fossil age, but this is neither logical nor necessary. Keeley et al. (2007) concluded that their study reveals long-distance dispersal, but this is based on treating the new oldest fossil date – a minimum – as a maximum date.

The major problems with the literal reading of the record have been ignored or glossed over; there is even a book entitled *The Adequacy of the Fossil Record* (Donovan and Paul, 1998a). This is a response to Darwin's (1859) chapter 'On the imperfection of the geological record'. In their introduction, Donovan and Paul (1998b) referred to Darwin's 'bias' in his argument that the fossil record must be very incomplete, but it may be the authors themselves, rather than Darwin, who are biased. What is the reason for this? Darlington (1966: 320) proposed that the fossil record 'allows an almost magical view into the past' and Briggs (1974: 249) used the same words. This approach, a virtual cult of the fossil, proposed that the age of the earliest known fossil of a group is the age of the group, that the location of that fossil is the group's centre of origin, and that the fossil itself is the group's ancestor. None of these need be correct and the data from paleontology are no more magical than any other data. Instead of reading the fossil record of a group literally, a distinction can be drawn between the *age of a fossil* and the *age of its group* (Craw *et al.*, 1999).

The transmogrification of minimum (fossil-based) ages into maximum ages and the use of these to rule out earlier vicariance

In many studies on molecular phylogeny the age of at least one node is equated with the age of an oldest known fossil. Sometimes more 'oldest fossil' dates are added to different parts of the tree but this does not address the fundamental problem with the approach – without further manipulation, the oldest fossils of a clade only give minimum ages (at best) for clades, not absolute ages. It is now often acknowledged that the age of the oldest fossil of a group is less, to much less, than the age of the group (Smith and Peterson, 2002; Forest, 2009). Yet while many published studies acknowledge the idea in their introductions, in their actual analyses most overlook it and shift from citing ages as minimum estimates to citing them as absolute dates. In this way estimates of minimum ages are transmogrified into estimates of maximum clade ages.

Many modern groups, including many orders of mammals and birds, have their oldest known fossils in the Cenozoic. A literal reading of the fossil record would indicate that the groups evolved in the Cenozoic. Treating the fossil-based dates in this way as *maximum* ages of the clades means that Mesozoic events, such as continental rifting and the breakup of Gondwana, can be ruled out as irrelevant, but this line of reasoning is not logical.

In one case, based on a fossil-calibrated phylogeny of South African Proteaceae, Sauquet *et al.* (2009b) concluded that the clade Leucadendrinae evolved '*no earlier* than 46 Myr ago . . . '. [Italics added.] Again, this assumes that fossil-calibrated dates give actual clade ages and so eliminates the possibility of vicariance. Yet supporters of the vicariance model have always stressed limitations in the fossil record.

Authors have used dates based on Cenozoic fossils but treated as maximum dates to rule out Mesozoic vicariance in groups such as the plant family Atherospermataceae (Renner *et al.*, 2000), *Myosotis* (Boraginaceae; Winkworth *et al.*, 2002a), South African dragonflies (Ware *et al.*, 2009), the orchids (Bouetard *et al.*, 2010), and many others.

For the South Pacific *Hebe* complex (Plantaginaceae), Wagstaff *et al.* (2002) calibrated a clock based on mid-Miocene fossils (15 Ma) and proposed divergence estimates of the 'Australasian species clade' at 9.9 Ma and the '*Hebe* clade' at 3.9 Ma. Although these were estimates of minimum, not absolute, age, the authors suggested that 'it would be inconsistent with this [fossil] record to assume that divergence . . . occurred in Gondwanan (Cretaceous) or earlier times'. This is not correct; inferring *older* events as causative would not be inconsistent with the record; only the use of *younger* events would be inconsistent.

For the danthonioid grasses in Africa, Australasia and South America, discussed in the last chapter, Pirie *et al.* (2009a) ruled out vicariance and instead inferred a complicated series of dispersal events. Although the calibration was crucial, the paper did not mention its precise nature. The calibration, derived from Christin *et al.* (2008), was in fact based on fossils. Christin *et al.* (2008: S1) admitted that 'Unfortunately, fossils only give a lower bound (i.e. minimal age)', but in practice they used fossils to set both lower and upper bounds. Pirie *et al.* (2009a) were only able to rule out vicariance in Danthonioideae by using this minimum date in its transmogrified form as a maximum.

Early, informal transmogrification of fossil-calibrated ages into maximum clade ages

In many studies, fossil ages – estimates of minimum clade age – have been converted into estimates of maximum clade age. There was no discussion or even mention of this process, and so it can be referred to as a transmogrification (Heads, 2012b). In earlier studies the transmogrification was informal. Authors accepted that the fossil record gave a more or less accurate representation of evolution and so they equated the age of a clade with the age of its earliest known fossil, 'perhaps adding a safety margin of a few million years' (Soligo *et al.*, 2007: 30). In this approach, groups that have their oldest known fossil in, for example, the Eocene, such as bats or modern primates (not including plesiadapiforms) could, at a stretch, have evolved in the Paleocene, but a Cretaceous origin would be ruled out. One literalist study estimated that the probability of primates existing at 80 Ma (Late Cretaceous) was one in 200 million (Gingerich and Uhen, 1994).

Another group with oldest known fossils in the Eocene (55 Ma) are the horses. The oldest known fossils of their sister-group, the rhino–tapir lineage, are about the same age.

Waddell *et al.* (1999: 125) wrote:

This is about as good as fossil calibration points get, in that there exist multiple good fossils representing both sister lineages, which appear in appropriate chronological order . . . To account for the *already differentiated fossils of the two lineages appearing rather suddenly* (probably from migration to the fossil sites), we consider the split could be as much as 58 mybp (a conservative estimate; e.g., D. Archibald, pers. comm.). So we have a conservative 55 mybp calibration point (SE ~1.5). [Italics added.]

The estimate of such a small standard error reflects the assumption that oldest known fossil age is more or less equivalent to clade age.

Although Benton and Donoghue (2007: 26) agreed that 'paleontological data can provide good estimates only for minimum constraints on the timing of lineage divergence events', they also proposed transmogrified maximum ages for a selection of clades and these were also based on the fossil record. They were derived simply by 'bracketing' (based on the maximum ages of sister-groups – established with fossils) and 'bounding' (based on the age of the youngest fossiliferous formation that lacks a fossil of the clade) (Donoghue and Benton, 2007). In a similar way, Goswami and Upchurch (2010) argued that: (a) fossil dates give minimum ages. (b) The oldest known fossil of true primates is dated at 56 Ma; the oldest known eutherian mammal at 125 Ma. Therefore, (c) 'it seems probable that the first true primate originated somewhere between 56 and 125 Ma' (p. 407). This is not logical, as C does not follow from A and B.

Transmogrification of clade ages in a Bayesian framework

Following the earlier, informal transmogrifications of clade age, formal transmogrifi-cation is now often carried out in a Bayesian framework. Many papers annotate fossil-calibrated phylogenies with minimum and maximum estimates of clade ages, given in the form of 95% credibility intervals. In other words, fossil-calibrated ages have been converted from minimum estimates into maximum estimates that have statistical sup-port. How exactly is this achieved? In these studies, the transmogrification is carried out in a Bayesian framework, using programs such as BEAST (Drummond and Rambaut, 2007). The key point is that specific prior probability distributions (priors) are assigned to the calibrations before any analysis is carried out.

A clade's oldest known fossil is often used to calibrate the phylogeny of a clade. The actual age of the clade, or at least the probabilities of different actual ages, are specified in Bayesian analysis. Given an oldest fossil age of, say, 10 Ma, possible clade ages and their probabilities could be specified as, say, 10 Ma (90%), 20 Ma (50%) and 30 Ma (10%). As an alternative, a steeper probability/age curve could be set, prior to any analysis, with clade ages of 10 Ma (99%), 11 Ma (10%) and 12 Ma (0.5%).

The question for Bayesian analysis is: how are the priors selected? For a given fossil age, normal, log normal, gamma or exponential curves are often specified as priors for probability/age curves and these give rapidly decreasing probabilities for older clade ages. An exponential prior will provide particularly young clade ages. It assigns the highest probability to a clade age that is the same as that of its oldest known fossil age, with the decreasing probabilities for older ages following an exponential curve. Ho and Phillips (2009: 372) warned that exponential priors should be used only 'when there is strong expectation that the oldest fossil lies very close to the divergence event'. Nevertheless, the traditional approach – a literal reading of the fossil record – *always* expects clade age to reflect oldest known fossil age (although this has been refuted in many individual cases) and so many authors now adopt exponential priors (review in Heads, 2012b).

The priors, or prior probability curves, for clade ages in Bayesian analyses represent 'sources of knowledge such as expert interpretation of the fossil record' (Drummond and Rambaut, 2007). Yet they can also introduce error, by incorporating literal interpretations of the fossil record; in this approach, the oldest fossil age provides the most likely clade age. By selecting appropriate priors, young clade ages with narrow credibility intervals can be generated.

A few authors have recognized the problem with the priors. Parham *et al.* (2012: 352) observed: 'Most studies use a Bayesian framework for estimating divergence dates with probability curves between minimum and maximum bounds . . . but there is presently no practical way to estimate curve parameters'. A review of recent studies showed that the parameters are usually not justified (Warnock *et al.*, 2012). Lee and Skinner (2011: 540) noted that: 'current practice often consists of little more than educated guesswork'.

Parham *et al.* (2012: 352) wrote: 'the fact that a widely applied methodology is subjected to such ambiguous assumptions that have a major impact on results . . . is a major limitation of molecular divergence dating studies'. As these authors concluded, authors should adopt maximum bounds that are 'soft and liberal . . . ' (p. 352), but many studies fail to do so. Using the softest, most liberal, bounds means treating the fossil dates as minimum ages.

Wilkinson *et al.* (2011: 28) also criticized the current approach, as seen in studies on primates (Chatterjee *et al.*, 2009), and wrote:

[Their] young age estimate may be due to the use of two exponential distributions . . . These distributions . . . implicitly assume that the true age is close to the minima and unlikely to be much older than those minima. This assumption, we feel, is unlikely to be warranted, as it does not take account of the sizable gaps that exist in the primate fossil record.

The gaps are not infinite, though, and possible maximum ages can be suggested, based on tectonics. For example, in the primates, the oldest known fossils are from the Paleogene, and molecular clocks suggest a Cretaceous origin, but tectonic-vicariance calibration (assuming the Madagascar and American endemic clades are due to rifting) suggests a Jurassic age (Heads, 2012a).

Bayesian analyses that stipulate appropriate priors will 'validate' young ages for clades, a key component of Modern Synthesis biogeography. These clade ages can then be used to 'rule out' earlier vicariance. This whole process is then said to provide 'evidence' supporting a centre of origin/dispersal model. As in traditional transmogrification, the age of the oldest known fossil in a clade is converted from a *minimum* clade age into an estimate of *maximum* clade age, and the Bayesian framework adds a gloss of respectability to the process. The Bayesian credibility intervals, or highest posterior density (HPD) intervals, provide a false illusion of statistical support, and the calibrations, together with the maximum clade ages based on them, are likely to be gross underestimates of clade age.

The following studies concern groups represented in Australasia (others are given in Heads, 2012b); they provide typical examples of Bayesian transmogrification and the problem of the priors. In all cases, clade ages estimated using Cenozoic fossils and steep priors are used to rule out Mesozoic clade ages and vicariance associated with Gondwana breakup.

Monimiaceae

Based on the oldest known fossils of Monimiaceae, dated at 87–83 Ma and 83–71 Ma, Renner *et al.* (2010) placed a normally distributed prior of 83 Ma with a standard deviation of 1.5 m.y. on the crown group node. This gave an age for the group (95% HPD interval) of 80.5–85.5 Ma. Based on fossils dated at 34–28 Ma they also placed a normally distributed prior of 30 Ma on the divergence of *Xymalos*, with a standard deviation of 1.5 m.y. This gave an age (95% HPD interval) of 33–28 Ma (Oligocene). Using these priors Renner *et al.* (2010) calculated that the Australasian *Wilkiea–Kibara–Kairoa* clade and the tropical American *Mollinedia* clade diverged at only 28 Ma. This is long after Gondwanan rifting and the authors inferred trans-Pacific dispersal from Australasia (Renner *et al.*, 2010). Likewise, the New Zealand–New Caledonia clade *Hedycarya arborea* + *Kibaropsis* separated from its sister-group (remaining

Hedycarya + *Levieria* in New Guinea, eastern Australia and New Caledonia) at 24 Ma (95% HPD: 35–14 Ma). This rules out pre-breakup vicariance.

Annonaceae

Pseuduvaria (Annonaceae) is distributed from Burma to Australia, with maximum species diversity in New Guinea. The Annonaceae have a poor fossil record and Su and Saunders (2009) used the only two fossils regarded as reliable enough for calibration: *Futabanthus* from 89 Ma, to date the Annonaceae (except *Anaxagorea*), and *Archaeanthus* from 98 Ma, to date the Magnoliaceae stem group. Instead of treating the fossil dates as estimates of minimum clade ages, exponential priors were assigned to both calibrations. This resulted in an age for *Pseuduvaria* of just 10–20 Ma (95% HPD). Based on this date and a DIVA analysis, the authors were able to support the traditional model of a centre of origin in mainland Asia/Sumatra, followed by eastward dispersal through Malesia to New Guinea and Queensland.

Uvaria (Annonaceae) ranges from Africa to Australasia. In their study of the genus, Zhou *et al.* (2012) calibrated a phylogeny for Annonaceae with *Futabanthus*, dated at 89 Ma (see above). In addition, a recently discovered fossil, *Endressinia*, dated at 115 Ma was used to date the clade Himantandraceae + Degeneriaceae + Eupomatiaceae + Annonaceae. Exponential priors were assigned to both calibrations and this gave a divergence date for *Uvaria* of just 38–24 Ma (95% HPD). Based on this date and a DIVA analysis the authors wrote that 'the data provide . . . convincing evidence' for a centre of origin of *Uvaria* in Africa, followed by dispersal to Asia and Australasia. Nevertheless, the result is only convincing if the transmogrification (the selection of priors) is accepted.

Arecaceae

Clade ages in the palm family have been calculated by Baker and Couvreur (2013a). They suggested that: 'in the case of fossil evidence the exponential prior distribution is generally preferred', although, as discussed above, exponential priors can only be justified in groups that have exceptional fossil records. Palms do not and so it is not surprising that young ages were estimated for the clades. A DEC analysis indicated a centre of origin for palms in Laurasia and so long-distance oceanic dispersal was invoked: 'The importance of dispersal events in the biogeographical history of palms is unequivocal' (Baker and Couvreur, 2013b). For example, the tribe Areceae includes many south-west Pacific endemics in Australia, New Zealand, New Caledonia and elsewhere, but was dated as originating in the Eocene (mean: 41 Ma, 95% HPD: 32–52 Ma) (Baker and Couvreur, 2013a).

Begoniaceae

Thomas *et al.* (2012a) studied the pantropical *Begonia* (Begoniaceae) in Asia and Malesia. They used fossil calibrations and treated the eudicot oldest fossil age as a maximum constraint. Other fossils, such as the oldest Fagales fossil, were assigned exponential priors, 'reflecting the assumption that, based on the good fossil record . . . the age of the oldest relevant fossils is relatively close to the actual divergence date'. Using the calculated clade ages and results from centre of origin programs (DIVA and DEC), the authors were able to support the standard model of an Asian centre of origin followed by eastward dispersal into Malesia and New Guinea in the Late Miocene to Pleistocene.

Bees

The bee family Colletidae shows many phylogenetic connections among the southern continents (*Hylaeus* was discussed on p. 10). A Bayesian study ruled out vicariance resulting from Gondwana breakup and instead supported trans-oceanic dispersal in the Cenozoic (Almeida *et al.*, 2012). Nevertheless, this result was determined by the use of log normal priors stipulating that clades could be no more than 10–15 m.y. older than their earliest fossils.

Specifying Bayesian priors and their parameters

Huelsenbeck *et al.* (2002: 684) wrote that:

> The use of a prior probability distribution on trees can be viewed as either a strength or a weakness of the method. It seems a strength when the systematist has prior information about the phylogeny of a group. Why not incorporate such information when it is available? However, when the systematist does not have strong prior beliefs, specifying a prior seems more difficult.

Even more serious problems can arise when the systematist does have strong prior beliefs, as these can be imposed as priors even if they are wrong. Earlier beliefs of the Modern Synthesis are often employed as priors. In Bayesian analyses, the specification of exponential, log normal and normal probability curves as priors, and the use of small standard deviations, incorporates the Modern Synthesis view that the terrestrial fossil record gives a more or less accurate representation of maximum clade ages.

Jennings and Edwards (2005) agreed that the priors can be a strength or weakness, 'depending on how informative (or misleading) a given set of priors are relative to how much information exists in the data (likelihood)'. These authors dated a phylogeny in Australian grass finches, *Poephila*, and concluded: 'As is likely to be the case with this type of study we did not have good prior information about our parameters. Faced with such a problem, Bayesians often employ the use of "vague" prior distributions . . . , a strategy that reduces the prior's influence on the resulting posterior probability distributions . . . '. Nevertheless, in many cases Bayesian studies employ very precise priors, which will influence the result. This is the approach now used in fossil-calibrated analyses of evolution, as seen in the studies cited above.

Anyone reviewing the Bayesian studies cited above would conclude that few, if any, geographical disjunctions date back to the breakup of Gondwana, at least for genera and tribes. Nevertheless, this consensus view has not developed because of any new data or analysis. Instead, it reflects the imposition of a prior belief – that fossil age more or less equals clade age. In these recent studies, fossil-calibrated minimum clade ages are converted into maximum ages (with good statistical support), but only by decree, and the potential magnitude of the gaps in the record is swept under the carpet. The alternative method advocated here instead integrates data from tectonics, biogeography and the fossil record, with fossil data used to provide minimum ages.

The comments given here are not meant as a rejection of Bayesian analysis per se, but of the selection of priors in current biogeographic work. Treating fossil ages only as minimum clade ages, as suggested here, is equivalent to using flat priors. The imposition of steep, non-flat priors for fossil-based clade ages is not justified, is unnecessary and leads to erroneous conclusions about the formation of biogeographic patterns. The impact of these conclusions on ecological and evolutionary interpretations has been profound. Authors have felt obliged to reject simple tectonic explanations for general distribution patterns and instead invoke chance processes and unknown ecological factors.

Estimating sampling error in the fossil record

Critical assessments of molecular dating have often focused on fossil calibration error 'because this error is least well understood and nearly universally disregarded' (Van Tuinen and Hadly, 2004). The authors discussed potential error in dating strata and in placing fossils on a phylogeny, although they did not deal with sampling error in the fossil record itself.

Various methods have been proposed for calculating confidence intervals of clade ages derived from fossils by examining the details of the fossil record (Magallón, 2004). These analyses produce estimates of maximum clade age, but they assess the fossil record using the fossil record itself and the ages proposed are often only minor extensions of the oldest fossil age. Givnish and Renner (2004a) were not enthusiastic: 'The construction of stratigraphic confidence intervals from the temporal distribution and abundance of known fossils may, to a limited extent, help compensate for inherent uncertainties in the record'. Brochu (2004) pointed out that methods for extracting an actual divergence date from the fossil record or assessing the confidence limits of a fossil time range 'require information about stratigraphic sampling that we simply do not have for most groups . . . and these methods can themselves be very sensitive to a priori assumptions'. In general, attempts to estimate error in the fossil record using data from the fossil record itself are not convincing.

As mentioned already, there is a widespread assumption that a clade could be older than its oldest fossil age (or fossil-calibrated age), but not by too much. Yet this conflicts with the logical principle that a fossil age gives a minimum age. A clade might be estimated as 1 m.y. older than its oldest fossil, but why not 10 m.y. or 100 m.y.? Gaps of this extent are well known in the fossil record; for example, extant groups whose only known fossils are from the Cretaceous.

There are many possible reasons for a group's absence from the record. At most places there is no rock record for most times, let alone any fossil record. On a smaller scale, presences and absences in the record often reflect local phenomena of ecology and differential preservation of different communities rather than biogeographic patterns. For example, an apparent 'transition' from one species to another over time in the fossil record often represents a phase of ecological change, rather than phylogenetic events or biogeographic dispersal.

Ho and Philips (2009: 374) discussed the assignment of priors and wrote:

> By what period of time could we reasonably expect the age of a node to predate the age of the oldest fossil on either of its descendent lineages? Answering this question is an exceptionally difficult task... It is evident that estimating the level of uncertainty might simply be impossible for the majority of fossil calibrations.

On the other hand, if phylogenetic breaks were correlated with tectonic breaks, estimating the error in the fossil record would be straightforward. For example, one main phylogenetic break in primates (between lemurs on Madagascar and lorises in Africa and Asia) occurs at the Mozambique Channel. If this break is attributed to the formation of the channel, at 160 Ma, each of the clades originated at 160 Ma. Another main break in primates (between the New World platyrrhines and the Old World catarrhines) occurs at the Atlantic Ocean, with the rifting and thus the clades dated at 120–130 Ma. These tectonic/biogeographic estimates of clade ages can be compared with the age of the oldest fossils in the clades to give an estimate of error in the fossil record.

The fossil record and the rock record

Other problems of interpreting the fossil record are evident in studies on biodiversity levels through time. This is often portrayed as a gradual increase, but Peters and Foote (2001) concluded that variation in marine diversity instead reflects variation in the amount of rock available for study. These authors found that extinction rates within marine animals can be predicted on the basis of temporal variation in the amount of exposed rock (Peters and Foote, 2002). This raises the possibility that there has been no substantial increase in taxonomic diversity since the early Paleozoic.

Peters (2005) concluded: 'Many features of the highly variable record of taxonomic first and last occurrences in the marine animal fossil record, including the major mass extinctions, the frequency distribution of genus longevities, and short- and long-term patterns of genus diversity, can be predicted on the basis of the temporal continuity and quantity of preserved sedimentary rock'. Peters suggested that variation in the available rock record has distorted macroevolutionary patterns in the fossil record. If this applies to the marine record, it probably also applies to the terrestrial fossil record.

The assignment of fossils on phylogenies: another problem in molecular dating

Molecular work has shown that many morphology-based groups are dubious. Nevertheless, molecular work often relies on fossil calibration, which in turn relies on identification of fossils and their assignment to a position in a phylogeny. This is done using morphology. Yet morphological analyses of living taxa have been wrong in countless cases, as the molecular results indicate. If analysis of morphology in living groups is difficult, in fossils the problems are even more obscure and the results more controversial, so basing the molecular clock calibration on these is not justified.

Nothofagus is a tree of Australasia and South America. Sauquet *et al.* (2012) studied the timeline of evolution and showed that using different calibrations led to estimates for the crown group age of *Nothofagus* that varied from 13 to 113 Ma. This shows that fossil data alone cannot resolve the problem of dating, even in a group such as *Nothofagus* with its 'rich and well-studied fossil record' (Sauquet *et al.*, 2012). Using younger, more safely identified fossil calibrations gave young ages consistent with previous molecular dating studies. These studies implied that the geographic disjunctions in *Nothofagus* were caused by long-distance dispersal rather than vicariance. In contrast, when older, risky fossils were used for calibration, the estimated ages were compatible with vicariance. Sauquet *et al.* (2012) wrote that several alternative explanations could weaken the inferences of long-distance dispersal made in previous studies on *Nothofagus* and in the 'safe but late' scenarios of their own study. 'First, the maximum age constraint of 125 Ma on the root (node U) [the eudicot clade] might be an incorrect assumption . . . Second, there might have been systematic changes in the rates of evolution, with generally higher rates of evolution early in the diversification of the group than at later stages. Third, the risky fossils might have provided a more accurate calibration of the phylogeny'.

Phylogenetic and taxonomic analysis can be biased by evolutionary and biogeographic preconceptions. For example, authors who assumed traditional ideas on dispersal have accepted that island populations of coastal, sea-dispersed trees cannot be distinct species, while tree populations in montane habitats of islands that are far apart must be distinct species (Heads, 2006a). Neither of these assumptions is justified. In the same way, biologists have often assumed that fossil taxa must be primitive and basal to living relatives. Many Cenozoic fossils are probably closer to extant clades than is often acknowledged and the clades themselves are correspondingly older. Pennington *et al.* (2004b) noted 'a tendency in many studies' to assign fossils to the stem of the clade to which they belong. As they emphasized, this will lead to underestimates of divergence times. Smith *et al.* (2010: 5897) also described 'the default practice of assigning fossils to the stem of the most inclusive crown clade to which they probably belong, thereby possibly biasing estimated ages (possibly throughout the tree) to be younger'. There is no evidence that many fossil groups assumed to be basal or even ancestral just because they are 'old' have such a special status. A typical example concerns the geckos (Gekkota) and their oldest fossil, the mid-Cretaceous *Cretaceogekko*. Studies in Australia (Pepper *et al.*, 2011a) and New Zealand (Nielsen *et al.*, 2011) have used this fossil to calibrate the base of the gecko tree. Nevertheless, while *Cretaceogekko* is the oldest gecko fossil, the only study of the genus (Arnold and Poinar, 2008) gave no indication that it is basal in the group.

The fossil record and its interpretation

Calibrating a phylogeny with fossils is an explicit, logical and testable approach. A problem only arises when the minimum dates are transmogrified into maximum dates. In some cases the fossil record is excellent – in certain geographic areas, at certain times, some organs of some taxonomic groups of particular habitat types can be well preserved. But the fossil record cannot be expected to give a detailed account of the development of life on Earth, or even single lineages through long periods of time. For example, the freshwater cichlid fishes, as with many groups, have their oldest *fossils* in the Eocene, but show major disjunctions among the Gondwanan landmasses. In the traditional approach, a literal reading of the fossil record indicated Eocene *origins* of these groups, and so trans-oceanic dispersal, not vicariance, was inferred. On the other hand, Genner *et al.* (2007) concluded that the 'dispersal implied by the cichlid fossil record may be due to its incompleteness' and supported earlier vicariance, as did Chakrabarty (2004), Sparks (2004), Sparks and Smith (2004a, 2005), and Azuma *et al.* (2008). In practice, most terrestrial groups have a fossil record that is poor to non-existent and most groups with a 'good' fossil record are only good in comparison with these.

Although many molecular clock studies stake everything on a minor and unreliable aspect of the fossil record – the oldest individual fossil – few have shown much interest in what the record as a whole indicates about evolutionary rates and clocks. Fossil taxa often exhibit both short-term evolutionary dynamics and long-term stasis, and the patterns of stasis constitute a genuine problem for evolutionary theory (Eldredge *et al.*, 2005). Advocates of a clock, or relaxed-clock, model of evolution overlook this aspect of the fossil record and only use selected fossil evidence. A valid synthesis must account for all the evidence from paleontology, phylogeny and biogeography, and this suggests that evolution can take place by short, rapid phases of modernization followed by tens of millions of years or more of relative stasis. In this model, evolution is not at all clock-like.

Fossils can be of great biogeographic significance. For example, Worthy *et al.* (2006) discovered a mammal in Miocene rocks of New Zealand and identified it as a clade outside the marsupial + placental group. It is unknown in the extensive Australian fossil record and is one of the most interesting discoveries in New Zealand biology since the nineteenth century. Other fossil groups with great potential relevance for modern biogeography include the extinct orders of South American mammals (Notoungulata, Litopterna and others). But despite the great significance of the fossil record it cannot be taken at face value. It does not make sense to base evolutionary analysis of a group on the absence of fossils, when the record is so poor for most groups and when the available information on extant distribution and differentiation is so extensive. Comparative biogeography should aim at integrating living and fossil records whenever possible, but paleontological monographs seldom include geographic distribution maps of taxa and are more concerned with their stratigraphic distribution. Panbiogeographic analyses of paleofaunas are now appearing, though; for example, that of Gallo *et al.* (2007) on Turonian marine groups.

Using the age of islands or strata to date their endemic clades

Many groups have a poor fossil record. The snakes are a typical example and Burbrink and Lawson (2007) emphasized that fossils of modern groups earlier than the Miocene are 'limited'. The authors argued that this underscores the need for a biogeographic methodology that uses extant taxa, as argued here. Fossil evidence is not the only means of identifying the area and date of origin for modern groups of snakes.

For bryophytes, Vanderpoorten *et al.* (2010) suggested: 'In view of the scarcity of the fossil record..., one promising solution is to use instances of island neoendemic speciation to provide geographic calibration points'. Many biogeographers have accepted that the age of an island, as indicated by the age of the exposed strata, 'indirectly places a maximum age limit on any endemic plants that have evolved *in situ*' (Baldwin *et al.*, 1998). This approach accepts that the age of endemic taxa on volcanic islands can be 'precisely measured' by dating the volcanic strata (Richardson *et al.*, 2001), but the method is flawed as it assumes that the species have only ever lived on that particular volcanic stratum.

Old clades on young islands

While some molecular studies do suggest that clades are younger than their islands, many others have concluded that island endemics are older or much older than their islands and some examples are cited next (see Heads, 2011, for more details). In most cases the studies used fossil calibrations and so the cited ages are estimates of minimum age.

In the south-west Indian Ocean, several taxa endemic to the Mascarene Islands have been dated as older than the islands. These groups include the shrub *Monimia* (Monimiaceae; Renner *et al.*, 2010), the palm *Hyophorbe* (Cuenca *et al.*, 2008), stick insects (Buckley *et al.*, 2010), snakes (the family Bolyeridae; Vidal *et al.*, 2009) and birds (the dodo *Raphus* and the solitaire *Pezophaps*, both in Columbidae (Pereira *et al.*, 2007)).

The Lord Howe Island group, in the Tasman Sea, includes among its endemics the stick insect *Dryococelus* (now restricted to Ball's Pyramid). *Dryococelus* was dated at 22 Ma or 25–50 Ma (Buckley *et al.*, 2009, 2010), while Lord Howe Island itself is dated at 7 Ma. Buckley *et al.* (2009) suggested that before 7 Ma *Dryococelus* inhabited former islands in the vicinity that are now submerged seamounts.

Norfolk Island lies east of Lord Howe Isand in the Tasman Sea. The skink *Oligosoma* of New Zealand has a sister-group *Cyclodina*, dated at 25 Ma, endemic to Lord Howe Island (7 Ma) and Phillip Island, 6 km off Norfolk Island (3 Ma) (Chapple *et al.*, 2009). The authors gave a similar explanation to that of Buckley *et al.* (2009) and concluded that island-endemic clades can be older than their islands.

The Chatham Islands east of New Zealand emerged at ~2.5 Ma (Campbell, 2008; Campbell *et al.*, 2009). Nevertheless, several endemic plants there have been dated as older than this (*Hymenanthera chathamica*: 5–7 Ma; *Embergeria grandifolia*: 6–13 Ma; *Sporadanthus traversii*: 11–13 Ma; *Myosotidium hortensium*: 10–22 Ma; Heenan *et al.*, 2010). An unnamed stag beetle, a species of *Geodorcus* (Lucanidae), is endemic to the Chatham Islands and has been dated at 6 Ma (Trewick, 2000) and the endemic skink, *Oligosoma nigriplantare nigriplantare*, was dated at 5.9–7.3 Ma (Liggins *et al.*, 2008a). The Chatham Islands region has a long history of volcanism dating back to the Mesozoic and so it is possible that these plants and animals survived on prior islands.

Other possible examples of old taxa on young islands occur in the tropical Pacific. Wall (2005) calibrated a phylogeny of the Old World moss *Mitthyridium* using endemic species on Samoa and the Society Islands. He wrote (p. 1421): 'The higher islands of the

Samoan archipelago and the Society Islands (including Bora Bora) were not present until after about 3 mya, which likely represents the earliest conceivable time of establishment of *Mitthyridium* lineages on these islands'. The exposed rock on the *current* islands is young, but there is no need to assume that the current islands were the first ones in the area, whether the archipelagos are caused by hotspots or by propagating fissures.

Murienne *et al.* (2005) used the age of the Loyalty Islands, the young (2 Ma) atolls off New Caledonia that cap the Loyalty Ridge, to calibrate a clock for endemic insects there. Yet the Loyalty Ridge itself is known to have been an active volcanic chain in the Eocene and some models propose it was already active in the Cretaceous at the main south-west Pacific plate margin (Picard *et al.*, 2002; Sdrolias *et al.*, 2003).

In Fiji, Monaghan *et al.* (2006) calibrated a phylogeny with a suggested age of Kadavu Island (1.5–2.5 Ma), based on the age of exposed volcanic strata there. These strata belong to the current phase of volcanism, in which ocean island basalts (typical of intraplate volcanism) have been erupted and emplaced over earlier rocks. Before the latest volcanism, a prior arc that passed through Fiji instead produced andesites, typical of subduction zones. Exposed rocks of this earlier phase are dated from Eocene to Miocene (Colley and Hindle, 1984; Cronin *et al.*, 2003), but regional tectonic models propose that the arc dates back to the Cretaceous (Schellart *et al.*, 2006). Thus the long history of earlier islands is more important than the age of the current islands. Most islands are young but have formed at subduction zones, mid-ocean ridges, hotspots and fissures that have existed for tens of millions of years longer than the individual islands.

Other young Pacific islands with older endemics include the Hawaiian Islands. In a discussion of this critical topic, the 2000-, 4000- and 5000-m isobaths of the central Pacific were mapped (Heads, 2012a). O'Grady *et al.* (2012) suggested that this was 'disingenuous', because sea level has not dropped by more than 100 m or so. But the authors overlooked the *thousands* of metres of subsidence that the Pacific seafloor itself has undergone. As the seafloor has drifted away from the East Pacific Rise – the spreading ridge that produced it – it has cooled (increasing its density) over tens of millions of years and has subsided by these large amounts. This has led to the submergence of most of the islands that were perched on it (the current high islands are new ones). Evidence for the subsidence is seen in the numerous atolls of the region, formed by coral reefs that have grown as the seafloor subsided. The many flat-topped seamounts (guyots) located north, south, east and west of Hawaii are former high islands that were eroded to sea level before being submerged with the tectonic subsidence.

Other young islands with old endemic clades include the Revillagigedo Islands off Mexico and also the Galapagos, where endemics dated as older than the islands include weevils, the giant tortoise *Geochelone nigra*, marine and land iguanas, and geckos (Heads, 2011). The Juan Fernández Islands off central Chile were formed at 4 Ma, but include endemics such as the plant *Lactoris* (Aristolochiaceae). *Lactoris* is dated to the Late Cretaceous from fossil pollen and to the Early Cretaceous (125 Ma) in molecular clock studies (Wikström *et al.*, 2001). Records of fossil *Lactoris* pollen have a circumglobal distribution, with the youngest being from the Early Miocene of Patagonia (~20 Ma; Gamerro and Barreda, 2008). *Lactoris* pollen is distinctive (it is the only angiosperm pollen with sacci) and is good evidence that the plant is older than the island. It is possible that other Juan Fernández endemics also have more widespread, early pollen, but because their pollen is not as distinctive as that of *Lactoris* it has not been recognized.

Island geology and biology

Some island endemics probably had a similar history to *Lactoris* and in the past occurred on a neighbouring mainland. Other old island endemics would have inhabited the region around the island on former, nearby islands that have been submerged or subducted. Whatever the case, using the age of islands to date their endemic taxa will often give unreliable results with unpredictable and sometimes massive errors. This is sometimes acknowledged, at least in theory, for individual islands in 'conveyor belt' hotspot systems, such as the Hawaiian and Galapagos Islands. In these settings, populations can persist by colonizing younger islands from older ones, before the older ones submerge. The same process can occur in other tectonic settings, such as island arc complexes (for example, the Solomon Islands) or uplifted parts of accretionary wedges (for example, Barbados).

The idea that an endemic on a volcanic island cannot be any older than its island is not a new one. After 'landbridges' were discredited in the early twentieth century, biologists assumed that this left long-distance dispersal, either from a mainland or from other, current islands, as the only way that taxa could establish on 'oceanic' islands. This overlooks the fact that volcanic islands are always forming at divergent plate margins (for example, Ascension and Saint Helena Islands in the Atlantic), convergent plate margins (for example, the Lesser Antilles or the south-west Pacific islands), and intraplate centres of volcanism (such as Mauritius and Réunion). The current volcanic islands at these sites are not the first to be formed there – they are just the latest in a long history of island formation and disappearance. Older strata or islands are now eroded, buried, subsided or subducted. The detailed paleogeography of archipelagos and islands is often not known and probably never will be; small islands and hotspots come and go, and geologists have little expertise or interest in deducing whether small areas of land were above water at a particular time. On the other hand, living taxa, at least in certain groups, can persist *in situ* on rejuvenating systems of small islands as dynamic populations. Although the distribution patterns of taxa of small islands are seldom studied in detail (because of the underlying assumption of chance dispersal), they have the potential to provide more information on paleogeography than the ages of exposed volcanic strata.

Old regional metapopulations surviving on young islands at old geological features

Islands do not form at random localities in the middle of an empty ocean. The vast majority are built up in the vicinity of other, prior islands around island-producing structures such as subduction zones, spreading ridges and centres of intraplate volcanism. The individual islands or volcanoes are often short-lived compared to the age of the volcanism in the general area. Taxa persist as metapopulations (populations of populations) that survive by colonizing new islands from older islands in the vicinity. In this way, an old group with suitable ecology can persist indefinitely on young islands or strata within a region. The ability to fly or swim, or the possession of wind-dispersed or sea-dispersed seed, means that reef fishes, Diptera, seed plants and others can survive as metapopulations on dynamic archipelagos, whereas animals such as cursorial mammals cannot. But the means of survival utilized by island metapopulations do not explain the patterns of biogeographic differentiation among different archipelagos or why the groups are there in the first place.

The idea that clade age must be less than or equal to the age of the clade's current island has often been used as evidence for long-distance dispersal. The islands in Hawaii, Tahiti and the Galapagos, for example, are young and so dispersal theorists have argued that their biotas have resulted from long-distance dispersal, from a mainland. Dispersal theory suggests that, because individual volcanic islands were not joined at any stage, later separating, there cannot have been a vicariance history and groups must have spread and evolved by chance dispersal. Woo *et al.* (2011: 434) proposed that: 'Dispersal patterns and processes can be most clearly investigated on oceanic island chains ..., whereas older and larger continental islands have more complex histories and more diverse processes, with the added possibility of vicariance to explain current distributions'. Yet if island populations behave as metapopulations they can also be subject to vicariance, for example, when active volcanic island arcs are rifted apart (Fig. 2.1). This has been suggested for the Vitiaz arc, an arc in the south-west Pacific that rifted apart to form the islands of Melanesia (Bismarck Archipelago, Solomon Islands, Vanuatu, Fiji and others).

The standard theory of island biogeography rejects the ordinary, local dispersal among the islands and islets in a region – the process by which old metapopulations persist there – but at the same time the theory supports extraordinary, long-distance dispersal (founder speciation) from much further away. An alternative model proposed here accepts local dispersal among populations on islands in a region (with constant colonization of new islands) and vicariance between metapopulations, by rifting and strike-slip displacement of island chains. In this model founder speciation is unnecessary. Ordinary ecological movement permits metapopulation survival, but does not in itself lead to differentiation.

If island taxa exist through time as dynamic metapopulations, regional and island endemics can be much older than the individual islands they inhabit at present. This has now been indicated in many groups, as in the clock studies cited above. In this model there is no difference between the biogeography of oceanic islands and the biogeography of habitat islands on a continent.

The archipelagos of Vanuatu and Fiji are typical island arc systems and provide a good example of metapopulation biogeography. Together Vanuatu and Fiji form an important centre of endemism that is now being rifted apart by the opening of the North Fiji basin. Both archipelagos lie along an active plate margin. In Vanuatu, islands such as Tanna are currently being built up by active volcanism, while older islands have disappeared in historical times by sliding down-slope into interarc rifts (Nunn *et al.*, 2006). In Fiji, recent volcanism (beginning at 0.8 Ma) has built the island of Taveuni, while considerable

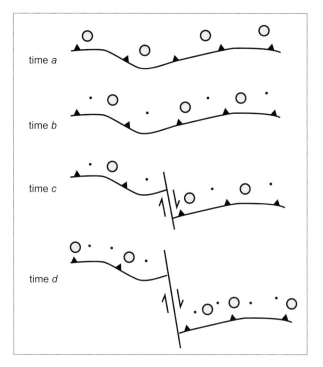

Fig. 2.1 Persistence and vicariance of a metapopulation on a rifted island arc. Lines = active convergent plate margins with subduction zones and island arcs (barbs on over-riding plate). Large grey dots = populations on recently formed, high islands; small black dots = subsided islands (atolls and seamounts) where populations have gone extinct. At times *a*, *b*, *c* and *d*, a metapopulation survives along an active island arc by continuously colonizing new islands. During times *c* and *d*, the arc and the metapopulation are rifted apart. Each of the two separate metapopulations continues to survive within its own region by colonizing young islands produced by the continued volcanism.

subsidence has occurred south of Taveuni, in the Lau group (Heads, 2006a). Many groups in the Vanuatu and Fiji archipelagos exist as dynamic metapopulations that were once adjacent but have been separated by ~800 km of seafloor spreading.

It is sometimes assumed that vicariance requires a continuous landmass to begin with, but metapopulations surviving in dynamic island systems that are being rifted apart are just as likely to undergo vicariance as populations on rifting continents. A metapopulation history resolves discrepancies in the usual model, as this proposes dispersal to Hawaii from the continental mainland, over thousands of kilometres, but denies dispersal to Hawaii from the Line Islands (now all atolls) or the Musicians Seamounts (now all submerged), over hundreds of kilometres. A metapopulation model also allows the integration of evolutionary ecology with geological time and tectonic processes. Losos and Ricklefs (2009) wrote: 'Because many islands are young and have relatively few species, evolutionary adaptation and species proliferation are obvious and easy to study'. Yet the model of diversification on islands that they proposed does not mention geological change.

Palms and other plants on Lord Howe Island: sympatric evolution or secondary overlap?

Lord Howe Island, with rocks dated at 7 Ma, is an emergent part of the Lord Howe Rise, a mainly submerged strip of continental crust that lies east of Australia. The Lord Howe Islands have a fascinating biota, including the stick insect *Dryococelus*, dated as older than the islands (see p. 45). The palm genus *Howea* is endemic to Lord Howe Island and has two species there. Savolainen *et al.* (2006) calculated that *Howea* originated ~5 Ma, based on a palm phylogeny that was calibrated with a Cretaceous fossil and four palm genera endemic to the Mascarene Islands. It was assumed that these last four genera can be no older than the young strata they now inhabit, although this is now doubtful; as indicated above, at least some Mascarenes endemics are much older than their island. Savolainen *et al.* (2006) concluded that the two species of *Howea* evolved in sympatry on Lord Howe Island, but this was only because of their inferred young ages and because Lord Howe in its present form is 'so small that geographical isolation cannot realistically occur'. Nevertheless, as Stuessy (2006) pointed out, the geography of the island and its environs has changed over time. Ball's Pyramid, the only known locality of the stick insect *Dryococelus*, is a good example. It is obvious that this island, an impressive rock stack 23 km off Lord Howe Island and 562 m high, is the relic of a larger island and there are many other now-submerged seamounts on the Lord Howe Rise (Mortimer, 2004). The Lord Howe Rise itself is a thinned, submerged block of continental crust that has had a history of formation, accretion to Gondwana, separation by rifting from Gondwana, and later separation by rifting from the Norfolk Ridge to the east. The ages of the current islands probably have little relevance to the historical biogeography of the plants and animals that survive there.

A period of compression, perhaps related to overthrusting in New Caledonia, occurred in the Eocene and led to uplift of the northern Lord Howe Rise. By the Oligocene, the area was again in bathyal depths, but by this time there were two seamount chains active to the west of the Lord Howe Rise. The Lord Howe seamount chain is 1000 km long and includes Lord Howe Island along with 30 large guyots (Gifford seamount etc.). These are former high islands that have subsided and been eroded flat by the sea before submerging, and so at one stage would have supported terrestrial life. The chain extends north to Chesterfield Plateau and the oldest rocks, in the north, formed at ~30 Ma (Early Oligocene) (Knesel *et al.*, 2008). Between the Lord Howe seamount chain and Australia there is another line of seamounts, the Tasmantid chain. The oldest rocks, again in the north, formed at ~35 Ma (Late Eocene).

The sister-group of *Howea* is *Laccospadix* of north-eastern Queensland (Townsville–Cairns), 2000 km away from Lord Howe Island (Savolainen *et al.*, 2006). The connection would often be attributed to chance, overwater dispersal to the island. Yet if long-distance dispersal can explain one species of *Howea* on Lord Howe, double invasion followed by chance extinction of the ancestor in north-eastern Queensland can explain two species there. Thus sympatric divergence is unnecessary in a dispersal-based biogeographic model. If the Queensland–Lord Howe connection is instead due to vicariance of a widespread Tasman Sea group, with a long subaerial history of different islands on and around the Lord Howe Rise, sympatric speciation on Lord Howe Island is again unnecessary.

In morphological studies, the stick insect *Dryococelus* of Lord Howe Island forms a pair with *Papuacocelus* of eastern New Guinea (Mumeng and Wau, at the northern margin of the Owen Stanley terrane) (Hennemann and Conle, 2006a). The pair is related to *Thaumatobactron* of New Guinea and *Eurycantha* of New Guinea, Bismarck Archipelago, Solomon Islands, New Caledonia, Loyalty Islands and Vanuatu.

Papadopulos *et al.* (2011) proposed that nine Lord Howe Island plant species, in addition to the two *Howea* palms, have arisen by sympatric speciation; five in *Coprosma* (Rubiaceae), two in *Metrosideros* (Myrtaceae) and two in the fern *Polystichum*. Nevertheless, there is no real evidence for sympatric speciation apart from the fact that sister-species currently occur on the same island. The biota of the Lord Howe Island region is probably much older than the island. For example, the stick insect *Dryococelus* on Ball's Pyramid is dated at 22–50 Ma, suggesting that Lord Howe Island itself, dated at 7 Ma, has little direct relevance to the evolution and biogeography of its endemic biota, apart from providing a current 'perch'. Before 7 Ma, *Dryococelus*, the *Howea* palms and the plants mentioned by Papadopulos *et al.* (2011) would all have inhabited former islands in the region that are now submerged seamounts. Subsidence and submergence have occurred due to erosion and also crustal extension. This occurred along Lord Howe Rise and in the rest of the New Zealand Plateau (Zealandia), before, during and after the rifting of the Tasman Sea.

Mountains as islands

Continental mountains provide good examples of habitat islands and, again, illustrate the general principle that old taxa are often endemic to much younger strata. The giant stratovolcano Mount Kilimanjaro in Tanzania began to erupt at ~2.5 Ma (Nonnotte *et al.*, 2008). Yet the mountain's fauna includes endemics such as the katydid *Monticolaria kilimandjarica* (Tettigoniidae: Orthoptera), dated as 7–8 m.y. old (Voje *et al.*, 2009). Populations on large volcanic edifices can survive *in situ* during the construction of the mountain if they are able to colonize younger lava flows and ash beds from older ones before the latter are buried.

Old taxa can persist within a zone of volcanism or uplift by having a weedy, pioneer ecology; for example, one that favours disturbed sites with young soils. Many old taxa survive on very young coastal sands, upland scarps or alpine stream banks. A particular cliff, valley or volcano may be recent, but this does not mean that the habitat *type* is a new development in the region. Volcanism on land, as in the ocean, often persists over long periods in the same area and so the direct ancestors of taxa endemic to Quaternary volcanoes could have existed within their region for tens of millions of years, always inhabiting volcanics less than 2 m.y. old. Examples include the katydids in East Africa, birds of paradise in the New Guinea Highlands and land snails and plants in New Zealand.

Habitats as islands

Habitats other than mountains can form ecological 'islands' and the ages of some of these habitats have been used to date clades. For example, the shrub *Hebe stenophylla*

(Plantaginaceae) inhabits 'islands' of suitable ecology in northern New Zealand, and Bayly *et al.* (2000) wrote that 'The distribution of *Hebe stenophylla* in the North Island cannot predate the formation of the landscapes that it occupies'. This view emphasizes the current topography and stratigraphy, but the underlying tectonic history is more significant. *Hebe stenophylla* occurs in open, rocky habitats, often along rivers and coastlines on calcium-rich marine sediments (Bayly *et al.*, 2000), and these habitat *types* – not the current landscapes themselves – have been available throughout New Zealand's history.

Many island plants and animals occupy habitats that have formed in recent times, including beaches with coral rubble, uplifted carbonate platforms, lava flows, volcanic sediments, alluvial deposits and young surfaces created by rapid erosion in recent mountains. Again, the particular beach, landslide or lava flow that is inhabited is probably quite young, but that does not mean that the habitat type has not existed in the region for a long time. By constant colonization of new landslides and so on, old populations can 'float' on younger stratigraphy. Of course, life cannot survive on molten lava, but individual lava flows are often narrow. Many recent ones on the volcanic islands north of New Guinea resemble straight, sealed roads running through the rainforest. The forest is eliminated along the path of the flow but soon regenerates. Individual lava flows often do not cover an entire island all at once, but over thousands or millions of years, eruption by eruption. This allows the biota to survive *in situ*, by constantly colonizing younger flows from older ones. With continued eruptions, the entire island can end up being covered by younger strata and older communities that have persisted in the area (cf. Craw *et al.*, 1999, Figures 2–5). Many plants of active margins show impressive tolerance of volcanic activity. After eruptions on the volcanic islands off northern New Guinea, palm fruits germinate in the steaming ash following the first rain, and other plant families, such as Thymelaeaceae, Symplocaceae and Ericaceae, also thrive around active craters in the Asia-Pacific region (personal observations).

Other possible cases of old taxa on young islands include fig species (*Ficus*, Moraceae) that live on seashores and coral rock of central Pacific islands. Corner (1963) wrote that these 'may be the pioneers on new coral islands, or they may be the remnants of a disappearing flora'. If the taxa have survived as metapopulations they could be both weedy pioneers *and* ancient relics. The same applies to local endemics such as 'fireweeds' in New Zealand *Celmisia* (Asteraceae; Allan, 1961: 656) that become dominant after burning, and endemic 'volcano weeds' along the Kermadec–Tonga subduction zone, such as *Scaevola gracilis* (Goodeniaceae). Even these weedy, pioneer groups are probably much older than any of the individual, ephemeral islands that they inhabit at present.

Corroboration of fossil-calibrated dates with dates calibrated using island age

The two methods of calibration discussed so far, using fossil age and island age, are sometimes used together in the hope that this will increase the reliability of the dating. This approach is limited though; both methods give underestimates and so any mutual corroboration is likely to be an artefact.

One group that has been studied using fossil age and island age calibrations is the sub-shrub *Abrotanella* (Asteraceae), with a disjunct distribution in Australasia and Patagonia. The fossil evidence can be considered first. Wagstaff *et al.* (2006) noted that fossil dates only give minimum ages, but transmogrified the fossil-calibrated dates for *Abrotanella* and concluded that the genus 'initially diverged during [not 'before'] the Miocene' (p. 100), with secondary species radiations 'about [not 'before about'] 3.1 million years ago'. Using these dates as maximums they ruled out earlier vicariance, concluding that the disjunct distribution '*must* reflect long-distance dispersal' (p. 104) and (in the Abstract) that the species radiations '*undoubtedly* reflect long-distance dispersal . . .'. [Italics added.] For three alpine species in New Zealand, Wagstaff *et al.* (2006) wrote: 'based on our molecular evidence', i.e. the fossil-calibrated rate, 'these three species are very closely related and must have diverged during [not 'before'] the past 500 000 years' (p. 102). Again, this is a minimum age only.

Wagstaff *et al.* (2006) compared their fossil-calibrated rates in *Abrotanella* with rates estimated for the genera *Dendroseris* and *Robinsonia*, also members of Asteraceae and both endemic to the Juan Fernández Islands. They found that the rates calculated for these genera (Sang *et al.*, 1994) were 'only slightly slower' than the rates calculated for *Abrotanella*. However, the rates calculated for the Juan Fernández genera are probably much too fast, as the genera were assumed to have differentiated from their respective ancestors only after the formation of the islands (Sang *et al.*, 1994). Other taxa endemic to Juan Fernández, such as *Lactoris* (dated as Cretaceous; see above), show that this assumption is unwarranted. Thus the fossil-calibrated dates and the island-age calibrated dates for *Abrotanella* are both likely to be underestimates of actual age and so the corroboration between them is unconvincing.

To summarize: many young islands and, in more general terms, young strata around the world host endemic clades that are older than the islands or strata themselves. Authors have often overlooked this principle and applied the idea that island age equals clade age to other areas of evolutionary ecology. Verboom *et al.* (2009: 44) suggested that 'Like island-endemic taxa, whose origins are expected to postdate the appearance of the islands on which they occur, biome-endemic taxa should be younger than the biomes to which they are endemic'. Nevertheless, at least some clades are much older than their islands, strata or biomes.

Using tectonics to date clades

After biologists have estimated a phylogeny for their study group, they often aim to place the biogeographic events within a geological context. Everyone agrees that this is desirable, but how it should be done is controversial. Phylogeneticists sometimes suggest that, to integrate biogeography and geology, divergence times for nodes must be estimated (Fries *et al.*, 2007). Condamine *et al.* (2013a) wrote that phylogenies should be based on adequate sampling and 'should also be well connected with a temporal framework to allow relevant comparisons with known past geological events'. Yet there is no need to connect the phylogeny with a temporal framework to compare it with geology. The comparison of the phylogeny and geology can be done more simply and directly by using the geography of the molecular clades, rather than dates based on their fossils, and this comparison will itself allow the connection with a temporal framework. Examples of this approach are discussed in the next section. The dates of biological divergences are controversial, whereas the geography of molecular clades is often clear-cut and unambiguous, suggesting that geography has a special significance in evolution.

The traditional view maintains that fossils are necessary to calibrate the time course of evolution. For example, Donoghue and Benton (2007: 424) wrote that: 'directly or indirectly, all molecular clock analyses rely on paleontological data for calibration . . . molecular clocks require fossil calibration'. Nevertheless, many modern clock studies do not use fossils at all. Instead they use island age calibrations, as discussed above, or more realistic calibrations based on tectonic events in Earth history, discussed next. Forest (2009: 790) wrote 'There is general consensus that the fossil record provides by far the best information with which to transform relative time estimates into absolute ages', yet this is not accepted by the many authors who use alternatives to the fossil record to date a phylogeny.

Some authors have suggested that molecular clocks 'inherently tend to overestimate' ages (Vences, 2004: 224, cf. Rodríguez-Trelles et al., 2002), but this is not true for clocks calibrated using fossils and islands, as in most cases both will give underestimates. Instead, the time course of evolution can be calibrated by relating the distribution of molecular clades to the tectonic structures and events they coincide with spatially. Tectonic calibration is often not straightforward as many tectonic features, such as major fault zones, have been reactivated at different times. Nevertheless, many events are distinct and well dated, and the general approach seems to hold more promise than the last two methods.

Tectonic calibrations in cichlids, onychophorans, passerines and skinks

Most taxa do not have a good fossil record, but most do have molecular clades with distinctive, geographic distributions, and these include boundaries that can be related to aspects of regional tectonics. Many tectonic events have been studied in detail by geologists and are dated with a high degree of confidence. Cichlid fishes in African lakes are often cited as the classic example of recent 'explosive radiation'. Yet the family has one main clade in America, Africa and Madagascar, with the other in Madagascar and India, and interesting studies using tectonic calibrations have proposed that the group is much older than its fossil record suggests (Genner et al., 2007; Azuma et al., 2008).

In Australasia, disjunction across the Tasman Sea is a classic problem in biogeography. A study of Onychophora utilized tectonic calibrations (the split between Africa and America) and found that an Australia–New Zealand disjunction is compatible with vicariance (Allwood et al., 2010). In other words, a rifting origin for the trans-Atlantic break is compatible with a similar origin for the trans-Tasman break, and a coherent evolutionary/tectonic history is possible – chance dispersal is not needed. In other studies based on tectonic calibration, breakup dates for Gondwana have been used to calibrate phylogenies of Australian carabid beetles (Sota *et al.*, 2005) and passerine birds (Shepherd and Lambert, 2007; Moyle *et al.*, 2012).

In an example of trans-Tasman Basin vertebrates, the *Oligosoma* skinks and allies are endemic to Zealandia (New Zealand, Lord Howe Island, Norfolk Island and New Caledonia) and have their sister-group in Australia (Chapple *et al.*, 2009). One option for calibrating the phylogeny would be to utilize the start of rifting between the two regions – Zealandia and the main part of Gondwana – dated at ~85 Ma (Late Cretaceous). This even has some support in the fossil record, as a member of Scincoidea has been described from the Late Cretaceous of Madagascar (Krause *et al.*, 2003) and a fossil-calibrated molecular clock study placed the origin of Scincidae in the Middle Jurassic (Hedges and Vidal, 2009). The oldest fossil specimens of New Zealand lizards are Miocene in age (16–19 Ma) and are 'very similar to (if not indistinguishable from)' extant New Zealand skinks and geckos (Lee *et al.*, 2009). The authors inferred 'long-term conservatism of the New Zealand reptile fauna'.

Problems with tectonic calibration

Calibrating phylogeny with tectonic events avoids relying on the fossil record to give maximum clade ages, but it is not always straightforward. One of the best known tectonic calibrations attributes the break between Pacific and Atlantic clades to the rise of the Panama isthmus, but many authors have recognized that this is problematic (Heads, 2012a). Geological processes are often episodic and even singular events are not instantaneous. For example, uplift of the Andes began in the Cretaceous, with phases of rapid uplift in the Eocene and the Miocene. The first phases of Andean uplift, in which land was raised out of the sea and low hills formed, would probably have been the most important for the lower elevation groups in the region. In another example, seafloor spreading in the southern Tasman basin began at ~84 Ma and proceeded northward, but some land connections between Australia and Zealandia persisted in the northern Tasman region until ~55 Ma. Which date should be used to estimate divergence of sister-groups in Australia and Zealandia? Nattier *et al.* (2011) used the latter date and this would be logical if Zealandia was new land. However, the continental land of Zealandia formed tens of millions of years before seafloor spreading began, in the Early Cretaceous orogeny. By the time of seafloor spreading, significant dispersal between Zealandia and Australia had probably ceased, as each area would have developed its own biota. In many trans-Tasman groups discussed below, the pre-breakup crustal extension and volcanism has been more important for phylogeny than the later seafloor spreading.

Using multiple tectonic events to calibrate nodes on a phylogeny

The birds of paradise, members of Paradisaeidae, occur in New Guinea and parts of eastern Australia. In their study of the family, Irestedt *et al.* (2009) avoided relying on oldest-fossil age or island age to calibrate the phylogeny and instead used a tectonic event. They assumed that the basal differentiation in passerines (New Zealand Acanthisittidae versus all other passerines) was caused by the split of New Zealand from Antarctica at ~76 Ma. This is a useful minimum estimate, although it is possible that the passerines are older than this. In any case the tectonic calibration already suggests a time course of the phylogeny in which the birds of paradise have an 'unexpectedly long history' (Irestedt *et al.*, 2009). The study proposed, for example, that most of the speciation occurred before the Pleistocene (cf. Heads, 2001a). In this example and the others cited above, biologists are moving beyond the illogical and dangerous reliance on fossils and island ages to calibrate clocks, and are instead using well-dated tectonic events.

The study by Irestedt *et al.* (2009) dated a group using a single tectonic event and a related node (the basal node in passerines) that are distant in space and phylogeny from the group studied. If this is valid it would also make sense to use many tectonic events to date many individual nodes within the group (a standard technique used with fossils, when they are available). This proposal would involve a tectonic/biogeographic calibration of each node, not just one, and a detailed, systematic engagement of phylogeny with tectonics. This approach has been used in studies of birds of paradise (Heads, 2001a,b) and primates (Heads, 2012a).

Calibration of phylogenies with tectonics is still in its infancy, but this approach combines the best of molecular biology (the clades and their distributions) with hard-rock geology and tectonics (rather than stratigraphy) and avoids the many problems of fossil calibration. Until recent times, biologists have been reluctant to rely on tectonics and geographic distribution to calibrate phylogeny. This is because the distribution of the living populations has been perceived as fluid and transient compared with the evidence of the fossil record. Nevertheless, molecular studies now show that distribution has special significance.

The significance of geography in molecular phylogeny

The importance of geography is a recurring theme in current molecular studies. Many authors are finding clades that are incongruent with earlier, morphological classifica-tions, but show distinct geographic structure, often much more so than in the earlier arrangements (e.g. Uit de Weerd *et al.*, 2004; Noonan and Chippindale, 2006; Ste-fanović *et al.*, 2007; Larsen *et al.*, 2007; Stadelmann *et al.*, 2007; Rawlings *et al.*, 2008; Chintauan-Marquier *et al.*, 2011). In this sense, geographic distribution has turned out to be the 'character of characters' for molecular phylogeny in general. Most large groups show simple allopatry or at least clear geographic structure among the main molecular clades. This discovery has already impressed authors such as Avise (2007), but its full implications remain to be explored. It provides some of the strongest evidence yet that geographic distribution has a determined, genetic relationship with phylogeny and that a group's distribution is not the result of chance dispersal after the group's origin. Recog-nition of this relationship is encouraging authors to map their clades and leading to a renewed interest in comparative biogeography.

Extant mammals are an example of a large group in which the phylogeny has a strong geographic component (Springer *et al.*, 2011). The basal group includes the egg-laying monotremes of Australia and New Guinea (platypus and echidnas). In the remaining groups, extant marsupials are in Australasia and America, and are sister to placentals.

In placentals the basal group is a trans-tropical Atlantic clade, the Atlantogenata. This comprises the Afrotheria (elephants, hyraxes, tenrecs, etc.), mainly in Africa (exceptions are the Asian elephant and sirenians), and its sister-group the Xenarthra (armadillos and sloths), restricted to the Americas.

In the phylogeny of mammals and many other groups the precision and depth of the geographic structure is becoming more and more evident with increased sampling. This great progress in describing the distributions has, in many cases, led to an increased understanding of the processes that gave rise to them. In mammals, traditional scenarios were based on oldest-fossil calibrations and chance dispersal but, based on the new phylogeny, Springer *et al.* (2011) recognized the importance of vicariance among the main clades.

A phylogeny is not just a topology as it also involves values for branch lengths (degree of differentiation), but the meaning of branch length is not obvious. It could be proportional to the age of the group, to the time involved in its differentiation, to both of these, or to neither. In contrast, the spatial significance of the topology is less ambiguous and in most cases the geographic distributions of molecular clades show clear-cut patterns.

Craw *et al.* (1999) suggested that life is the uppermost geological layer. It differs from other strata through its 'stickiness' or 'weediness' and its resistance to erosion. Because of this it can evolve together with the geological strata it survives on as metapopulations. A community will be redeposited onto lower strata in areas of erosion and elevated during phases of uplift. In areas of horizontal tectonic movement a community will undergo local or regional 'deformation' and 'metamorphism'. In this way the geography of a group can represent inherited information that can be preserved and passed on, with minor modifications, for tens of millions of years.

Mode of evolution: is evolution more or less clock-like?

Problems with calibrating the time course of phylogenies are well known and often discussed among biogeographers. In addition, different models of genetic substitution will also give different estimates of phylogenies and distances between clades. Edwards (2009) commented: 'the most commonly employed models of DNA substitution do not adequately describe the complexities of DNA sequence evolution'.

With respect to evolutionary rate, Smith *et al.* (2010: 5901) wrote 'It is increasingly clear that there may be extreme differences in molecular rate [of evolution] . . . and current methods may be unable to cope'. The method of dating used in this book – fitting tectonic events (rather than fossils) to a phylogeny – does not assume an evolutionary clock, even a relaxed one. Instead, it accepts that rates of evolution can show extreme changes in lineages and genes at different times and places. Despite the biogeographic evidence for extreme rate changes, many biogeographers still make the implicit assumption that evolution is more or less clock-like. Yet evolutionists did not always think this way. Nineteenth-century evolutionists were well aware that the different groups sharing a similar biogeographic pattern, such as the Australasia–Patagonia disjunction, show different degrees of differentiation (as species, subgenera and genera etc.). Based on this information and a good knowledge of geology, Hutton (1872) was able to conclude that 'differentiation of form, even in closely allied species, is evidently a very fallacious guide in judging of lapse of time'. Hutton recognized that, if evolution is not clock-like, a single geographic pattern could have developed at the same time in the different groups with different taxonomic ranks.

In contrast with these early workers, the mid-twentieth century Modern Synthesis provided a spatial analysis based on the idea of centre of origin and a chronological analysis based on a fossil-calibrated evolutionary clock. This model accepted that the different groups in a biogeographic pattern have all evolved at different times, depending on their rank. This in turn means that a single origin, for example by community-wide vicariance, can be rejected. In the key text of the new synthesis, Matthew (1915) concluded that 'the Malagasy mammals point to a number of colonizations of the island by single species of animals at different times.' Following Matthew's lead, Mayr (1931: 9) discussed the avifauna of Rennell Island, in the Solomon Islands, and wrote: 'The different degree of speciation suggests that the time of immigration has not been the same for all the species'. For the birds of New Caledonia, Hawaii and other islands, he wrote that 'Strikingly different degrees of differentiation indicate colonization at different ages' (Mayr, 1944: 186). For the birds of the Pantepui mountains in Venezuela, Mayr and Phelps (1967: 290) wrote: 'We now know that populations of different species and on islands of different size may differ rather widely in the rate at which they diverge morphologically under the influence of the same degree of isolation. *Nevertheless, all other things being equal*, degree of differentiation of an isolated population reflects length of isolation' (italics added). But among the 96 Pantepui birds 'all other things' are probably not equal and there are likely to be clade-specific differences in the genetic potential of the different taxa to evolve and diverge.

Later authors in the Modern Synthesis have continued to accept the evolutionary clock idea. Ehrendorfer and Samuel (2000) discussed the pattern of disjunction between South America and New Zealand seen in many groups. They wrote that: 'Judged by their morphological (and molecular) divergence, these disjunctions, which range from the infraspecific (e.g. in *Hebe*) to the specific (e.g. in *Anemone*), sectional or even subgeneric level (e.g. in *Fuchsia* or *Nothofagus* p.p.), must be of very different ages ... '. This interpretation assumes that rates of evolution, even in different orders, are similar, and that the taxonomic rank of a clade supplies its age.

Matthew's (1915) idea of a single rate and an 'evolutionary clock' became an integral part of the Modern Synthesis; later it was applied to molecular data and termed the 'molecular clock'. Nevertheless, adopting this idea is problematic in light of the biogeographic evidence and the wide differences in evolutionary rates that Mayr and Phelps (1967) themselves accepted (cf. the start of the passage quoted above).

Avise's paradox

Avise (1992: 63) described a structure that has turned out to be widespread: 'Concordant phylogeographic patterns among independently evolving species provide evidence of similar vicariant histories . . . However, the heterogeneity of observed genetic distances and inferred speciation times are difficult to accommodate under a uniform molecular clock'. A typical example of Avise's paradox is seen in a group of ovenbirds: 'Despite the high congruence among the spatial patterns identified, the variance in divergence times suggests multiple speciation events occurring independently across the same barrier, and a role for [chance] dispersal' (d'Horta *et al.*, 2013).

All important spatial patterns are shared by groups that show significant geographic concordance, indicating general vicariance, but all patterns are shared by different taxa showing many different degrees of differentiation, indicating separate histories and dispersal. The paradox only arises if degree of differentiation is related to time since divergence. In evolutionary clock theory, evolutionary rate is more or less continuous and so the degree of divergence is related to *time* since the divergence. An endemic genus will be older than an endemic subgenus, at least within the same family. Nevertheless, in many cases the degree of differentiation is probably not related to time elapsed and is instead caused by *prior aspects of genome architecture* that determine the evolvability or evolutionary propensity of a group. This explanation for branch length accounts for Avise's paradox.

In a few cases there is some evidence for a molecular clock, at least at local scales. A study of beetles separated by a single break – the mid-Aegean trench – found similar branch lengths for several pairs of separated sister-species (Papadopoulou *et al.*, 2010). The study assumed that the break was caused by tectonics rather than chance dispersal. The results are interesting and support clock-like evolution, but they are all from one family (Tenebrionidae), the break is recent (9–12 Ma) and the sample was not large (six pairs). The results are probably not typical for multiple pairs at a single break.

The likelihood of extreme variability in evolutionary rate

In the clock model of evolution, a degree of variation in evolutionary rate is often accepted. In other words the clock can be somewhat relaxed. Yet tectonic calibrations suggest instead that evolution is not at all clock-like and that extreme changes in evolutionary rate are common. The different groups showing breaks at the same biogeographic boundary usually have different degrees of divergence there, but the break can still reflect a single, ancient tectonic event that affected the entire community. In the plant family Melastomataceae, Renner (2004: 1485) observed that:

Divergence events thought to date back to well-understood Gondwanan events, for example the break-up of South America and Africa, occur at very different distances from the phylogenetic trees' roots . . . Accordingly, hypotheses of trans-oceanic long-distance dispersal were put forward to explain the shallowest geographical disjunctions. *Explaining them other than by different absolute ages* [for example, with a single vicariance event] *would have required assuming tremendous rate heterogeneity.* [Italics added.]

This last point is valid and important, but great-rate heterogeneity (rather than chance dispersal) is accepted here as the explanation.

In another example, the plant *Scleranthus* (Caryophyllaceae) has a disjunct distribution in Europe, Australia and New Zealand. Smissen *et al.* (2003) calibrated nodes in a phylogeny with fossil ages, transmogrified the minimum dates into maximum ages, and concluded that the genus 'diverged within the last 10 million years'. Based on this, the authors concluded that: 'Clearly, *Scleranthus* is capable of long-distance dispersal [from Europe to Australasia, direct, and from Australia to New Zealand] despite lacking any obvious adaptations to facilitate it'. Smissen *et al.* (2005) suggested that trans-Tasman or trans-Tethyan vicariance in *Scleranthus* would imply a very old age for the genus and an age for the family Caryophyllaceae older than the planet. This would be true only if the molecular evolutionary rate were constant over tens of millions of years. Evidence from biogeography suggests that there was a rapid phase of differentiation in the mid to late Mesozoic that produced many angiosperm clades of different rank in and around Caryophyllaceae.

In a similar way, Wagstaff *et al.* (2010b) suggested that if trans-Tasman sister affinities between *Gynatrix* (Malvaceae) in Australia and *Plagianthus* in New Zealand were due to vicariance (rifting of the Tasman Sea at 83 Ma), their relative *Hoheria* in New Zealand must be much older and *Lawrencia*, sister to all three, must have diversified in Australia hundreds of millions of years ago. Instead of contradicting a vicariance model, though, the pattern probably indicates that there have been significant changes in evolutionary rate.

All distribution patterns include clades of different rank, so are all patterns the result of chance dispersal?

Twentieth century dispersal theorists such as Mayr (1965) accepted that most allopatry was caused by dispersal rather than vicariance and most molecular studies on angiosperms and vertebrates agree – vicariance is very rare. Dates produced using fossil or island age calibrations give systematic underestimates but are cited in dispersal theory as evidence that most biogeographic distributions are caused by separate, Cenozoic events of chance dispersal for all the clades in a community, rather than a single, community-wide vicariance event. Many molecular studies now attribute all biogeographic patterns in a group to dispersal and none to vicariance (e.g. Wikström *et al.*, 2010, on the Rubiaceae of Madagascar, a group with 95 genera and 650 species).

Molecular studies of multiple groups that share a particular biogeographic pattern always report a great range of branch lengths and (assuming a molecular clock) ages for the taxa involved. This has been proposed for South African plants (Galley and Linder, 2006), Africa/Madagascar mayflies (Monaghan *et al.*, 2005), western/eastern Pacific fishes (Lessios and Robertson, 2006), south-western/south-eastern Australian groups (Crisp and Cook, 2007, Fig. 2; Morgan *et al.*, 2007), fishes east and west of the Torres Strait (Mirams *et al.* 2011), butterflies on both sides of a biogeographic boundary in Peru (Whinnett *et al.*, 2005), groups divided by the Mississippi River (Pyron and Burbrink, 2010), and birds on both sides of the Bering Strait region (Humphries and Winker, 2011). This work confirms earlier conclusions based on morphology: all patterns include taxa of differing rank. This means that if an evolutionary clock is assumed, no biogeographic pattern would ever be explained by vicariance. All patterns would be interpreted as pseudopatterns, with each of the component taxa being the result of an unlikely event that was repeated, by chance, in many groups at different times. This centre of origin/evolutionary clock model is simple and popular but always concludes with a mystery – how does the process work? Why has each (monophyletic) group in the pattern dispersed only once? Why exactly does dispersal start and then stop?

To summarize: spatial analyses of evolution often assume that basal clades and grades indicate centres of origin, while chronological analyses often assume that evolution is more or less clock-like and that fossil-calibrated ages can be treated as maximum dates. Neither of these assumptions is valid or necessary.

The deep historical roots of some ideas in modern phylogeography

Early dispersal theory focused on finding the centre of origin and deducing a chronology of evolution from the fossil record. Many contemporary studies rely on just one or two popular programs to analyse their data. Using DEC in LAGRANGE for spatial analysis ensures that a centre of origin will be found and multiple area/multiple area vicariance rejected; using BEAST for chronology ensures that clade ages will not be much older than the fossil record. Using both programs, students can find a Modern Synthesis model for their data.

In this way, much current work maintains the core of the earlier research program. Behind the obvious technical advances, data interpretation is still based on the concepts of the Modern Synthesis: centre of origin, dispersal and adaptation. For example, the modern idea that a 'basal' clade or grade occupies the centre of origin is derived from the idea that the primitive member of a group occurs at the centre of origin (Mayr, 1942; Hennig, 1966). In its methods of interpretation, 'Phylogeography has re-invented dispersal biogeography . . . ' (Ebach and Humphries, 2002); the approach favoured by Matthew (1915) 'is revived in molecular systematics of the present, in the search for ancestors and centres of origin' (Nelson, 2004).

With respect to branch length and evolutionary rate, cladistic biogeography was an attempt to avoid relying on degree of difference; it emphasized monophyletic clades rather than their rank. Yet in modern molecular studies, degree of difference, or branch length, has again assumed fundamental importance. By relying on branch length (rather than the topology of the phylogeny) the Modern Synthesis and its recent applications in molecular work have concluded that each group has had a different history. The fundamental process is chance dispersal and any shared aspect of biogeography is assumed to be a pseudopattern caused by chance congruence.

Modern approaches to calibration are also minor variations of earlier ideas. Givnish and Renner (2004a) introduced a special issue on intercontinental tropical disjunctions (Givnish and Renner, 2004b) and commented that all the contributions calibrated molecular trees using the age of the oldest known fossil of the group. As for the results of the symposium, they found it 'surprising' that most studies concluded that long-distance dispersal was pervasive. But, given the fossil-based methodology, the conclusion of trans-oceanic dispersal is inevitable. Many groups have a record that begins only in the Cenozoic, well after the tectonic revolutions of the Cretaceous. Modern Synthesis biogeography was based on a literal reading of the fossil record and concluded by supporting trans-oceanic dispersal; molecular biogeography that calibrates trees using a literal reading of the fossil record is bound to reach the same conclusion.

Case-studies in evolutionary chronology

The significance of the Pleistocene ice ages for evolution

One of Matthew's (1915) main concepts was that the Pleistocene ice ages were of fundamental importance for global evolution and biogeography. The Pleistocene was seen as a time when most extant species evolved and the distributions of any prior species were transformed. This idea has been adopted in studies on New Zealand (Fleming, 1979) and Melanesia (Mayr and Diamond, 2001). Nevertheless, biogeographic work has questioned the idea that most speciation took place in the Pleistocene, and ages produced in molecular clock work are also leading to a critical reassessment of the idea. The glaciations that began with the Pleistocene at 2.6 Ma modified distributions and caused extinction, but molecular evidence indicates that they have not had the profound evolutionary effects that were suggested earlier. A typical example is seen in the tree *Nothofagus* (Nothofagaceae), that occurs in Australasia and South America. The main phylogenetic and geographic break *within* a single species, *N. pumilio* of Patagonia (at the latitude of Chiloé Island), has been dated at 40 Ma, in the Eocene (Mathiasen and Premoli, 2010). The authors concluded that *N. pumilio* survived Pleistocene glaciations by persisting in multiple, small refugia, and following postglacial warming the species re-expanded out from these. In this way, patterns of variation in *N. pumilio* sequences 'were controlled by ancient paleogeography rather than more recent Neogene events such as Pleistocene glaciations'. Similar results have been found in many studies in the tropics and even in the north temperate zone.

Large areas in the northern and southern temperate zones were glaciated in the Pleistocene. Species that are widespread there have been assumed to be the result of reinvasion after the last glacial maximum. Because of this history the widespread species were predicted to show little genetic variation. A similar prediction made for marine clades was discussed above; populations were thought to be uniform over their range. Nevertheless, sequencing studies have shown distinctive geographic structure in many marine taxa. The same pattern – clear phylogenetic structure – has also been found in surveys of many north temperate taxa, even in regions that were glaciated.

At the end of the nineteenth century, the paleobotanist Clement Reid observed that there are oak trees in northern Britain, despite the fact that they would have had to have travelled ~950 km from southern refugia since the last glacial maximum ended ~10 000 years ago (Provan and Bennett, 2008). Reid believed that this rate exceeded the dispersal capabilities of oaks, and the observation, known as Reid's Paradox, was explained by rare, long-distance dispersal events. Later paleobotanical and molecular studies have instead suggested that the species persisted throughout the last glacial maximum in unknown, cryptic refugia at higher latitudes.

Provan and Bennett (2008) cited northern, cryptic refugia for vertebrate lineages in the Carpathians, the Canadian Arctic and elsewhere. Refugia for plants in the Arctic have also been proposed (Abbott *et al.*, 2000). In central Europe, Kotlík *et al.* (2006) found that a woodland mammal, the bank vole (*Clethrionomys glareolus*), survived the height of the last glaciation (~25 000–10 000 years B.P.) in the vicinity of the Carpathians. These mountains lie well to the north of the Iberian, Italian and Balkan peninsulas, often regarded as glacial refugia for temperate species. In Europe, multiple glacial refugia located north of the Alps, further north than was thought possible, have now been proposed for plants (Michl *et al.*, 2010), beetles (Dress *et al.*, 2010) and mammals (Vega *et al.*, 2010). Horsák *et al.* (2010) found that the landscapes of central Europe at the glacial maximum were not all dominated by open, dry steppe, but included areas of shrubby vegetation, wet habitat and woodland. In North America, Rowe *et al.* (2004) proposed northern refugia for the forest chipmunk, *Tamias striatus*.

This new work supports the idea that life is 'stickier' and more resistant to 'erosion' than has been thought. The studies on northern refugia (nunataks are just one form) show that clades can survive *in situ*, even in the face of major environmental change, by persisting as a metapopulation in microrefugia. Following glaciations, many taxa have expanded from local microrefugia in the region itself, rather than invading from warmer regions in the south.

These recent studies suggest that during phases of environmental adversity, populations retreat, not to other regions, but to multiple microrefugia within the region. In eastern North America biogeographers postulated the retreat of many plant and animal clades into southern refugia during the ice ages, followed by northward range expansion after the last glacial maximum. Yet plants of the Gulf and Atlantic Coastal Plain demonstrate complex, recurrent distributional patterns that cannot be explained by this model (Wall *et al.*, 2010).

For organisms on the rocky shores of the north-eastern Pacific, 'regional persistence during the last glacial maximum appears a common biogeographic history' (Marko *et al.*, 2010). The authors concluded that in this region and also in the Atlantic, events predating the last glacial maximum have been more important in shaping the demographic histories of rocky-shore species than the glaciations.

In China, earlier, literal readings of fossil pollen data suggested that all temperate forest species migrated from north to south during the last glacial maximum. Instead, sequencing studies of temperate forest Betulaceae indicated multiple refugia in both southern and northern China (Tian *et al.*, 2009). On the Tibetan Plateau, molecular studies of plants have revealed a high degree of geographic structuring. They suggest that *Juniperus* species (Cupressaceae) survived the coldest phases of the Pleistocene in microrefugia above 3500 m elevation (Opgenoorth *et al.*, 2010) and that *Hippophae tibetana* (Elaeagnaceae) survived in microrefugia at 4000 m (Wang *et al.*, 2010). The authors cited the ever-increasing number of microrefugia that have been detected in the region.

Arctic clades include groups with local endemism and poor means of dispersal. For these reasons, biogeographers have long suspected that some arctic groups survived the glaciations in local ice-free sites within the limits of the northern European ice sheets. Now there is good molecular evidence corroborating this. Westergaard *et al.* (2011) presented evidence for *in-situ* glacial persistence of two arctic-alpine plants, *Sagina caespitosa* and *Arenaria humifusa* (both in Caryophyllaceae). Both species occur in North America and Europe and belong to the 'west-arctic element' of amphi-Atlantic disjuncts. Their only European localities occur well within the limits of the last glaciations. Two distinct and diverse genetic groups, one North American and one European, were detected in each species. The groups showed reciprocal monophyly, and this excluded postglacial dispersal from North America as an explanation for the European occurrences. The authors concluded that populations persisted through the glaciations in East Greenland and/or Svalbard (*A. humifusa*), and in southern Scandinavia (*S. caespitosa*).

High alpine habitat round the world maintains many distinctive clades and it has often been debated whether these survived the ice ages *in situ*, on small, ice-free islands of habitat (nunataks) or on larger 'massifs de refuge' on the periphery of the main mountain ranges. Lohse *et al.* (2011) sampled the high alpine carabid beetle *Trechus* in the Orobian Alps of northern Italy, where the genus shows local diversity. During the last glacial maximum the southern areas of the range remained unglaciated, while summits along the northern ridge formed nunataks surrounded by the icesheet. Mitochondrial node ages (estimates of minimum age) suggested that the beetles persisted even on the northern ridge through the last ice age.

Thus taxa in glaciated regions have survived *in situ* on nunataks and other sites in the same way that terrestrial taxa survive on ephemeral islands in the ocean: as metapopulations persisting *in situ* at multiple, local sites. ('Refugia' is the usual term for these, although it implies that groups have *fled* there, from somewhere else, and this will not always be the case.) Many taxa will be wiped out during glaciation, marine transgression, mountain uplift and any other geological upheaval, but in many cases at least some living taxa will persist in microrefugia. In this way, life, as the uppermost geological stratum, can often avoid complete erosion.

The paradigm shift in studies of groups of the glaciated north is similar to the shift in marine biology and freshwater microbiology. All have come about through molecular studies showing high levels of geographic structure that are incompatible with the traditional model.

The time course of evolution in bryophytes

Mishler (2001) reviewed bryophyte biology and stressed three points. First, the fossil record indicates that ancient forms were very similar to modern ones. The oldest bryophyte fossil, the thalloid liverwort *Pallaviciniites devonicus*, is Devonian but, despite this great age, resembles extant genera such as *Pallavicinia* or *Symphyogyna* (Vanderpoorten *et al.*, 2010). For liverworts, 'Tertiary fossils are, in general, similar to the modern flora: virtually all are attributable to extant genera and some even to extant species' (Vanderpoorten *et al.*, 2010).

Mishler's (2001) second point was that bryophytes, in their biogeography, tend to follow the same characteristic disjunctions that are seen in tracheophytes. For example, the distributions of liverwort genera can be categorized as Laurasian or Gondwanan (Vanderpoorten *et al.*, 2010), as in many tracheophytes. In Laurasia the authors cited 'the striking similarity between the bryophyte floras of North America and Europe', with a common species pool that contributes up to 70% of the flora in areas such as the Ozark Plateau. In addition to the Laurasia/Gondwana allopatry, other patterns in liverworts, including breaks at Wallace's line and the South Atlantic, have often been documented and interpreted in terms of vicariance. 'These observations contrast with the idea that, in spore-dispersed organisms like bryophytes and pteridophytes, dispersal obscures evidence of vicariance' (Vanderpoorten *et al.*, 2010: 471). This conclusion implies that bryophyte distributions are valuable biogeographic data and reflect tectonic events of the Cenozoic, Mesozoic and even Paleozoic.

Mishler's (2001) third point noted that, while the geographic patterns in bryophytes resemble those of tracheophytes, they are manifest in groups of lower taxonomic rank. For example, species distributions in bryophytes often resemble genus distributions in tracheophytes, suggesting that the different levels reflect phylogenetic constraints in evolution. In a similar way, angiosperms often show much more local endemism than bryophytes. For example, on the tepui mountains of southern Venezuela, 42% of the angiosperm species are endemic, but only 10% of the hepatics and probably fewer in mosses (Désamoré *et al.*, 2010). For Hawaii, the figures for the three groups are 90%, 49% and 30%. Désamoré *et al.* (2010) attributed the differences to the greater dispersal ability of bryophytes, but they could also reflect phylogenetic constraints, with bryophytes having had lower ancestral potential for differentiation than angiosperms. This would also account for Mishler's (2001) second point – bryophyte clades have similar, characteristic disjunctions to those of tracheophytes – while greater dispersal ability would contradict it.

Vanderpoorten *et al.* (2010) followed Mishler (2001) in stressing that bryophyte species often show similar disjunctions to higher category taxa in angiosperms. They wrote (p. 12): 'The corollary of the acceptance of an ancient origin for the disjunctions currently observed amongst the world bryophyte floras is that the low proportion of endemics in the liverwort flora as compared to angiosperms must originate from the slow evolution rates of the former...'. Vanderpoorten *et al.* (2010) cited the case of the disjunct southern hemisphere moss, *Pyrrhobryum mnioides*, in which molecular dating studies attributed the disjunctions to continental drift (McDaniel and Shaw, 2003). Vanderpoorten *et al.* (2010) inferred long-term 'resistance to speciation' in bryophytes and concluded that their evolution has involved periods of stasis over tens or even hundreds of millions of years.

To summarize, although the mosses and liverworts are older, as groups, than angiosperms, individual clades in all three show the same series of biogeographic patterns, reflecting evolutionary responses to the same tectonic events.

The evolution of angiosperms

From the 1960s until the last few years it was assumed that angiosperms evolved in the Early Cretaceous. This is the age of the oldest undisputed fossils, pollen dated at ~136 Ma. Angiosperms were thought to be nested in one of the gymnosperm groups. However, molecular studies have instead concluded that angiosperms are sister either to all extant gymnosperms (Soltis *et al.*, 2011), to all extant gymnosperms except gnetophytes, or to cycads (Mathews, 2009). Of the extant gymnosperm groups, conifers are known from fossils back to the late Carboniferous, and gnetophytes and cycads back to the Permian (~280 Ma; Renner, 2009). Even if angiosperms are sister to cycads alone, cycads, and thus angiosperms, existed in the Permian. This implies that there was a period of ~130–150 m.y., from the Permian to the Cretaceous, when angiosperms existed, although fossils are either unknown or not recognized. If groups known only from fossils are considered, angiosperms are likely to be sister to corystosperms (Upper Permian to Eocene) or Caytoniales (Triassic–Cretaceous), and each of these relationships would be consistent with a Permian origin.

Smith *et al.* (2010) calibrated a land plant phylogeny with 33 fossils and proposed an origin of crown-group (extant clade) angiosperms in the Triassic (~217 Ma), 60 m.y. before the oldest undisputed angiosperm fossils in the Cretaceous. The new Triassic dating is significant as it places the group's origin before Gondwana breakup began in the Jurassic. Smith *et al.* (2010) suggested that the only living lineages that existed between the Triassic and Cretaceous are in the basal ANA grade of angiosperms (Amborellales, Nymphaeales and Austrobaileyales). Nevertheless, the other clade of angiosperms, comprising all the rest, is sister to Austrobaileyales and so is the same age as that order (estimated as ~190 Ma, Early Jurassic, in Smith *et al.*, 2010, Fig. 1).

Molecular studies have estimated that the crown (extant) angiosperm clade originated between the Early Permian and Late Triassic (Magallón, 2010). With this conclusion the subject returns to the point reached by earlier writers (such as Croizat, 1964), who supported a Permian date before the detour of 'Cretaceous origins' began in the 1960s. Angiosperms as a group are probably much older than the Cretaceous. Extant genera are known from the Cretaceous fossil record, and fossil-calibrated clock studies indicate that pantropical genera such as *Piper* and *Peperomia* (Piperaceae) date back to the Cretaceous, at least (Smith *et al.*, 2008).

As the molecular studies supported earlier dates for angiosperms than the fossils, paleobotanists re-examined pre-Cretaceous fossils to see if any of these might be angiosperms. Doyle (2005) wrote that the Triassic pollen fossils termed *Crinopolles* have 'remarkably angiosperm-like' reticulate-columellar structure and could fit on the angiosperm stem lineage. Earlier, Cornet (1989) had concluded: 'If pollen with reticulate-columellate-footlayer ectexinal and non-laminated endexinal structure proves to be an angiosperm autapomorphy . . ., then the Triassic *Crinopolles* Group probably evolved within the angiosperms rather than within a pre-angiospermous anthophyte group . . . '. It is likely that early angiosperm pollen would not be recognized for what it is. Many other early lineages contemporary with the *Crinopolles* group and related to it have either not been found as fossils or not recognized as related.

The eudicots

Apart from the basal groups – the ANA grade families, magnoliids and monocots – the angiosperms comprise one large clade, the eudicots, that includes about 75% of all the species. In the pre-molecular model, magnoliids were thought to be primitive. Early Cretaceous fossil sequences were interpreted as showing the evolution of magnoliid-type (monosulcate) pollen into typical eudicot (tricolpate) pollen. This model for the origin of eudicots is based on the appearance of tricolpate pollen in the Early Cretaceous (~125 Ma). Although this can only provide a minimum date for the clade, in many studies it is treated as a maximum age and used to calibrate the time course of phylogenies (e.g. Buerki *et al.*, 2011). Forest (2009: 792) justified this, suggesting that:

An exception to the rule of using fossils as minimum constraints can be applied to fossilized pollen grains. Pollen grains have a much higher fossilization potential than any other plant organs, but not all plant groups will have an extensive pollen fossil record or possess palynological features assigned with confidence to extant taxa. Tricolpate pollen grains (those with three apertures or colpi), for example, are unique to the eudicots in plants, and age estimates for these fossils place them in the Barremian and Aptian of the Early Cretaceous (130–112 Mya); earlier occurrence is thought to be very unlikely. The abundance and widespread distribution of early tricolpate pollen fossils coupled to their easily identified features has led to their frequent use as a maximum constraint or fixed age in molecular dating of angiosperms (e.g. Anderson *et al.*, 2005; Magallón and Castillo, 2009). It is only in such rare cases that fossils can be used as maximum constraints or fixed ages without serious risk of underestimating molecular ages.

Smith *et al.* (2010: 5897) disagreed with this reasoning and concluded instead:

Using the first appearance of tricolpate pollen as a fixed calibration may underestimate the origin of eudicots and, by extension, other age estimates that have relied on this constraint. Tricolpate grains first appear in separated geographical areas and the grains themselves are not uniform in morphology . . . , both observations implying that the tricolpate clade originated some time before its appearance in the fossil record . . . .

Although the oldest tricolpate (eudicot-type) pollen is dated as ~125 Ma, excellent macrofossil material from 123 to 126 Ma shows closest affinities with the eudicot family Ranunculaceae (Sun *et al.*, 2011), suggesting that modern eudicot families, or their immediate precursors, already existed by then.

A further problem with the dating of eudicots from fossils concerns the fact that, while most eudicots have tricolpate or 'tricolpate-derived' pollen, some do not. One example is *Duparquetia*, a liane from West Africa. It is near-basal in the phylogeny of legumes and has bizarre pollen that in some features is 'unique in the Fabales and eudicot clades, resembling more closely the monosulcate pollen found in monocots and basal angiosperms' (Banks *et al.*, 2006: 107). It is doubtful that this pollen, if known only from fossils, would be identified as eudicot.

In their clock study of eudicots, Anderson *et al.* (2005) showed that removing a single fossil can change the results by a magnitude of tens of millions of years. They noted that the tricolpate pollen of eudicots may have evolved *after* the split of eudicots from their sister-group, for example, by parallel evolution in the main eudicot lineages. In another clock study, Smith *et al.* (2010) estimated that eudicots evolved in the Late Jurassic.

They concluded (p. 5900): 'we favour the use of 125 Myr as a *minimum* age for the origin of the eudicot crown clade'. [Italics added.] They wrote: 'our results suggest that the first appearance of tricolpate grains at ca. 125 Myr underestimates the origin of the tricolpate clade by perhaps 3 to 22 Myr. This finding is problematic because the record of fossil pollen is judged to be very good through this time period (Friis *et al.*, 2006)'. But the pollen record is very good only in relation to the fossil record of other plant parts, not in an absolute sense. If angiosperms are so much older than their acknowledged fossil record (see last section), perhaps eudicots are as well. This is consistent with results from panbiogeography and the dates of Smith *et al.* (2010). The study by Smith *et al.* was fossil-calibrated and the age of eudicots they suggested is still only a minimum. Yet even extending the age of eudicots from Early Cretaceous to Late Jurassic means that eudicot groups would have been exposed to important tectonic and paleogeographic events. These include the marine flooding of central Australia and the Jurassic–Early Cretaceous Rangitata orogeny of New Zealand.

Angiosperm families

The global phylogeny of several large, pantropical plant families was considered by molecular systematists in a volume edited by Pennington *et al.* (2004a). In the introduction, Pennington *et al.* (2004b: 1457) observed that 'calibration is potentially the largest source of error in the dating'. In addition, treatment of fossil-calibrated minimum ages as maximum ages can produce large errors. Pennington *et al.* (2004b) (citing Renner, 2004) wrote that endemic radiations of Melastomataceae on Madagascar date only from the Miocene and so are due to long-distance dispersal, but these dates were based on calibrations from fossils and so are all minimum, not absolute, dates.

Pennington *et al.* (2004b: 1458–1460) concluded their introduction: 'A clear message emerging from all these studies is that long-distance, trans-oceanic dispersal has been a major force determining plant distributions'. They argued that 'dated phylogenies show clear evidence of recent long-distance dispersal events', that 'it is clear that long-distance dispersal must have had a substantial influence' on plant evolution, and that 'recent rapid speciation has clearly played a role'. Nevertheless, these conclusions all depended on the transmogrification of fossil-calibrated dates into maximum clade ages.

Renner (2004: 1485) used chance dispersal to account for trans-Atlantic disjunctions in different Melastomataceae, as cited already, because vicariance would have required 'tremendous rate heterogeneity.' Renner (2004) also studied trans-Indian Ocean disjunctions (Madagascar–India) in Melastomataceae and calibrated a tree using oldest fossils. Nodes were assigned minimum ages but were then treated as maximum ages. In this way the trans-Indian Ocean disjunctions were shown to be 'too young' for Cretaceous vicariance and so they were attributed to multiple dispersal events, as with the trans-Atlantic patterns. Again, a vicariance model would require 'tremendous rate variation'. Renner (2004: 1491) concluded: 'molecular data continuously bring to light new examples of trans-oceanic long-distance dispersal in groups traditionally thought to be poor dispersers'. Nevertheless, the detailed similarities of these repeated intercontinental affinities suggests they are not the result of chance events, but instead reflect vicariance and great variations in evolutionary rate.

In the families Rhamnaceae and Annonaceae, Richardson *et al.* (2004) calibrated phylogenies based on fossil material and in their 'Material and Methods' section 'emphasized that all timings are therefore minimum ages'. Nevertheless, following transmogrification of the dates, the authors were able to conclude that 'Rhamnaceae and most lineages within Annonaceae are *too young* to have had their distribution patterns influenced by break-up of previously connected Gondwanan landmasses . . . long-distance dispersal appears to have played a more significant role . . . than had previously been assumed'. (p. 1495; italics added.) A subsequent study of New Caledonian Annonaceae accepted the transmogrification, with the authors writing: 'The exceptionally young age of only 3.6–4.8 (±1.5) Mya recently postulated for *Goniothalamus* by Richardson *et al.* (2004) *clearly indicates* that the genus cannot represent an ancient lineage derived before the separation of New Caledonia from Australia' (Saunders and Munzinger, 2007: 502; italics added).

An account of biogeography in Annonaceae was again based on the treatment of fossil-calibrated dates as absolute ages (Couvreur *et al.*, 2011). For example, the authors argued that most of the intercontinental splits found in one of the two main clades (the 'long branch clade') originated 'during [not before] the Eocene'. Based on this reasoning they accepted that distribution in the family has been caused by chance, trans-oceanic dispersal.

Richardson *et al.* (2004) argued that Africa–South America disjunctions in both Rhamnaceae and Annonaceae are too recent for migration by land routes and so they proposed long-distance dispersal. They suggested that 'long-distance dispersal does occur, as evidenced by the molecular trees and the presence of Annonaceae on volcanic islands in the Antilles' (p. 1502). These islands occur at a subduction zone and this feature is more important for the history of the group than the ephemeral, individual islands that it produces. As discussed already, taxa survive on islands around subduction zones as metapopulations and colonize new islands from nearby older ones by ordinary means of survival, not by extraordinary long-distance dispersal from a mainland.

In Myrtaceae, seven genera comprise the eucalypt clade. Crisp *et al.* (2004: 1562) wrote that 'given the uncertainty of any eucalypt fossil before the Miocene . . . , it would be reasonable to conclude that a Cretaceous date for the basal node is too old'. Again, this result depends on the use of a minimum (fossil-based) date to rule out earlier ages for a clade. The fossil record of eucalypts is poor overall (Gandolfo *et al.*, 2011), but eucalypt fossils that are already unambiguous are documented from the Early Eocene (cf. Hermsen *et al.*, 2012).

Pennington and Dick (2004: 1616) accepted that Gondwanan explanations for trans-Atlantic Melastomataceae and Malpighiaceae 'have been refuted by molecular phylogenetic studies coupled to fossil-calibrated molecular clock analyses' and that trans-Atlantic Lauraceae are 'clearly' the result of recent radiation. Fossil-calibrated phylogenies indicate that 'waif' or 'sweepstakes' dispersal across the Atlantic Ocean 'has indeed occurred in multiple taxa and explains disjunctions at species, generic and higher taxonomic levels'. Pennington and Dick (2004: 1618) also observed: 'It is remarkable that some of these examples are of plants that show little adaptation for over-water dispersal, such as *Symphonia globulifera* . . . whose large, recalcitrant seeds cannot survive immersion in sea-water . . . ' and this was taken as further evidence of the great powers of chance dispersal.

Pennington and Dick (2004: 1615) admitted that pantropical distribution in angiosperm orders such as Lamiales, Gentianales and Boraginales 'is both tempting and parsimonious to explain by Gondwanic vicariance'. Nevertheless, they concluded that 'conflicting data' – tree topology and the absence of older fossils – indicate recent origin. Yet tree topology by itself provides no information on age and the fossil record only conflicts with older clade ages if the fossil dates are treated as absolute ages. A young fossil record shows no true conflict with an older clade age, only with a younger clade age.

3 Global affinities of Australasian groups

The last two chapters discussed problems interpreting evolution in time and space. These problems can be overcome, though, and the wealth of new molecular data allows a revision of biogeography in Australasia, as presented in the rest of this book. This chapter discusses the broader affinities of the Australasian biota in the global context.

Widespread Australasian groups with global sisters

Many Australasian groups have no immediate affinity with any particular part of the world, as their sister-group is more or less global. This interesting pattern is seen in terrestrial groups and coastal marine groups such as seahorses. As mentioned already, nearshore marine groups need substrate close to sea level and share many biogeographic patterns with terrestrial groups rather than groups of the open ocean. Seahorses (*Hippocampus*: Syngnathidae) are a worldwide, shallow-water group in which the basal clade includes *Hippocampus breviceps* and *H. abdominalis* of New Zealand, south-eastern Australia, south-western Australia and the Torres Strait region (Teske *et al.*, 2004).

A similar pattern occurs in the brown algae of Australasian coasts, as the genus *Notheia*, recorded from south-western Australia to New Zealand, is basal in the diverse, cosmopolitan order Fucales (*Sargassum*, *Durvillaea*, etc.). Another clade of Fucales has a phylogeny (Sanderson, 1997; Cho *et al.*, 2006):

> SW Australia to New Zealand (*Hormosira*).
>
>> SE Australia (Victoria to near Kangaroo I.; Tasmania) and New Zealand (*Xiphophora*).
>>
>> Northern hemisphere, widespread (Fucaceae).

Standard theory suggests that the paraphyletic basal grade in Australasia represents a centre of origin, but the phylogeny is also consistent with a cosmopolitan ancestor undergoing two consecutive breaks in Australasia.

Australasian groups such as *Hormosira* and *Xiphophora* that have global sister-groups are of special interest for biogeography and are discussed below (Chapter 5), but they cannot help in locating particular areas to which the Australasian biota is related. The following sections indicate some of the global and intercontinental affinities of the biota.

Australasian groups basal in panaustral complexes

Panaustral groups occur in all or most of the Gondwanan regions. Many panaustral clades have basal groups in the Tasman Sea region and some examples of this are cited next. The pattern implies evolutionary differentiation around the Tasman region before the breakup of Gondwana.

In monocots, *Phormium* (Xanthorrhoeaceae subfam. Hemerocallidoideae) occurs in New Zealand and Norfolk Island. Its sister-group ranges from southern Africa, via Asia and the Pacific, to South America (Devey *et al.*, 2006). The main genus in this widespread group is *Dianella*, which ranges from the Zimbabwe mountains (by the Lebombo monocline) to the Tasman region and the central Pacific (Marquesas and Hawaiian Islands). *Dianella* has its sister-group in the northern Andes (*Eccremis*: Bolivia to Venezuela) (Wurdack, 2009). Other genera in the subfamily include the trans-Tasman *Herpolirion* (Tasmania, south-eastern Australia and New Zealand).

The tree tribe Cunonieae (Cunoniaceae; Bradford, 2002; Heads, 2012a, Fig. 9.6) has the phylogeny:

> Central east coast Australia: McPherson–Macleay Overlap (*Vesselowskya*).
>
> New Caledonia (*Pancheria*).
>
> South Africa, Madagascar, Mascarenes east to the central Pacific and South America (*Weinmannia* and *Cunonia*).

Here *Vesselowskya* and *Pancheria* are each equivalent to *Phormium* – they are Tasman groups with panaustral sisters. (The McPherson–Macleay Overlap is an important biogeographic centre on the central east coast of Australia, at the Queensland/New South Wales border. It extends from the McPherson Range to the Macleay River.)

In tenebrionoid beetles, the Pyrochroidae subfam. Pilipalpinae occurs in Madagascar, southern Australia, New Zealand and Chile. The basal group, *Paromarteon*, is endemic to eastern Australia. The genus has one species from Melbourne to the McPherson–Macleay Overlap, two endemic to Bribie Island near Brisbane (by the McPherson–Macleay Overlap), and one in north-eastern Queensland (Pollock, 1995).

Placostylus is a genus of large land snails (up to 11 cm long) from the Tasman region and is often discussed by biogeographers. It belongs to the family Bothryembryontidae, which has the following phylogeny and distribution (Breure, 1979; Herbert and Mitchell, 2008: 203; Breure and Romero, 2012) (main regions in bold):

> **Eastern Tasman**: northern New Zealand, Lord Howe Island, New Caledonia, Vanuatu, Fiji and the Solomon Islands (*Placostylus*), sister to Solomon Islands. *Placocharis* + *Eumecostylus*.
>
> **Australia** (SW Western Australia and Tasmania: *Bothriembryon*), sister to **South Africa** (*Prestonella*) + **South America** (central-eastern Argentina: *Discoleus*; northern Chile: *Plectostylus*).

Again, the Tasman clade, in this case lying east of Australia, is sister to a panaustral complex and cannot be simply derived from any particular member of it. Herbert and

Mitchell (2008: 203) concluded that 'either the origin of the family must at the least predate the separation of Africa and South America in the Mid Cretaceous (under a vicariance scenario) or there must have been subsequent dispersal between the isolated Gondwanan fragments. In view of the limited dispersal ability of terrestrial snails, we consider the former more likely'.

In the mite harvestmen (Opiliones suborder Cyphophthalmi), *Neopurcellia* is endemic to the western South Island, New Zealand, and is sister to a clade of South Africa, Sri Lanka, Australia, New Zealand and Chile (Boyer and Giribet, 2007).

Australasia–Indian Ocean groups

In traditional theory, the biogeographic regions for terrestrial organisms correspond to the continents. Africa, for example, represents the 'Ethiopian Region'. Further studies analysing the terrestrial biotas have shown that the continental faunas are not unitary, but instead decompose into different elements distributed around the main ocean basins. For example, some 'African' groups (often in West Africa) have closest relatives in South America and the Caribbean, and together these form trans-Atlantic clades, while others (often in East Africa) are shared with Australia, forming Indian Ocean groups. The large biogeographic regions for terrestrial groups correspond not to the continents but to the ocean basins (Craw *et al.*, 1999). The reason for this surprising result is not known but the patterns coincide with the major tectonic features, the spreading centres, that caused the breakup of Gondwana. The rest of this chapter discusses the trans-oceanic areas of endemism that include Australasia and straddle the basins (and spreading centres) of the Indian Ocean, the former Tethys Ocean and the South and central Pacific Ocean. These intercontinental patterns provide the context for most distributions within Australasia; the 'globally basal' groups cited above are of special interest but are unusual.

Distributions based around the Indian Ocean can be considered first. These groups have records in localities such as Australasia, Madagascar, Africa and, often, southern Asia. Plate tectonics reconstructions suggest that the biogeographic links among these regions also involved parts of East Antarctica, as it was juxtaposed next to them in the Mesozoic. Indian Ocean groups are often interpreted in terms of dispersal across the basin, with the polarity of the dispersal determined from the phylogeny. For example, the butterfly subtribe Mycalesina (Nymphalidae: Satyrinae) has a standard range from Africa through Asia to New Guinea, Australia and the Solomon Islands. Kodandaramaiah *et al.* (2010: 8) retrieved phylogenies in which the African *Bicyclus* is sister to the other five genera and wrote: 'This would point to an African origin ... With the poorly supported basal relationships, however, we are unable to distinguish between hypotheses of African and Asian origin. Nonetheless, it is clear that there has been at least one dispersal event between the Afro-Malagasy and Asian regions ... '. Instead of being derived from a centre of origin in Africa or Asia, the Australasian group could be derived more simply by *in-situ* evolution of a trans-Indian Ocean ancestor. The phylogeny, with the basal group in Africa, indicates the polarity of the wave of differentiation that has passed through this ancestor from west to east.

The Crustacean suborder Phreatoicidea includes freshwater isopods that often inhabit underground freshwater or freshwater/saline systems. They are most diverse in Australia but there are also members in South Africa, India and New Zealand, and fossils in Antarctica, giving a standard Indian Ocean distribution pattern (Wilson, 2008). Morphological studies suggested that the Indian clade belongs to Hypsimetopidae, also in Western Australia (Perth and Pilbara), Victoria and Tasmania. The New Zealand members belong to Phreatoicinae, also present in south-eastern Australia (New South Wales and Victoria). The South African clade is sister to an Australia–New Zealand group. There is a basal grade in Australia, but instead of interpreting this as a centre of origin, with dispersal to India, New Zealand and South Africa, Wilson (2008) accepted a simple vicariance history. There is fossil material from the Triassic and Wilson suggested that overseas dispersal was unlikely.

Other Indian Ocean groups show direct (disjunct) connections between Australasia, and Africa/Madagascar. In the beetle family Boganiidae, *Paracucujus* (Australia) and *Metacucujus* (Africa) are sister taxa both associated with cycads, and *Boganium* (Australia) and *Afroboganium* (Africa) are sister taxa both associated with angiosperms (A.D. Austin *et al.*, 2004). In other beetles, the family Chaetosomatidae is known only from New Zealand and Madagascar (Leschen *et al.*, 2003).

The monocot family Xanthorrhoeaceae includes at least three trans-Indian Ocean affinities:

> South Africa, Madagascar and Australia: south-western, south-eastern and northern regions (*Caesia*).
> South Africa and Australia: south-western, south-eastern and northern regions (*Bulbine*).
> Africa (*Kniphofia*) and New Zealand (*Bulbinella*) (Chase *et al.*, 2000).

Other Indian Ocean groups in Asparagales include *Arthropodium* (Asparagaceae s.lat.): Madagascar, New Guinea, Australia, New Caledonia and New Zealand, and the plant genus *Dietes* (Iridaceae), in Lord Howe Island and Africa (South Africa to Ethiopia) (Goldblatt *et al.*, 2008). Another monocot, *Ficinia* (Cyperaceae), occurs in South Africa, southern Australia and New Zealand ('*Desmoschoenus*') (Muasya *et al.*, 2009).

Examples of groups in the southern Indian Ocean include the plant *Hectorella* (Montiaceae) of the South Island, New Zealand, and its sister, *Lyallia* of Kerguelen Island. This pair is sister to a circumglobal clade (*Montia*, *Claytonia* and others) (Wagstaff and Hennion, 2007). Kerguelen and Zealandia are the Earth's two largest areas of submerged continental crust.

Crossosomatales

Another clade with direct, disjunct connections between the Tasman region and Africa/Madagascar includes *Ixerba* and allies in the eudicot order Crossosomatales (Fig. 3.1; Heads, 2010b). This Indian Ocean group is sister to the remaining clade of Crossosomatales, made up of four families that occur further north (Stevens, 2012). The simplest explanation for the global pattern is vicariance between the Indian Ocean group

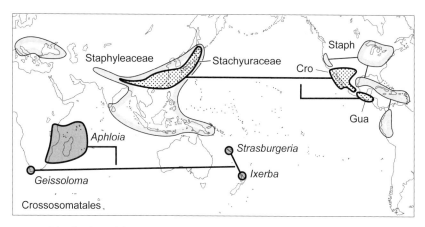

Fig. 3.1 Distribution of the Crossosomatales. Cro = Crossosomataceae, Gua = Guamatelaceae, Staph = Staphyleaceae. (From Heads, 2010b, reprinted by permission of John Wiley & Sons.)

and the northern clade, followed by vicariance within both groups. The only dispersal that is required is range expansion in China and Japan, which is needed to explain the overlap of Stachyuraceae and Staphyleaceae there. Part of the initial fracture between the Indian Ocean clade and the northern clade occurred between New Caledonia and New Guinea, and similar breaks in this locality are an important feature of the regional biogeography.

Some endemic weed trees in Malvaceae

Entelea (Malvaceae) is a monotypic genus endemic to New Zealand (from the Three Kings Islands to the north of South Island). It is never found more than 8 km from the sea and is an opportunist invader, a light-demanding weed tree up to 15 m tall (Millener, 1946). In parts of northern New Zealand it is the first woody colonizer of slips, treefall gaps in forest, burnt areas, rocky outcrops and consolidated sand dunes. The populations are transient; following a great initial increase in numbers, there is a sudden collapse and within 25 years the dense population has died out and been replaced. Despite its pioneer ecology and effective means of dispersal it is a New Zealand endemic. It is sister to an African genus, *Sparrmannia*, giving a typical Indian Ocean clade, as in the Crossosomatales. The phylogeny of the complex (from Brunken and Muellner, 2012) is:

Brazil and Bolivia to Mexico (*Apeiba*).

New Zealand (Three Kings, North and northern South Islands) (*Entelea*).

southern and eastern Africa, Madagascar (*Sparrmannia*).

Based on morphology, this group also includes *Clappertonia* and *Ancistrocarpus* of tropical Africa. These genera were not included in the molecular analysis, but they will probably not disrupt an Indian Ocean affinity of some sort.

Asteraceae tribe Gnaphalieae

Members of the tribe Gnaphalieae are found worldwide but have their greatest diversity in southern Africa and Australia. The basal members are South African and are concentrated in the Cape region (Bayer *et al.*, 2006; Bergh and Linder, 2009). In the Australian members (84 genera), there is massive species diversity in south-western Australia, but the basal clade is the monotypic *Parantennaria*, restricted to mountains in south-eastern Australia (on the border of Victoria and New South Wales). This genus is a dioecious cushion-plant and in traditional morphology would be regarded as advanced and derived. Nevertheless, other basal clades in the Australian group are also dominated by perennial shrubs and alpine cushion-plants from south-eastern Australia ('clade E' in Bayer *et al.*, 2002), while the Western Australian clades, including many species of annual herbs, are nested higher up in the phylogeny.

Bayer *et al.* (2002) suggested that the phylogenetic sequence seen in Gnaphalieae (South Africa, south-eastern Australia, then south-western Australia) was attained by chance dispersal. Instead, the standard Indian Ocean distribution suggests a sequence of differentiation in a widespread ancestor, first around South Africa, then around south-eastern Australia (Tasman region), and then south-western Australia, where a final phase of extensive, low-rank evolution has taken place. Another direct 'jump' in the sequence of differentiation, from Africa to the Tasman region, occurs in Droseraceae, discussed next.

Drosera, with a basal grade around the Indian Ocean

The sundew family Droseraceae is the basal member of the carnivorous plant clade (Stevens, 2012). *Drosera* itself is found worldwide and, while about a third of the 110 species occur in south-western Australia, this is not the primary node, as the basal groups in the genus are from South Africa and the Tasman region. In this way the pattern resembles that of Gnaphalieae. The phylogeny of Droseraceae (Rivadavia *et al.*, 2003) is:

> Europe (*Aldrovanda*) + south-eastern US (*Dionaea*).
>> South Africa: western Cape (*Drosera regia*) (this sometimes groups with *Aldrovanda* + *Dionaea*).
>>> New Zealand (mainland, montane to subalpine bogs, near sea level in the south), SE Australia (mountains) and Tasmania (*Drosera arcturi*).
>>> Global clade (all other *Drosera* species).

The last, global clade of *Drosera* includes a trans-South Pacific pair: *D. stenopetala* of New Zealand + *D. uniflora* of Patagonia, that is also useful for dating the group.

Cymbonotus and Arctotidinae (Asteraceae tribe Arctotideae)

The clade Arctotidinae comprises a basal grade of three southern African clades, *Cymbonotus* of south-western and south-eastern Australia, and two other southern African genera (Fig. 1.7; Funk *et al.*, 2007; McKenzie and Barker, 2008). Funk *et al.* (2007: 7) suggested: 'given that the family is relatively young (less than 50 million years old . . .), it seems certain that *Cymbonotus* is the result of a long-distance dispersal event from southern Africa, long after contact between the two continents ceased'. Yet the actual age of the family itself (not just its fossil members) is unresolved; the 50 million year date is fossil-based and so is a minimum age only. (Keeley *et al.* (2007) extended the oldest known fossil date to ~60 Ma.) Funk *et al.* (2007) interpreted *Cymbonotus* and similar cases as evidence of extreme long-distance dispersal, but the phylogeny topology does not necessarily mean that southern Africa is a centre of origin (cf. Fig. 1.6). *Cymbonotus* is not an endemic Australian genus embedded in a southern African clade as the authors inferred; instead, it is an Australian genus embedded in a southern African–Australian clade. Rather than being a centre of origin, southern Africa, in particular south-eastern Africa, is a fracture zone around which an African/Australian ancestral complex has broken up into several subclades. One of the subclades included populations located in areas that became Africa and Australia. Among the most important Mesozoic events in this region were the Early Jurassic flexing and intense volcanism on the Lebombo monocline (a volcanic rifted margin) and the subsequent mid-Jurassic to Cretaceous breakup of Gondwana.

After splitting from its diverse African sister-group, the primary division in *Cymbonotus* is between one clade of south-western and south-eastern Australia and one at the McPherson–Macleay Overlap. Overall, evolution in Arctotidinae involved differentiation around two main nodes, one in the south-western Indian Ocean and one in southern Australia.

McKenzie and Barker (2008) also studied the Arctotidinae and provided divergence dates, although they emphasized that these must be 'interpreted with caution'. They noted that the 'poor fossil record' for the group means that obtaining 'reliable' calibration points is a 'widespread problem'. Nevertheless, based on fossil evidence they inferred that dispersal from southern Africa to Australia has taken place in the Pliocene. This leads to a problem, as the fruits in Arctotidinae lack the family's usual adaptations for long-distance dispersal and so 'the mode of dispersal is difficult to envisage'. The problem only arises in a dispersal model based on converting minimum fossil-based dates into maximum ages and assuming that a paraphyletic basal grade occupies a centre of origin.

Calostemmateae (Amaryllidaceae: Asparagales)

In the monocots, the tribe Calostemmateae is endemic to Australia and New Guinea, and provides a parallel case to that of *Cymbonotus*, as it is nested in an otherwise African group. Meerow and Snijman (2006) found a phylogeny:

Africa, mainly South Africa (Cyrtantheae).

Africa, mainly South Africa (Haemantheae).

Australia and New Guinea (Calostemmateae).

Meerow *et al.* (1999) suggested that the Australasian group was isolated as Australia separated from Gondwana (Africa).

Begonia (Begoniaceae)

The plant genus *Begonia* is almost pantropical, but there are no indigenous species in the Hawaiian Islands (several introduced species are naturalized there). In contrast, the sister-group of *Begonia*, *Hillebrandia*, is endemic to the Hawaiian Islands. In the traditional model, this pattern would be explained by chance dispersal to or from Hawaii. Yet this would not explain why the dispersal occurred only once or why it occurred so early in the history of the group, and a simple vicariance event would account for both these features. In this model, *Begonia* originated as a simple vicariant over its present range rather than spreading out from a centre of origin.

Within *Begonia* itself, Thomas *et al.* (2012a) found a phylogeny:

Africa.

    Africa and South America.

      Africa.

       Socotra, India, Sri Lanka, China, SE Asia, Malesia to Fiji (Vanua Levu).

The authors located a centre of origin in Africa (because of the basal grade there), with dispersal to Asia, Malesia and the Coral Sea region. Nevertheless, this does not account for the intriguing break between Africa and Socotra–Asia, a break that recurs in other groups such as *Exacum* (Gentianaceae) (Yuan *et al.*, 2005). Within Malesia, there is also large-scale allopatry in *Begonia*; for example, the New Guinea–Philippines clade is sister to one from the Lesser Sunda Islands (Thomas *et al.*, 2012a).

Hypoxidaceae (Asparagales)

In the *Pauridia* clade (Hypoxidaceae), Kocyan *et al.* (2011) found an Australia/New Zealand clade (*Hypoxis* sect. *Ianthe*) nested in an otherwise African group, and they inferred long-distance dispersal from Africa to Australia. They compared the pattern with other trans-Indian Ocean monocots such as *Caesia* and *Bulbine* (cited above).

In the last three clades described – Calostemmateae, *Begonia* and the *Pauridia* clade – differentiation began around the south-west Indian Ocean (South Africa, Madagascar, etc.) and breaks around the Tasman region took place later. Rifting between Africa and Madagascar occurred in the Jurassic, but rifting east of Madagascar (for example, between Madagascar and the Mascarenes) developed in the Cretaceous, at about the same time as seafloor spreading began in the Tasman region. Pre-breakup tectonics in the Tasman region are also important and so phylogeny that developed through the Cretaceous can sometimes appear to 'jump' between the south-west Indian Ocean and south-west Pacific (cf. Fig. 1.11, showing the same process between south-west and south-east Australia).

The spider family Archaeidae

In the archaeid spiders, Mesozoic fossils are widespread globally and are very similar to modern forms. The extant groups have a phylogeny (Rix and Harvey, 2012):

Madagascar (southern African members not sampled).
  SW Australia to Victoria.
    New South Wales to SE Queensland.
    NE Queensland.

The sequence of breaks proceeds from between Africa and Australia, to the Victoria/New South Wales border, and then to a break between south-eastern and north-eastern Queensland (the 'St Lawrence gap').

Indian Ocean groups with basal members in the Tasman region

In the Indian Ocean groups cited so far, the basal group is in Africa or the islands of the south-west Indian Ocean and differentiation proceeds eastwards. In other Indian Ocean groups differentiation begins in the east. For example, in fruit bats (Pteropodidae), *Melonycteris* and *Pteralopex* of the Bismarck Archipelago and the Solomon Islands form a clade with *Desmalopex* of the Philippines. This clade's sister-group is widespread from the East African coast to the Pacific (*Pteropus*, *Acerodon* and *Styloctenium*) (Almeida *et al.*, 2011).

Another example is the New Guinea pigeon *Otidiphaps*, which is basal in a group ranging from the Mascarenes to Samoa (including the dodo, *Raphus*) (Pereira *et al.*, 2007). The Crustacea suborder Phreatoicidea is an Indian Ocean group, mentioned already, from South Africa, India and Australasia; a morphological study indicated that the basal clade is a Tasman endemic (Phreatoicopsidae) (Wilson, 2008). In beetles, the Scarabaeinae include the following clade (Monaghan *et al.*, 2007): New Caledonia: *Pseudonthobium* etc. (New Zealand: *Saphobius* (Australia and South Africa: *Pseudignambia* etc.)).

The different phylogenetic orientations of trans-Indian Ocean tracks, westward and eastward, suggest a phase of more or less simultaneous breaks around southern Africa/Madagascar and eastern Australia/Zealandia. These coincide with rifted margins in the first area that were active in the Jurassic/Cretaceous and with the Rangitata orogeny that developed in the second area at the same time.

Divaricate shrubs and ratite birds

Indian Ocean distribution is seen in large numbers of clades, but also in features that do not define clades and instead represent convergences or parallelisms across many families. One example concerns the shrubs and small trees of New Zealand that local botanists term 'divaricate'. These show a distinctive shoot architecture that is characterized by small foliage borne on shoots differentiated into brachyblasts (short-, or spur-shoots), and determinate long shoots in which the apices abort (Heads, 1998a). Apart from these features, divaricate shoot architecture is diverse. Shoot axes can all be plagiotropic (more or less horizontal, e.g. *Coprosma acerosa*), all orthotropic (more or less vertical, e.g. *Pittosporum* species), orthotropic and then plagiotropic, all by primary growth (e.g. many *Coprosma* species), or plagiotropic by primary growth with basal sections becoming orthotropic by secondary growth (e.g. *Sophora microphylla*).

The divaricate syndrome occurs in a few species in each of many different families and the group is nothing like monophyletic. Despite this, it has a distinctive geographic range. Most divaricates occur around the south-west Indian Ocean (East Africa, South Africa and Madagascar) and the south-west Pacific (Australia and New Zealand), and the architecture is very rare in New Guinea, the central Pacific islands and West Africa (personal observations). Some plants with the architecture also occur in parts of the Americas (e.g. *Enriquebeltrania*: Euphorbiaceae). The distribution of this parallelism is a standard Indian (–Atlantic) Ocean pattern, as seen in many monophyletic clades.

The distinctive architecture of the divaricates means they are not just 'normal' plants in which the size of the leaves and stems has been reduced. Nevertheless, the architectural differences can still be explained by minor morphogenetic changes. In many small-leaved shrubs, the original vegetative shoot system is more or less entirely suppressed (it may be represented by phases of 'juvenile' or 'reversion' foliage) and the plant equivalent to an inflorescence. The example of the Australian tremands was mentioned in Chapter 1. This process explains the cymose (sympodial), often helicoid, branching of divaricates, and the racemose (monopodial) branching of ericoid shrubs. The other features of divaricate and ericoid architecture can be accounted for by a second development, in which most of the 'inflorescence-plant' has been sterilized, and so the foliage is composed of sterile floral bracts (Heads, 1994a). These simple processes – suppression of the vegetative shoot and sterilization of most of the inflorescence – explain many aspects of morphology but involve quite minor morphogenetic variants. Another possible example of this reduction is seen in *Chamaesyce*, one of the four main clades in the giant genus *Euphorbia* (Euphorbiaceae) and a diverse, cosmopolitan group in its own right. The branching architecture of plants of *Chamaesyce* shows close resemblances to that of an inflorescence (Yang and Berry, 2007) because the plant *is* an inflorescence, largely sterilized.

The traditional explanation for a parallelism that occurs in a particular region is adaptation to local conditions by selection. Authors have suggested that, because divaricates are diverse in New Zealand, Australia and Madagascar, their morphology could be the result of selection pressure exerted by the browsing of ratites and other giant birds that have occurred in these regions, rather than ungulates (Bond *et al.*, 2004; Bond and Silander, 2007). The idea is based on biogeography, but Dempewolf and Reiseberg (2007) pointed to what they called a 'fly in the ointment' for the theory – the presence of divaricates in Patagonia where there are both ratites and ungulates. In addition, divaricates are abundant in south-eastern Africa (e.g. in Combretaceae) where there are many ungulates. Also, divaricates are more or less absent in the diverse New Guinea flora, although ratites (cassowaries) are widespread there. Thus the evolutionary process that has led to ratites and divaricates sharing many aspects of their distribution probably did not involve one (ratites) being the cause of the other (divaricates). Instead, a common cause has led to the distribution of both. The distribution indicates a general phase of modernization of already polymorphic, worldwide groups, with primary differentiation around the south-west Indian and south-west Pacific Oceans. This produced new recombinations of characters, monophyletic groups, paraphyletic basal groups and hybrid swarms, with high levels of parallelism, convergence and incongruence. The phylogenetic and geographic distribution of divaricate plants suggests that 'divaricate genes' were widespread around the Indian Ocean and south-west Pacific (but not New Guinea or the central Pacific) long before the origin of the modern clades in which they are now maintained.

Toads, woodpeckers and centres of absence in the southern Indian Ocean

Most important areas of endemism, such as Madagascar + Australasia, are also significant as areas of absence. For example, the true toads, Bufonidae, are almost cosmopolitan but are absent (living and fossil), from Madagascar, Australia and New Guinea. (Pramuk *et al.* (2007) described this absence as confounding the hypothesis of a Gondwanan distribution, but this would only be true if the biota of Gondwana was homogeneous.) Pramuk *et al.* (2007) observed that 'this apparently enigmatic absence is not unique to bufonids' and cited several widespread groups that are absent in Madagascar and Australasia. Others include woodpeckers (Picidae).

Australasia–Tethys affinities

The last sections introduced Australasian groups with relatives located around the Indian Ocean basin, in classic Gondwanan patterns. In other Australasian groups the affinities lie with Tethys and Pacific regions. Many Tasman clades have Tethyan, not Gondwanan, affinities that occur along the former Tethys Sea basins, now preserved from south-east Asia to central Asia and the Mediterranean (for the closure of these basins, see Heads, 2012a, Fig. 5.13). These groups are often absent from the cratonic cores of Gondwana in western Australia, Africa and eastern South America. Examples include the following.

The monocot family Posidoniaceae (Alismatales) occurs in Australia (south-west to south-east) and the Mediterranean (Stevens, 2012).

In Caryophyllaceae, *Scleranthus* occurs in the Tasman region (New Zealand, south-eastern Australia and eastern New Guinea; three species) and also in Europe and the Mediterranean (three species). *Pentastemonodiscus* of Afghanistan is possibly related (Smissen *et al.*, 2003) and would fill the disjunction.

In Ranunculaceae, *Ceratocephala* has three species, one in New Zealand and two others from Kashmir and Pakistan to Spain and North Africa (Smissen *et al.*, 2003).

In Apiaceae, the tribe Aciphylleae of New Zealand and south-eastern Australia forms a clade with the *Acrotrema* group of China and the Himalayas, and the tribe Smyrnieae of Iran to Europe and the Mediterranean (Spalik *et al.*, 2010).

A group in *Solanum* (Solanaceae) comprises the 'Archaesolanum clade' of Australia, Tasmania, New Guinea and New Zealand, and the 'Normania clade' of north-western Africa, the adjacent Iberian Peninsula and the Macaronesian islands (Weese and Bohs, 2010).

In Boraginaceae, *Myosotidium* of the Chatham Islands, New Zealand, is sister to *Omphalodes nitida* of Spain and Portugal (Heenan *et al.*, 2010).

In a sample of Asteraceae tribe Cichorieae, the clade *Kirkianella* of New Zealand + *Embergeria* of the Chatham Islands is sister to *Sonchus aquatilis* of the Mediterranean (Kim *et al.*, 2007).

The gastropod group Otinoidea has two clades, both found in intertidal habitat on rocky shores (Bouchet and Rocroi, 2005): Smeagolidae in central New Zealand (Kaikoura Peninsula, Wellington Peninsula), Tasmania and Victoria, and Otinidae of Western Europe (Sardinia to Britain and Belgium).

Two major, Mesozoic oceanic domains are involved in these tracks: Tethys to the north and west of Australia, and the Paleo-Pacific east of Australia. The tectonic relationship between the two has remained cryptic. This is because the old plate boundaries between Eurasia and eastern Australia have been destroyed during Gondwana breakup and seafloor spreading in the Pacific (Gaina *et al.*, 2002). The relationship of Tethys and the Pacific is of special interest for biogeography as Tethys affinities, distinct from both Gondwanan and Pacific connections, are so important in the modern biota.

Eastern Tethys

Many groups occur around north-western margins of the Pacific basin (Indochina, China, Japan) and also south-western margins (Australasia), and these groups often reflect a Tethyan history. A typical example is the gastropod family Batillariidae (Cerithioidea). It inhabits mudflats and rocky shores from Australia and New Zealand to China and Japan and fossil forms occur through to the Mediterranean region. The family has been interpreted as a Tethyan relict, with Australasia representing 'a refugium of Tethyan fauna' (Ozawa *et al.*, 2009). Fossils of the family are known back to the Late Cretaceous and three extinct genera became diverse in the Tethyan realm before the group disappeared from Europe at the end of the Miocene. Ozawa *et al.* (2009) recognized four extant genera: *Zeacumantus* (Australia and New Zealand; basal), *Pyrazus* and *Velacumantus* (both Australia) and *Batillaria* (Australia, Philippines, Malaysia, China and Japan).

These hard-shelled, mudflat snails fossilize well and there is a rich fossil record, although identification is difficult and controversial. Many other groups with a China–Australasia distribution do not have such a good fossil record and their distributions are often explained by dispersal; however, this is only because the groups are assumed to be young.

The standard Tethyan range: New Zealand–southern Europe, as seen in Batillariidae, also occurs in other groups of cerithioid gastropods such as the freshwater *Melanopsis* (Melanopsidae). This genus also has intermediate records from New Caledonia, Asia Minor and North Africa (Powell, 1979). A vertebrate example of an eastern Tethys group is the snake family Homalopsidae (= Colubridae subfam. Homalopsinae). Homalopsidae are widespread in mangroves, mudflats and lakes from India and eastern China through

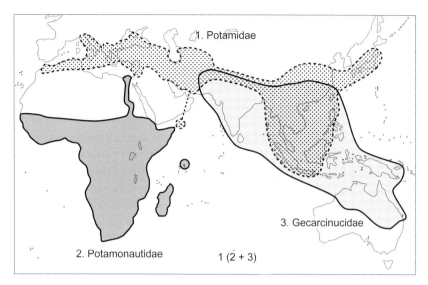

Fig. 3.2 A Gondwanan clade (Potamonautidae + Gecarcinucidae) and a Tethyan clade (Potamidae) of freshwater crabs, with the putative phylogeny indicated (Klaus *et al.*, 2011).

Malesia to southern New Guinea and northern Australia (Alfaro *et al.*, 2008). The family has been proposed as sister to the more or less cosmopolitan clade Colubridae + Elapidae (Yan and Zhou, 2008).

Tethys + Indian Ocean affinities

Tethys and Indian Ocean affinities both occur in many groups, as illustrated above for *Lordhowea* (Fig. 1.14). Another example concerns an Indian Ocean (Gondwanan) clade of freshwater crabs with two components (Fig. 3.2; Klaus *et al.*, 2011):

Africa, Madagascar and the Seychelles (Potamonautidae).

India to McPherson–Macleay Overlap (Gecarcinucidae).

This Gondwanan clade is closely related, and possibly sister, to a typical Tethyan family, Potamidae, that ranges from Borneo to Gibraltar. Only the Gondwanan clade is present in India, while only the Tethyan clade is in Japan, Socotra and the Mediterranean, and both overlap in Southeast Asia. These are the three Old World families of 'primary freshwater crabs'. The phylogeny is not yet fully resolved and the Pseudothelphusidae of the New World tropics may also belong here (Klaus *et al.*, 2011).

Australasia–Pacific groups

Many groups are endemic to Australasia plus southern South America; one of the best-known examples is the tree *Nothofagus*, also present as fossils in Antarctica. The disjunction has intrigued biogeographers for more than a century. Marine paleontologists have recognized an equivalent centre of endemism, the 'Weddellian region', that is best-documented in shallow continental shelf groups of south-eastern Australia, New Zealand, West Antarctica and Patagonia. This region forms the margin of the South Pacific (it does not include southern Africa or India). This same area of endemism is evident in many fossil groups; for example, mosasaurs, a group of large marine reptiles known from Cretaceous rocks. Mosasaurs are squamates, sister to either varanoid lizards or snakes; and one, *Taniwhasaurus* of New Zealand (Haumuri Bluff), now includes *Lakumasaurus* of the Antarctic Peninsula (Martin and Fernández, 2007). *Moanasaurus* is another Pacific mosasaur endemic to New Zealand and the Antarctic Peninsula.

This pattern in mosasaurs is matched in plesiosaurs. This Jurassic–Cretaceous order of marine reptiles is sister to squamates plus Rhynchocephalia (tuatara). In Late Cretaceous plesiosaurs, *Aristonectes* of Chile, Argentina and Antarctica appears similar to *Kaiwhekea* from New Zealand. *Mauisaurus* occurs in the Antarctic Peninsula and New Zealand (Martin and Fernández, 2007). All these affinities span the Pacific Southern Ocean and appear to have diverged from their sister-groups elsewhere before (perhaps soon before) they were themselves rifted apart by Gondwana breakup.

Crame (1999) concluded that the present-day benthic faunas from Magellan and Antarctic provinces have origins lying as far back as the Early Cretaceous. At that time extensive marine incursions were developing around the world, while southern continents were being isolated by seafloor spreading. This was also the probable time of formation of the Weddellian Province and there is fossil and phylogenetic evidence that a number of the living families and genera can be traced back to the Late Cretaceous, at least (Crame, 1999).

These Cretaceous marine vertebrates have a standard area of endemism shared with large numbers of other groups that are still extant. For example, the bryophyte family Acrocladiaceae occurs in south-eastern Australia, New Zealand and southern South America (Tangney *et al.*, 2010; for other moss examples, see Tangney, 2007).

Australasia–South America connections in craneflies

In craneflies (Diptera: Tipulomorpha), Ribeiro and Eterovic (2011) analysed the 15 morphological and molecular clades found only in Australasia and South America; the groups include a total of ~700 species. The clades (most are genera and subgenera) illustrate four main distribution patterns:

> South America and New Zealand (four clades).
> South America and Australia (three clades).
> South America, Australia and New Zealand (five clades).
> South America, Australia, New Zealand, New Caledonia and New Guinea (three clades).

The faunas of each geographic region (South America etc.) are not monophyletic and so the pattern has not been caused simply by the rifting of a homogeneous fauna. Instead of explaining the pattern by long-distance dispersal after rifting, Ribeiro and Eterovic

(2011) concluded that the four patterns had already differentiated from each other *before* Gondwana breakup.

The first pattern is well known in many groups. Direct New Zealand–South America connections that do not involve Australia already existed in the early Mesozoic, as Murihiku terrane fossil faunas of New Zealand and New Caledonia, along with faunas of Argentina, form a centre of endemism (Damborenea and Manceñido, 1992).

The second pattern in the craneflies involves south-eastern Australia and South America, but not New Zealand. In a similar example, the bird family Pedionomidae (Charadriiformes) of south-eastern Australia is sister to Thinocoridae, from Patagonia to the Peruvian Andes (Fain and Houde, 2007). This South Pacific pair is sister to a pantropical clade, Jacanidae + Rostratulidae. Likewise, in dicot plants, the order Berberidopsidales is restricted to eastern Australia (the McPherson–Macleay Overlap) and central Chile, and is sister to a cosmopolitan clade (see Chapter 5). It is likely that the Australia–Patagonia connection ran south of New Zealand, approximating the great circle (shortest) route and including parts of West Antarctica.

A variant of the Australia–Patagonia connection misses mainland New Zealand but involves the New Zealand subantarctic islands. For example, *Nothotrichocera* (Diptera: Trichoceridae) occurs in south-western and south-eastern Australia, Tasmania, the New Zealand subantarctic islands (Auckland and Campbell Islands) and Chile (by Chiloé Island) (Krzeminska, 2001). In the same way, a clade of carabid beetles comprises groups of Australia and Patagonia, and also *Bountya*, endemic to the New Zealand subantarctic islands (Liebherr *et al.*, 2011).

The third cranefly pattern links South America, Australia and New Zealand, and is well known in many groups. The fourth pattern involves a separate New Guinea–New Caledonia sector.

Chironomidae (Diptera) and the Australia–New Zealand–South America problem

This family is well known among biogeographers for its putative Gondwanan distributions, as monographed by Brundin (1965, 1966). In one chironomid subfamily, the Podonominae, Brundin found consistent patterns of relationships among taxa from Australia, New Zealand and South America. Within each clade, New Zealand taxa always possessed a sister-group in South America or South America–Australia, and there were no direct phylogenetic connections between Australia and New Zealand. Cranston *et al.* (2010) provided preliminary molecular support for Brundin's idea; none of the sampled New Zealand taxa showed closer relationships to Australian taxa than to South American groups.

If a group is in Australia, New Zealand and South America, but the groups in each area are not monophyletic, as in craneflies and Podonominae, the phylogeny cannot have been caused simply by the rifting of a homogeneous population. Instead, the pattern suggests that differentiation developed in a long phase of intracontinental rifting that was a precursor to Gondwana breakup with seafloor spreading. The whole breakup process was drawn out and included early phases of 'failed' rifting; this occurred at the West Antarctic Rift System about the same time (Late Cretaceous) as the successful

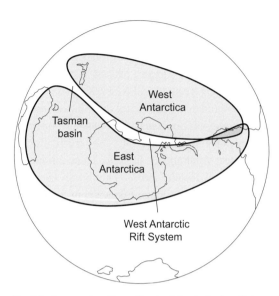

Fig. 3.3 A reconstruction of typical Australasian–Patagonian distributions before glaciation and extinction in Antarctica.

seafloor spreading in the Tasman basin. West Antarctica, part of the circum-Pacific mobile belt, is an ensemble of blocks that have moved independently of each other and of cratonic East Antarctica (Veevers, 2012, Fig. 3). The West Antarctic Rift System is one of the largest zones of continental extension on Earth.

Brundin (1966) observed that before Gondwana breakup, New Zealand was connected to South America via West Antarctica, and Australia was connected to South America via East Antarctica (Fig. 3.3). This suggests that the break seen in craneflies and Podonominae – between New Zealand + South America groups, and Australia + South America groups – was associated with Jurassic–Cretaceous extension in the West Antarctic Rift System, not the seafloor spreading that later broke up Gondwana.

In another chironomid subfamily, the Orthocladiinae, Krosch *et al.* (2011) found a more complex pattern than in Podonominae. The phylogeny is as follows (Af = Africa, Aus = Australia, NZ = New Zealand, SAm = South America):

SAm, NZ.

    Af + (Aus, SAm).

        Af + cosmopolitan.

        SAm.

            Aus.

                NZ, SAm.

                Aus, NZ.

                SAm.

                    (Aus, NZ) (Aus, SAm).

The basal node was a break around the South Pacific, and took place between New Zealand + South America, and the rest of the world. This was even earlier than the breaks around Africa, and may have happened long before the New Zealand versus South America split.

For the areas Australia, New Zealand and South America, all three pairwise tracks (Aus–NZ, NZ–SAm and SAm–Aus) are defined by endemic clades of Orthocladiinae. There are at least two Australia–New Zealand clades, a pattern that does not occur at all in Podonominae. Again, if the Gondwanan ancestor was undifferentiated, the pattern could not have been caused simply by vicariance with seafloor spreading. New Zealand was rifted from Gondwana before Australia and then South America were, so the phylogeny would be: NZ (Aus + SAm). Incongruence between taxon phylogenies and the geological sequence of continental fragmentation has been taken as evidence for long-distance dispersal. Instead, it may indicate that the Gondwanan chironomid fauna was already diverse at the time of rifting. The phylogeny of Orthocladiinae suggests that repeated phases of 'breakup and overlap' in a global ancestor have occurred in each of Australia, New Zealand and South America, over the same period. Gondwana breakup occurred later and rifted apart many clades that had originated during the extension and magmatism of the prior 10–20 m.y.

In this model, the last main evolutionary event in New Zealand was not caused by seafloor spreading and the breakup of Gondwana in the Late Cretaceous (~85 Ma), but by the prior events that led up to this. In New Zealand, for example, these include the Rangitata orogeny of the Jurassic–Early Cretaceous (which formed proto-New Zealand), the associated magmatism of the Whitsunday province and Median batholith, widespread extensional faulting with graben formation and exhumation of metamorphic core complexes (see Glossary). The incongruence among the South Pacific phylogenies is compatible with evolution mediated by these large-scale, pre-breakup events and their equivalents in Australia and South America. There is no need to assume that if phylogenies were generated by geology they must follow the split sequence; there is more to the geological history of Gondwana than its final breakup.

Monocot lianes of the South Pacific: *Rhipogonum* (Liliales)

The standard pattern: South America, Australia and New Zealand, occurs in lianes such as Rhipogonaceae (Liliales; Fig. 3.4). The single genus, *Rhipogonum*, comprises six species that inhabit eastern Australian (east of the Great Dividing Range), New Guinea and New Zealand. In New Guinea, *Rhipogonum* is in the east only. In New Zealand, *Rhipogonum* is widespread in the three main islands but, despite its bird-dispersed fruit, it is not on the Chatham Islands. This suggests a more southern connection between Australasia and South America. *Rhipogonum* is most diverse around the McPherson–Macleay Overlap, where all five Australian species occur and only the New Zealand species is absent (Conran and Clifford, 1985).

Rhipogonum is sister to the Philesiaceae, with two monotypic Chilean genera (*Philesia* and *Lapageria*). The South Pacific Rhipogonaceae + Philesiaceae is sister to the

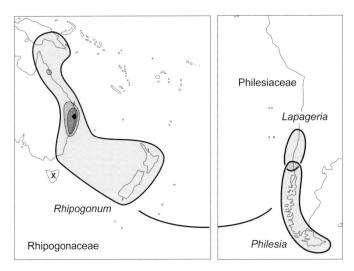

Fig. 3.4 Distribution of Philesiaceae and Rhipogonaceae, showing increase in species numbers towards the McPherson–Macleay Overlap in *Rhipogonum*; areas with 1, 3, 4 and 5 species are indicated with increasing depths of grey. X = Early Eocene fossil from Tasmania (Conran *et al.*, 2009).

cosmopolitan Liliaceae + Smilacaceae (Liliales), indicating early breaks around Australasia and South America.

The Pacific margin of Gondwana, including the Tasman region, Antarctica, and southern and western South America, includes a major orogenic belt in which active terrane accretion took place from Neoproterozoic to late Mesozoic time (Vaughan *et al.*, 2005). This belt, termed the Australides, is comparable in scale to the orogenic belt in western North America (Cordillera) that extends from Mexico to Alaska. The Australides region is well known to biogeographers as a major centre of endemism that does *not* involve core areas of Gondwana such as Western Australia, India, Africa or cratonic (eastern) South America (Gibbs, 2006). So, although some biogeographers have considered Australides ('Pacific') distributions to form a simple subset of Gondwanan patterns, they are distinct from the 'core Gondwana', Indian Ocean patterns that were discussed above.

Peloridiidae (Hemiptera)

The family Peloridiidae (17 genera, 37 species; Burckhardt, 2009; Burckhardt *et al.*, 2011) has been described as the rarest and most remarkable of all the Hemiptera. The members live in wet moss in the cooler rain forests of Australasia and southern Chile. All the species except one are flightless and the family is often cited, along with *Nothofagus*, as a typical 'Gondwanan' element. Neither Peloridiidae nor *Nothofagus* is in Africa or India, though, and they can both be described as South Pacific groups. Molecular studies indicate that the South Pacific peloridiids are the sister-group of the Heteroptera, a diverse, worldwide group. This suggests that they are not simply derived from a group in, say, Australia or New Guinea.

A morphological analysis of the Peloridiidae (Burckhardt, 2009) proposed two main clades:

New Zealand and New Caledonia.

East Australia (Tasmania to north-eastern Queensland), Lord Howe Island, southern Chile.

Apart from the position of the endemic Lord Howe genus, *Howeria* (Burckhardt, 2009), this is consistent with the breakup of Gondwana causing the diversification. (Lord Howe Island is located on the western margin of Zealandia.)

South Pacific brachiopods

As discussed above, sequencing studies have shown that many marine clades have deep and intricate geographic structure, and nearshore organisms repeat many of the patterns seen in true terrestrial groups. In brachiopods, Cohen *et al.* (2011) found that a clade endemic to New Zealand (*Neothyris* etc.) is sister to one in Patagonia (*Magellania* etc.). Using a clock calibrated with Triassic brachiopods, the authors found that the two clades diverged at ~82 Ma and they concluded that the break was due to vicariance. The coincidence of the clade age and the age of the geological event suggests that in this case the fossil record reflects the actual age of the groups. This would not be too surprising, as brachiopods are well known for their rich fossil record.

Nothofagus (basal Fagales)

Southern beech, *Nothofagus*, is the only genus in Nothofagaceae. The trees dominate many cooler rain forests around the margins of the South Pacific and fossils extend the range to south-western Australia and Antarctica (Fig. 3.5; Heads, 2006b; Moreira-Muñoz, 2011). *Nothofagus* and Nothofagaceae are sister to the rest of the Fagales and are allopatric with most of them.

    Nothofagus is a wind-pollinated tree that is dominant over large areas and produces large amounts of pollen. It is not surprising that it has an exceptional fossil record. The four extant subgenera are all known from Cretaceous fossils and the total ranges of the subgenera show geographic overlap at different times in history. Nevertheless, the main species concentrations of the four subgenera are allopatric in four localities (Fig. 3.5), and these are the same, whether extant species only or all species (extant plus fossil) are counted. Each of the subgenera has a total range (living and fossil) that spans the South Pacific and each has an extensive fossil record back to the Cretaceous. For example, subgenus *Brassospora* has its centre of total species diversity in New Guinea and New Caledonia, and also has a small number of extinct species in New Zealand, Australia, Antarctica and South America. In the last four areas it overlaps with other subgenera, but in its centre of diversity, no other subgenera are known, living or fossil. This evidence all suggests that the subgenera originated prior to Gondwana breakup.

    A traditional analysis of a group such as *Nothofagus*, ranging from New Guinea to Chile, would locate a centre of origin in one part of this region (or elsewhere) by

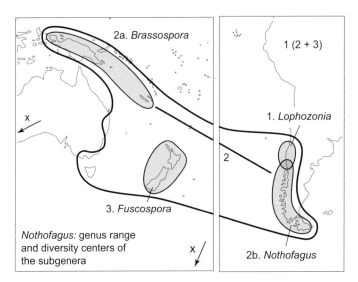

Fig. 3.5 Extant distribution of *Nothofagus*, basal in the cosmopolitan group Fagales. The phylogeny of the subgenera is indicated. Extralimital fossils occur in Antarctica and Western Australia (× symbols and arrows). Grey = main centres of diversity in the four subgenera. (From Heads, 2006b.)

examining the fossils and the phylogeny, looking for basal paraphyletic grades, and so on. Instead of investigating a group's origin in this way, by examining the group itself, the group's spatial relationship with its relatives can be considered. *Nothofagus* is sister to the rest of Fagales (Stevens, 2012), a large group with seven families found in most parts of the world where *Nothofagus* does not occur, and with lower diversity in areas where the two overlap. This indicates that *Nothofagus* evolved more or less *in situ*, as a vicariant of its sister-group. Analyses using centre of origin programs such as DIVA often invoke dispersal, but a DIVA analysis of *Nothofagus* concluded instead that its ancestral area was 'a broad realm almost including the total extant distribution pattern of the genus', not a narrow centre of origin (Zhang, 2011).

Although *Nothofagus* is widespread and often dominant in the southern Andes and New Guinea, it has a conspicuous absence in the northern Andes (Bolivia–Venezuela) and the mountains of Borneo, two major areas of endemism and centres of diversity. There is no obvious ecological explanation for this and, because these areas are occupied by other Fagales (Fagaceae, Betulaceae, Myricaceae, Juglandaceae and others), the absences are likely to be phylogenetic, the result of vicariance.

Although most debate about *Nothofagus* has focused on its trans-Pacific and trans-Tasman disjunctions, a more fundamental issue is the cause of the break between *Nothofagus* and the related families. There is no evidence for the range of *Nothofagus*, or any point in it, being the centre of origin for the order. Instead, a break has developed *between* *Nothofagus* and the other Fagales. The geographic location of this break – somewhere around the margins of *Nothofagus* – indicates the site of differentiation.

Nothofagaceae show some overlap with Casuarinaceae, but this group has its centre of diversity in Australia, unlike *Nothofagus*, and the overlap is probably the result of secondary range expansion. Casuarinaceae are known in New Zealand and South America, but only from a few fossil species. Overlap in New Guinea and, in particular, New Caledonia is more significant and may have occurred in an earlier phase.

Some South Pacific fungi (Hysterangiales)

An order of Basidiomycetes, the Hysterangiales, comprises two clades. One is a widespread global family, Phallogastraceae, and the other is distributed as follows (Hosaka *et al.*, 2008): (Aus = Australia, NZ = New Zealand, NC = New Caledonia, NG = New Guinea, SAm = southern South America).

Aus, NC, NZ, SAm (Gallaceaceae).

Aus (Mesophelliaceae).

Aus, NG, NC, NZ, SAm (*Hysterangium* II).

Northern and southern hemispheres, widespread (*Hysterangium* I and *Aroramyces*).

In traditional models this pattern would be attributed to a South Pacific centre of origin, followed by dispersal of the last clade to other areas of the world. The pattern is also consistent with dispersal occurring earlier in the phylogeny, with the phylogeny reflecting differentiation in a worldwide ancestor. The differentiation has begun around Australasia–southern South America, as in groups such as the worldwide Fagales, where the basal clade is Nothofagaceae. In Hysterangiales, the phylogeny indicates three successive differentiation events around Australasia–southern South America, and, as with the *Nothofagus* subgenera, the trans-Pacific distributions indicate that this took place before Gondwana breakup.

A 'Gondwanan' clade of copepods

The significance of pre-breakup rifting and magmatism for phylogeny has been discussed for several groups. It is also suggested in the copepod family Centropagidae. This has two main clades. One is widespread in the open ocean but also occurs at inland sites at the McPherson–Macleay Overlap (Brisbane) and in the northern hemisphere. The other clade is restricted to Australia, New Zealand, Antarctica and South America, and inhabits coastal, estuarine and inland sites (Adamowicz *et al.*, 2010). The authors concluded: 'Members of this southern continental clade were evidently both widespread and already diversified in at least some parts of Gondwanaland at the time of the separation of Australia and South America' (p. 427). This accounts for different affinities in the southern group that are incongruent with each other and with the sequence of breakup.

The Gondwanan biota is sometimes assumed to have been homogeneous throughout the supercontinent, but the southern centropagid clade was diverse in pre-breakup Gondwana, and only occurs in *parts* of Gondwana. It is absent from Africa and India, and instead is based around the South Pacific. Within this group, both *Boeckella* and *Parabroteas/Calamoecia* are disjunct between Australasia and South America, and Adamowicz *et al.* (2010) wrote that: 'The topology taken alone is suggestive of parallel vicariant cladogenesis' in the two clades. They suggested that the southern clade invaded the continents from marine ancestors, after Pangaea breakup but before final Gondwana breakup, and this period coincides with the mid-Cretaceous marine transgressions.

Brachyglottis and allies (Asteraceae: Senecioneae)

A diverse clade of South Pacific shrubs in Asteraceae tribe Senecioneae has two subgroups (Pelser *et al.*, 2010):

1. New Zealand: South and Stewart Islands, montane to alpine (*Dolichoglottis*).
2. New Guinea (*Papuacalia*), New Zealand (*Brachyglottis, Haastia, Urostemon, Traversia*), south-eastern Australia and Tasmania (*Bedfordia, Centropappus*) and central Chile (*Acrisione*). (The Australian clade is sister to the Chilean clade based on plastid data and to the New Guinea clade based on nuclear data, not to the New Zealand clade.)

Recent, rapid evolution in the group (including *Dolichoglottis*) has been suggested because of the short branch lengths (Wagstaff and Breitwieser, 2004), but branch length is not always related to clade age. The phylogeny indicates that the break, somewhere around Zealandia, between *Dolichoglottis* and its larger sister clade occurred before the trans-Pacific and trans-Tasman breaks in the latter. These trans-oceanic breaks suggest the primary differentiation in the group was Mesozoic and *Dolichoglottis* is probably much older than the Neogene mountains it now inhabits. In this way it matches other groups endemic to alpine New Zealand, such as the plant *Hectorella* (Montiaceae), cited above with its sister-group from Kerguelen (p. 74).

Ourisia (Plantaginaceae)

Ourisia is another group of montane plants found around the margins of the South Pacific. There are two main clades (Fig. 3.6; Meudt and Simpson, 2006). The first, subgen. *Suffruticosae*, includes three species of woody subshrubs found in the Coast Range of central Chile and also adjacent parts of the Andes, between Coquimbo and Concepción (30–37°). The Coast Range is lower and less well known than the Andes, but its diverse and interesting biota (Elórtegui Francioli and Moreira Muñoz, 2002) has special significance for Pacific biogeography. In another example, *Pitavia* (Rutaceae) is endemic to the Coast Range and is sister to a clade (*Flindersia* plus *Lunasia*) ranging from the Philippines to New Caledonia and New South Wales (Groppo *et al.*, 2012).

The second clade in *Ourisia*, subgen. *Ourisia*, is much more widespread than the Coast Range subshrub group. It includes 20 species of herbs in the northern Andes (the standard Bolivia–Venezuela sector), the southern Andes (Chile and Argentina south of Concepción), New Zealand and Tasmania.

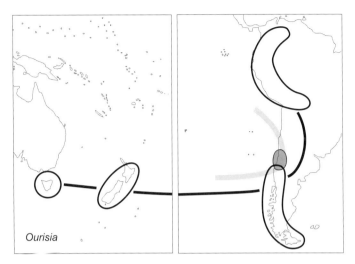

Fig. 3.6 Distribution of *Ourisia* (Orobanchaceae), showing the two clades. The subshrub clade is mapped as a grey ellipse and its possible prior distribution is indicated by the grey line. (From Heads, 2009b, reprinted by permission of John Wiley & Sons.)

The two clades of *Ourisia* are allopatric through most of their range but show minor overlap in central Chile. In Meudt and Simpson's (2006) analysis, the smaller (basal) group was assumed to occupy the centre of origin. The authors suggested that the genus dispersed from central Chile to the southern Andes, and from there north to the northern Andes (around or over the basal clade), and also west (against the prevailing winds) to New Zealand and Tasmania. Instead, the two sister-groups could have resulted from a single, simple vicariance event and the main question is: why does the genus show a split between a small part of central Chile on one hand and the great arc, Tasmania–Venezuela, on the other? One possibility (Fig. 3.6) is that the subgenera occupied two parallel arcs, with extinction offshore (perhaps during subduction) in what is now the localized Coast Range clade. The endemism and diversity of the Coast Range biota may be even higher than that of the Andes at the same latitude and altitude (Smith-Ramírez, 2004; Smith-Ramírez *et al.*, 2005). For example, the butterflies of Chile reach their highest species diversity in the Coast Range near Concepción (Biobío to Maule) (Samaniego and Marquet, 2009). This diversity anomaly suggests a tectonic origin and there are indications that the Jurassic–Lower Cretaceous island arc rocks forming the Coast Range terranes are allochthonous (Moores *et al.*, 2002). In any case, the pre-breakup tectonics of the Coast Range region involved a switch at the Early Cretaceous/Late Cretaceous boundary (100 Ma); Arancibia (2004) inferred a major change from an extensional tectonic regime with basin formation, to a contractional regime with crustal shortening and uplift. A regional scale, contractional shear zone occurs at the western boundary of the Coastal Range, between Valparaíso and Coquimbo. Arancibia (2004) dated the ductile deformation at ~100 Ma (mid-Cretaceous), coeval with uplift of the Coastal Range. The timing and geography of this activity suggest it is the cause of the pre-breakup node in *Ourisia*.

In *Ourisia*, the subshrub group on the Coast Range is closer to the central Pacific than is the herbaceous group. This reflects a trend towards woodiness and a tree habit in central Pacific angiosperm groups, compared with their mainland relatives. This is seen in plants of Juan Fernández, Hawaii, south-eastern Polynesia and the Chatham Islands, New Zealand (Heads, 2012a).

Basal differentiation occurs around nodes in central Chile in *Ourisia*, Rhipogonaceae (Fig. 3.4) and *Nothofagus* (Fig. 3.5). In the monocot family Bromeliaceae, both the two main clades (Puyoideae and Bromelioideae) have basal clades concentrated in central Chile (Jabaily and Sytsma, 2010), again indicating early breaks in the region.

Calceolariaceae (Lamiales)

This trans-South Pacific family comprises two groups (Cosacov *et al.*, 2009; Nylinder *et al.*, 2012):

1. Mexico to Tierra del Fuego, mainly in the Andes (250 species). (*Calceolaria*, incl. *Porodittia*.)
2. Central Chile (two species), from about Concepción to Valdivia (36–41° S lat., Moreira-Muñoz, 2007) and New Zealand (four species) (*Jovellana*).

As in *Ourisia*, the main break is in central Chile and was active prior to the break in the South Pacific.

Laurales

Trees in the order Laurales are abundant and diverse in most tropical and subtropical rainforests. There are three main clades (Fig. 3.7): one in the north (basal), one mainly in the south, and one widespread throughout the tropics and warmer areas (Lauraceae, Monimiaceae, etc., not mapped here).

The basal clade, Calycanthaceae, occurs mainly in the northern hemisphere but has its own basal group, *Idiospermum*, in north-eastern Queensland (Stevens, 2012).

The second split in Laurales is between the mainly southern clade (extending north to Mexico; Siparunaceae, Gomortegaceae and Atherospermataceae) and the rest of the order (the pantropical group, not shown). This implies later secondary overlap in the southern hemisphere between the three southern families and the pantropical group.

Within the southern group, the first split is between the trans-Atlantic family Siparunaceae (tropical America and West Africa) and a Pacific group: Gomortegaceae of central Chile (Coast Range) plus Atherospermataceae of southern Chile and Australasia (Renner *et al.*, 2000). The split, somewhere between Peru/Bolivia and the Chilean Coast Range, developed before the split *within* Siparunaceae at the Atlantic (opening in the Cretaceous). Figure 3.7 suggests a possible former range of Gomortegaceae in the eastern Pacific, for example on islands that have been subducted or accreted. A similar history would also explain why *Paschalocarpus*, an extinct palm of Easter Island, is most similar to *Jubaea* of the Coast Range of central Chile (31–35°; Moreira-Muñoz, 2007;

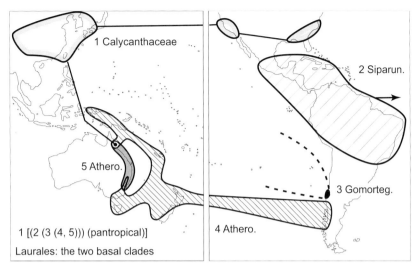

Fig. 3.7 The two basal clades of Laurales. The nested sequence of numbers indicates the phylogeny. 1 = Calycanthaceae with an outlier in north-eastern Queensland (*Idiospermum*); 2 = Siparun: Siparunaceae, also in Cameroon; 3 = Gomorteg: Gomortegaceae (possible former range indicated with dashed line); 4 = Athero: the main clade of Atherospermataceae extending north to the Barrington Tops (not north of the Macleay River in the McPherson–Macleay Overlap); 5 = Athero: the basal clade of Atherospermataceae (*Doryphora* and *Daphnandra*; both at the McPherson–Macleay Overlap). (From Heads, 2009b, reprinted by permission of John Wiley & Sons.)

Gonzalez *et al.*, 2009; Dransfield, 2012). Both *Gomortega* (35–38°) and the palms can be compared with *Ourisia* subgen. *Suffruticosae* (30–37°).

  Within the South Pacific Atherospermataceae there are two clades:

  1. Widespread in southern Chile and Australasia, often in swamp forest. The New Zealand member, *Laurelia*, is a large, rainforest tree with pneumatophores (breathing roots; Dawson, 1993, Figs 10 and 11), consistent with a former ecology as a mangrove associate.
  2. Eastern Australia, in rainforest. Both genera in the clade (*Doryphora* and *Daphnandra*) occur at the McPherson–Macleay Overlap, where the first clade is absent.

To summarize, for the southern group of basal Laurales, the sequence of differentiation proceeds westward from a node between Peru/Bolivia (Siparunaceae) and central Chile (Gomortegaceae/Atherospermataceae), to a node between central Chile (Gomortegaceae) and southern Chile (Atherospermataceae), to a node in eastern Queensland.

Marsupials and Chiloé Island

Extant marsupial mammals are another group highlighting the importance of central Chile in South Pacific distribution. The marsupials are restricted to the Americas and Australasia, and have the phylogeny (Fig. 3.8; Nilsson *et al.*, 2010):

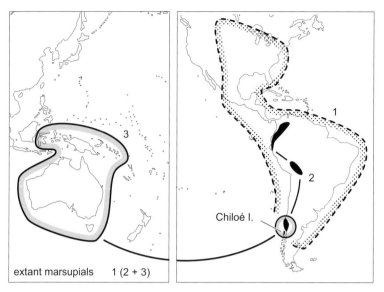

Fig. 3.8 Distribution of extant marsupials. The nested sequence of numbers indicates the phylogeny. (From Heads, 2009b, reprinted by permission of John Wiley & Sons.)

Central Chile north to Mexico (65 species), plus one species in the eastern US (introduced in the western US) (Didelphidae).

Central Chile and northern Andes (Caenolestidae).

Central Chile and Australasia (all remaining families).

The basal vicariance in the second clade, between the caenolestids and their sister-group, does not involve a split in the South Pacific itself, but at a node in central Chile, south of the break in the last examples. The Caenolestidae occur along the standard Andean track: Bolivia–Venezuela, but also have a disjunct genus, *Rhyncholestes*, in central Chile: Chiloé Island–Concepción, and adjacent Argentina. *Dromiciops*, of the otherwise Australasian clade, has an almost identical distribution to that of *Rhyncholestes*. Didelphidae also reach their southern limit at a point between Chiloé Island and Concepción. The overlap of three clades here, at the southern end of the Coast Range, reflects a fracture zone where there has been subsequent local overlap.

Through its marsupials the Chiloé Island area has a close biogeographic relationship with Australasia. Chiloé has already been mentioned; the fly *Nothotrichocera* (Trichoceridae) is in southern Australasia and Chile, by Chiloé Island, and the hymenopteran family Rotoitidae is in New Zealand, Chiloé and the adjacent mainland. The main phylogenetic and geographic break within *Nothofagus pumilio* (subgen. *Nothofagus*) is in Chile at the latitude of Chiloé, and has been dated as Eocene (40 Ma) (Mathiasen and Premoli, 2010). Earlier breaks in the same area are also possible; for example, it is the southern limit of *Nothofagus* subgen. *Lophozonia* (Moreira-Muñoz, 2011).

In other groups, the beetle family Priasilphidae (Coleoptera: Cucujoidea) is regarded as a Gondwanan relict (Leschen *et al.*, 2005; Leschen and Michaux, 2005) and has the following phylogeny (NZ = New Zealand):

Chile: Chiloé I. and adjacent parts of the mainland in Los Lagos province (*Chileosilpha*).

   Tasmania (*Priastichus*).

      NZ: South Island, mainly west of the Alpine fault (*Priasilpha* clade).

      NZ: (North, South and Auckland Islands, east and west of the Alpine fault) (*Priasilpha* clade).

Again, differentiation at Chiloé has been the earliest event. Leschen and Michaux (2005) concluded that the original break between the two New Zealand clades corresponded with the Alpine fault, and the overlap is a secondary development.

A node near Chiloé Island is also important for near-shore marine groups. On the southern coast of Chile the main genetic break in the brown seaweed *Macrocystis pyrifera* is at Chiloé (Macaya and Zuccarello, 2010). The southern group extends north to inland of Chiloé, the northern group extends south to the seaward side of the island.

Merluccius (hakes, Merlucciidae) are marine fishes that occur mainly on the continental shelf and upper continental slope of Pacific and Atlantic coasts. They have the following phylogeny (Campo *et al.*, 2007):

Atlantic Ocean (eastern US, all western coast of Africa and Mediterranean).

   West coast of the Americas from Canada south to Chiloé I. (*M. productus, M. angustimanus, M. gayi*).

   New Zealand and southern South America, north in Chile to Chiloé I. (*M. australis, M. hubbsii*). Caribbean (*M. albidus*).

In similar South Pacific–Americas patterns, Tetrachondraceae (Lamiales) comprise the herb *Tetrachondra* of New Zealand and Chile, and *Polypremum*, recorded from the south-eastern United States to Paraguay (Wagstaff *et al.*, 2000; Stevens, 2012). Here the split between the two genera has developed not in central Chile but somewhere around north-western Argentina, before the trans-Pacific connection was broken.

In another variant, *Discaria* s.str. (Rhamnaceae) of southern South America (Patagonia to south-eastern Brazil), New Zealand and Australia is sister to *Adolphia* of California, Texas and Mexico (Aagesen *et al.*, 2005). Again, a break in the Americas (this time between Brazil and Mexico) occurred before the break in the South Pacific, between Patagonia and Australasia.

Oreomyrrhis (Apiaceae)

Oreomyrrhis (Fig. 3.9) differs from the previous examples of Pacific endemics as it is so widespread around the Pacific margin. (In earlier work it was regarded as a genus but it is now treated as an unranked, unnamed clade in the genus *Chaerophyllum*; Chung

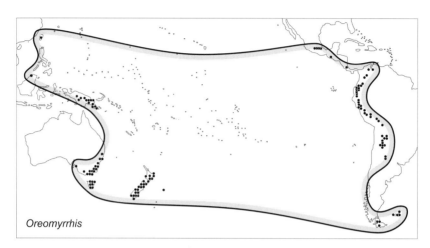

Oreomyrrhis

Fig. 3.9 Distribution of *Oreomyrrhis* (= *Chaerophyllum* 'Oreomyrrhis clade') (Apiaceae) (Chung *et al.*, 2005).

et al., 2005; Chung, 2007; Spalik *et al.*, 2010.) It is often cited for its circum-Pacific distribution – it is absent in the far north, but still has a huge range, from the far south north to Taiwan and the Trans-Mexican Volcanic Belt. These last regions are dominated by Pacific tectonics and biogeography, but they have never been part of Gondwana. Nevertheless, the distributions indicate that this group has been affected by Gondwanan rifting around Australasia and South America. The phylogeny (from Spalik *et al.*, 2010) is: Borneo and Taiwan (New Guinea (New Zealand (Australia and America))), with the first three nodes following a Pacific arc that encircles Australia.

Santalaceae, the sandalwood family

In Santalaceae, a phylogeny and distribution data are available (Der and Nickrent, 2008; Nickrent, 2012). One of the main clades of Santalaceae has 29 genera and has a global distribution. The basal group in this widespread clade is a South Pacific pair:

New Zealand and Juan Fernández Islands (*Mida*).

Patagonia, Tierra del Fuego, Falkland Islands (*Nanodea*).

The break between Juan Fernández and mainland South America is prior to the trans-South Pacific break in *Mida*.

The three other groups in the widespread clade are each widespread:

1. '*Santalum* clade'. The basal clade is the central Pacific genus *Exocarpos*: Malesia to New Zealand, Hawaii and SE Polynesia, + *Omphacomeria*: SE Australia. Sister to nine other genera, more or less worldwide.
2. '*Amphorogyne* clade'. The basal clade is *Amphorogyne* and *Daenikera* of New Caledonia. Sister to seven other genera, Himalayas to Australia.

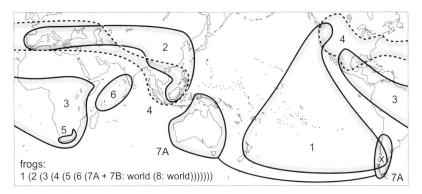

Fig. 3.10 The seven basal clades of frogs (Pyron and Wiens, 2011). The last two groups of frogs, Hyloidea (7B) and Ranoidea (8), are each distributed globally and are not mapped. The nested sequence of numbers indicates the phylogeny.

3. '*Viscum* clade'. The basal clade is the Indo–Pacific *Korthalsella*: eastern Africa to Hawaii and SE Polynesia. There are six other genera, more or less worldwide.

Differentiation at the basal nodes in each of these three clades and at the basal node of the whole group was concentrated around breaks in the south, central and south-west Pacific. This suggests that the three last clades were already widespread by the time they differentiated.

Leiopelmatidae: basal frogs

The basal frogs, Leiopelmatidae s.lat., form a Pacific basin affinity. *Leiopelma* of New Zealand and *Ascaphus* of western North America are sister genera and the only other members of the family are two Jurassic fossil genera of Argentina (*Vieraella* and *Notobatrachus*) (Fig. 3.10; Báez and Basso, 1996; Pyron and Wiens, 2011). The family is sister to all other modern frogs (Early Triassic fossil frogs represent another group). With its western North American records, Leiopelmatidae cannot be Gondwanan.

San Mauro *et al.* (2005) described the leiopelmatid genera as 'merely relicts', but this does not explain the overall allopatry between the family, straddling the central Pacific basin, and all other frogs, which are absent from the Pacific east of Fiji. The absence of the main clade of frogs from New Zealand and eastern Polynesia is probably for phylogenetic reasons. In contrast, the trans-Pacific affinities of the Leiopelmatidae suggest that the clade's absence in eastern Polynesia has been caused by secondary, ecological factors, such as the phases of large-scale volcanism and subsidence that occurred in the Cretaceous. The split between the Pacific Leiopelmatidae and all other frogs dates back to the Paleozoic or early Mesozoic (San Mauro *et al.*, 2005), before the Jurassic formation of the Pacific plate and the Pacific basin.

As San Mauro *et al.* (2005) wrote, their proposed older time frame for amphibian evolution explains 'heretofore paradoxical distribution' (p. 590) in the group. One of the most paradoxical aspects is the trans-central Pacific distribution of the basal clade, but this can be explained with reference to Mesozoic terrane accretion in New Zealand, western North America and South America (Adams, 2010), followed by extinction in the central Pacific.

Frog phylogeny: the basal nodes

In frogs, Pyron and Wiens (2011) proposed eight main clades and the distributions of these show a distinct pattern (Fig. 3.10). The first six show a high degree of allopatry, while the seventh comprises a South Pacific group (7A) and its sister (7B), the worldwide Hyloidea. Clade 8 is another worldwide group, Ranoidea. The phylogeny is: (1 (2 (3 (4 (5 (6 (7A + 7B: world (8: world)))))))). The first seven clades are mapped in Figure 3.10, but the worldwide clades 7B and 8 are not shown. The groups have the following distributions:

1. *Central Pacific*. Leiopelmatidae: New Zealand and western USA. Putative fossil relatives in Patagonia.
2. *Tethys*. Discoglossidae and Alytidae: Mediterranean. Bombinatoridae: Philippines, Borneo, Eurasia.
3. *Trans-Atlantic*. Rhinophrynidae: Costa Rica to USA. Pipidae: South America and sub-Saharan Africa.
4. *Tethys*. Pelodytidae: Europe to Caucasus. Scaphiopodidae: southern Mexico to southern Canada. Pelobatidae: Mediterranean to western Asia. Megophryidae: Himalayas to Philippines.
5. *South Africa* (west of the Lebombo monocline). Heleophrynidae.
6. *Seychelles and India (Western Ghats)*. Sooglossidae (incl. Nasikabatrachidae).
7. *South Pacific (7A) and global (7B)*. 7A. Myobatrachidae: Australia and New Guinea; and Calyptocephalellidae: central and southern Chile. 7B. Hyloidea: global (except Madagascar) (Hylidae, Bufonidae, many American families).
8. *World*. The global group Ranoidea.

The large complex formed by clades 7 and 8 has a more or less global distribution, but is absent from the central Pacific, New Zealand (where clade 1 is present) and the Seychelles (where clade 6 is present). The latter absence may reflect the original allopatry between clade 6 and clades 7 + 8.

The six basal groups of frogs (numbers 1–6 in Fig. 3.10) occur in most parts of the world, but are absent from Madagascar, Sulawesi, New Guinea and Australia, as well as parts of Africa and South America. It is possible that these gaps indicate the original distribution of clades 7 and 8.

The evolutionary sequence in the main clades of frogs has moved from initial breaks around the Pacific (clade 1), to breaks around the Tethys (2 and 4), Atlantic (3) and Indian (5 and 6) basins. The geography of the break between the clades 7 and 8 is obscure, but within clade 7 the main break is between the South Pacific clade 7A and its global sister.

The diverse, worldwide clade 8, Ranoidea, is mainly an Old World group, with many lineages in Africa and Asia. One family, the Microhylidae, extends to the New World and also contributes the majority of the frog species in Madagascar and New Guinea.

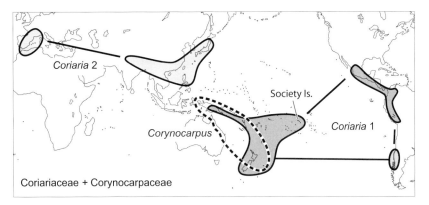

Fig. 3.11 Distribution of Coriariaceae (*Coriaria*) and their sister-group Corynocarpaceae (*Corynocarpus*) (Stevens, 2012).

Yet in Australia members of Ranoidea are restricted to Queensland and Arnhem Land (there is only one genus of Ranidae and two of Microhylidae). It might be suspected that Ranoidea are absent from most of Australia because they are a 'tropical' group. Nevertheless, many members occur outside the tropics or at high elevations in the tropics, and so the absence could instead reflect an old, phylogenetic break with clade 7 (Hyloidea and Myobatrachidae). The South Pacific clade (Myobatrachidae) makes up the bulk of Australia's amphibians (20 genera out of 26) and there are also three genera of Hylidae.

Pacific + Tethys groups in Australasia

Both Pacific and Tethyan tracks are evident in many Australasian groups. For example, in Plantaginaceae, the South Pacific *Hebe* complex of Australasia (150 species) and Patagonia (two species) is sister to *Veronica* subgen. *Triangulicapsula*, comprising *V. grisebachii* of Turkey and Bulgaria, and *V. chamaepithyoides* of Spain (Albach and Meudt, 2010). Other examples include the following.

Coriariaceae and Corynocarpaceae (Cucurbitales)

The trees and shrubs in *Coriaria* have two main groups, one distributed around the Pacific, the other around Tethys (Fig. 3.11; Yokoyama *et al.*, 2000). Pacific + Tethys distributions, as seen here in an extant angiosperm, also occur in groups such as the Mesozoic bivalve *Monotis* (Craw *et al.*, 1999, Fig. 6.8). *Coriaria* as a whole is sister to *Corynocarpus*, a Tasman–Coral Sea endemic, and together they are sister to a diverse, cosmopolitan clade including Cucurbitaceae, Begoniaceae and others (Stevens, 2012).

Corynocarpus and *Coriaria* show significant allopatry, as *Corynocarpus* occurs alone in western New Guinea, Australia, New Caledonia and Norfolk Island (Wagstaff and

Dawson, 2000). The two groups overlap only along the eastern strip: New Zealand–Vanuatu–eastern New Guinea, indicating the original fracture between the two. Christenhusz and Chase (2013) suggested that Coriariaceae have such a wide distribution that it must have required long-distance dispersal, but they did not consider the high degree of allopatry with the family's sister-group, Corynocarpaceae. The two probably evolved by vicariance, with dispersal occurring before the origin of the two, in a common ancestor, and after their origin, causing the overlap in New Zealand.

The total distribution of *Coriaria* and *Corynocarpus* along the Tethys and circum-Pacific (including Caribbean) belts corresponds to the world's active subduction zones. *Coriaria* and *Corynocarpus* are subduction zone weeds and have maintained this ecology over geological time scales, surviving the high levels of disturbance caused by volcanism, earthquakes, uplift, rifting and subsidence. The two plants are also weedy pioneers in the ecological time scale, and are often conspicuous colonists of disturbed habitat around cliffs, gullies and landslides. *Coriaria* has interesting records in the central Pacific (east to the summit of Tahiti, Society Islands), a region characterized by high levels of volcanism since the formation of the Pacific plate and the emplacement of the large igneous plateaus.

Myosotis (Boraginaceae)

The plant genus *Myosotis* has a circumglobal distribution and notable diversity in the New Zealand mountains. One clade has two subgroups (Winkworth *et al.*, 2002a):

1. South Pacific: New Guinea, Australia, New Zealand and South America (sect. *Exarrhena* 'austral group').
2. Mediterranean: Europe and North Africa to Ethiopia (sect. *Exarrhena* 'discolor group').

The pattern indicates a former connection of the South Pacific clade with the Mediterranean clade somewhere between New Guinea and Ethiopia, probably along a Tethys track. Later there was a break between the two clades, with one evolving in the north and one around the South Pacific. Finally, breaks occurred in the South Pacific clade between Australasian and South American members. The pattern repeats that of *Coriaria*.

Winkworth *et al.* (2002a) calibrated the *Myosotis* phylogeny using the oldest fossil pollen of the genus from New Zealand (2 Ma), but concluded that the extent of sequence divergence suggests that *Myosotis* was in the southern hemisphere longer than is indicated by the fossil record. The authors inferred a centre of origin for *Myosotis* in the northern hemisphere because of a paraphyletic grade there (but see Fig. 1.6), but this does not explain key features: why is there only one South Pacific-based clade in *Myosotis*, and why does its distribution conform to a common, global pattern, vicariant with its Mediterranean sister (cf. the *Hebe* complex)? Within the South Pacific clade of *Myosotis* the New Zealand species are basal, but again a New Zealand centre of origin for the group is not necessary if the group occupied the South Pacific by evolving there, rather than migrating there.

Gentianella (Gentianaceae)

Gentianella is a group of herbs widespread in cooler habitats around the Pacific and Tethys basins. The two primary clades are distributed as follows (Hagen and Kadereit, 2001):

1. Eurasia and North America (26 species).
2. Eurasia and North America (16 species), South America (170), New Zealand (30) and Australia (2).

The sister-group of *Gentianella* is an East Asian clade (*Swertia*, in part). Hagen and Kadereit (2001) suggested this means the progenitor of *Gentianella* and clade 2 is likely to have originated in East or Central Asia. They drew the migration of clade 2 from Asia, via the Bering land bridge, down through North and South America, and finally across the South Pacific to Australia/New Zealand. As usual, this sequence can be interpreted instead as a sequence of differentiation and this would be more compatible with the high degree of allopatry among the 10 components of clade 2. These are found, respectively, in Eurasia, North America, North and South America, South America (seven clades), and Australasia.

Other plant genera that are diverse in South America but are thought to have a northern hemisphere origin include *Berberis*, *Ribes*, *Potentilla*, *Lupinus*, *Salvia* and *Draba* (Hagen and Kadereit, 2001). Studies of fossil pollen records show that these appear at different times in the southern fossil record, but this result is probably an artefact of preservation and discovery; as Hagen and Kadereit (2001) wrote, 'many taxa, including *Gentianella*, are not well suited for palynological investigations because their pollen is only rarely found'. These are small, alpine, insect-pollinated herbs that produce minute amounts of pollen and fossil preservation in mountain ranges is unusual. Nevertheless, Hagen and Kadereit (2001) used the age of the earliest documented *Gentianella* fossil in South America (1.6 Ma) to calibrate their molecular tree. They also employed a second calibration, based on the first appearance of high alpine habitat in South America at 3 Ma. At present *Gentianella* occurs in *tropical* South America only in the high alpine zone, and the authors argued it could not have occurred in the continent before high alpine habitat existed. Yet populations in southern South America are recorded from lower elevations and before the ice ages these could have inhabited areas of open vegetation caused by extreme soil type or microclimate. Based on the two calibrations, Hagen and Kadereit (2001) proposed that the last common ancestor of *Gentianella* dates back only to 3–5 Ma, but it is possible that the actual age is much older than this.

Morphological similarities indicate a possible trans-Pacific affinity between the Australasian *Gentianella* species and some of the South American species (this has not yet been resolved with molecular data). To explain this possible connection, Hagen and Kadereit (2001) suggested trans-Pacific wind dispersal from South America to Australasia, although they noted that prevailing winds are in the wrong direction for this. They also suggested seabirds as possible vectors, with seeds either attached to the body or in the gut, but they noted that 'such behaviour ... is not known for alpine birds of South America, the potential dispersal agents of *Gentianella*'. 'In summary', the authors wrote, 'we cannot answer the question of how *Gentianella* dispersed from South America to Australia/New Zealand'. Perhaps this dispersal event never occurred; the dating calibrations that support it relied on some doubtful assumptions.

Gentianella has its greatest species diversity in the Andes, and Hagen and Kadereit (2001) suggested that this 'probably resulted from the availability of a very large alpine area'. Nevertheless, there is a much larger area of alpine habitat in North America and Eurasia, where the genus occurs but with many fewer species. As in the Andes, species numbers per unit area in Australasia (mainly in New Zealand) are also much greater than in the northern hemisphere.

Glenny (2004) suggested that the New Zealand clades are closest to species of Ecuador, Peru and Bolivia. The earliest *Gentianella* pollen in New Zealand is dated at 2.6–1.6 Ma and Glenny suggested dispersal from South America and New Zealand. Nevertheless, if the group was in the region since before the rifting of New Zealand from Gondwana at ~85 Ma, this would account for both the intercontinental patterns and also key aspects of distribution within New Zealand. These include massive disjunction between the sister species *G. saxosa* and *G. scopulorum* along the Alpine fault (maps in Glenny, 2004), where strike-slip displacement began in the Miocene (see Chapter 7).

Muehlenbeckia (Polygonaceae)

A clade in Polygonaceae (Schuster *et al.*, 2011a) is made up of three allopatric genera:

Japan, Sakhalin (*Reynoutria*).

North America, Europe, North Africa, central Asia (*Fallopia*).

Papua New Guinea, Solomon Islands, Australia, Tasmania, New Zealand, South and Central America (*Muehlenbeckia*).

The pattern is consistent with a widespread Pacific–Tethys ancestor.

Australasia–central Pacific groups

The last sections presented some of the biogeographic connections that Australasia has with other parts of the Indian, Tethys and South Pacific regions. One of these groups, *Coriaria* (Fig. 3.11) is also represented in the central Pacific (French Polynesia), far from the Pacific plate margin. Other groups in Australasia have immediate affinities in the central Pacific.

Alseuosmiaceae and allies (Asterales)

A clade made up of three families of trees and shrubs, Alseuosmiaceae, Phellinaceae and Argophyllaceae, is distributed around the northern Tasman Sea region and also on Rapa Island in south-eastern Polynesia (Fig. 3.12). Only New Caledonia has all three groups. Phellinaceae are endemic in New Caledonia, with possible fossils in New

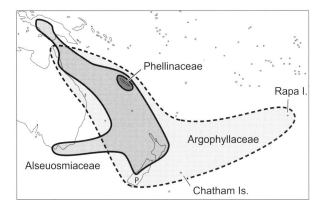

Fig. 3.12 Distribution of the *Phelline* complex: Phellinaceae, Argophyllaceae and Alscuosmiaccac. Distributions from Stcvcns (2012) and *Australia's Virtual Herbarium* (2011). P = fossil record of *Phelline* in New Zealand (Pole, 2010). (From Heads, 2010b, reprinted by permission of John Wiley & Sons.)

Zealand. Alseuosmiaceae are more widespread around the Tasman–Coral Sea region, but have significant gaps at the McPherson–Macleay Overlap and Tasmania, and in New Zealand extend south only to southern Westland (Mount Ellery; personal observations). The third clade in the complex, Argophyllaceae, extends from Cape York east to the Chatham Islands and Rapa Island. Alseuosmiaceae and allies are not well known but are important in phylogeny as they are sister to a large, cosmopolitan group comprising Asteraceae, Goodeniaceae, Stylidiaceae and others (Tank and Donoghue, 2010).

Other Tasman Sea groups with outliers on Rapa Island include the *Hebe* complex (Plantaginaceae) and *Olearia* (Asteraceae). (The biogeography of Rapa and the central Pacific is discussed elsewhere; Heads, 2012a, Chapter 6.) Similar Australasia–Polynesia connections are seen in the blue-pollen group of *Fuchsia* (Onagraceae), present in New Zealand and Tahiti (Society Islands), north of Rapa. Its sister-group (sections *Quelusia* and *Kierschlegeria*) is in southern and central Chile, Argentina and south-eastern Brazil (Berry *et al.*, 2004). Mosses in the genus *Weymouthia* (Lembophyllaceae) occur in Australasia, Tahiti and southern Chile (Frey *et al.*, 2010).

Sophora section *Edwardsia* and *S. tomentosa* (Fabaceae)

The trees and shrubs in *Sophora* section *Edwardsia* (Fig. 3.13) occur around a central Pacific triangle: New Zealand–Hawaii–Juan Fernández, and are also on the Mascarenes and Gough Island in the South Atlantic (Hurr *et al.*, 1999; Tassin *et al.*, 2004). (Some authors have cited *Edwardsia* from the Marquesas Islands, but no specimens are known; *S. tomentosa* is indigenous east only to the Society and Austral Islands; W. Wagner and D. Lorence, personal communication, 3 June 2010.) In the Tasman region, section *Edwardsia* is on Lord Howe Island but is absent from Australia and New Caledonia.

Hurr *et al.* (1999) calculated ages for section *Edwardsia* based on a fossil of the sister-group, *S. tomentosa*, dated at 30 Ma. They concluded that this is a 'clear indication' of

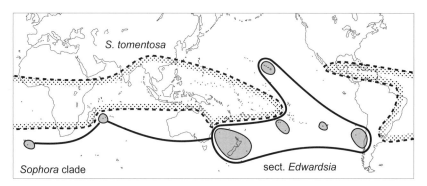

Fig. 3.13 *Sophora tomentosa* (dashed, stippled line) and its sister-group *S.* sect. *Edwardsia* (grey and solid line) (Hurr *et al.*, 1999).

the recent origin of *Edwardsia* and that long-distance dispersal of buoyant seeds must account for the distribution. Recent studies have also proposed centres of origin for *Edwardsia* (Hurr *et al.*, 1999; Peña *et al.*, 2000; Mitchell and Heenan, 2008), but these have not referred to or explained the precise allopatry between sect. *Edwardsia* and its sister, *S. tomentosa* (Fig. 3.13). The allopatry of the two widespread groups does not simply follow the latitudes and ecology does not explain the boundary. The break could reflect early phylogeny and this would account for the anomalous absence of section *Edwardsia* in Madagascar, Africa, Australia and New Caledonia. The main breaks between *Edwardsia* and its sister-group occur at Madagascar and the Mascarenes, at the McPherson–Macleay Overlap and around the Society Islands, and these correspond to zones of late Mesozoic rifting, epicontinental seas and volcanism, respectively.

Members of section *Edwardsia* are widespread inland and form a single trunk, whereas *Sophora tomentosa*, a beach plant, branches from ground level (Heenan *et al.*, 2004). This is a variant of a common pattern in which the central Pacific members of an otherwise non-arborescent group develop a single trunk (see notes on *Ourisia*, p. 92). The pattern is consistent with *Edwardsia* representing former mangroves that have been stranded inland, with the plants having had all but one of the multiple shoot axes suppressed.

Trans-tropical Pacific connections

New Zealand–Polynesia connections, seen in groups such as Argophyllaceae, *Fuchsia* and *Sophora* sect. *Edwardsia*, are related to complete trans-tropical Pacific affinities; for example, between Australasia and the Caribbean. These tropical affinities are less well known than the trans-Pacific connections further south, but the patterns are common and examples are given below. The central Pacific biota has been involved in the Earth's greatest volcanic event, in which vast volcanic plateaus were formed in Polynesia during the mid-Cretaceous. These were thought to have been erupted under water, but parts of the plateaus were erupted in a subaerial environment and fossil wood has been recovered from the Ontong Java plateau. In addition, while the plateaus are now all submerged, they include many seamounts with flat tops, indicating that they were former islands eroded to sea level. (The plateaus are discussed in more detail in Chapter 6.) Following their eruption, perhaps as a single structure, the plateaus broke up and some dispersed to the western and eastern margins of the Pacific; the Manihiki Plateau has remained in central Polynesia (Cook Islands). The islands on the plateaus and around other volcanic structures have allowed survival of metapopulations in the Pacific, and the rifting of the plateaus and other island groups has probably led to their vicariance (Heads, 2012a, Chapters 6–8).

Liverworts, mosses and ferns

The liverwort *Symphyogyna hymenophyllum* of south-west Australia, south-east Australia and New Zealand is sister to *S. bogotensis*, endemic to the standard northern Andean track: Bolivia to Venezuela (Frey *et al.*, 2010). It is replaced in Patagonia, central America, Brazil and Africa by its sister-group *S. podophylla* and so the trans-Pacific connection appears to be across the tropics, not via the well known Australasia–Patagonia track.

The liverwort genus *Lobatiriccardia* (Aneuraceae) is diverse and well documented along the west Pacific margin, from Japan to New Zealand. Molecular studies now show that it also occurs in Ecuador (Preußing *et al.*, 2010).

The moss *Elmeriobryum* (Hypnaceae) comprises one species in Taiwan and the Philippines (Luzon), one in the Papua New Guinea mountains and one in Central America (Guatemala, El Salvador and Costa Rica) (Buck and Tan, 2008).

The New Zealand endemic fern *Loxsoma* (Loxsomataceae) is sister to *Loxsomopsis* of the standard Andean track: Bolivia to Colombia/Venezuela border, also Costa Rica (Perrie and Brownsey, 2007).

Orchids

The orchid *Paphiopedilum* is native from India to the Solomon Islands and is sister to *Mexipedium* and *Phragmipedium*, extending from Mexico to south-eastern Brazil (Guo *et al.*, 2012).

Parsonsia and allies – a Pacific group of lianes (Apocynaceae)

A trans-Pacific clade of lianes is composed of the predominantly Australasian *Parsonsia* and *Artia*, plus the tropical American *Prestonia* p.p. (Fig. 3.14; Livshultz *et al.*, 2007). The group is mainly tropical and subtropical; for example, in New Caledonia there are 15 species of *Parsonsia* and five of *Artia*, while New Zealand only has three species of *Parsonsia*. *Artia lifuana* of New Caledonia is a typical example. It is found on calcareous, ultramafic and schist substrates in lowland to mid-altitude rainforest, in the Loyalty Islands (an extinct, accreted arc) and central Grande Terre (another accreted arc) (Morat *et al.*, 2001).

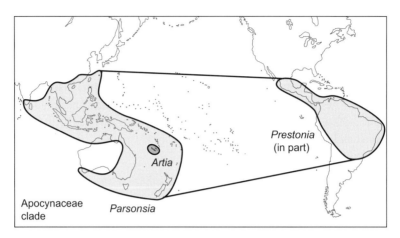

Fig. 3.14 Distribution of a clade in Apocynaceae (Livshultz *et al.*, 2007, 2011).

Members of the coffee family, Rubiaceae

A clade in Rubiaceae has the following phylogeny (Barrabé *et al.*, 2012):

> Tropical Africa (*Chazaliella*).
>> Burma to SE Polynesia (Society Islands) ('Clade G': *Psychotria* p.p., *Hodgkinsonia*, *Readea*).
>> Tropical America (*Margaritopsis*).

The allopatry is consistent with the origin of the trans-tropical Pacific clade by simple vicariance with its African sister-group; there is no need for dispersal across the Pacific in the modern group.

Snails, millipedes, ants and beetles

Pachychilid river snails include a typical trans-tropical Pacific clade, present in Sulawesi (*Tylomelaenia*), Torres Strait islands (*Pseudopotamis*) and tropical America (*Pachychilus*) (Köhler and Glaubrecht, 2007).

Based on morphological studies, the millipede *Agathodesmus* includes species from New South Wales, Queensland and New Caledonia, and also a little known species from Jamaica (Jeekel, 1986; Mesibov, 2009). Some authors had suggested human transport to New Caledonia and Jamaica, but the New Caledonian species is endemic (Mesibov, 2009), and Hoffmann (1999) reported a second Jamaican species in the genus.

In ants, *Nasutitermes polygynus* is found throughout New Guinea and forms a clade with Neotropical species (Miura *et al.*, 2000). The authors regarded this as 'most surprising' and wrote that 'how the ancestors of *N. polygynus* could cross the Pacific Ocean remains an enigma'.

In the beetle *Arrhipis* (Eucnemidae), Brüstle *et al.* (2010) gave the following phylogeny: Southeast Asia (Southeast Asia (Africa (Australia + South and Central America))). Given the allopatry of the clades, the authors concluded that 'vicariance

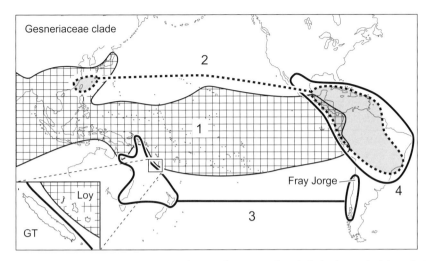

Fig. 3.15 Gesneriaceae except *Peltanthera* and *Sanango* (basal clades in tropical America). 1 = Didymocarpoideae + Epithematoideae; 2–4 = Gesnerioideae: 2 = Napeantheae and Beslerieae, 3 = Coronanthereae, 4 = Sinningieae, Gesnerieae, etc. (Perret *et al.*, 2013). Inset shows New Caledonia; GT = Grande Terre, L = Loyalty Islands.

seems to be the logical explanation', with possible overlap in Southeast Asia the result of local dispersal. Brüstle *et al.* (2010) described *Arrhipis* as 'pantropical', implying that the Australia–America affinity occurs at tropical latitudes.

Based on morphological evidence, the stag beetle *Syndesus* (Lucanidae) is recognized in Australia, New Caledonia, New Guinea, Ecuador and Brazil (Holloway, 2007). There is also a fossil species from Dominican Republic amber which is 'remarkably similar' to the Australian *S. cornutus* (Woodruff, 2009).

Harpy eagles (Accipitridae)

The harpy eagle *Harpyopsis* is widespread in the mountains of New Guinea and is sister to *Harpia* + *Morphnus* of tropical America (Lerner and Mindell, 2005). *Harpia* is the largest raptor in America. The trans-tropical Pacific clade is well supported and may be sister to a global group, the main clade of Accipitridae (*Accipiter, Circus, Aquila, Haliaeetus*, etc.).

Gesneriaceae

The pantropical plant family Gesneriaceae has two basal genera in tropical America, *Peltanthera* from Costa Rica to Peru, and *Sanango* in Ecuador and Peru (Perret *et al.*, 2013). The rest of the family forms a global group, with one clade in the North Pacific, one in the South Pacific and one across the tropical Pacific, in Asia and in Africa (Fig. 3.15). Weber (2004a) observed that an explanation for the 'enigmatic' trans-tropical Pacific clade is still needed, but the pattern in the global clade overall can be explained by simple vicariance of all three groups, with overlap restricted to tropical

America, around Taiwan, and around the Coral Sea. The clades form belts that coincide, in part, with the latitudes, although these do not correlate with the present climate. For example, the southern clade is well into the tropics in the northern Solomon Islands, while the central Pacific clade is present in America only along the western margin and is replaced through the rest of the American tropics by its sister-group. This indicates a standard break between the cratonic part of South America and the allochthonous part, accreted in the mid-Cretaceous. Two earlier breaks around the same margin, followed by overlap, account for the two basal genera.

Where the two widespread clades of Gesneriaceae meet in the south-west Pacific, overlap is restricted to the northern Solomon Islands and near Cairns. Simple allopatry occurs in New Caledonia, for example, between Grande Terre and the Loyalty Islands. This is a common biogeographic break discussed in Chapter 8.

The break between clade 3 (Coronanthereae) and 4 (Sinningieae, Gesnerieae, etc.) is not located in the Pacific or Southern Oceans, but at an earlier break between central Chile (where Coronanthereae range north to Fray Jorge; Moreira-Muñoz, 2007, 2011) and northern Argentina. This region underwent large-scale faulting before and during Gondwana breakup, as South America rotated away from Africa.

Old World suboscines and the enigmatic *Sapayoa*

Trans-tropical Pacific clades with limits in western Colombia, as in Gesneriaceae, are rare in passerines but do occur in the basal clades. In passerine birds, the basal node separates the New Zealand wrens from all the others. In 'the others', the first node separates a smaller group, the suboscines, from the main group, the oscines. The singing abilities of suboscines are more limited than in the oscines owing to differences in syrinx musculature. Suboscines are a worldwide group, with two subgroups (Irestedt *et al.*, 2006):

1. Old World–Pacific suboscines (50 species): through Africa and Asia to the Solomon Islands, but with one disjunct member, *Sapayoa aenigma*, in Panama, western Colombia (Chocó) and north-western Ecuador.
2. New World suboscines (1150 species): South and Central America, a few extending to North America.

The two clades overlap only in the Chocó region of Colombia and so, overall, are separated by a break in western Colombia near the Romeral fault zone. This is the boundary between the craton in the east and the accreted terranes to the west.

The pattern in Gesneriaceae and suboscines is similar to one in snakes: pythons range through Africa and Asia to the Solomon Islands, and have one member, *Loxocemus*, in western Mexico to Costa Rica. Otherwise they are replaced in America by their sister-group, the boas.

Gaultheria (Ericaceae subfam. Vaccinioideae), with an introduction to the Ericaceae

The circum-Pacific distribution of *Gaultheria* (Fig. 3.16) has long intrigued biogeographers, and molecular work is now resolving the phylogeny within the group. In the broad

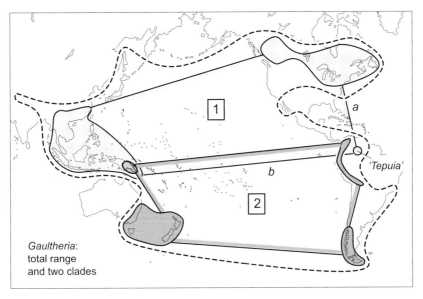

Fig. 3.16 Distribution of *Gaultheria* and two of its main clades (Ericaceae) (Fritsch *et al.*, 2011).

sense the genus has 130 species of shrubs, which range north to Sakhalin and Canada, and south to New Zealand and Patagonia. There are six main clades (Fritsch *et al.*, 2011), including several named earlier as genera. Two of the clades have widespread, intercontinental distributions and are shown here, along with the total range of the genus. One of the widespread clades (the basal clade of the genus), is in the north Pacific, while the other widespread clade is in the South Pacific. Of the four clades not mapped here, one extends from southern South America to Mexico (Oaxaca terrane boundary), one is in Japan and western US/south-western Canada, and two are in Asia.

In the north Pacific clade of *Gaultheria*, the former genus *Tepuia* of the southern Venezuelan mountains is sister to a clade comprising *Diplycosia* of Malesia (Sumatra to New Guinea), two species of Southeast Asia, and *Gaultheria hispidula* of the United States and Canada. Thus *Tepuia* has potential affinities across the Caribbean in North America (track *a* in Fig. 3.16) and across the tropical Pacific in Malesia (track *b*). Tracks *a* and *b* both relate to Early Cretaceous events – for *a*, seafloor spreading and the emplacement of a large igneous province (Caribbean plate) between North and South America, and for *b*, seafloor spreading and the emplacement of a large igneous province (Ontong Java–Hikurangi–Manihiki plateau) in the central Pacific. These Caribbean and central Pacific events have been considered as part of a single process by geologists and together coincide with the affinities of *Tepuia*.

Another genus of Ericaceae from the southern Venezuelan mountains, *Ledothamnus*, has a trans-North Pacific affinity, as its sister-group, *Bryanthus*, is endemic to Japan (Gillespie and Kron, 2012). The usual explanation for the trans-Pacific patterns is dispersal via the Beringia land bridge, followed by extinction there owing to cooling, but these trans-Pacific Ericaceae are montane and alpine plants that thrive in cold conditions. In addition, the *Gaultheria* clades already have close relatives (other members of their subfamily, Vaccinioideae) around Beringia and there is no evidence that they themselves ever existed there.

In the South Pacific clade of *Gaultheria*, an Australia–New Zealand clade (*G. tasmanica* etc.) is grouped with species of southern South America (*G. pumila* etc.) and the northern Andes (along the usual Bolivia–Venezuela track; Fig. 3.16). *G. mundula*, of New Guinea, has not been sequenced, but morphological studies agree that it belongs to this group. As an alternative to chance dispersal, it is possible that the trans-oceanic connections already existed prior to breakup. The southern Pacific connection is well known and the disjunction can be explained by Gondwana breakup, but the intercontinental boundary and connections that span the tropical Pacific cannot, and instead these implicate the tectonic history of the central Pacific.

Fossil data for *Gaultheria* (Vaccinioideae) are scarce and in most cases are too recent to be useful for estimating divergence times in the genus (Bush *et al.*, 2009). In other groups of Ericaceae, fossil records for *Rhododendron* (Ericoideae) are more plentiful and the oldest are seeds from the Early Paleocene (~65 Ma). Bush *et al.* (2009) proposed that Ericoideae originated 'much earlier' than the Vaccinioideae (they gave no evidence for this but it is probably based on the fossil record). They reasoned, 'It would therefore be conservative to assume that *Gaultheria* evolved less than 80 million years ago' (the age of Gondwana breakup). This is a flawed, risky way of calculating the date, though, not a conservative approach, as it treats fossil-based estimates (minimum ages) as maximum clade ages. In any case, a centre of origin model does not account for many aspects of the distribution; for example, the existence of allopatric groups spanning the North and South Pacific, and the conspicuous absence of *Gaultheria* and its subfamily, Vaccinioideae, from most of Australia. This local phenomenon is not due to chance, but instead is a function of the global relationships among higher clades in the family. Several different groups are cited in this book and so it is worthwhile summarizing the phylogeny of the family here.

Ericaceae are hardy plants able to survive extreme cold, poor drainage, acid soil and volcanic activity such as ash falls. They survive where many other plants cannot and so they are valuable as biogeographic indicators. The family comprises three basal clades in the northern hemisphere (Enkianthoideae, Monotropoideae and Arbutoideae) and a diverse clade with two branches, each widespread in the northern and southern hemispheres:

1. **Cassiopoideae**: circumboreal. **Ericoideae**: widespread globally, but in Australia only in the Cairns area (two species of *Rhododendron*). Absent from the Pacific islands (except the Solomon Islands).
2. **Harrimanelloideae**: circumboreal. **Vaccinioideae**: widespread globally but in Australia only in the Cairns area (one species of *Agapetes*) and south-eastern Australia (eight species of *Gaultheria*). **Styphelioideae**: Australasia (throughout Australia with about 30 genera and 350 species), Malesia and the Pacific islands.

Given the 'basal' groups Enkianthoideae, Monotropoideae and Arbutoideae in the north, along with Cassiopoideae and Harrimanelloideae, it is not surprising that a DIVA analysis gave centre of origin of Ericaceae in Laurasia (Kron and Luteyn, 2005), but the overlap of the groups in the north is probably secondary. In particular, the two widespread clades listed above could have been allopatric in the north as well as in the south. The main biogeographic problem is accounting for these two widespread groups and their distinct distributions which overlap but show significant differences in Australia and the Pacific islands. The distributions around Australia can be considered first.

The absence of Ericoideae + Cassiopoideae from all Australia (and New Zealand) except Cairns is striking, especially as there are 29 naturalized species of *Erica* in Australia, including the south-west, and they are also invasive weeds in New Zealand. The simplest explanation is that the Ericoideae + Cassiopoideae group is represented through the rest of Australia by its sister-group, Harrimanelloideae + Vaccinioideae + Styphelioideae, and that the different distributions reflect a vicariance event, not ecology or chance dispersal.

As with Ericoideae, Vaccinioideae also show a striking absence from most of Australia (they are present in the Cairns area and along the south-eastern seaboard). The simplest explanation is that they are represented throughout most of Australia by their sister-group, Styphelioideae, and the considerable allopatry reflects vicariance. In this model, overlap along the eastern seaboard is secondary.

Outside Australia, the two global clades in Ericaceae show extensive overlap in the northern hemisphere but, apart from this, their distributions are quite different, with major centres of diversity in different regions:

> Ericoideae and Cassiopoideae: most diverse in **South Africa** (including 680 species of *Erica* in the Cape region), **Burma/southern China** (500 species of *Rhododendron* in China) and **Malesia** (300 species of *Rhododendron*). Absent from the central Pacific. Mainly an Indian Ocean group.

>> Vaccinioideae and Harrimanelloideae: most diverse in the **Andes, central America and the Antilles** (800 species), in and around the **Pacific basin**, **Malesia** (450 species) and southern China. Few members in Africa (absent in the Cape region, but present in south-eastern South Africa). Mainly a Pacific group.

>> Styphelioideae: mainly in **Australia and New Zealand**.

The impressive centres of diversity for the main clades overlap in Malesia and in the far north, but otherwise they show high levels of allopatry, consistent with their derivation by vicariance. The overlap in the northern hemisphere involves relatively few species and could have been a young development (even if the groups themselves were not), while the overlap in Malesia and Burma/southern China involves hundreds of species, as well as many higher level clades, and could have been caused by earlier events.

Bracken fern (*Pteridium*: Dennstaedtiaceae)

The bracken fern *Pteridium* is cosmopolitan with two primary clades (Der *et al.*, 2009):

1. Widespread, but rare around the South Pacific: South Africa to Asia (south to India, Indonesia and Queensland), Hawaii and North America, south to Costa Rica and Colombia.

2. South Pacific: Malaysia, western and eastern Australia, New Caledonia, New
Zealand and South America north to Mexico.

The 'widespread' clade includes potential trans-tropical Pacific connections north of
the South Pacific group. The only overlap between the two clades occurs in Malesia/
Queensland and central America. A line linking these two areas separates the South
Pacific clade from the 'widespread' clade and represents part of the margin where
differentiation occurred, as in *Gaultheria*.

Elaeocarpaceae

In the widespread tree family Elaeocarpaceae the basal clade has three genera (Crayn
et al., 2006):

Widespread Indo–Pacific: Madagascar, Yunnan to Sydney and New Caledonia,
also tropical America (Mexico to SE Brazil) (*Sloanea*). Fossils are known from the
northern hemisphere (US, Greenland and Europe) and are dated as Paleogene or
Late Cretaceous (Manchester and Kvaček, 2009).

South Pacific: SW and SE Australia (north to Armidale), Tasmania, New
Zealand, and central Chile (31–43°S) (*Aristotelia*).

Northern Andes: Bolivia/NW Argentina border to Venezuela (*Vallea*).

As with *Pteridium*, the group has one clade around Africa/Madagascar, Asia and across
the tropical Pacific (New Caledonia and Colombia etc.), and one spanning the South
Pacific. The two show local overlap around the McPherson–Macleay Overlap and in the
northern Andes. In addition to the trans-tropical Pacific link, *Sloanea* also has possible
trans-Pacific connections at northern latitudes (cf. the fossils from North America).

Sloanea and the widespread clade of *Pteridium* show a common pattern, occurring
in the south of the Indian Ocean (Madagascar), but with the distribution 'lifted up'
northwards in the Pacific. *Aristotelia* + *Vallea* form a typical southern Pacific affinity, with a main break not in the Pacific, but between central Chile and north-western
Argentina. *Sloanea*, disjunct across the tropical and North Pacific, and its sister-group,
disjunct across the South Pacific, probably differentiated from each other by vicariance. If the southern disjunction is Cretaceous, perhaps the tropical/northern one is as
well.

Crayn *et al.* (2006) concluded that the trans-Pacific genus *Aristotelia* 'arrived in New
Zealand from Australia *at least* 6–7 million yr ago' [italics added], and, if this is treated
as an estimate of minimum age, Gondwanan vicariance is possible. The high level of
allopatry among *Aristotelia*, *Vallea* and *Sloanea* is compatible with vicariance.

The three elaeocarp genera form a clade that is pantropical, except for its absence on
mainland Africa. Its sister-group comprises the remaining seven genera in the family
and has a similar, widespread distribution. Again, the Pacific basin is significant and the
four basal groups are all located around its southern margins:

New Guinea (Morobe), NE Queensland (*Peripentadenia*).

South America: Chile, Argentina, Bolivia (*Crinodendron*).

Maluku Islands (Moluccas), New Guinea, northern Australia, New Caledonia (*Dubouzetia*).

Southern Australia: south-western Australia, Adelaide, Sydney ('Tremandraceae': *Tremandra*, *Platytheca* and *Tetratheca*) (Fig. 1.9).

Widespread Indo–Pacific, from Madagascar to Brazil (*Elaeocarpus* incl. *Sericolea* and *Aceratium*).

Distribution in a Pacific triangle: Australasia–western North America–southern South America

Many Australasian groups show direct trans-Pacific connections with accreted terranes of western North America. For example, in the plant *Montia* (Montiaceae), a group of Australian and New Zealand species have their sister-group (*M. howellii*) in California and Oregon (O'Quinn and Hufford, 2005). In other groups this Australasia–western North America affinity is supplemented with records in southern South America, giving a Pacific triangle of distribution. This was illustrated above in the basal frogs (the New Zealand *Leiopelma*, the western North American *Ascaphus* and the Patagonian fossil genera) (Fig. 3.10). The fishfly *Protochauliodes* (Megaloptera) has similar distribution, occurring in eastern Australia, Chile and the western United States (Liu *et al.*, 2012). The pattern is also seen in the following groups.

Blennospermatinae (Asteraceae: Senecioneae)

This well-supported, disjunct clade of small herbs has a phylogeny (Wagstaff *et al.*, 2006; Pelser *et al.*, 2010):

North-western US (*Crocidium*).

California and central Chile (Valparaíso to Concepción) (*Blennosperma*).

New Guinea (*Ischnea*).

Microseris (Asteraceae: Cichorieae)

Microseris (Vijverberg *et al.*, 1999; Vijverberg, 2001) occurs in the same three, disjunct regions:

Western North America (California to Alaska: one clade of perennials and another of annuals).

South America (Chile and Peru: one species, in the annual clade).

Australasia (southern Australia, Tasmania and New Zealand: two sister species of perennials).

The traditional model proposes a centre of origin in North America. Yet in the South American clade, Peruvian populations are nested in a Chilean paraphyletic group, conflicting with southward 'stepping-stone' dispersal along the Andes (Lohwasser *et al.*, 2004).

Based on karyotypes and morphology, Chambers (1955) suggested that the Australasian *Microseris* group arose by hybridization between an annual and a perennial species. Because these occur together only in North America, he proposed this as the centre of origin (by hybridization). This was followed by long-distance dispersal to Australasia and extinction of the hybrid in North America.

In their nuclear DNA, Australasian members of *Microseris* showed features of both the annual and perennial American species, while chloroplast sequences placed it at the base of the annuals (Vijverberg *et al.*, 1999). These molecular results are consistent with a hybrid origin for the Australasian clade, but the allopatric distribution suggests it is not derived from them by simple hybridism and dispersal. Instead, the *Microseris* groups probably had a common ancestral complex that already occurred on both sides of the Pacific and was already variable before the modern clades existed. The chloroplast DNA phylogeny (Vijverberg *et al.*, 1999) indicates that the Australasian hybrid lineage arose once, early in the evolution of the North American annuals. In addition, no other perennial × annual hybrids are known. The dispersal scenario leaves several unanswered questions: Why did the hybridism occur only once? Why did it occur so early in the evolution of the group? Why did the plant disperse from North America to Australasia but nowhere else, and why only once? Why did it then go extinct in America? These aspects would all be accounted for if the pattern was caused by vicariance in a polymorphic ancestor.

With respect to its broader affinities, *Microseris* is related to clades in the following localities (distributions from Bremer, 1994; phylogeny from Lee *et al.*, 2003).

Western US/northern Mexico (*Uropappus* (= *Microseris lindleyi*); sister to *Microseris*).

Central and western North America (*Nothocalais*).

Western North America, Argentina and Chile (*Agoseris*).

South-western US (*Stebbinsoseris*).

(*Picrosia* is distributed from Peru to south-eastern Brazil and was placed in Microseridinae by Bremer, 1994, but was not sampled by Lee *et al.*, 2003.) A dispersal analysis would suggest a centre of origin for *Microseris* and its allies in or around California, as most of the genera, including a local endemic, occur there. Instead, the Pacific distribution of the group – western North America, southern South America and Australasia – is a standard triangular pattern seen in many groups, and so the diversity in California could represent the first breaks in a widespread Pacific ancestor rather than a centre of origin (as in Fig. 1.6).

Nicotiana (Solanaceae)

Trans-Pacific affinities also occur in the tobacco genus, *Nicotiana* (Solanaceae), which is composed of the following groups (Knapp *et al.*, 2004; Marks *et al.*, 2011):

Twelve sections in the Americas (including Juan Fernández and Revillagigedo Islands).

One Indo–Pacific section, *Suaveolentes*, with three clades in an unresolved trichotomy:

Southern Africa (Namibia).

Australia (widespread), Lord Howe Island and New Caledonia.

Central Pacific: New Caledonia to the Marquesas Islands.

Authors have suggested that the Indo-Pacific sect. *Suaveolentes* formed as a hybrid derived from modern groups in South America, dispersed to Australia, and then died out in South America (Aoki and Ito, 2000; Chase *et al.*, 2003; Knapp *et al.*, 2004). As with *Microseris*, this does not explain the section's standard (Indian Ocean/Pacific Ocean) distribution, its precise allopatry with the other sections in America (at a break between the Marquesas and Juan Fernández/Revillagigedo), the fact that the hybridism and dispersal only occurred once, or the extinction of the hybrid group in America. A simple vicariance model accounts for all these and also explains the allopatry of the three clades *within* sect. *Suaveolentes* (with overlap restricted to New Caledonia).

In considering the origin of any group and its distribution, it is useful to examine the phylogenetic context. Although *Nicotiana* is, for the most part, an American genus, it belongs to a diverse clade, subfamily Nicotianoideae, that is otherwise Australasian and has a centre of diversity in south-western Australia (Olmstead *et al.*, 2008). So in dispersal theory, *Nicotiana* dispersed from Australasia to America, just once, at its origin, and then back again, just once. On the other hand, Nicotianoideae is the only subfamily in Solanaceae that is not endemic to America (except for one in America + Madagascar), so in dispersal theory there would have been dispersal from America to Australasia (Nicotianoideae), then back to America (*Nicotiana*), then back again to Australasia (*Nicotiana* section *Suaveolentes*).

Most large clades show a similar pattern, with repeated differentiation in a small number of centres that are often distant from each other (Fig. 1.11). This has been explained by proposing very unusual, intercontinental leaps; just three times across the Pacific in the history of Nicotianoideae, and just once in *Microseris* and Blennospermatinae. Instead, the distributions suggest that pre-breakup and breakup phases of episodic vicariance have fractured a widespread ancestor at nodes in America and Australasia. If these nodes were active over the same time period, the phylogeny will give the impression of intercontinental 'jumps'.

A similar process of trans-Pacific dispersal and ad hoc extinction has also been invoked for allopatric hybrids in *Lepidium* (Brassicaceae) (Mummenhoff *et al.*, 2001, 2004). It is not a question of denying the hybrid origin of the 'hybrid clades', but it is unlikely that these were hybrids of the modern clades. Instead, the allopatric distributions of the hybrids along with their multiple affinities suggest that they evolved by simple vicariance with the others. Another example of an allopatric hybrid is *Santalum boninense* of the Bonin Islands, interpreted as a hybrid of two Hawaiian species (Harbaugh, 2008), but possibly an allopatric vicariant of a widespread north-west Pacific ancestor.

Geum subgen. *Oncostylus* (Rosaceae)

This group of herbs occurs in Australasia and southern South America. *G. schofieldii* from the Queen Charlotte Islands, off western Canada, was also placed here by Smedmark and Eriksson (2002). The authors wrote that the standard South Pacific connection is 'somewhat disturbed' by *G. schofieldii*, but the Pacific triangle distribution of the group is itself a standard pattern.

Macrocystis (Laminariaceae), the giant kelp

Macrocystis is the largest brown alga. It forms intertidal forests along the coasts of western North American and the Southern Ocean, including the Tasman region. The distribution has been attributed to dispersal from the northern to the southern hemisphere (Coyer *et al.*, 2001), but only because of a paraphyletic, basal grade in western America and a vicariance history is also possible (Chin *et al.*, 1991).

    Macrocystis forms a clade with *Pelagophycus* (southern California to Baja California), *Postelsia* (California) and *Nereocystis* (central California to Alaska) (Lane *et al.*, 2006). This California/Alaska–Southern Ocean clade is sister to a North Pacific–North Atlantic group, distributed from Japan to California and Scandinavia (*Hedophyllum*, *Kjellmaniella* and *Laminaria* p.p., now treated as *Saccharina*; distributions from www.algaebase.org). In their global distributions the two clades are largely allopatric, indicating vicariance with secondary overlap between Alaska and California. This would provide a simple explanation for the absence of *Macrocystis* as such from Japan and the North Atlantic.

Indian + Pacific Ocean groups

The groups discussed in the last sections have affinities that span either the Indian or Pacific Ocean. Both these connections occur together in many Indo–Pacific clades, marine, freshwater and terrestrial.

Pythons

Pythons (Pythonidae) are mainly an Old World group with an outlier in western tropical America, as in Gesneriaceae and suboscine birds. One python genus occurs in Africa and tropical Asia, and there are seven others in New Guinea and Australia. The New Guinea region is the centre of diversity and there is an endemic genus, *Bothrochilus*, in the Bismarck Archipelago. In addition to the Old World pythons, there is one monotypic American genus, *Loxocemus*. As mentioned already, this is found along the Pacific coast from Mexico to Costa Rica, in a belt of accreted terranes.

    In Pythonidae, after the American *Loxocemus* and the Southeast Asian *Xenopeltis* branch off, the phylogeny is as follows (from Rawlings *et al.*, 2008):

Africa, tropical Asia to Sulawesi (*Python*, in part).

Nicobars to the Philippines (*Python*, in part).

New Guinea, the Bismarck Archipelago, and Australia (seven genera).

Rawlings *et al*. (2008) suggested a centre of origin in Africa or Asia with dispersal to Australia and New Guinea, but in a simple vicariance model the only movement required in the three modern clades is between the Nicobars and Sulawesi. In this model, the direct ancestors of the python complex already occupied all its current areas, or at least the terranes that eventually formed them, at the time that true pythons became recognizable. This is suggested by the complementary geographic relationship between the pythons (plus some smaller groups) and the boas or Boidae (Noonan and Chippindale, 2006). The two clades, both non-venomous, have long been regarded as sister-groups and are sometimes treated as subfamilies of Boidae s.lat. (Wiens *et al*., 2008, portrayed Pythonidae, Boidae and the Southeast Asian Uropeltidae in an unresolved trichotomy.) Unlike the Old World pythons, the boas (including Ungaliophiidae) are pantropical, but they are most diverse in the New World. They are in Africa and Eurasia, but there is only one genus (*Candoia*) in Melanesia (Sulawesi to Samoa) and none in Australia. The great differences in their distributions mean that the mainly Old World pythons and the largely New World boas could have originated as vicariants in a widespread precursor, followed by overlap in parts of Africa and Asia. Noonan and Chippindale (2006) found that the Pythonidae–Boidae break occurred in the Early Cretaceous. The phylogeny was calibrated with the oldest snake fossil, from earlier in the Early Cretaceous.

To summarize: there is no need for pythons to have invaded Australasia by physical movement, as they could have evolved there (and in Africa and western Central America) as vicariants of the mainly New World boids.

Iguanid lizards (Squamata)

In lizards, the widespread Old World group of agamids plus chameleonids (Acrodonta) is diverse in Africa, Asia and Australia, and occurs east to the Solomon Islands (see 'Ecological centres of origin', p. 31). The sister-group, Iguanidae s.lat. (Pleurodonta), is found mainly in America, but has a few species across the tropical Pacific (Galapagos, Fiji and Tonga) and in Madagascar. This whole complex, Acrodonta plus Pleurodonta, is found throughout the tropics, and makes up ∼28% of all lizards (non-snake squamates), but its two main clades overlap only in Madagascar. The allopatry of the two (best seen on a South Polar projection) suggests that Iguanidae were present in both East and West Antarctica before glaciation. A simple vicariance origin implies that agamid lizards, like pythons, did not invade Australia by migration, but evolved there. In this model, the direct ancestors of pythons and agamids were present in Australia before pythons, agamids, or Australia existed as such.

The iguanid genus *Brachylophus* (living in Fiji, fossil in Tonga) is sister to *Dipsosaurus* of Baja California and adjacent parts of the south-western USA and Mexico (Noonan

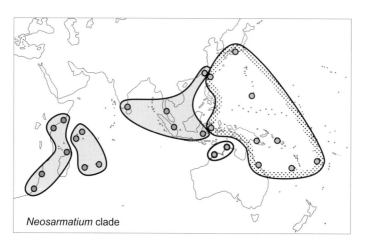

Neosarmatium clade

Fig. 3.17 Distribution of a clade in *Neosarmatium* (Sesarmidae) (Schubart and Ng, 2002; Ragionieri *et al.*, 2009).

and Sites, 2010). This gives a typical trans-tropical Pacific connection. Other iguanids occur in the Galapagos, the Caribbean and South and Central America, with a strong concentration in the west.

Mangrove crabs

A clade in the mangrove crab *Neosarmatium* (Sesarmidae) has a typical Indo–west Pacific distribution (Fig. 3.17, based on Ragionieri *et al.*, 2009; unsequenced records from New Caledonia, New Guinea and Sulawesi are from Schubart and Ng, 2002). There are two main subclades, one each in the Indian and west Pacific Oceans. Within the Indian Ocean 'clade (the *Neosarmatium meinerti* complex)' there is a clear phylogenetic structure, notwithstanding a larval phase that lives for about 1 month in the marine plankton (Ragionieri *et al.*, 2009). The Indian Ocean/Pacific Ocean split includes a break between eastern and western Australia, and this continues through Sulawesi, Philippines and Taiwan. Within the Indian Ocean group there is a split between one clade in Africa–Madagascar and one in the Seychelles and the Mascarenes, a pattern also seen in terrestrial groups (Heads, 2012a).

If the Indian and Pacific Ocean clades were to develop local overlap in the Philippines and Sulawesi, or more extensive overlap through this region, the result would be a standard pattern – an Indo–Pacific range with highest diversity in Southeast Asia. For example, a clade of bees comprises an Indian Ocean group (Africa and Southeast Asia: *Ceratina* subgen. *Pithitus*: Apidae) sister to a trans-Pacific group (Southeast Asia, Japan, New Guinea: subgen. *Lioceratina* and subgen. *Ceratinidia*, plus the Americas: subgen. *Zadontomerus* etc.). (This phylogeny, from Rehan *et al.*, 2010, surveyed 15 of the 21 described subgenera.)

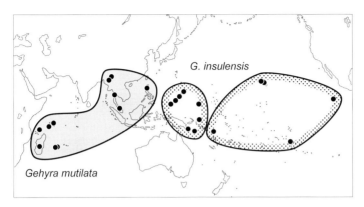

Fig. 3.18 An Indo–Pacific clade in *Gehyra* (Gekkonidae) (Rocha *et al.*, 2009). *G. insulensis* is introduced in California and South America, possibly native in western Mexico.

A group of bees

In the southern bee family Colletidae, the following pair forms one of the four main clades (Almeida *et al.*, 2012):

1. Africa (southern Africa to Kenya) (Scrapterinae) + Australia (Euryglossinae).
2. Australasia (apart from one of the seven clades in *Hylaeus*, which is worldwide) (Hylaeinae) + South America (Xeromelissinae).

This gives a pattern in which an Indian basin clade is sister to a Pacific basin (plus world) clade, compatible with vicariance at a break somewhere in Australia. Michener (2007: 105) wrote that: 'Because bees fly well one might think that they would be rather successful at crossing barriers ... Nonetheless, distributional data suggest that most groups of bees are not particularly good at crossing major barriers ... Thus for the majority of kinds of bees, dispersal has been by slow spread across continents or nearby landmasses, or by transport on moving continents'.

A group of 'endemic weeds' in the gecko *Gehyra*

Gehyra (Gekkonidae: Gekkoninae) has most of its species in mainland Australia, but also occurs in Madagascar and surrounding islands, Southeast Asia and the Pacific islands. It has a full Indo–Pacific range, straddling the Indian Ridge and the East Pacific Rise, unlike Indo–west Pacific groups such as the *Neosarmatium* clade in Figure 3.17 (both patterns are common). The basal clade in *Gehyra* is *G. mutilata* plus *G. insulensis* (Fig. 3.18), found throughout the extra-Australian range, although not in Australia itself (Rocha *et al.*, 2009; Heinicke *et al.*, 2011). The clade is a commensal of humans and was thought to be a single species that had spread from a centre of origin in Asia. Instead, molecular studies found evidence for at least two species, one in the Indian Ocean and the other in the Pacific. The break occurs at the Philippine Sea plate, a basin that has opened up between the Eurasian and Pacific plates. The Pacific clade itself has two subclades with a break at the Solomon Islands and the Ontong Java plateau. The absence of this weedy group from Africa is another interesting feature of the pattern. This same absence occurs in many Madagascar–Indo–Pacific clades cited already, such as the plants *Arthropodium*, *Elaeocarpus* and *Sloanea*.

Iridaceae (Asparagales)

This monocot family has three main clades (Goldblatt *et al.*, 2008), one in Tasmania, one around the Indian Ocean and one around the Indian and Pacific Oceans.

1. **Tasmania** (*Isophysis*).
2. **Indian Ocean** (incl. Crocoideae etc.) (35 genera):

> Australia (wide), New Guinea, Borneo, Sumatra (*Patersonia*).
>> Madagascar (*Geosiris*).
>>> Madagascar, Africa (*Aristea*).
>>>> Cape region of South Africa (*Nivenia*, *Klattia*, *Witsenia*).
>>>> Mainly Africa, other genera in Eurasia and North America (Crocoideae: 29 genera).

3. **Indian and Pacific Oceans** (Iridoideae, 29 genera):

> Tasmania and SE Australia (*Diplarrena*).
>> Indian Ocean (Irideae). *Dietes* in Africa and Lord Howe Island, remaining genera mainly in Africa (some in Eurasia and North America).
>> Pacific Ocean (Sisyrinchieae etc.). Australasia and America (mainly South America to southern North America). The groups in Australasia are *Libertia*: SE Australia, New Guinea, New Zealand, and South America (Andes), and *Orthrosanthus*: SW and SE Australia, Mexico to Bolivia.

The basal breaks around Tasmania and Australasia, in which *Isophysis*, *Patersonia* and *Diplarrena* each differentiated from their sister-group, indicate repeated fracturing in the region early in the group's evolution. At about the same time there has been repeated differentiation in the south-western Indian Ocean (between *Geosiris*, *Aristea*, etc. and their sister-groups). Goldblatt *et al.* (2008) attributed the basal parts of the phylogeny to long-distance dispersal from a centre of origin in Australasia to Africa, while in a vicariance model these areas instead represent centres of differentiation. In either case, these basal events took place prior to South Pacific disjunctions in *Libertia* and *Orthrosanthus*.

Monimiaceae (Laurales)

The phylogeny of Monimiaceae (Renner *et al.*, 2010; Heads, 2012a, Fig. 9.7) is of special interest as it shows:

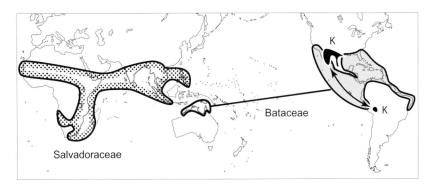

Fig. 3.19 The related plant families Bataceae, Salvadoraceae and Koeberliniaceae (K) (Brassicales) (Stevens, 2012).

> Trans-Indian Ocean (Gondwana) connections.
> Southern South Pacific (Gondwana) connections (between Australasia and Patagonia).
> Trans-tropical Pacific connections (between Australasia and tropical America).

Renner *et al.* (2010) suggested that the tropical American *Mollinedia* clade originated at 28–16 Ma (a fossil-calibrated minimum age) and must have arrived via trans-Pacific dispersal from Australasia (~15 000 km). The authors admitted that the young date is 'most surprising', but it is a minimum age and the actual clade age could be much older. The authors also commented that 'Given the family's distinct pollen (in some genera), it is surprising that there are no pollen fossils . . . ', yet many groups have no fossil record at all. The lack of Monimiaceae fossil pollen is probably a sampling artefact of the record.

Bataceae and allies (Brassicales)

Members of the plant family Bataceae grow in salt marsh and have a trans-tropical Pacific distribution, disjunct between the Arafura Sea–Torres Strait area and tropical America (Fig. 3.19; Stevens, 2012). Their sister-group is the Indian Ocean family Salvadoraceae. The sister of these two families is Koeberliniaceae, disjunct between California and Bolivia and vicariant, for the most part, with Bataceae. This suggests a global group that split apart, first at breaks in the east Pacific, into Koeberliniaceae and everything else, and then around western Malesia and the Atlantic, into Bataceae and Salvadoraceae. The only dispersal required in the families themselves is local range expansion in northern Mexico. In the past, it is likely that pre- or proto-Bataceae occupied the mid-Cretaceous basalt plateaus and their islands in the central Pacific.

The family Koeberliniaceae is an inland equivalent of the coastal Bataceae and earlier in its history probably included coastal weeds. The distributions suggest that populations have been stranded inland in Mexico and Bolivia following Andean orogeny and marine regressions in the Cretaceous. The disjunction between North and South America can be explained by the Cretaceous separation of the continents.

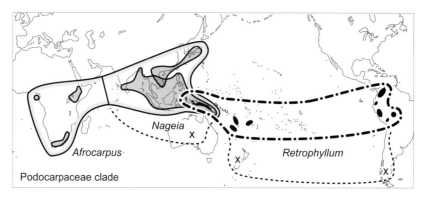

Fig. 3.20 Distribution of a clade in *Podocarpus* (Podocarpaceae) (de Laubenfels, 1988; Biffin *et al.*, 2011; Knopf *et al.*, 2012). X symbols in fine dashed lines = fossil localities.

Podocarp trees

The conifer family Podocarpaceae comprises five basal clades located around the Pacific, and one pantropical clade, *Podocarpus*. *Podocarpus* is made up of two pantropical groups, typical *Podocarpus* and the '*Polypodiopsis* clade'. The latter is made up of a trans-Indian Ocean clade (*Afrocarpus* and *Nageia*) plus a trans-Pacific clade (*Retrophyllum*) (Fig. 3.20; de Laubenfels, 1988; Biffin *et al.*, 2011; Knopf *et al.*, 2012). *Nageia* fossils have been identified in Australia, while *Retrophyllum* fossils have been identified in Australia, New Zealand and southern Chile (Hill and Pole, 1992). The Indian Ocean and Pacific Ocean clades overlap only in the Maluku Islands and mainland New Guinea; the Pacific clade replaces the Indian clade in New Britain, Vanuatu and other Pacific islands.

Ratite birds

The ratites include the largest birds and are the sister-group of all other extant birds. A phylogeny of ratites shows the ostriches of Africa as basal, with the other two clades spanning the Indian and Pacific basins respectively. The distribution is as follows (Fig. 3.21, phylogeny from Smith *et al.*, 2013; membership of Aepyornithidae in the Indian Ocean clade is from Phillips *et al.*, 2010):

> Africa (fossil records extend to China): ostriches (Struthionidae).
>
> **Pacific basin**: moas (Dinornithidae) of New Zealand (extinct in historical times), tinamous (Tinamidae) and rheas (Rheidae) of South America to Mexico.
>
> **Indian Ocean basin**: elephant birds (Aepyornithidae) of Madagascar (extinct in historical times), emus of Australia plus cassowaries of New Guinea and Queensland (Casuariidae) and kiwis (Apterygidae) of New Zealand.

The three main clades are allopatric everywhere except New Zealand, consistent with an origin by vicariance. The primary break is between Africa and Madagascar (Middle

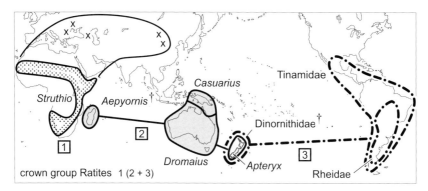

Fig. 3.21 Distribution of the ratite birds (Phillips *et al.*, 2010). The numbers indicate the main clades and their phylogeny. X symbols indicate fossil localities. Extinct groups are indicated with dagger symbols.

Jurassic) and the Atlantic Ocean (Early Cretaceous). The second break, between Pacific and Indian Ocean clades, lies somewhere in the New Zealand region, where the extant kiwis (Apterygidae) overlap with the extinct moas (Dinornithidae). The fossil record of both families extends only to the Miocene, but the groups themselves are probably much older than that. The distribution of the moa–tinamou–rhea clade is similar to that of *Coriaria* clade 1 (Fig. 3.11) and the *Penthesilenula incae* group of ostracods, restricted to New Zealand, South America and central America (Martens and Rossetti, 2002).

Phillips *et al.* (2010) did not discuss the Indian Ocean/Pacific Ocean allopatry in ratites. Instead, they examined the evolution of the group by estimating clade ages, using fossil calibrations. By transmogrifying fossil-based ages into maximum ages they ruled out the presence of the birds in Gondwana (and a simple vicariance history), and suggested that the moas had flown into New Zealand from South America. Yet chance dispersal does not explain the intriguing Indian Ocean–Pacific Ocean allopatry and there is no real need for intercontinental dispersal in the ratite clades themselves. Similar Indian Ocean–Pacific Ocean distributions in ratites and in their sister-group, 'other birds', indicates that the global overlap between ratite birds and the others probably took place before the differentiation of the ratite clades. The ratite phylogeny indicates that the break in the New Zealand region occurred after the Mozambique Channel/Atlantic Ocean break (160–120 Ma), but before rifting in the Tasman and southern Pacific basins (85 Ma). This means it overlapped in time with the major event in New Zealand at that time, the Rangitata orogeny (Late Jurassic–Early Cretaceous, last phase ~100 Ma). The geographic breaks between New Zealand and South America, and between New Zealand and Australia, developed later in the Late Cretaceous, with the final phases of Gondwana breakup.

Birdwing butterflies and their allies (Lepidoptera: Papilionidae: Troidini)

In swallowtail butterflies (Papilionidae), the tribe Troidini includes the largest butterflies in the world – *Ornithoptera alexandrae* and *O. goliath* of northern New Guinea. The

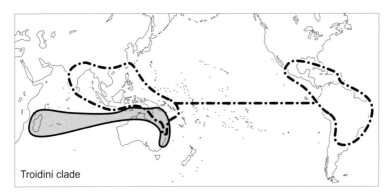

Troidini clade

Fig. 3.22 Distribution of the Indo–Pacific clade of Troidini (Papilionidae) (Condamine *et al.*, 2013a; Figs 2 and 3).

tribe has the following phylogeny (Condamine *et al.*, 2013a; cf. Braby *et al.*, 2005; Simonsen *et al.*, 2011):

> US – Brazil (*Battus*).
>> India to NE Australia (*Pachliopta* incl. *Losaria*, + *Atrophaneura*).
>>> **Indian Ocean:** Madagascar (*Pharmacophagus*), Timor, southern New Guinea, northern and eastern Australia (*Cressida*).
>>> **Pacific Ocean:**
>>>> Sri Lanka and north-eastern India to Taiwan, New Guinea, eastern Australia and the Solomon Islands (*Trogonoptera, Troides, Ornithoptera*).
>>>> Northern Argentina to Mexico (not in Chile) (*Euryades* + *Parides*).

The Indian and Pacific clades are shown in Figure 3.22. The tribe's biogeographic history is controversial, with one molecular study supporting an origin by vicariance in Gondwana (Braby *et al.*, 2005). The authors wrote that there was no reason to assume the Troidini became extinct in Africa, especially as the food plant family, Aristolochiaceae, is widespread there. They identified the 'Africa/Madagascar anomaly' as a crucial feature indicating a major weakness in the traditional model, a northern hemisphere centre of origin in the Cenozoic. Another molecular study supported a centre of origin/dispersal model, but this relied on the use of a centre of origin program (Bayesian DIVA) and the conversion of minimum fossil-calibrated dates into maximum clade ages (Condamine *et al.*, 2013a).

The dove-tailing vicariance of the Indian Ocean and Pacific Ocean clades of Troidini has been neglected. Condamine *et al.* (2013a) explained the distribution of the Indian Ocean clade by dispersal from northern Asia to Madagascar (*Pharmacophagus*) and Australasia (*Cressida*). For the trans-Pacific clade, they proposed that the Australasian genera (then in eastern Siberia) and the South American genera (then in North America) originated by vicariance along a boundary: North Pole–Bering Strait at 31 Ma. The three northern groups then migrated south to the tropical regions and went extinct in the north. The migrations are unnecessary, though, and it is difficult to imagine the two main groups migrating so far – from the Pole to the Equator – while maintaining almost complete allopatry.

Simonsen *et al.* (2011) calibrated their phylogeny of Papilionidae with fossils rather than vicariance. They wrote that '*a priori* acceptance of vicariance as the main (or only) explanation of current distribution patterns is problematic as it ignores the mounting evidence for long-distance dispersal events as important factors in biogeography' (p. 129). A biogeography based on vicariance alone would be impossible – phases of range expansion are needed to explain overlap. Yet there is no real need for chance dispersal as a mode of speciation. The only data that support dispersal are the young clade ages and these are all produced either by transmogrifying fossil-calibrated ages into maximum age or calibrating phylogenies with island age. The end result of using chance to explain all general patterns is that the patterns themselves are neglected, and Indian Ocean–Pacific Ocean vicariance is an example of this.

The Indian Ocean and Pacific Ocean clades in Troidini straddle the spreading ridges in their respective ocean basins. They overlap only in Queensland and southern Papua New Guinea, and this margin between the two groups is equivalent to the Coral Sea region. The break that took place here in Troidini coincides with the Paleogene Coral Sea basin, which is probably too young to be involved. The mid-Cretaceous Whitsunday magmatic belt is a large-scale feature that could provide a possible mechanism for vicariance.

Winteraceae (Canellales)

In the tree family Winteraceae a trans-Pacific clade is distributed from the Philippines to South America. Its sister is in northern Madagascar, giving an Indo–Pacific group (Marquínez *et al.*, 2009; Heads, 2012a, Fig. 9.5). There is no need to propose any intercontinental dispersal in Winteraceae, as it shows almost complete allopatry with its sister-group, Canellaceae, in southern Madagascar, Africa and eastern South America.

Rytidosperma and its allies: an Indo–Pacific group in the grass subfamily Danthonioideae

The basal clades in this subfamily, Danthonioideae (*Merxmuellera*, *Chionochloa*, etc.) were discussed in Chapter 1, as an example of Indian Ocean distribution. The group as a whole has a phylogeny: Africa (Africa (Africa (Tasman (Africa (world))))). The 'world' clade is made up of an Indo–Pacific group and its Pacific–Atlantic sister-group.

In the Indo–Pacific group, the grass *Rytidosperma* has a typical South Pacific disjunct range (Humphreys *et al.*, 2009; Linder *et al.*, 2010), while its closest allies are across the Indian Ocean in southern Africa. The authors proposed a phylogeny for the group:

South Africa via Ethiopia to the Himalayas (*Tenaxia*).

South Africa and southern Europe (*Schismus* incl. *Karoochloa*).

Southern Africa (*Tribolium*).

South Pacific: New Guinea, Australia, New Zealand and Chile (*Rytidosperma* incl. *Austrodanthonia*, *Notodanthonia* and *Joycea*).

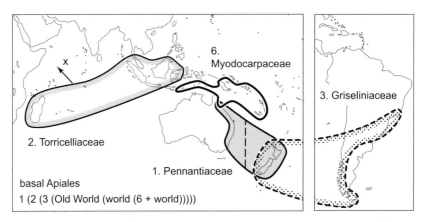

Fig. 3.23 Distribution of the three basal clades in Apiales. X symbol and arrow indicate possible fossils of Torricelliaceae in Europe and North America (Stevens, 2012). (From Heads, 2009b, reprinted by permission of John Wiley & Sons.)

The pattern could imply a single dispersal event out of Africa followed by a major radiation in Australasia (Humphreys *et al.*, 2009), or an ancestral distribution around the southern Indian, Atlantic and Pacific basins, with an early phase of differentiation around South Africa.

The Indo–Pacific *Tenaxia–Rytidosperma* clade has a Pacific–Atlantic sister-group (not in Africa):

New Zealand (*Austroderia*).

Blue Mts, near Sydney (*Notochloe*).

SE Australia coast: Victoria/NSW border to the McPherson–Macleay Overlap (*Plinthanthesis*).

South America: Tierra del Fuego to Colombia, not in the Amazon basin (*Cortaderia*).

New Guinea (*Chimaerochloa*).

Temperate South and North America, Europe (*Danthonia*).

Here, a Tasman basin clade (the first three genera) is sister to a widespread Pacific–Atlantic group.

Basal Apiales

Apiales are cosmopolitan plants. They comprise three, small, southern hemisphere families forming a basal grade (Fig. 3.23) and one much more diverse clade with a global distribution. The phylogenetic sequence (Tank and Donoghue, 2010) is:

1. Pennantiaceae.
2. Torricelliaceae.
3. Griseliniaceae.

4. Pittosporaceae: Old World trees (not shown on Fig. 3.23).
5. Araliaceae: pantropical trees (not shown on Fig. 3.23).
6. Myodocarpaceae.
7. Apiaceae: cosmopolitan herbs (not shown on Fig. 3.23).

Assuming a worldwide ancestor, the first three breaks are: Pennantiaceae (Tasman Sea region) and the rest; Torricelliaceae (Indian Ocean region) and the rest; and Griseliniaceae (South Pacific region plus south-eastern Brazil) and the rest, or, in short: Tasman basin, Indian Ocean, South Pacific, world. (In the Indian Ocean group, possible fossil material of *Torricellia* is recorded in Europe and North America; Manchester *et al.*, 2009.)

Apart from the overlap of *Pennantia* and *Griselinia* in New Zealand, the three basal clades in Apiales have remained allopatric after their formation, but are completely overlapped by the cosmopolitan clade (4 7). Given the allopatry among the basal clades, this overlap is probably a result of a phase of range expansion by the cosmopolitan clade and, given the allopatry among many of its own intercontinental subclades, this took place early in its history. Clade 6, Myodocarpaceae, exemplifies endemism around the northern end of the Tasman–Coral region.

Loranthaceae, the mistletoe family

In Loranthaceae there are three, small, basal clades and one diverse, more or less global clade (phylogeny from Vidal-Russell and Nickrent, 2008; distributions from Nickrent, 2012):

SW Australia (*Nuytsia*).
  SE Australia: Blue Mountains (*Atkinsonia*).
    Tropical America (*Gaiadendron*).
    Widespread (absent from many parts of the north temperate zone).

The basal breaks occur in Australia and tropical America. The widespread clade has three clades:

1. New Zealand (*Alepis*, *Peraxilla*), sister to a diverse Asia–Pacific group (Sri Lanka to Tahiti, six genera).
2. New Zealand (*Tupeia*), southern Chile (*Desmaria*), and a large group in South and Central America (11 genera). (The arrangement of the three clades is not resolved.)
3. New Zealand (*Ileostylus*) and eastern Australia (*Muellerina*), sister to a diverse Old World group (34 genera, mainly tropical).

All three clades have basal nodes around New Zealand, possibly indicating a single phase of differentiation. The second clade is based around the South Pacific basin, the third around the Indian Ocean basin, and the two overlap only in New Zealand.

Pacific + Atlantic groups

Bignoniaceae clade

A widespread group in the family Bignoniaceae has two clades, one trans-Pacific and the other trans-Atlantic (Olmstead *et al.*, 2009):

> **Pacific Ocean:**
>
>> East New Guinea (*Lamiodendron*) + Sumatra to New Caledonia (*Deplanchea*) (Heads, 2003, Figs 112, 129).
>>
>>> New Zealand, Queensland, New Guinea (*Tecomanthe*).
>>>
>>> East Australia, New Guinea, New Caledonia (*Pandorea*).
>>>
>>> Central Chile (*Campsidium*).
>
> **Atlantic Ocean:**
>
>> Warm America and Africa (*Tecomaria*) + South Africa (*Podranea*).

The break between the two clades occurs somewhere between central Chile and northern Argentina.

Poaceae clade

The grass subfamily Danthonioideae is a mainly southern group. As noted on p. 128, one clade is endemic in the Tasman region and its sister has a Pacific–Atlantic distribution (Linder *et al.*, 2010):

> **Tasman region:** New Zealand (*Austroderia*), Blue Mountains (*Notochloe*) and east coast Australia (*Plinthanthesis*).
>
> **Pacific and Atlantic Oceans:** New Guinea (*Chimaerochloa*), North and South America, Europe: *Danthonia* and *Cortaderia*).

Pacific + Indian + Atlantic groups

Many Australasian groups belong to clades that are pantropical, or pantropical plus south temperate, and show disjunctions across all three ocean basins.

Bryophytes

Frey *et al.* (2010) reviewed molecular studies on bryophytes with distributions disjunct among Africa, India, Australasia and South America. This work discussed many interesting examples of trans-Indian, trans-Atlantic and trans-Pacific patterns that show close parallels with those in other groups.

Phyllanthaceae (Euphorbiaceae s.lat.) tribe Poranthereae

The pantropical tribe Poranthereae showed the following pattern (Vorontsova *et al.*, 2007b, Fig. 5):

Atlantic Ocean: Mexico and Peru to Africa and Afghanistan (*Andrachne* p.p.).

Indian Ocean: Australia, New Zealand and South Africa (*Poranthera* etc.).

Atlantic Ocean: tropical America, Africa, Madagascar and Sri Lanka (*Meineckia*, *Zimmermannia* etc.).

Trans-tropical Pacific Ocean (Sri Lanka and China to Mexico and Alabama (*Actephila*, *Leptopus*, etc.).

The sample included all the genera in the tribe and 63 of 120 species. So far the phylogeny indicates a simple geographic pattern, with each of the four main clades centred around one of the ocean basins. The last three clades are located around the Indian, Atlantic and Pacific basins, respectively, and together form a global group. Dispersal (range expansion) is only required to account for marginal overlap among these three groups (for example, in Sri Lanka) and the more widespread overlap between these three and the basal group, around the Atlantic Ocean.

Given the presence of three of the four clades in different parts of Africa, a centre of origin analysis deduced a centre there, followed by trans-oceanic dispersal events in all four clades (Vorontsova *et al.*, 2007b). Instead of proposing primary breaks between the continents, a vicariance model accepts phylogenetic and tectonic breaks within what are now the separate continents, before final continental breakup rifted the distributions of the clades that had already evolved.

Notonemouridae (Plecoptera)

The stonefly family Notonemouridae has one clade in South Africa and the other in Australia, New Zealand and South America (McCulloch, 2010). The break between the two coincides with breaks in the Indian and Atlantic basins that extend into a break between West and East Antarctica. The sister of the family (the rest of the clade Euholognatha: Capniidae, Leuctridae, Nemouridae and Taeniopterygidae) is endemic to the northern hemisphere, and this is consistent with simple vicariance of a global ancestor and subsequent overlap of the northern groups.

Parastacidae (Crustacea: Decapoda)

The large, freshwater crayfish (Astacidea) break down into a widespread northern hemisphere clade (Astacidae + Cambaridae) and its disjunct, southern hemisphere sister (Parastacidae), often cited in biogeographic studies. The southern Appalachian Mountains have the highest diversity of freshwater crayfish, with south-eastern Australia having the second highest levels of diversity (Toon *et al.*, 2010). Parastacidae have three main groups (Toon *et al.*, 2010):

South America (disjunct between central Chile and south-eastern Brazil/Uruguay).

Indian Ocean: Madagascar, western Tasmania (*Spinastacoides*, *Ombrastacoides*), New Zealand.

Tasman region: New Guinea, Australia (widespread) and Tasmania (widespread).

The last two clades are southern and northern vicariants, overlapping only in western Tasmania. Together they form an Indian Ocean group separated from the South American clade by a break in the Pacific.

In part of the widespread Australian clade, the south-western Australian *Engaewa* is sister to a group of south-eastern Australia (including sister endemics at the McPherson–Macleay Overlap and Tasmania/Victoria). This south-west/south-east split was dated by Toon *et al.* (2010) as Early Cretaceous, compatible with a split by epicontinental seas. The dating was based on a rich fossil record that includes Early Cretaceous parastacids from south-eastern Australia (116–106 Ma).

Toon *et al.* (2010) concluded: 'The estimated phylogenetic relationships and time of divergence among the Southern Hemisphere crayfishes were consistent with an east–west pattern of Gondwanan divergence. The divergence between Australia [Tasmania] and New Zealand (109–160 Ma) pre-dated the rifting at around 80 Ma, suggesting that these lineages were established prior to the break-up'. Pre-breakup events included extension and magmatism and so it is likely that these caused the differentiation, as suggested already for several groups.

The origin of the southern hemisphere family Parastacidae itself is straightforward. Sinclair *et al.* (2004) concluded that the reciprocal monophyly of the Parastacidae and its northern hemisphere sister-group ' . . . is consistent with the break-up of Pangea into a southern continent Gondwana and a northern continent Laurasia during the Jurassic (200–140 Myr)'. Overall, the biogeography suggests that the Parastacidae are another example, along with the trans-Pacific brachiopods cited on p. 89, in which the fossil record reflects the actual age of the group.

Intercontinental affinities of Australasia: a summary

The patterns discussed in this chapter show that a group in, say, Africa and Australia did not necessarily disperse from one to the other. If the group has a sister in, say, New Zealand and America, it is just as likely that both groups arose by vicariance. The argument for dispersal is often based on timing, with groups thought to be too young to have undergone vicariance, but this is based on flawed dating techniques that 'transmogrify' minimum fossil-based dates into maximum clade ages.

Many of the groups cited show large-scale overlap between two or more of the member clades. These indicate range expansions that have taken place at different times in the phylogeny. But a phylogeny: Gondwana (Gondwana (Gondwana (Laurasia))) does not require any dispersal between Gondwana and Laurasia in the group itself to explain the pattern (range expansion in the ancestor is another question). The only requirement is range expansion within Gondwana, to account for overlap there.

4   Biogeography of Australia

The last three chapters introduced general topics. The rest of this book describes a biogeographic 'transect' from west to east across Australia, across the Tasman basin to New Zealand, and then north to New Caledonia, New Guinea and the Philippines.

Distribution in and around Australia

Is Australia a unit? The allodapine bees suggest it is not

Many biogeographic analyses assume that current geographic entities are also natural biogeographic regions. Australia is a geographic unit, but in terms of its biogeography, parts of it are often more closely related to external areas than to the rest of Australia. For example, the allodapine bees (Apidae: Allodapini) comprise three main clades (Chenoweth and Schwarz, 2011):

Africa and Madagascar (*Macrogalea*).

Africa, Madagascar, Middle East, southern and Southeast Asia, *northern* Australia.

Southern Australia.

As in many widespread Indian Ocean groups, the first break has occurred in Africa/Madagascar, and there has been subsequent overlap there. The second break has developed somewhere between northern and southern Australia. This break in central Australia coincides with marine transgressions there during Jurassic and Cretaceous time and these are discussed in more detail below. Chenoweth and Schwarz (2011) suggested a single colonization event from Africa to Australia at ~34 Ma – 'too late for Gondwanan vicariance models' – but this was a minimum (fossil-calibrated) age for the group and so it is possible that the only dispersal in the allodapine clades has occurred within Africa/Madagascar.

The allodapine bees show that distribution within Australia bears a close relationship with distribution beyond Australia, and that both must be considered together. In particular, it cannot be assumed that Australia is a simple biogeographic unit, as the bees indicate there was differentiation and complexity in the region before Australia existed as a discrete geographic entity.

Affinities in and around Australia

Australian groups show trans-Tasman Sea affinities as well as the intercontinental connections across the Indian, Tethys and Pacific Oceans already discussed for Australasia in general. For example, in Australian Diptera (Bickel, 2009), there are:

Many Tethyan groups with connections to the north-west.

Disjunct trans-Indian Ocean groups (the dolichopodid *Heteropsilopus*: Australia and India; Neminidae: Australia, New Guinea, Madagascar and South Africa).

Disjunct Tasman Sea groups (for example, Australimyzidae: southern Australia, Macquarie Island and New Zealand).

Disjunct trans-Pacific groups (Brachystomatidae: Australia/Tasmania, New Zealand and South America; Perissommatidae: Australia and South America).

With respect to the trans-Tasman affinities, Bickel (2009: 239) wrote that: 'Relatively recent trans-Tasman dispersal explains the distributions of some plants and animals on the southern continents and islands better than ancient Gondwanan vicariance . . . , however, we have no well-established examples in Diptera yet'. If dispersal is important, it would be strange if flies did not indicate trans-Tasman flight, but this is what Bickel (2009) proposed. This implies that the different trans-ocean-basin connections of Australian groups could be the result of tectonic history. Similar external patterns are seen in other Australian groups (Chapter 3). For example, Australian members of the water plant family Alismataceae (Chen *et al.*, 2012) belong to trans-Indian Ocean affinities:

Africa, Madagascar, India (*Wiesneria*), + Africa, Madagascar, India, Peninsular Malaysia (*Limnophyton*), + NE Australia (*Astonia*).

Tropical Africa, India, Indochina and Java, New Guinea and northern Australia (*Butomopsis*).

and Tethyan affinities:

Eastern and south-western Australia, NE India, Mediterranean region and North America (*Damasonium*).

An early vicariance model for the Australian flora

Hooker (1860) observed that many groups, such as Melastomataceae, are rich in tropical Asia and Africa, but absent or rare in Australia. Conversely, a relative of Melastomataceae, the family Myrtaceae, is diverse in Australia, but poorly represented in Africa. In other examples of higher level allopatry, Hooker (1860) noted that Campanulaceae (Asterales) are diverse in South Africa, while the related Goodeniaceae and Stylidiaceae are diverse in Australia, and that the Ericaceae subfamily Ericoideae of South Africa is represented in Australia by Ericaceae subfam. Styphelioideae (formerly Epacridaceae). In the light of these and similar patterns, Hooker (1860: liv) accepted a vicariance model and wrote that the Australian flora: ' . . . though manifestly more allied to the Indian than to any other, differs from it so organically, that it is impossible to look upon one as derived from the other, though both may have had a common parentage'. Hooker inter-preted the trans-Tasman relationship between Australia and New Zealand in a similar way.

Distribution within Australia: affinities between the coasts and the central deserts

Phylogenetic links between the coasts and the central deserts

The central, arid part of Australia (often referred to as the Eremean or Eyrean region) has a distinctive biota, and explaining its history has been one of the most vexed problems in Australian biogeography. Biologists have often described affinities beween the plants and animals of the desert and those on the coast, and Burbidge (1960: 106) wrote that many central Australian elements 'may have developed from species associated with coastal habitats.' The flora of central Australia includes many calcicoles and halophytes as well as xerophytes, and all of these can be derived from shore floras. Salt lakes are abundant in inland Australia, especially in the south-western and south-central areas, with the salt derived either from erosion of marine strata or from airborne sea salt. The saline lakes have long been a feature of the flat landscape and the halophytes around them show very high levels of diversity (Hopper, 2009). Endemism here occurs in typical halophyte families, such as Chenopodiaceae, but also in the dominant trees, such as individual series and species of eucalypts, and these communities around salt lake margins preserve ancient links with the coast. Most of the large rivers in south-western Australia have their sources in salt lakes and the riverine plants tolerate some salinity (George, 2009). The riverine trees include paperbarks (*Melaleuca*: Myrtaceae), flooded gums (*Eucalyptus rudis*: Myrtaceae) and sheoaks (*Casuarina*: Casuarinaceae). The fauna associated with the salt lakes is also diverse. For example, despite its lack of permanent rivers and lakes, the Australian arid zone has a higher number of autochthonous water and shore birds than any other part of Australia (Schodde, 1982). Most waterfowl in Australia visit salt lakes and feed on the abundant invertebrate populations. As with the plants, the distributions and phylogeny suggest that this ecology dates back to the Mesozoic and ancestral groups that were associated with inland seas.

The close affinity between Australian desert floras and coastal ones has been explained most often by dispersal inland, long after the inland seas had retreated. Crisp *et al.* (2004) suggested that molecular studies 'may reveal pathways between these habitats, perhaps along riverine floodplains'. Nevertheless, the dispersal of coastal, weedy groups far inland along valleys and flood plains probably occurred with the expansion of entire communities during the marine flooding of the Cretaceous, not in the Cenozoic. Phases of flooding and regression would have also led to differentiation.

The Mesozoic inland seas and their geological context

To explain the biogeographic patterns around the Indian Ocean, Hooker (1860) proposed former land west of Australia. Geology now accepts that former land existed to the

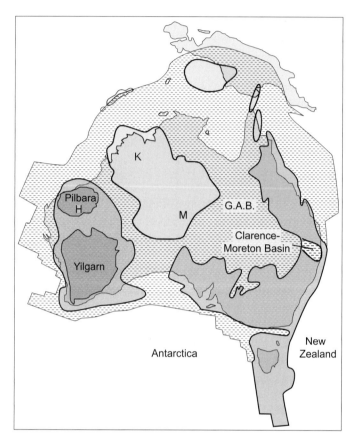

Fig. 4.1 The Pilbara and Yilgarn cratons (the two oldest Archean cratons of Australia), the Clarence–Moreton basin, where extension continued into the Early Cretaceous, and mid-Cretaceous (Aptian–Albian) coastlines in Australia. G.A.B. = Great Artesian Basin, H = Hamersley Range, M = MacDonnell Ranges, K = Kimberley. (Mainly from Veevers, 1988.)

west, south and east of Australia, until the breakup of Gondwana. Before breakup, Jurassic–Cretaceous marine transgressions developed through much of central Australia and peaked at the end of the Early Cretaceous (Aptian–Albian). During the sea level maxima, even the Great Western Plateau, that forms the western half of the country, was divided by the sea (Fig. 4.1; Veevers, 1988). In this mid-Cretaceous geography there was land where there is now sea (west, south and east of Australia), and sea where there is now land (in central Australia). The landscape revolutions that led to the modern geography also led to the modern biota.

Shallow seas spread over large parts of all the continents through the Mesozoic, before and about the same time as the breakup of Gondwana. Examples include the Guaraní Seaway of central South America (northern Argentina to south-eastern Brazil), and the Sundance (Jurassic) and Western Interior (Cretaceous) Seaways of North America

(Mexico to Alaska). The seas filled basins that are best known for trapping large volumes of oil, as in North America, and freshwater, as in the Guaraní aquifer and the Great Artesian Basin of Australia. The latter, with its associated opal fields, marks the site of a Mesozoic inland sea and the Earth's largest aquifer, with $65\,000$ km^3 of fresh water held in porous sandstone. Sediments deposited on the floor of the inland seas capped the aquifers and when the seas retreated the water remained trapped in the Great Artesian Basin.

The global sea level highs of the Cretaceous have never been reached again. The Cretaceous sea level increases were caused by tectonic changes and also global warming. Regional tectonic subsidence in Australia led to the flooding there being somewhat out of step with global changes in sea level. As Australia was moving eastward over slabs subducting westward, decreasing dip angles of the slabs induced widespread subsidence in eastern Australia (Matthews *et al.*, 2011). This meant that seas already dominated the Australian interior in the Early Cretaceous, prior to the global sea level maximum in the Late Cretaceous. Later, as global sea levels were reaching a maximum at ~80 Ma the inland seas in Australia were already retreating.

Earlier inland Australian seas developed as subsidence behind the Gondwana margin created foreland sedimentary basins. Foreland basins develop in a continent behind a marginal subduction zone and its associated orogen/volcanic arc, as in South America behind the Andes. The flexure of continental foreland basins is governed largely by the weight of the orogenic pile. The central Australian basins include the Cooper and Simpson Basins (Permo-Carboniferous to Triassic), Bowen and Gunnedah Basins (Early Permian to Middle Triassic), and Eromanga and Surat Basins in the Great Artesian Basin (Early Jurassic to Early Cretaceous).

The main mid-Cretaceous tectonic event in central and eastern Australia was a phase of basin inversion (a biologist might understand this process as 'eversion') that occurred at the start of the Late Cretaceous (Müller *et al.*, 2012). The basins that had formed earlier were inverted during phases of uplift that occurred in the Late Permian to Middle Triassic (New England Orogen) and the mid-Cretaceous (~95 Ma). This last phase is of direct interest for biogeography, and an important tectonic switch also occurred in New Zealand at this time, at the boundary between Early and Late Cretaceous (100 Ma).

The mid-Cretaceous basin inversion reflected plate tectonics. From 135 to 100 Ma, East Gondwana moved eastwards, but this slowed between 115 and 100 Ma (Müller *et al.*, 2012). Tectonic inversion in central and eastern Australia occurred when Gondwana stopped its eastward motion around 100 Ma. This led to the extensional collapse of the Zealandia Cordillera, the large, Andean-style orogen built along Gondwana's Pacific margin. The collapse of this large mountain range caused a period of compression landward of the mountain belt. In the foreland basins, the switch led to folding and reactivation of low-dip reverse (compressional) faults, uplift, basin inversion erosion and deep weathering (see Glossary for the distinction between normal and reverse faults). Some reactivation led to the formation of hydrocarbon traps in a number of basins, including the Cooper Basin.

The Mesozoic inland seas and the Key hypothesis of Australian biogeography

The idea that the great inland seas of Australia were important for modern biogeography was proposed by the entomologist K.H.L. Key, after four decades of research on African and Australian grasshoppers. Key (1976) related the modern distributions and centres of diversity in morabine grasshoppers (Eumastacidae: Morabinae) to areas that were dry land during the Cretaceous and Cenozoic marine incursions. It is surprising that this interesting idea has been neglected. Barker and Greenslade (1982) regretted that it was not mentioned at a symposium on the arid zone biota of Australia and they agreed that the origins of this biota have to be sought far back in time. Nevertheless, most biogeographic work in Australia has continued to overlook the Cretaceous and instead stresses Neogene events. In contrast, physical geographers have recognized the importance of the Cretaceous as a 'dividing point' in Australian landform history (Wasson, 1982). The epicontinental seas produced a diverse set of new habitats through central Australia and the littoral environment was widespread; the coastline was twice its current length.

Among the few biogeographic studies that have mentioned the Cretaceous seas is a paper on centipedes by Giribet and Edgecombe (2006b). They described a major phylogenetic break between Queensland and south-eastern Australia (the McPherson–Macleay Overlap) in *Paralamyctes* and attributed it to the Cretaceous seaway there (Fig. 4.1).

The Key hypothesis was also applied in an account of the tenebrionid beetles of arid Australia (Matthews, 2000). Matthews recognized Indo–Malayan, Austral and Tethyan elements in the fauna, and wrote: 'Tethyan groups are endemic in the arid zone at tribal level and have no forest-inhabiting relatives anywhere'(Matthews, 2000, p. 941). Because of their affinities with northern hemisphere groups, partial occurrence in coastal dunes, and apparently basal phylogenetic positions, arid Australian groups 'are surmised to have descended by vicariance from inhabitants of the coastal sand dunes of the Tethys Sea, probably in the Jurassic before there was an arid zone in Australia'. This is consistent with the beetles invading central Australia, along with their habitat (coastal sand dunes), and later undergoing vicariance when they were stranded there. This occurred when the seas retreated in the mid-Cretaceous. As the coastlines expanded inland, they would have brought an entire biota with them, not just beetles, and so this process could also explain the invasion of eucalypts, Chenopodiaceae and many others. With the retreat of the seas many of the groups were left behind, deposited like the ring in a bath.

Marine and freshwater groups can also shed light on the marine transgressions. Most clades in the fish family Atherinidae are marine, but one genus, *Craterocephalus*, includes 25 mainly freshwater species of Australia and New Guinea. The main break in *Craterocephalus* is between two coastal marine species of Australia and the rest (Unmack and Dowling, 2010). Crowley (1990) suggested that a freshwater invasion occurred during marine transgressions in the mid-Cretaceous, but Unmack and Dowling (2010) wrote 'it is unclear how or why marine transgressions would be a necessary condition for invasion of freshwater habitats since the same interface between freshwater and marine environments exists irrespective of sea level'. In theory this is correct – the

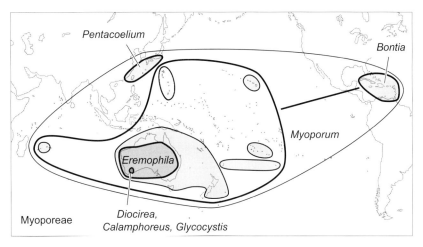

Fig. 4.2 Distribution of tribe Myoporeae (Scrophulariaceae) (Oxelman *et al.*, 2005; Chinnock, 2007).

seas should have encroached on the land and then left, taking all their fish with them – but in practice some populations may well have been left behind, stranded in deeper pools as at low tide. Many of these populations would die out but some would survive.

Central Australian clades can be grouped according to where their affinities intersect with the present coast, which reflects the clades' distributions prior to marine transgression. For example, the freshwater fish *Macquaria ambigua* is widespread through the eastern half of central Australia, the Murray-Darling Basin (these populations form a clade), and in addition the species has a basal grade near the coast, in the Fitzroy Basin of Queensland (Faulks *et al.*, 2010).

Two Indo–Pacific groups that are diverse in central Australia: Myoporeae (Scrophulariaceae) and wattles (*Acacia* s.str; Fabaceae)

The most diverse groups of shrubs and small trees in the Australian deserts are the tribe Myoporeae (Scrophulariaceae) and *Acacia* (the wattles) (Chinnock, 2007). Both of these characteristic central Australian groups share a similar intercontinental distribution.

Myoporeae (formerly Myoporaceae) have a typical Indo–Pacific range (Fig. 4.2). The presence of the Myoporeae in the Caribbean is a standard extension for Pacific groups and is congruent with the geological idea that the Caribbean plate had its origin in the Pacific. The relatives of Myoporeae are allopatric, giving a group with a total range from Madagascar to America and structured as follows (Oxelman *et al.*, 2005; Chinnock, 2007):

Madagascar (*Androya*).

Central America to Peru (Leucophylleae).

Mascarenes to the Caribbean (Myoporeae) (Fig. 4.2).

Rather than having a centre of origin in a restricted part of its range, the allopatry between Myoporeae and their relatives indicates that Myoporeae evolved *in situ*, by vicariance. This in turn suggests that the ancestor was already widespread in central Australia before the group itself had become recognizable. Within Myoporeae, the genera (as shown in Figure 4.2) and subgeneric groups are still based on morphological studies, but the tribe seems to be most diverse in Western Australia. In central Australia a significant break at the MacDonnell Ranges is evident in *Eremophila* sections *Sentis*, *Hygrophanae*, *Arenariae* and *Platycalyx* (Chinnock, 2007). This important locality at the margin of the Western Plateau is indicated in Figure 4.1, along with important north-western centres at the Hamersley Range and the Kimberley region.

Many members of Myoporeae are coastal plants and often grow in estuaries flooded by the tide, more or less as mangroves (Chinnock, 2007). Myoporeae are also diverse throughout arid, central Australia, where they are represented by *Eremophila* (215 species). Habitats in the region include sand dunes and the margins of salt lakes, and many species tolerate high levels of disturbance, for example, along roadsides. Some members, for example *Eremophila veronica*, are ericoid 'inflorescence plants', with foliage composed of sterile inflorescence bracts (cf. the Australian tremands, Chapter 1, and the New Zealand divaricates, Chapter 3). As a result of their morphological reduction, the Myoporeae are hardy plants and have strong chemical compounds; they are the most renowned medicinal plants among the local people and are also implicated in stock poisoning. *Myoporum montanum* is ubiquitous and variable, and the vegetation types that it inhabits – margins of mangrove, woodland with *Eucalyptus*, *Acacia*, *Casuarina* or *Callitris*, chenopod shrubland and open grassland – suggest that it retains at least parts of the diverse ancestral ecology.

The wattles, *Acacia* s.str. (Fabaceae), include ~1000 species (Murphy *et al.*, 2010). *Acacia* is the largest angiosperm genus in Australia and is almost confined there; only 19 species occur outside Australia and eight of these are also present in northern Australia. Overall, *Acacia* occurs in the Mascarenes (some authors also cite Madagascar), tropical Asia, Australia (most species) and the Pacific Islands to Hawaii. The distributions of the Myoporeae (Fig. 4.2) and the wattles are similar. Both groups show breaks in the south-west Indian Ocean region at Madagascar/Mascarenes, and are widespread between there and Hawaii (wattles) or Hawaii–Caribbean (Myoporeae). The trans-Indian Ocean connection is southern, while the trans-Pacific disjunction (complete in Myoporeae; only to Hawaii in the wattles) is in northern tropical latitudes.

Acacia is found throughout Australia, but is most diverse in the south-west and along the Great Dividing Range of eastern Australia. If the arid zone members were derived from mesic clades, they would be nested among them. Instead, though, the *A. victoriae*–*A. pyrifolia* clade, found predominantly in arid and semiarid central Australia, is sister to the rest (Murphy *et al.*, 2010). This indicates that central Australia was not colonized by the modern, mesic clades. Rather than a simple invasion of the centre, the phylogeny suggests an initial break between central Australia and the rest of the Indo–Pacific range, with subsequent overlap. The different phases of marine transgressions and regressions in the Mesozoic would have been complex in their local details, given the flat nature of the land overall, and the changing coasts would have created prime conditions for vicariance and overlap. With increasing aridity through the Neogene, pre-adapted forms in central Australia would have expanded their range.

Polygonaceae in central Australia

One of the most interesting members of Polygonaceae in Australia is *Duma* (Polygo-naceae). It is widespread through inland Australia (not Tasmania or northern Australia), reaching the coast in Western Australia and at Kangaroo Island. Rather than being nested in an Asian group, for example, as predicted in dispersal theory, it is sister to *Polygonum*, native on all continents (Schuster *et al*., 2011a; 2011b). *Duma* species are rhizomatous, much-branched shrubs, with abortive shoot apices that form thorns. The plants grow around freshwater lakes and tolerate high levels of soil salinity. *Polygonum* species are well known as weeds today, while the ancestors of *Duma* are likely to have been weeds in the Cretaceous, invading central Australia with the new coastlines.

Chenopodiaceae in central Australia

Carlquist (1974) observed that in Western Australia, typical beach genera such as *Franke-nia* (Frankeniaceae) also have numerous inland species. Other examples of this pattern are seen in Goodeniaceae, Aizoaceae and Chenopodiaceae, plant families that are well known from maritime and disturbed, open habitats around the world, but are also diverse in arid, central Australia. In Aizoaceae, the Indian Ocean genus *Zaleya* (Aizoaceae) is in Africa, Pakistan and India, and is also widespread through central Australia from the Pilbara region to New South Wales (Venning and Prescott, 1984).

The Chenopodiaceae are represented in Australia with about 300 species and most of these are endemic. Diversity is high along the coast and also in the inland desert, where populations can dominate large areas of saline soils. Martin (2006) listed the dominant plants in arid Australia as *Acacia*, Casuarinaceae, chenopods and grasses. Crisp *et al*. (2004: 1561) suggested that Australian Chenopodiaceae 'all probably originated as post-isolation immigrants, given the absence of fossils before that time'. Kadereit *et al*. (2005) also estimated Cenozoic ages for the Australian Chenopodiaceae, with the groups attributed to nine colonization events beginning at 42 Ma. In both these studies vicariance was ruled out, but only because minimum, fossil-calibrated ages were treated as maximum clade ages. Clades in two different chenopod subfamilies are examined next.

The coastal *Salicornia* and its central Australian sister-group (Chenopodiaceae: Salicornioideae)

Salicornia (including *Sarcocornia*) is a leafless, succulent plant found around coastlines in most parts of the world (Kadereit *et al*., 2006). It is often the dominant plant in saltmarsh that is flooded by the tides, and it also occurs in saline and alkaline soils above high tide level. In some areas the genus is found away from the coast; in South Africa, for example, *S. mossiana* occurs inland around salt pans derived from old lagoons. These were cut off from the sea during Pleistocene marine regressions (Steffen *et al.*, 2010). A similar 'ecological lag' can also account for anomalous populations and species trapped inland and at high elevation in the Andes, such as *S. andina* (2300–4200 m/7540– 13780 ft) and *S. pulvinata* (above 3500 m/11480 ft) (Alonso and Crespo, 2008). Their earlier location, around shores and basins that became uplifted to form the Andes, has determined their present alpine ecology, and a similar model of population uplift during orogeny has been proposed for Andean parrots (Ribas *et al.*, 2007).

Some *Salicornia* species occur in inland Australia (mainly in the south-west and the south-east), but in most parts of central Australia the genus is replaced by its sister-group, comprising *Halosarcia*, *Tecticornia*, *Tegicornia*, *Pachycornia* and *Sclerostegia* (Shepherd *et al.*, 2004; Kadereit *et al.*, 2006). These five genera (there are problems with their delimitation) all occur in central Australia and are abundant there. Four of the genera are endemic to Australia, while *Halosarcia* is also in Pakistan and tropical Africa, in a typical Indian Ocean group.There is no need for dispersal to or from Australia in *Salicornia* or its sister-group, as simple vicariance between them is indicated around the margins of Australia and Africa. The distributions suggest the ancestral complex was already global when it broke apart into one group on world coastlines and one in central Africa, Pakistan and Australia.

Tethys–Australia connections: Australian Camphorosmeae (Chenopodiaceae: Salsoloideae)

The Camphorosmeae are the most species-rich tribe of chenopods in Australia. The Australian members belong to a clade of subshrubs (*Sclerolaena*, *Maireana* and others) that occurs throughout mainland Australia (except the Kimberley and Cape York Peninsula), Tasmania and New Caledonia (Cabrera *et al.*, 2011). The plants are abundant in the central, arid zone of Australia and the species tolerate saline and alkaline soils that are high in sodium chloride or gypsum. The distribution of several species is linked to inland salt lakes.

Kadereit and Freitag (2011) suggested that Camphorosmeae evolved during the Late Eocene to Early Oligocene (a minimum, fossil-based age). The authors also suggested that the tribe originated in Eurasia and dispersed to southern Africa, North America and Australia, but this was only because of a paraphyletic basal grade ('early branching lineages') in Eurasia. The early branching lineages all had sister-groups the same age as themselves and these may have already been in Australia when differentiation began.

The *Sclerolaena* group of Australia–New Caledonia is sister to the central Asian *Grubovia*, found in steppes around the Tien Shan and Altai Mountains (Kadereit and Freitag, 2011; Cabrera *et al.*, 2011). This connection follows the Tethyan basins. Within Australia, clades in three western areas ('south-west', 'western desert' and 'Pilbara') form a basal grade, indicating primary centres of differentiation that were widespread through Western Australia. Overall the differentiation has followed the sequence: Tethys– Western Australia.

Some Australian species of Camphorosmeae have become weeds when introduced overseas and earlier authors invoked ongoing speciation associated with hybridization (see discussion in Kadereit and Freitag, 2011). Although the group has been seen as quite young, the distributions are consistent with an origin during a mid-Cretaceous phase of marine regression. Instead of being a recent development, the group could represent a Cretaceous hybrid swarm of weedy taxa that has been left stranded inland and more or less frozen in place, although when members are introduced elsewhere or cultivated together they can develop renewed mobilism and hybridism. *Hebe* (Plantaginaceae) provides a parallel case in New Zealand.

The south-western Australian biota

The biota of south-western Australia is well known for its high diversity and endemism. In angiosperms, Hopper and Gioia (2004) cited 6–11 families endemic there. The order Dasypogonales is almost endemic to the region; it comprises Dasypogonaceae, with four genera in south-western Australia, and a single species of one of these in South Australia and Victoria. The group has been placed as sister to either Commelinales + Zingiberales, to Arecales or to Poales (Stevens, 2012), indicating that basal nodes in global groups are present around south-western Australia (although they are more common in eastern Australia; Chapter 5).

Each of the following clades is endemic to south-western Australia and also has a global sister-group:

> *Emblingia* is sister to a global clade of Brassicales (Brassicaceae, Pentadiplandraceae, etc.) (Stevens, 2012).
> *Nuytsia* is basal in the mistletoe family Loranthaceae, a large, subcosmopolitan group (Stevens, 2012).
> In *Daphnia*, a worldwide genus of freshwater Crustacea, *D. occidentalis* of south-western Australia is sister to all the rest (92 out of 107 species sampled; Adamowicz *et al.*, 2009).
> The freshwater fish *Lepidogalaxias* is sister to a diverse, global clade comprising Neoteleostei (eight orders), Galaxiiformes, and five other orders (Li *et al.*, 2010).

Other interesting endemics in south-western Australia include *Campanile*, a marine gastropod. Fossils of the genus include the largest known gastropod and are widespread in shallow Tethyan seas, through to Mexico.

In plants, Haemodoraceae (Commelinales) have two clades (Stevens, 2012):

> SW Australia (Conostylidoideae; six genera).
> South Africa, Australia (eastern, northern and south-western parts), southern New Guinea, North and South America (Haemodoroideae, eight genera).

The pattern is consistent with a break around south-western Australia in a widespread, mainly southern hemisphere group.

External affinities related to distributions within Australia: Southern Africa–south-western Australia–central Australia

Within Australia, Hooker (1860) was most impressed with the flora of the south-western region and interpreted it as a relic of the original Australian flora. Hooker (1860: lv) further reasoned that '. . . the antecedents of the peculiar Australian flora may have inhabited an area to the westward of the present Australian continent, and that the curious analogies which the latter presents with the South African flora, and which are so much more conspicuous in the south-west quarter, may be connected with such a prior state of things'. In this way Hooker integrated the biogeographic affinities between the Australian biota and areas outside the country with distribution patterns within the country.

Hooker's (1860) ideas are confirmed by Indian Ocean groups, such as the chironomid fly *Austrochlus*, restricted to granite outcrops in south-west Western Australia and the MacDonnell Ranges in central Australia. Its sister-group is *Archaeochlus* + *Afrochlus* of southern Africa (Drakensberg, Namibia, etc.; Cranston *et al.*, 2002, 2010, 2012). Biogeographic evidence and Early Cretaceous chironomid fossils from Lebanese amber date the group to the Upper Jurassic, at least (Cranston *et al.*, 1987).

South-western Australia and south-eastern Australia: a 'curious case of great differences'

Hooker (1860) recognized the great biogeographic difference between eastern and western Australia as an important problem in global biogeography. He wrote: 'In studying the extra-tropical flora of Australia, the first phenomenon that attracts attention is the remarkable difference between the eastern and western quarters' (p. l), a difference that is 'without a parallel in the geography of plants' (p. li). Hooker even suggested the deconstruction of Australia as a biogeographic entity: 'There is a greater specific difference between two quarters of Australia (south-eastern and south-western) than between Australia and the rest of the globe' (p. xxviii). Hooker concluded that 'This curious case of great differences . . . will no doubt eventually be found to offer the best means of testing whatever theory of creation and distribution may be established' (p. liii). This alluded to Darwin's *Origin of Species*, published the year before, and it also stressed the importance of biogeography in interpreting evolution.

Although Hooker (1860) adopted a vicariance interpretation for the Australian flora as a whole (see above), for distribution within Australia he argued for a centre of origin and dispersal. He saw the south-western region as the 'centrum' of Australian vegetation, with a flora of 'purely Australian' plants and he suggested that this flora had migrated eastward. Instead, it is suggested here that western Australia and the Tasman region (including eastern Australia) are two distinct biogeographic regions in their own right, and that the biota of each one is not derived from the other (cf. Fig. 1.11).

South-western and south-eastern Australia are phylogenetic and geographic nodes of global significance and have intercontinental affinities with many areas. South-western Australia shows many connections with Africa, while eastern Australia has many connections with New Zealand and South America. It is often assumed that the split between the south-west and the south-east occurred with the formation of the limestone Nullarbor Plain by uplift at 15 Ma. This event has been used, for example, to calibrate a clock for the African–Australian fungus *Auritella* (Matheny and Bougher, 2006; Matheny *et al.*, 2009). However, a consideration of the east/west breaks in a broader context suggests earlier dates for the differentiation. For example, a clade of fishflies (Megaloptera) (Liu *et al.*, 2012) has a phylogeny:

> South-western Australia (*Apochauliodes*).
>
> Eastern Australia (Victoria to Cape York), New Zealand and Chile (*Archichauliodes*).

The plant genus *Empodisma* (Restionaceae) comprises three species with the phylogeny (Wagstaff and Clarkson, 2012):

> South-western Australia (*E. gracillimum*).
>
> Tasmania, south-eastern Australia and southern New Zealand (*E. minus*), + northern New Zealand (*E. robustum*).

A similar pattern is seen in the following clade of Ericaceae (Wagstaff *et al.*, 2010a):

> South-western Australia (*Sphenotoma*).
>
> Eastern Australia (Tasmania to north-eastern Queensland) and New Zealand (*Dracophyllum* incl. *Richea*).

These patterns indicate that the break between south-western Australia and eastern Australia/New Zealand is older than the Late Cretaceous opening of the Tasman basin. This implicates the inland epicontinental seas of the mid-Cretaceous that separated western and eastern Australia as a possible mechanism.

South-western Australia is well known for its diverse, bizarre plants. Carlquist (1974) gave a well-illustrated introduction and described the speciation in the area as 'rather fantastic'. He attributed this to the local variation in rainfall, topography and soil type. As usual, the taxa within the region sort themselves out according to the current ecological variations yet these do not explain the origin of the biota, which includes connections with Africa, India and South America, as well as endemics with global sisters. The high levels of genetic diversity in south-western Australia probably pre-date the modern species themselves and the intercontinental affinities of the older groups suggest that these date back to Mesozoic times.

The Cretaceous inland seas divided Australia into several areas of land that were isolated from each other while they were still adjacent to other parts of Gondwana. Thus it is not surprising that the east/west break within Australia can often be related to intercontinental patterns outside the country. For example, in the fly *Empis macror-rhyncha* and its relatives (Empididae) there are two main clades, one in south-western Australia and Patagonia, the other in south-eastern Australia and Patagonia (Daugeron *et al.*, 2009). The split between south-western and south-eastern Australia is prior to the break between Australia and Patagonia (probably involving Antarctica) that occurs in both clades.

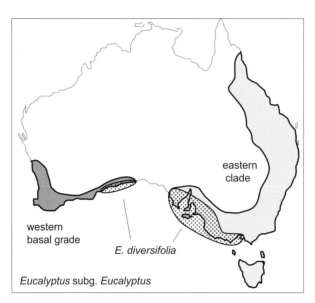

Fig. 4.3 Distribution of *Eucalyptus* subgenus *Eucalyptus* (Myrtaceae; Ladiges *et al.*, 2010).

East/west vicariance in *Eucalyptus* (Myrtaceae)

The trees called monocalypts (*Eucalyptus* subgenus *Eucalyptus*: Myrtaceae) are a classic case of vicariance between eastern and western Australia (Fig. 4.3; Ladiges *et al.*, 2010). All the species in the east (except *E. diversifolia*) form a monophyletic clade of 85 species. The other geographic group is a paraphyletic basal grade of 26 species which are all, except for *E. diversifolia*, restricted to south-western Australia. The distribution of *E. diversifolia* overlaps that of the western grade and the eastern clade. The basal grade in the west indicates that the first differentiation events occurred there. In the first break, for example, a widespread southern and eastern Australian group evolved into one western endemic and its sister-group, widespread in the west and the east. The final break, between east and west, coincides with the area of Cretaceous marine incursions.

Species of monocalypts in wetter forest include tall trees such as *E. marginata* (jarrah) in south-western Australia and *E. regnans*, the tallest angiosperm, in the southeast. In areas between the wetter regions of the southern coasts and the central arid zone, many eucalypts have a mallee growth form. Mallees are giant, multileadered shrubs with large, underground lignotubers, and grow on sandy and calcareous substrates. The only monocalypt species that occurs in both the south-west and south-east, *E. diversifolia* in the basal paraphyletic grade, is a mallee and grows in coastal calcareous soils over limestone. This suggests that the species inherited a sector of the ancestral distribution, around the Great Australian Bight, along with part of the ancestral habitat, coastal limestone. Many other members of the Myrtaceae have a maritime ecology and *Osbornia* (northern Australia, New Guinea, the Philippines and Borneo) is a true mangrove. In many plant groups there are close phylogenetic links between saline habitat and calcareous habitat. For example, in Fiji, 39 indigenous plant species are recorded in

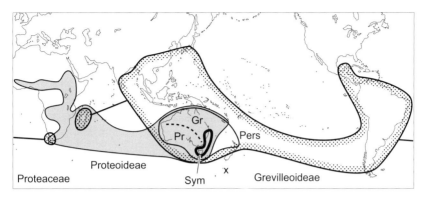

Fig. 4.4 The four main clades in Proteaceae. Sym = Symphionematoideae, Pers = Persoonioideae + Bellendenoideae (Weston, 2007). Gr = main massing of Australian Grevilleoideae, Pr = main massing of Australian Proteoideae. X = fossil pollen described as *Beauprea*-like and *Knightia*-like.

or around mangrove and also on limestone (Heads, 2006a). Soil calcium can reduce the negative effects of salinity on plant growth and the evidence from biogeography, ecology and physiology suggests that the mangrove–limestone connection is an old one.

Paleocene fossil pollen (*Myrtaceidites eucalyptoides*) has been described as 'eucalypt-like', and Early Eocene macrofossils have been attributed to *Eucalyptus* (Ladiges *et al.*, 2010; Gandolfo *et al.*, 2011). If these fossil dates are treated as possible minimum ages they are compatible with an origin of *Eucalyptus* before the Aptian-Albian flooding of Australia.

Proteaceae: an Indian Ocean group diverse in south-western Australia, and a Pacific group diverse in north-eastern Australia

The chironomid flies cited above (*Austrochlus* and relatives) have a southern Africa–southwestern/central Australia connection that does not involve eastern Australia. Similar patterns occur in many other groups, including the plant family Proteaceae. Proteaceae are often cited as a classic southern or Gondwanan group and occur in Africa, Southeast Asia, Australasia and South America (Fig. 4.4). The family's sister-group, Platanaceae, is a northern vicariant, present in Southeast Asia, the Mediterranean region and North America (Stevens, 2012). This is consistent with the two groups having split apart along a Tethyan margin. Platanaceae have Early Cretaceous fossils (Stevens, 2012) and these give a useful minimum age for the two families. In their study of Proteaceae, Sauquet *et al.* (2009a, b) did not refer to the vicariance with Platanaceae and instead adopted a dispersal model, with a centre of origin for Proteaceae in Australia. They attributed the great diversity in South Africa and south-west Australia to the development of a 'Mediterranean climate' there, but this does not explain why Proteaceae are absent from the Mediterranean itself and from California. The simplest explanation is that the group is replaced in these regions by Platanaceae as a result of allopatric evolution. In this process the Proteaceae would inherit their trans-Indian, trans-South Pacific and trans-Atlantic distribution from the common, global ancestor and there is no requirement for trans-oceanic dispersal after the Proteaceae/Platanaceae split.

Within Proteaceae the main clades are an Indian Ocean group (Proteoideae) and a South Pacific group (Grevilleoideae) (Barker *et al.*, 2007). Through most of Africa, Southeast Asia and South America, only one of the subfamilies is present. In contrast, both overlap throughout Australia. There is also overlap in Madagascar and the south-western Cape region, but with only one genus of Grevilleoideae in each. Despite the overlap, the high level of allopatry between the Proteoideae and Grevilleoideae is the most striking large-scale aspect of the family's distribution and needs to be explained.

In addition to their extensive allopatry, the two subfamilies have different centres of diversity within Australia and this is illustrated here in a brief breakdown of the genus distributions (from McCarthy, 1995, Wilson, 1999, Makinson, 2000 and Weston, 2007).

The distributions within Australia are described with reference to the four quarters lying north-west, south-west, north-east (Qld) and south-east of a point in the centre of Australia (200 km south-west of Alice Springs, between the MacDonnell Ranges and Uluru/Ayers Rock).

Subfamily Persoonioideae + subfamily Bellendenoideae

This Australia–Tasman Sea group is the basal clade in Proteaceae.

> *Bellendena*: SE (Tasmania) (= subfam. Bellendenoideae)
> *Placospermum*: Qld
> *Acidonia*: SW
> *Persoonia*: All quarters (most species in SW or SE)
> *Toronia*: New Zealand
> *Garnieria*: New Caledonia.

Subfamily Symphionematoideae

This is found at the margins of the Tasman basin.

> *Agastachys*: SE (Tasmania)
> *Symphionema*: SE.

Subfamily Proteoideae

There are 13 genera in Africa, mainly South Africa (*Aulax*, *Faurea* and *Protea*, and a clade of 10 genera including *Leucadendron* etc.). The subfamily is also in Madagascar, Australia and New Caledonia (*Beauprea/Beaupreopsis*). Wanntorp *et al.* (2011) recorded 'Beauprea-like' fossil pollen in New Zealand. The Australian genera are:

> *Eidothea*: Qld
> *Cenarrhenes*: SE (Tas)
> *Stirlingia*: SW
> *Petrophile*: SW, SE and one species extends into Qld
> *Isopogon*: SW and SE

Conospermum: SW, SE and two species just extend into Qld
Synaphea: SW
Franklandia: SW
Adenanthos: SW and two species barely in SE, i.e. eastern SA.

Of the nine Australian genera, south-western Australia has seven and three of these are endemic, while Queensland has only three genera, with one endemic. This concentration of diversity in the south-west fits with the group's extra-Australian distribution in Africa.

Subfamily Grevilleoideae

This subfamily mainly occurs around the Pacific basin, with a single South African genus (related to a South American/Australia one) and single Madagascan genus (related to an Australian one).

The group includes the following genera not present in Australia:

Malagasia: Madagascar
Heliciopsis: Burma to western Malesia
Finschia: New Guinea to Vanuatu and Palau
Eucarpha: New Caledonia
Virotia: New Caledonia
Sleumerodendron: New Caledonia
Kermadecia: New Caledonia
Knightia: New Caledonia and New Zealand
Turrillia: Vanuatu and Fiji
Euplassa: tropical South America
Roupala: South and central America
Gevuina: Chile and Argentina
Embothrium: Chile, Argentina
Oreocallis: Peru and Ecuador
Panopsis: South and Central America
Brabejum: South Africa (south-western Cape region). This is sister to *Nothorites* (NE
  Queensland) and *Panopsis* (South and Central America) (Mast *et al.*, 2008).

The Australian genera of Grevilleoideae are as follows:

Sphalmium: Qld
Carnarvonia: Qld
Orites: SE and Qld, also Chile and Argentina
Nothorites: Qld (Mast *et al.*, 2008)
Lasjia: Qld and Sulawesi (Mast *et al.*, 2008)
Neorites: Qld
Megahertzia: Qld
Darlingia: Qld
Cardwellia: Qld
Strangea: SW and SE
Stenocarpus: SE, Qld, NW, also New Guinea and New Caledonia

Buckinghamia: Qld
Opisthiolepis: Qld
Lomatia: SE and Qld
Alloxylon: SE and Qld
Telopea: SE
Hollandaea: Qld
Helicia: SE, Qld and NW, also India and China to Japan and Malesia
Xylomelum: SW, SE and Qld
Triunia: SE and Qld
Hicksbeachia: SE and Qld
Bleasdalea: Qld, also in New Guinea
Athertonia: Qld
Catalepidia: Qld
Floydia: SE
Macadamia s.str.: SE and Qld
Lambertia: SW and SE
Grevillea: All quarters, also in New Caledonia and New Guinea
Hakea: All quarters
Musgravea: Qld
Austromuellera: Qld
Banksia (including *Dryandra*; Mast and Thiele, 2007): all quarters of Australia, and a fossil species in New Zealand (Carpenter *et al.*, 2010)

Unlike the Proteoideae, the Grevilleoideae are concentrated in north-eastern and south-eastern Australia, with much less diversity in the south-west. Of the 30 genera, 25 occur in the Queensland quarter, 16 in the south-east, and only five (none endemic) in the south-west. In contrast, the Indian Ocean Proteoideae are concentrated in western and south-eastern Australia and have much poorer representation in north-eastern Australia, where their Pacific-based sister-group, Grevilleoideae, is most diverse. The main massings of the two subfamilies indicate that the original split between the groups took place around a boundary: north-western Australia–central Australia–south-eastern Australia. Assuming that the allopatry was caused by vicarance, this would have occurred before the breakup of Gondwana and the rifting of Indian Ocean and Pacific Ocean clades. Mast *et al.* (2008) proposed a centre of origin/dispersal history for the Proteaceae, but this was based on the treatment of the oldest tricolpate fossil pollen age as a maximum age for eudicots (see Chapter 2).

The Proteaceae are ubiquitous in Australia but show low levels of endemism in the north-western quarter. This area is often involved with Tethyan tracks and the lack of endemism here is consistent with the southern, rather than Tethyan, overall distribution of the family. Despite their distribution in Mediterranean climates Proteaceae are absent from the Mediterranean itself and from most parts of the Tethys track that do not have direct involvement with Indian Ocean or Pacific Ocean distribution. They are replaced in these regions by their sister-group, Platanaceae.

Wanntorp *et al.* (2011) reported 12 fossil species of proteaceous pollen, mostly *Proteacidites* and *Beaupreaidites*, from the Maastrichtian–Paleocene of Campbell Island, southern New Zealand. This pollen is described as *Knightia*-like and *Beauprea*-like and indicates that Proteaceae were already diverse in the region by the end of the Cretaceous. Wanntorp *et al.* (2011) concluded that the family's presence in Zealandia can be explained in terms of 'relictual vicariant distributions'.

North-western Australia

Many clades in southern Australia are distinctive, isolated endemics. In contrast, plants and animals in northern Australia are often allied with Asian groups and have been attributed to recent immigration from Asia. A typical example is the land snail family Camaenidae s.lat., distributed from Eurasia to Australia. In Australia it is the most diverse land snail family and has 80 genera and 400 species. These are widespread in rainforest and in desert, but the group is a northern one and is absent from south-western Australia and Tasmania. The traditional model for the Australian contingent proposes Miocene dispersal from a Laurasian centre of origin (Hugall and Stanisic, 2011). This model depends on an evolutionary clock and interprets the locality of a basal, paraphyletic group as a centre of origin. But instead of invading Australia by dispersal, it is possible that Camaenidae 'invaded' the region by evolving there. The geographic connections of the family with relatives in Southeast Asia and elsewhere would then represent widespread ancestral distributions and reflect aspects of former geography. Several groups in north-western Australia have been interpreted in this way and examples of camaenid distribution in north-western Australia are cited below.

Subterranean endemism around the Hamersley Range (Pilbara region) and Tethyan connections

One of the most interesting recent developments in Australian biogeography has been the discovery of a diverse, subterranean fauna in the north-west of the country (Eberhard *et al.*, 2005; Humphreys, 2008). The fauna comprises a wide range of groups, including fish, but is best known for its diversity of unusual Crustacea. Many of the subterranean clades are stygobionts living in the groundwater (subterranean, air-breathing species are termed troglobionts). The fauna was first discovered at Cape Range and Barrow Island, and then further inland, at the Hamersley Range and other parts of the Pilbara region. Most of Australia's iron ore is mined in this area, especially in and around the Hamersley Range. The local fauna and flora, above and below ground, are important for understanding the evolutionary history of Australia, but the stygofauna is threatened by groundwater abstraction and the large-scale dewatering of aquifers that is carried out before mining.

Apart from its unexpected, great diversity, a striking feature of the stygobiont fauna is the high proportion of species with Tethyan or Gondwanan affinities (Eberhard *et al.*, 2005). The Tethyan clades have sister-groups in coastal caves of the Mediterranean, the Canary Islands and the Caribbean; biologists working on the fauna often refer to distribution extending from Australia via the Mediterranean to the Caribbean as the 'full Tethyan track' (Jaume *et al.*, 2001; Jaume, 2008; Humphreys *et al.*, 2009; Karanovic and Eberhard, 2009).

The north-western stygofauna occupies sandstone or limestone aquifers that are anchialine – there is a subterranean connection with the sea. The aquifers are salinity-stratified and are affected by marine tides, and so they have been described as underground estuaries. Page *et al.* (2008) explained the origin of the cave fauna by a process in which local marine species are stranded in new, terrestrial limestone habitat as it emerges. This process can occur either by uplift or marine regression and is one of the most significant causes of vicariance in stygobiotic faunas (Craw *et al.*, 1999; Culver and Pipan, 2009).

A sample from the stygofauna: some Crustacea

Typical examples of the north-western stygofauna include the five following clades of Crustacea.

Crustacean class Remipedia occurs in anchialine caves in north-western Australia at Cape Range, the Canary Islands and the Caribbean (with most diversity) (Neiber *et al.*, 2011).

The crustacean order Thermosbaenacea is known in Australia only in the north-west, with records of *Halosbaena* near Cape Range. Jaume (2008) mapped the order and showed that the global distribution 'matches precisely the area covered by the ancient Tethys Sea or its coastlines. [The members] are most probably relicts of a once widespread shallow-water marine Tethyan fauna stranded in interstitial or crevicular groundwater during marine regressions'. Jaume wrote that the full Tethyan track displayed by *Halosbaena* (Australia, the Canary Islands and the Caribbean) suggests its origin dates back to at least the maximum extent of the Tethys sea (120 Ma).

Members of suborder Phreatoicidea (Isopoda) live in freshwaters of South Africa, India, Australia and New Zealand (fossil in Antarctica), in a typical Gondwanan pattern. Many are subterranean. In a morphological study, the basal members were groups from south-eastern Australia (*Phreatoicopsis* and *Synamphisopus*, both from Victoria) (Wilson, 2008). Another clade has the phylogeny:

Pilbara (*Pilbarophreatoicus*).

India (*Nichollsia*).

SE Australia (*Phreatoicoides* and *Hypsimetopus*).

Wilson and Johnson (1999) attributed the intercontinental patterns in the suborder to differentiation prior to and during Gondwanan breakup. They explained the many local genera in Australia by differentiation caused by mid-Cretaceous marine transgressions.

The copepod *Speleophryia* is found in Cape Range, the Mediterranean and the Caribbean, and also extends the Tethyan track to southern Australia in the Nullarbor Plain (Karanovic and Eberhard, 2009).

The shrimp *Stygiocaris* (Decapoda: Atyidae) is endemic to coastal caves at Barrow Island and Cape Range. It is closely related to *Typhlatya*, a genus with a centre of diversity in the Mediterranean, North Atlantic and Caribbean (Page *et al.*, 2008). The authors suggested that *Stygiocaris* and *Typhlatya* may descend from a common ancestor that lived in coastal marine habitat of the Tethys Sea, and the two genera were separated by continental drift. Further sampling by Botello *et al.* (2012) gave a phylogeny:

NW Australia (*Stygiocaris*) + circum-Caribbean: Hispaniola, Puerto Rico, Barbuda, Curaçao, San Andrés I. (*Typhlatya monae*).

W Mediterranean (*Typhlatya* species).

Mid-Atlantic, Bahamas, Cuba, Yucatán (*Typhlatya* species).

The boundary between *Typhlatya monae* and the other West Indian *Typhlatya* species to its north coincides with the margin of the Caribbean plate, a Cretaceous structure, and this supports the vicariance model proposed by Page *et al.* (2008). In contrast, Botello *et al.* (2012) found that the distribution cannot be explained by vicariance as the clades are too young, but they calibrated the phylogeny by assuming that *Stygiocaris* diverged only after the emergence of the Cape Range anticline (7–10 Ma).

A subterranean fish of north-western Australia: *Milyeringa*

Milyeringa, a gobiiform fish, is an example of a north-western Australian group that has a trans-Indian Ocean connection rather than a Tethys one. The sister-group affinity, comprising the family Milyeringidae, is as follows (Chakrabarty *et al.*, 2012):

Milyeringa: north-western Australia, in groundwaters of Cape Range and Barrow Island (in freshwater caves and in seawater in anchialine systems; Chakrabarty, 2010).

Typhleotris: Madagascar, in subterranean freshwaters.

Chakrabarty *et al.* (2012) concluded that vicariance is the simplest explanation for the distribution and that *Milyeringa* and *Typhleotris* owe their origins to the Mesozoic break-up of Gondwana. The pair is sister to Eleotridae, widespread in tropical–subtropical seas and freshwaters. Connections between north-western Australia and Madagascar are also documented in plants such as *Adansonia* (see the section on 'The Kimberley', below).

Endemics around the Hamersley Range (Pilbara region) and their widespread Australian sister-groups

In the widespread Australian gecko *Lucasium stenodactylum* (formerly treated in *Diplodactylus*), a clade endemic to the Hamersley Range and the Pilbara craton (Fig. 4.1) has an allopatric sister clade that is widespread through the central desert (to the MacDonnell Ranges), and the two are separated by a 'deep and ancient' split (Pepper *et al.*, 2006). Pepper *et al.* (2011a) dated the divergence at 15.5 Ma and, as this was based on a fossil calibration, the actual age could be much older. The authors used the fossil genus *Cretaceogekko* (mid-Cretaceous, 97–110 Ma) to calibrate the root of the Gekkota phylogeny, but the only analysis of *Cretaceogekko* (Arnold and Poinar, 2008) gave no indication that the genus is basal in Gekkota and so, again, *Lucasium stenodactylum* could much be older than suggested.

Pepper *et al.* (2006) pointed out that the break between the Pilbara and central Australian clades in *Lucasium stenodactylum* corresponds to the mid-Cretaceous coastline where the marine sediments begin. They suggested that the distributional boundary reflects ecology, as different substrate (sand rather than rock) is found to the east, but the original phylogenetic break could have been caused by an event – the marine transgression itself.

In a similar pattern, the gecko *Underwoodisaurus* (Carphodactylidae) comprises one species of the Hamersley area and one widespread through mainland Australia (rare in the north) (Doughty and Oliver, 2011). Three species in the frog genus *Uperoleia* (Myobatrachidae) are endemic to the Pilbara area and one of these, *U. glandulosa*, is sister to a clade of four species that is widespread from the Kimberley to north-easterrn Queensland (Catullo *et al.*, 2011). These breaks and the Tethyan connections indicate that endemism in the north-west is not just a local development but has a broader continental and intercontinental significance.

Gerygone tenebrosa (Acanthizidae) is a passerine bird that inhabits mangrove forest along the north-western coast of Australia, from Shark Bay to the Kimberley. Its sistergroup has five species and is widespread throughout the rest of Australia (except Tasmania), New Zealand, Norfolk Island and New Guinea (in some analyses the sister-group also includes *G. flavolateralis* of New Caledonia, Vanuatu and the Solomon Islands) (Nyári and Joseph, 2012). The pattern is compatible with the break between *G. tenebrosa* and the rest occurring before the opening of the Tasman Sea in the Late Cretaceous.

In *Nicotiana*, the tobacco genus, the Australian members form a clade. The basal node separates a Tasman basin group from the others, while the second node separates a north-western, paraphyletic grade centred around the Hamersley area from a widespread Australian group (Ladiges *et al.*, 2011):

Eastern Australia east of the main divide (Sydney to Cairns), Lord Howe Island, New Caledonia (*N. forsteri*).

A paraphyletic, unresolved grade of five species concentrated in north-western Australia: three around the Hamersley area (one extending to the Kimberley, and one extending through central Austalia); one inland from Geraldton; one near Darwin.

A widespread Australian clade, west of the main divide (four species).

The phylogenies of *Gerygone* and *Nicotiana* indicate that endemism in north-western Australia is not a recent, local development, but has involved breaks in widespread groups that occurred before the opening of the Tasman basin. The exact mechanism is not yet known, but studies at a local scale are also revealing interesting patterns that are relevant to the problem. For example, the east Asian–Australian land snail family Camaenidae

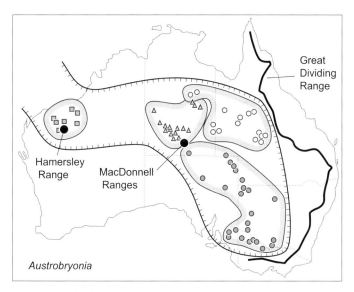

Fig. 4.5 Distribution of *Austrobryonia* (Cucurbitaceae), a genus with Tethyan affinities (Schaefer *et al.*, 2008; Schaefer and Renner, 2011; the different symbols indicate the four species).

was mentioned above. Within Australia, the group has its maximum diversity in the Kimberley region (186 species; Hugall and Stanisic, 2011). One of the genera, *Rhagada*, is diverse and endemic to north-western Australia. It extends from Shark Bay to the Kimberley (like *Gerygone tenebrosa*) and dominates the land snail fauna in the Pilbara region. Mainland species of the genus have large, non-overlapping geographic ranges. In contrast, in the Dampier Archipelago, off-shore from the Hamersley Range, there are several locally endemic species with intermingled ranges (Johnson *et al.*, 2004). Two species endemic there, on Barrow Island (cf. *Stygiocaris*, above), are genetically the most distinctive of all *Rhagada* species examined (Johnson *et al.*, 2006). This diversity off-shore from the Hamersley area is probably related not to the current geography of islands and mainland, but to former off-shore rifting.

A Tethyan group in arid Australia: *Austrobryonia* (Cucurbitaceae)

The Tethyan stygofauna of north-western and central Australia has equivalents above ground, as seen in desert plants of the gourd family. The tribe Bryonieae is a Tethyan group that occurs in Australia, central and south-western Asia, the Mediterranean region and the Canary Islands (Schaefer and Renner, 2011). In Australia the tribe is represented only by its basal member, the endemic *Austrobryonia*. The genus ranges through central Australia from the Hamersley Range and lower-lying parts of the Pilbara craton via the MacDonnell Ranges to north-western Victoria (Fig. 4.5; Schaefer *et al.*, 2008). Its eastern limit coincides with the Great Dividing Range. Although *Austrobryonia* is widespread in arid, central Australia, it is not found throughout this region; for example, while it

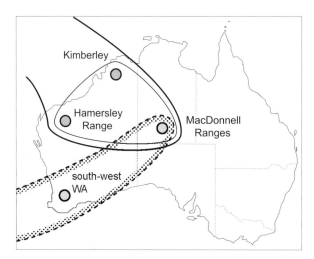

Fig. 4.6 The Hamersley–Kimberley–MacDonnell centre of endemism (with its potential Tethyan connections) and the southern Africa–south-western Australia–MacDonnell Ranges track.

is in Victoria, it is absent from the south-western quarter of Australia. This indicates that the pattern is not caused by ecology alone and that biogeographic factors have been important. The distribution suggests that ancestral Bryonieae invaded Australia with marine transgressions of the Tethys seas, along with the Tethyan stygofauna of north-western Australia. The divergence between Bryonieae and other, eastern Cucurbitaceae can be related to the mid-Cretaceous uplift of the Great Dividing Range.

Links between western and central Australia

Austrobryonia links central and north-western Australia with Tethyan regions. Another connection links the biota of central Australia with Africa. For example, the chironomid fly *Austrochlus*, mentioned already, connects the MacDonnell node with south-western Australia and southern Africa (Fig. 4.6). The MacDonnell node has strategic significance for Australian and global biogeography as both a centre and a boundary of endemism, and is located by a mid-Cretaceous coastline (Fig. 4.1).

Despite groups such as *Austrobryonia* and *Austrochlus*, central Australia has low local endemism. This is probably because the biota has been decimated by the aridity that began in the Miocene. Locally endemic species of Australian seed plants (those known from only 1–4 cells in a 1° × 1° grid) are instead concentrated in a mesic belt around the arid centre, in eight main regions: south-western Australia, Kangaroo Island, Tasmania, Sydney, McPherson–Macleay Overlap, Cairns, Cape York and Arnhem Land (Crisp *et al.*, 2001). Neither the Hamersley, MacDonnell nor the more humid Kimberley regions feature in this list. Yet the Hamersley–MacDonnell–Kimberley region (Fig. 4.6) is important as a larger area of endemism and is discussed in the next sections. It combines parts of arid, central Australia with the Kimberley, which has a monsoon climate.

The Kimberley

The Kimberley region biota includes endemics such as the scaly-tailed possum *Wyulda* (Phalangeridae), also present on the offshore Bonaparte Archipelago. *Wyulda* is the sister-group of the brush-tailed possum *Trichosurus*, widespread throughout Australia and Tasmania, except for the driest deserts (Meredith *et al.*, 2009). Kimberley groups illustrate southern, Gondwanan affinities with Madagascar, as well as Tethyan connections. For example, there are two main clades of baobab trees (*Adansonia*: Malvaceae), one in Africa and Madagascar, the other in the Kimberley (Baum *et al.*, 1998).

Within the Kimberley region, the offshore Bonaparte Archipelago is an important centre of endemism for groups such as the camaenid land snails (Köhler, 2011). In the camaenid genus *Amplirhagada*, a Kimberley endemic, Johnson *et al.* (2010) sequenced samples from two mainland areas and 16 islands in the archipelago. Four major clades corresponded to the major geographic groupings, separated by gaps of 10–160 km. Distinct lineages were also found on islands that are only a few kilometres apart, and the authors concluded that 'The large differences indicate that the lineages are much older than the islands themselves'. Another camaenid, *Australocosmica*, is a Kimberley endemic in which most of the species are narrow-range endemics restricted to single islands, although some extend to the adjacent mainland coast (Köhler, 2010). Again, this suggests that the endemism is older than the islands and is instead compatible with off-shore rifting. A similar process may explain endemism in the camaenid *Rhagada* on the Dampier Archipelago.

Hamersley–Kimberley distribution

This pattern has been cited for *Rhagada* land snails and the bird *Gerygone tenebrosa*. Another example is the centipede *Pilbarascutigera*, endemic to the Hamersley and Kimberley regions. Its sister-group is *Thereuopoda*, widespread from Australia to Japan (Edgecombe and Barrow, 2007). *Seychellonema* from the Seychelles has been allied with *Thereuopoda* (Butler *et al.*, 2010) and so the group as a whole may represent a disjunct, Indian Ocean complex.

Squamate diversity in arid Australia

Squamates – lizards and snakes – are diverse in the Australian deserts. Regions of spinifex (*Triodia*) grassland in arid Australia are thought to contain the highest diversity of squamate species in the world (Wilson and Swan, 2003). Between 40 and 50 species of lizards can be found in local sympatry and all of the Australian lizard families are involved (Jennings *et al.*, 2003). The diversity is intriguing as there are few geographic barriers in arid and semiarid Australia (Chapple *et al.*, 2004), and Jennings *et al.* (2003) observed that 'we still know little about why or when these radiations occurred'.

In some parts of arid Australia, lizard species show a high degree of habitat specificity. Based on this observation, Pianka (1972) suggested that the speciation has been driven by ecology, with lizards becoming adapted to particular habitats within the arid zone.

Yet molecular studies have not corroborated this idea. Chapple *et al.* (2004) wrote that the main prediction of Pianka's (1972) hypothesis is that generalist arid-zone species such as the skink *Egernia inornata* should show phylogeographic structure concordant with the major desert habitat types. Chapple *et al.* found considerable phylogeographic structure within *E. inornata*, with six major clades, but the clades' distributions were not concordant with habitat types. The authors concluded that their data 'do not support the main predictions of Pianka's (1972) hypothesis'.

If Neogene ecology cannot explain the diversity, perhaps earlier events can. Many studies have suggested that the arid zone lizard diversity has developed with the aridification in the Miocene, but the biogeography suggests an older history, with the origin and overlap of squamate clades in central Australia related to Paleogene and Mesozoic events.

Australian lizard families and their different centres of diversity

Overall, Australian lizards have maximum species diversity in a region extending from central Australia (MacDonnell Ranges and Alice Springs) west to the Hamersley Range (Powney *et al.*, 2010). There are secondary diversity centres around the Kimberley and the Cairns area. Powney *et al.* (2010) contrasted this pattern with those of mammals and amphibians (both most diverse in the north and east), and birds (most diverse in the east). Thus the concentration of lizard diversity along the MacDonnell–Hamersley sector is matched by a low diversity of the other groups there and so this could reflect early phylogenetic history as well as ecology. An analysis of snake diversity would be interesting. Hotspots of species diversity for the individual lizard families are:

Gekkonidae s.lat.: Hamersley Range.
Agamidae: central Australia and parts of Western Australia (not Hamersley Range).
Varanidae: Kimberley/Arnhem Land.
Scincidae: north-eastern Australia (McPherson–Macleay Overlap and Cairns) (cf. Hutchinson, 1993).

The allopatry of the main areas of diversity is clear-cut and the differentiation among most of them is congruent with Mesozoic seaways.

A north-western centre in Australian skinks

Although Australian skinks have maximum species diversity in the north-east, evolution around the north-west has also been important. The 'Australian sphenomorphines' (15 genera, 232 species) are the largest clade of Australian skinks, with 60% of the species. In the group, *Notoscincus* is centred in north-western Australia and is sister to all the rest (Rabosky *et al.*, 2007). *Notoscincus* has two species: *N. butleri* endemic to the Pilbara, and *N. ornatus*: Pilbara–Top End–MacDonnell Ranges, with rare records in northern Queensland (*Atlas of Living Australia*, 2012).

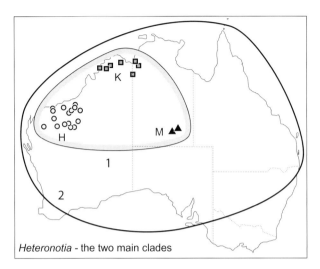

Heteronotia - the two main clades

Fig. 4.7 The two main clades in *Heteronotia* (Gekkonidae s.str.); one in north-western Australia, one widespread in Australia (Pepper *et al.*, 2011b; the different symbols indicate the three species in the north-western clade).

Hamersley–MacDonnell–Kimberley endemism in *Heteronotia* (Gekkonidae s.str.)

The prickly geckos (*Heteronotia*: Gekkonidae s.str.) have one clade in the north-western quarter of Australia, and the other, *H. binoei*, widespread in Australia (Fig. 4.7; Pepper *et al.*, 2011b). The north-western clade has a phylogeny: Kimberley (Hamersley + MacDonnell). The phylogeny was calibrated by assuming that the rise of the Iranian Plateau at 9 Ma divided northern and southern populations of agamid lizards. Instead, the internal node between the Hamersley and MacDonnell Ranges can be related to the incursion and retreat of the sea between these areas during Aptian–Albian time.

Other lizard clades that have their main breaks around the MacDonnell Ranges include the gekkonid *Rhynchoedura* and a clade in *Tympanocryptis* (Agamidae) (Shoo *et al.*, 2008). In *Rhynchoedura*, the first node separates a Kimberley/Top End clade from a diverse clade that is widespread in drier Australia (Pepper *et al.*, 2011c). The first break in this widespread clade is a west/east split near the MacDonnell node, corresponding to the Cretaceous coast. The eastern clade is divided into three groups whose distributions correspond to the Eyre, Bulloo–Bancannia, and Murray–Darling basins. These are current landscape features but their immediate precursors are thought to have been established in the Cretaceous (Pepper *et al.*, 2011c).

Crenadactylus and the diplodactyloid geckos: endemism in the western half of Australia

Three families of geckos (sometimes treated as subfamilies of Gekkonidae s.lat.) make up a large Australasian complex, the diplodactyloids, with the following phylogeny (Oliver and Sanders, 2009):

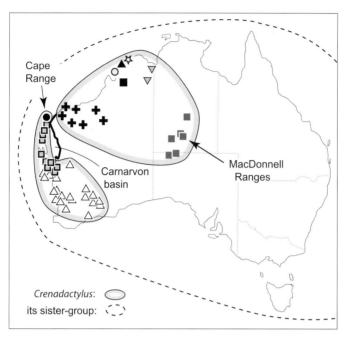

Fig. 4.8 Distribution of *Crenadactylus* and its sister-group (Diplodactylidae; Oliver *et al.*, 2010; the symbols indicate the 10 lineages in *Crenadactylus*).

Australia (throughout, but not in Tasmania) and New Guinea (Pygopodidae).

Australia (throughout, but not in Tasmania) (Carphodactylidae).

Australia (throughout), Tasmania, New Caledonia and New Zealand (Diplodactylidae).

The overlap of the three families in Australia implies basal differentiation of a widespread Australasian group around nodes somewhere in Australia, followed by overlap there. Even using fossil calibrations, Oliver and Sanders (2009) found that 'basal divergences within the diplodactyloids significantly pre-date the final break-up of East Gondwana'.

In Diplodactylidae, the basal group is *Crenadactylus*, endemic to north-western and south-western Australia, and with an eastern boundary at the MacDonnell Ranges (Fig. 4.8). Its sister, the rest of the family, is widespread throughout Australia, New Caledonia and New Zealand. In traditional taxonomy *Crenadactylus* has been treated as a single species, but Oliver *et al.* (2010) identified at least 10 lineages that show deep divergence, and five of these are endemic to the Kimberley. Four of the Kimberley lineages are known from areas of less than 100 km².

Oliver *et al.* (2010) concluded, using a fossil calibration, that *Crenadactylus* began to diversify in the Late Oligocene/Early Miocene (~20–30 Ma). Other lizard phylogenies have been calibrated with a combination of fossil dates and dates of geological/biogeographic events (Gamble *et al.*, 2010). Local correlations of geology and phylogeny are possible in *Crenadactylus*, as the 'Pilbara clade' (Hamersley node) is separated from its sister-group, the 'central ranges clade' (MacDonnell node) by the Aptian–Albian seaway (Fig. 4.1). Other early events in the group's history (including the basal break) have occurred around the Cape Range/Barrow Island area, where the endemic Crustacea cited above (*Stygiocaris*, etc.) are related to Tethyan forms.

In *Crenadactylus*, Oliver *et al.* (2010) found 'strong evidence for . . . the persistence of multiple lineages that pre-date the estimated onset of severe aridification [Miocene] . . . Highly divergent allopatric lineages are restricted to putative refugia across arid and semi-arid Australia . . . '. Following the widespread aridification that took place through Australia in the Miocene, populations of *Crenadactylus* have survived only in 'refugia', while the distribution of the genus and its sister-group indicates a Cretaceous origin of the two, prior to the opening of the Tasman.

The Pilbara–Yilgarn break

Humphreys (2008) discussed the groundwater fauna of Western Australia and noted: 'Typically there is fine spatial scale endemicity of species associated with local aquifers, but there are inexplicable regional differences, such as the change of fauna between the Yilgarn and Pilbara' (see Fig. 4.1). The break here, between the two Archaean cratons, is sometimes attributed to the climatic difference between the dry, summer monsoon climate in the north-east and the mesic, winter-rainfall climate in the south-west. However, the boundary is often precise and is evident in Tethyan and Gondwanan groups that are probably much older than the current climate. The northern part of the break coincides with the Late Cretaceous coast (the inland margin of the Carnarvon basin; Fig. 4.8). Further to the south-east, gentle tectonic displacement could have been involved. On flat terrain, even small vertical movements can affect drainage and have important ecological consequences. Quigley *et al.* (2010: 243) wrote that: 'The Australian continent is actively deforming in response to far-field stresses generated by plate boundary interactions and buoyancy forces associated with mantle dynamics . . . Although Australia is often regarded as tectonically and geomorphologically quiescent, Neogene to Recent tectonically induced landscape evolution has occurred across the continent, with geomorphological expressions ranging from mild to dramatic'. The northern shelf is submerging, the southern shelf is emerging. Australia has experienced southwest-up, northeast-down tilting of ∼300 m since the Late Eocene (Sandiford *et al.*, 2009) and, in addition, has undergone a broad-scale subsidence since the Late Cretaceous (DiCaprio *et al.*, 2011).

Arnhem Land: the shrubs and trees of Gonystyloideae

Among the distinctive endemics of Arnhem Land (Northern Territory) is the plant genus *Arnhemia* (Thymelaeaceae) (Fig. 4.9). Its sister-group is *Gonystylus* of Malesia and Fiji (Beaumont *et al.*, 2009). (*Amyxa* and *Aëtoxylon* are monotypic genera from Borneo and have not been sequenced. They are usually placed with *Gonystylus* and the three genera share a type of pollen that is unique in angiosperms; Nowicke *et al.*, 1985.) *Gonystylus* is

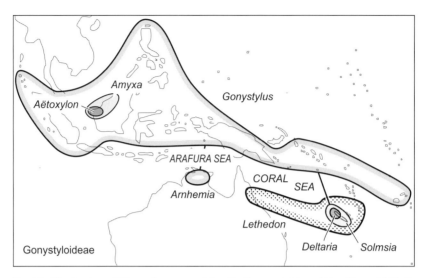

Fig. 4.9 Distribution of Thymelaeaceae subfamily Gonystyloideae (Airy Shaw, 1953; Beaumont *et al.*, 2009).

a typical tree of lowland rainforest, while *Arnhemia* is a subshrub and does not produce a trunk; instead, many prostrate shoots sprout from a woody, underground stock. The two patterns of architecture represent the two extremes in terms of trunk formation. As biogeographic and morphogenetic vicariants, both are probably derived from a common ancestral structure, rather than one from the other.

The *Gonystylus* + *Arnhemia* clade is sister to *Deltaria* + *Solmsia* of New Caledonia, and the sister of all four genera is *Lethedon*: north-eastern Queensland to southern Vanuatu (Beaumont *et al.*, 2009). The phylogenetic/biogeographic nodes in the group correspond with the basins of the Arafura and Coral Seas. The whole clade mapped here comprises the subfamily Gonystyloideae and is sister to all the other Thymelaeaceae (except *Octolepis*), a diverse, worldwide group.

Sea turtles (Dermochelidae and Cheloniidae) comprise five more or less worldwide clades (Duchene *et al.*, 2012). One of these is made up of *Natator* (Cheloniidae), breeding along the coast of northern Australia (from Exmouth near Cape Range to Bundaberg) and its sister, the worldwide genus *Chelonia*. As with *Arnhemia* and relatives, this suggests early vicariance in widespread ancestors at breaks around the Arafura and Coral Seas.

Marsupials in Australasia

The biogeography of extant marsupials was referred to above (Fig. 3.8). The Australasian members make up a clade with five main groups (Meredith *et al.*, 2009). These groups, with the distributions of their basal clades highlighted, are as follows (NG = New Guinea):

Bandicoots, dasyurids etc. (all Australia, NG). Basal: *Notoryctes*: **central Western Australia**.

Wombats etc. (eastern Australia). Basal: *Phascolarctos*: **Eastern Australia (fossil records inland to Lake Eyre)**.

Brushtail possums etc. (all Australia, NG, Sulawesi, Solomon Islands). Basal: *Burramys* + *Cercartetus*: **Southern and eastern Australia, Tasmania, NG**.

Kangaroos etc. (all Australia, NG). Basal: *Hypsiprymnodon*: **Cairns**.

Ringtail possums etc. (all Australia, NG). Basal: *Tarsipes*: **South-western Australia**.

The five clades show extensive overlap (four are widespread through Australia and New Guinea), indicating range expansion. In contrast, the basal clades in each of the widespread groups show a high degree of allopatry and together cover all of Australia and New Guinea.

The basal group in the basal clade of these marsupials is the marsupial mole, *Notoryctes*, of central Western Australia. As with *Acacia* (see p. 140), this suggests that central Australian groups are not the result of invasion from the current mesic areas. Instead, Mesozoic marine transgressions and regressions could have caused the overlap of the five marsupial clades, the basal differentiation events in the clades, or both.

The Great Dividing Range

The tree kangaroos (*Dendrolagus*: Macropodidae) are widespread in New Guinea and around Cairns, while members of their sister-group, the rock wallabies (*Petrogale*), occur in steep, rocky habitat through mainland Australia (Fig. 4.10). *Petrogale* has the following phylogeny and distribution (Potter *et al.*, 2012):

1. Kimberley – Top End (*P. brachyotis* etc.).
2. Central Queensland coast, near Proserpine and the Whitsunday Islands (*P. persephone*).
3–6. The rest of mainland Australia (all other species).

The break between *Dendrolagus* and *Petrogale*, and the break between *P. persephone* and the last clade, are both located near the north-eastern Queensland coast. The node coincides with the mid-Cretaceous pre-breakup volcanism of the Whitsunday Volcanic Province.

The last, widespread clade of *Petrogale* has four groups, numbered 3–6 in Figure 4.10. The group is divided by one boundary around the Pilbara craton, one running from Eyre Peninsula to the Gulf of Carpentaria (line A in Figure 4.10, equivalent to the mid-Cretaceous sea), and one extending along the eastern seaboard of Australia (line B). This last line coincides with the Great Dividing Range (Fig. 4.5) and is an important biogeographic break. An episode of large-scale uplift and fault reactivation took place here

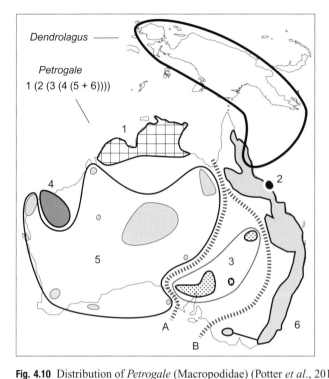

Fig. 4.10 Distribution of *Petrogale* (Macropodidae) (Potter *et al*., 2012; the six main clades in *Petrogale* and their phylogeny are indicated).

during the mid-Cretaceous (114–83 Ma), prior to breakup at 84 Ma (O'Sullivan *et al*., 1999; Kohn *et al*., 1999; Persano *et al*., 2006; Vandenberg, 2010). In some areas uplift continued into the Cenozoic. The mid-Cretaceous uplift was a precursor to the seafloor spreading that opened the Tasman basin. Coeval uplift occurred along much of the eastern margin of Gondwanaland (eastern Australia, New Zealand and Marie Byrd Land, Antarctica; Raza *et al*. 2009). Elsewhere, pre-breakup uplift at rift shoulders has had important consequences for biogeography in areas such as eastern Brazil and western Africa.

Petrogale illustrates the way in which the Great Dividing Range separates 'Tasman–Coral Sea' groups (6) from 'Australian' groups (5). In another example, the spider *Plebs rosemaryi* of Norfolk Island and the Cairns area is sister to *P. cyphoxis* of Western Australia to Kangaroo Island (Joseph and Framenau, 2012). Other groups with east–west breaks at the Dividing Range include *Nicotiana* (see 'North-western Australia', above) and the passerine family Cinclosomatidae (Corvida), with the following phylogeny (Toon *et al*., 2012):

New Guinea (*Ptilorrhoa*).

New Guinea (*Cinclosoma ajax*).

Australia, east and south of the Great Dividing Range (*C. punctatum*).

Australia, west of the Great Dividing Range, to the west coast of Australia (other *Cinclosoma* species).

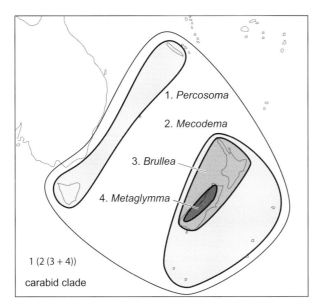

Fig. 4.11 Distribution of *Percosoma* and allies (Carabidae), with phylogeny indicated (Roig-Juñent, 2000; Liebherr *et al.*, 2011).

The pattern indicates that the first two nodes in an Australian/New Guinea ancestor have been active around New Guinea, while the third has developed at the Great Dividing Range.

The next sections introduce some of the main biogeographic nodes of eastern Australia, east of the Great Dividing Range. These are located in Tasmania, the McPherson–Macleay Overlap, north-eastern Queensland and Torres Strait/Coral Sea.

Tasmania

Tasmania has many endemics while other groups show strange absences there; Hooker (1860) commented on the relative paucity of *Eucalyptus*. Diverse Australian-endemic genera that occur in all the mainland states but not in Tasmania include *Templetonia* (Fabaceae), *Eremophila* (Scrophulariaceae), *Actinobole* and *Gnephosis* (both Asteraceae). The absence of mainland Australian groups in Tasmania is related to the existence of the Tasmania–Zealandia centre of endemism, as seen in the carabid beetle *Percosoma* and its relatives in New Zealand (Fig. 4.11; Roig-Juñent, 2000). (Disjunction between Tasmania and New Caledonia is also seen in the monocot family Campynemataceae, endemic to the two areas; Stevens, 2012.)

Taxa restricted to Tasmania and New Zealand include the basal centipede *Craterostigmus* (Edgecombe and Giribet, 2008), and the plants *Archeria* (Ericaceae: Styphelioideae), *Pachycladon* (incl. *Cheesemania*; Brassicaceae), *Liparophyllum* (Menyanthaceae) and '*Oreoporanthera*' (now included in *Poranthera*, Phyllanthaceae; only found in Nelson and south-western Tasmania; Orchard and Davies, 1985). Tasmania–New Zealand affinities in the fauna include the yponomeutoid moth *Proditrix*, present in New Zealand (feeding on monocots) and Tasmania (on *Dracophyllum* s.lat.; Ericaceae; McQuillan, 2003). Other yponomeutoids endemic to New Zealand and Tasmania are on conifers, for example *Chrysorthenches* (Dugdale, 1996). Other conifer-feeding taxa include a *Platycoelostoma* species (Hemiptera) in Tasmania with its sister in subalpine New Zealand (Gullan and Sjaarda, 2001).

The phylogeny in many groups indicates that the break between Tasmania and mainland Australian groups occurred before the separation of Tasmania and New Zealand. Tasmania is separated from mainland Australia by the Bass Strait, but the latest flooding of the Strait, at ~10 000 B.P., is probably not relevant for biogeography. Instead, the formation of the Bass Strait basins, now filled with sedimentary strata, was the critical event. The basins (from west to east, the Otway, Bass and Gippsland Basins) are Mesozoic to lower Cenozoic extensional structures. They were active in the Cretaceous, in association with the breakup of Gondwana, and represent a failed branch of the Southern Ocean rift that developed between Australia and Antarctica. The Gippsland basin is Australia's most prolific oil and gas province, and it is possible that the same tectonism that formed and trapped the hydrocarbons also separated Tasmanian and mainland groups. This occurred before New Zealand and Tasmanian populations were rifted apart by the seafloor spreading in the Tasman basin.

Apart from connections with New Zealand and New Caledonia, Tasmanian groups show intriguing disjunctions with the McPherson–Macleay Overlap (for example, in *Acradenia*: Rutaceae, in *Anopterus*: Escalloniaceae, between *Milligania* and *Neoastelia*: Asteliaceae; Birch *et al.*, 2012, and between clades of the centipede genus *Paralamyctes*; Giribet and Edgecombe, 2006a; 2006b).

The empidoid fly *Ceratomerus* occurs in Australia, New Zealand and South America. One species, *C. campbelli* of Tasmania, is sister to *C. athertonius* from the Atherton Tableland near Cairns, and a similar, disjunct species pair is also known in another empidoid, the genus *Clinocera* (Sinclair, 2003). Another common disjunction occurs between Tasmania and south-west Western Australia (e.g. Hydatellaceae).

Tasmania also has many connections with the south-western Indian Ocean. For example, the freshwater crayfish family Parastacidae includes one clade found in Madagascar, western Tasmania (*Spinastacoides*, *Ombrastacoides*) and New Zealand (Toon *et al.*, 2010).

To summarize, members of the Tasmanian biota have relatives in Madagascar, south-western Australia, the McPherson–Macleay Overlap, New Zealand, New Caledonia and South America. Tasmania itself often appears as a biogeographic hybrid, with eastern and western terranes having distinct geology and biology. In the Australasian–South American genus *Abrotanella* (Asteraceae), the eastern Tasmanian group is basal (Swenson *et al.*, 2012). The rest of the genus comprises one group in western Tasmania, south-eastern Australia, New Guinea and New Zealand, and the other in Stewart Island and South America. This pattern is compatible with the east Tasmania/west Tasmania split in *Abrotanella* having developed before rifting in the Southern Ocean separated Australia from New Zealand and South America.

McPherson–Macleay Overlap

The McPherson–Macleay Overlap (MMO), located where the Queensland/New South Wales border hits the coast, is an important area of overlap and diversity in many groups (Burbidge, 1960). It is not just an area of overlap, though, and it is an important centre of endemism in its own right. Burbidge (1960) recorded as many as 34 plant genera known only from there. The endemics are often not just local forms in a group with many others, but have sister-groups that are widespread. For example, the frog *Assa* (Myobatrachidae) is a local endemic at the MMO and its sister is *Geocrinia*, disjunct in south-western and south-eastern Australia (Read *et al.*, 2001). The liane *Nothoalsomitra* (Cucurbitaceae) of the MMO (near Brisbane) is sister to a diverse clade in eastern Asia and the Americas (Schaefer *et al.*, 2009). The aquatic monocot family Maundiaceae is endemic to the MMO and is sister to a circumglobal clade of Alismatales (Zosteraceae, Potamogetonaceae, etc.) (Stevens, 2012).

Many northern and southern groups overlap at the MMO. For example, the families and subfamilies of Australian butterflies have the following centres of diversity (Kitching and Dunn, 1999):

Cape York (Riodinidae).
Cape York and Cairns region (Nymphalidae, Lycaenidae).
Cairns region (Papilionidae, Pieridae, Hesperiidae subfam. Hesperiinae, Hesperiidae subfam. Pyrginae).
Cairns region and MMO (Hesperiidae).
Cape York, Cairns region, MMO (Hesperiidae subfam. Coeliadinae).
MMO to Victoria (Nymphalidae subfam. Satyrinae, Hesperiidae subfam. Trapezitinae).

As with all important nodes, the MMO is also a centre of absence. For example, the gecko *Diplodactylus* is almost a true Australian endemic (found only in Australia and throughout Australia), but is absent on the mainland from the MMO (data from *Atlas of Living Australia*, 2012).

Burbidge (1960) noted close affinities of MMO taxa with areas further south, including the Sydney/Blue Mountains area. The MMO in the narrow sense is illustrated by local endemics such as *Streptothamnus* (Berberidopsidaceae) and *Corokia whiteana* (Argophyllaceae), and, in the broader sense, including the Sydney region, by groups such as *Petermannia* (Petermanniaceae: Liliales) and *Blandfordia grandiflora* (Blandfordiaceae: Asparagales). The node in the strict sense lies south of Brisbane, but extending the MMO north of here defines a region with endemics such as *Durringtonia* (Rubiaceae) and associated geological structures such as the accreted Gympie terrane.

Apart from its high levels of endemism, diversity and absence, the MMO biota has far-flung disjunctions with Asia and South America (discussed below). The area has intercontinental phylogenetic significance and is more than just a local overlap of Queensland and south-eastern Australian groups (Ladiges, 1998).

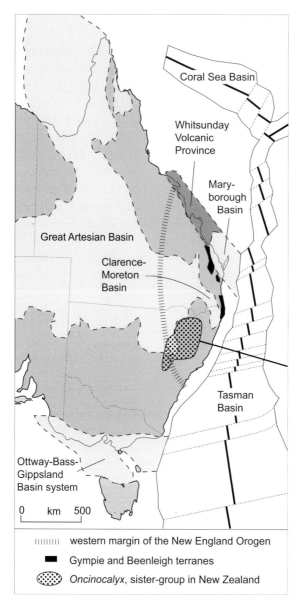

Fig. 4.12 Aspects of geology around the McPherson–Macleay Overlap (Bryan *et al.*, 2012). The distribution of *Oncinocalyx* (Lamiaceae), sister to *Teucridum* of New Zealand, is also shown (Cantino *et al.*, 1999). Solid line = spreading centre. (Geological map reprinted from *Episodes*, by permission of the International Union of Geological Sciences.)

The MMO region has a complex tectonic history (Fig. 4.12; Mortimer *et al.*, 2005; Bryan *et al.*, 2012). In early Mesozoic time it was involved in the New England orogeny and was the site of the last terrane accretion on the Australian mainland. The MMO part of the orogen later underwent Jurassic–Cretaceous extension, forming the

Clarence-Moreton basin. Through the Early Cretaceous a belt of pyroclastic volcanism, the Whitsunday volcanic province, developed along the eastern side of the MMO.

New England orogen

The Permian–Triassic New England (or Hunter–Bowen) orogen extends along eastern Australia from Sydney to just north of Mackay and includes the MMO region in its broadest sense – all of central east coast Australia. On its western margin, the New England orogen is flanked by the Sydney and Bowen foreland basins, well known for their huge coal reserves and *Glossopteris* fossils. Offshore, the New England orogen is recognized north to the Queensland plateau and south to the Lord Howe Rise (Mortimer *et al.*, 2008). The main phase of deformation in the orogen involved subduction, arc-accretion and back-arc basin formation.

The final event in the New England orogeny was the Triassic accretion of the arc-related Gympie terrane. This terrane has been equated with the Brook Street terrane in New Zealand (Mortimer *et al.*, 1999) and the Téremba terrane of New Caledonia. These three terranes are remnants of an extensive, intra-oceanic island arc system that formed in the Permian. The arc was accreted to the Gondwana margin in the Triassic/Jurassic. Dislocation of the arc rocks occurred during this process and again in the Cretaceous during Gondwana breakup (Spandler *et al.*, 2005b). Both events would have led to vicariance and disjunction in many communities.

Clarence-Moreton basin

The Great Artesian basin of the Australian central-eastern lowlands extends east to the coast as the Clarence-Moreton basin (Fig. 4.12; Turner *et al.*, 2009). This arm of the main basin was formed by Late Triassic to Early Cretaceous extension in the New England orogen. The Great Artesian basin floor trapped freshwater in porous aquifers, and there is coal in the Clarence-Moreton basin (for example, in the alluvial–lacustrine Grafton Formation of the Late Jurassic–Early Cretaceous). It is also likely that oil is present, and the location of oil belts coincides with centres of diversity and biogeographic boundaries in many areas, such as western Amazonia, central Africa and Indonesia (Heads, 2012a). The types of faulting that trap petroleum also lead to dynamic landscapes and biological evolution. The formation of the Clarence-Moreton basin would have created a barrier and phylogenetic break for some taxa, but a pathway for expansion and overlap in weedy, coastal groups. Giribet and Edgecombe (2006b) suggested that the Cretaceous seaways caused differentiation between centipedes of Queensland and their relatives in south-eastern Australia. The Clarence basin is significant for biogeography even at a local scale. For example, a group of leaf-tailed geckos in the genus *Saltuarius* (Diplodactylidae) breaks down into two clades (Couper *et al.*, 2008):

1. Three locally endemic species, from the Clarence River north to Brisbane.
2. One species, from the Clarence River south to Newcastle.

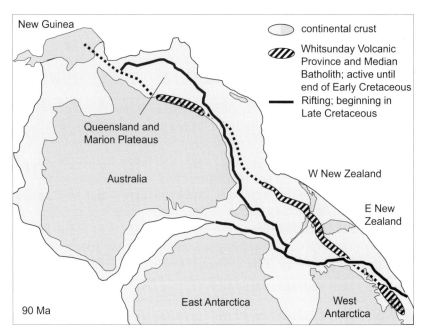

Fig. 4.13 A reconstruction for 90 Ma showing an Early Cretaceous silicic volcanic province and zones of Late Cretaceous rifting (Mortimer *et al.*, 2005).

Whitsunday volcanic province

Through Early and Late Cretaceous time, pre-breakup extension and magmatism developed in what became eastern Australia. This produced the silicic Whitsunday volcanic province (132–95 Ma) (Fig. 4.13; Campbell and Hutching, 2007). In the Maryborough basin part of the province (Fig. 4.12; Bryan *et al.*, 2002, 2012), intrusion, volcanism and deformation in the Late Jurassic–Early Cretaceous has been related to the New Zealand Rangitata orogeny (Harrington and Korsch, 1985). The last, Late Cretaceous deformation in the basin has been termed the Maryburian orogeny.

Silicic igneous rocks equivalent to those of the Whitsunday volcanic province occur on the other side of the Tasman basin, in the Lord Howe Rise, New Zealand and Marie Byrd Land (Antarctica), and the whole belt forms the Earth's largest silicic igneous province (Bryan and Ernst, 2007; Bryan *et al.*, 2012). It was emplaced about the same time (~120 Ma) that the mafic igneous province, the Ontong Java–Hikurangi–Manihiki plateau, was erupted further east.

Silicic volcanism is restricted to regions of continental crust and involves anatexis, or partial melting, of the crust. This produces the silicic (felsic) chemistry of the ejecta. Silicic volcanism can be a precursor to rifting, as in the Whitsunday volcanic province and in the Lebombo batholith of south-eastern Africa (composed of silicic rocks in the otherwise basaltic upper Karoo sequence). The Lebombo volcanism and associated crustal flexure at the Lebombo monocline was a precursor to the complete rifting of the crust that occurred in the Mozambique Channel. The volcanic rifted margin in south-eastern Africa appears as a biogeographic boundary in large numbers of groups, as does the volcanic rifted margin of eastern Australia. The Whitsunday volcanic province has close geological affinities with the New Zealand Median Batholith and so the northern end of the MMO (Maryborough basin) is connected with Nelson and other parts of New Zealand, as in the biota (cf. Fig. 6.3).

To summarize: protracted magmatism and continental rifting took place along the eastern Australian/New Zealand part of the Gondwana margin, before there was a complete rupture of the continent with the onset of seafloor spreading (Bryan *et al.*, 2002, 2012). The major events in the Cretaceous breakup of the margin were:

1. Silicic **magmatism** along the Whitsunday belt (Early Cretaceous, 130–95 Ma).
2a. Intracontinental **rifting** (mid-Cretaceous, 115–85 Ma).
2b. Kilometre-scale **uplift** along the eastern margin of Australia (mid-Cretaceous, 100–95 Ma), in New Zealand and in Marie Byrd Land (Raza *et al.*, 2009).
3. **Seafloor spreading** in the Tasman basin (Late Cretaceous–Eocene, 84–56 Ma), between the Gondwana mainland and Zealandia.

The sequence is a normal one. Rifts in the interior of continents that evolve to form large ocean basins often last for 30–80 m.y. before complete rupture of the continent and the onset of seafloor spreading (Umhoefer, 2011). The Whitsunday–Median Batholith magmatic belt and the seafloor spreading centre (the Tasman rift) both lie near the margin of Gondwana, but in different locations (Fig. 4.13). Each coincides with a biogeographic boundary.

The Whitsunday belt (rather than the later seafloor spreading) often divides clades in Australia–southern New Zealand (in the south-west Tasman) from sisters in New Guinea–northern New Zealand (north-east of the first clade). The second affinity suggests former connections east of Queensland. East of the Whitsunday volcanic province, the continental crust of the Queensland Plateau and Lord Howe Rise main-tained terrestrial habitats and biota until the crust thinned with continued extension and sank.

In New Zealand, the Median Batholith includes silicic intrusive rocks, mainly gran-ites, that match the silicic extrusives (mainly rhyolite) of the Whitsunday volcanic province, and the two are thought to represent a single belt of magmatism (Tulloch *et al.*, 2006; Mortimer, 2006). (Adams, 2010, instead mapped the Median Batholith as a continuation of the New England orogen.) Within the Median Batholith, Tulloch *et al.* (2006) recognized an older, eastern belt of plutons (Darran suite; 174–130 Ma) and a younger, western belt (Separation Point suite; 130–105 Ma). The younger magmatic arc was associated with thick crust and probably supported a large, Andean-style moun-tain range, 'Cordillera Zealandia'. Cessation of arc magmatism at ~105 Ma was fol-lowed by widespread intracontinental extension beginning at 102 Ma (forming the Great South Basin). This was prior to seafloor spreading which began in the Tasman basin at 85 Ma.

In the D'Entrecasteaux and Louisiade Islands, at the Woodlark rift in south-eastern Papua New Guinea, metamorphic rocks show isotopic similarities with rocks in the

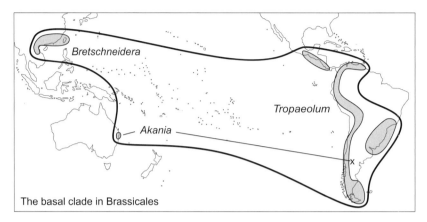

Fig. 4.14 The trans-Pacific, basal clade in Brassicales: Akaniaceae (including Bretschneideraceae) and Tropaeolaceae (Stevens, 2012). X = fossil record of *Akania*.

Whitsunday volcanic province (Zirakparvar *et al.*, 2013). The Woodlark rocks also contain inherited zircons with 90–100 Ma U–Pb ages that overlap the timing of magmatism in the Whitsunday belt. This information suggests that the volcaniclastic sediments generated in the Cretaceous Whitsunday volcanic province provided the protoliths for the metamorphic rocks in the Woodlark rift and an Early Cretaceous reconstruction shows the two regions adjacent to each other (Zirakparvar *et al.*, 2013, Fig. 8). Later they were rifted apart during the opening of the Coral Sea basin and New Guinea was rotated anti-clockwise.

The intercontinental connections of MMO clades

Several interesting MMO taxa have intercontinental affinities. The order Berberidopsidales, at the MMO and central Chile, was mentioned above. In centipedes, a clade of *Paralamyctes* occurs in western Tasmania, the MMO and Chile (Wellington Island, and Chiloé Island to Maule; Giribet and Edgecombe, 2006b).

Cadellia (Surianaceae) occurs at the MMO and also extends inland to central Queensland (Longreach). It is sister to *Recchia*, in southern Mexico and Costa Rica. This transtropical Pacific disjunct clade is basal in a circumglobal group, the Surianaceae (Forest, 2007b).

Akania is a tree endemic to the MMO (Fig. 4.14). Earlier botanists placed it in Sapindales, but the local name was 'turnip wood' and molecular studies have confirmed that the genus belongs in Brassicales. *Akania* forms a clade (Akaniaceae) with *Bretschneidera* of northern Vietnam, south-eastern China and Taiwan. This pair is sister to *Tropaeolum* (Tropaeolaceae) of Mexico to temperate South America, giving a trans-Pacific grouping. In addition to its extant records in Australia, *Akania* is known from fossil material in Chubut, southern Argentina (Gandalfo *et al.*, 1988), suggesting that the split between the genera occurred before the rifting in the Pacific basin.

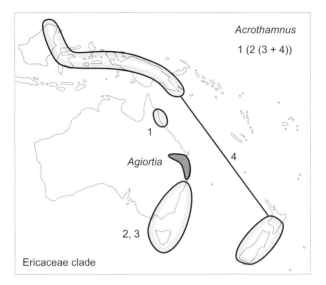

Fig. 4.15 Two sister clades in Ericaceae subfam. Styphelioideae (Quinn *et al.*, 2003, 2005). The line represents the affinity between the New Zealand *A. colensoi* and the New Guinea–Borneo *A. suaveolens* (Puente-Lelièvre *et al.*, 2013).

MMO and the Tasman region: two groups of heath (Ericaceae)

The following two clades of Ericaceae subfam. Styphelioideae are sister-groups (Fig. 4.15; Quinn *et al.*, 2003, 2005; Puente-Lelièvre *et al.*, 2013):

> MMO and inland along the Great Dividing Range into southern Queensland (*Agiortia*; 'clade L3' in Quinn *et al.*, 2003).
> Tasmania to south of the MMO, Cairns, New Guinea, Sulawesi, Borneo, New Zealand and (based on morphology) New Caledonia (*Acrothamnus*; 'clade L4' in Quinn *et al.*, 2003).

The New Zealand member of the group, *Acrothamnus colensoi*, is sister to *A. suaveolens* of New Guinea–Borneo. This New Zealand–New Guinea disjunction is a common one and coincides with the tectonic belts that run parallel with the Gondwana margin (Fig. 4.13); *Acrothamnus* has its southern boundary at Foveaux Strait and the Median Batholith. The distribution of the MMO group, *Agiortia*, is based around the Clarence-Moreton basin (Fig. 4.12).

Leptomyrmex (Formicidae): the MMO as a main node for Tasman groups

In the ant genus *Leptomyrmex*, Lucky (2011) found the following phylogeny for the 'macro' group (Fig. 4.16A):

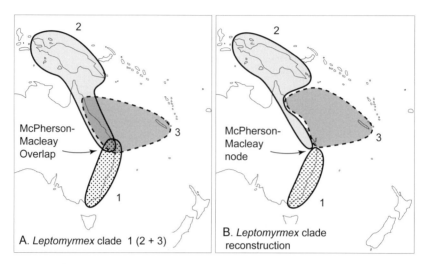

Fig. 4.16 Distribution of a clade in the ant genus *Leptomyrmex* (Lucky, 2011). A. Current distribution. B. Reconstruction of the original allopatry.

> MMO, New South Wales and Victoria.
>
> MMO, Queensland, New Caledonia.
>
> MMO, Queensland, New Guinea.

As Lucky (2011) noted, the three clades show strong correlations with geography. The simplest explanation for the considerable allopatry is that the group evolved by vicariance of a widespread Tasman–Coral Sea ancestor at a node around the MMO (one possible reconstruction is shown in Figure 4.16B). The subsequent overlap of all three clades at the MMO can be attributed to local range expansion. Trans-oceanic dispersal is unlikely; Lucky (2011) wrote that members of the 'macro' group are 'severely dispersal limited . . . colony reproduction presumably involves founding queens walking to new nest sites'.

The trans-Tasman connections of the MMO

Direct MMO–Tasmania disjunctions were mentioned above (under 'Tasmania'). To the north, many taxa are only known from the MMO and disjunct in north-eastern Queensland, often in the hills by Cairns. Examples include rainforest plants such as *Trimenia moorei* (Trimeniaceae: Austrobaileyales) and *Eupomatia bennettii* (Eupomatiaceae: Magnoliales).

Other MMO endemics show closer affinities with groups outside Australia than with any Australian taxa. Some of the most frequent connections are with New Zealand and New Caledonia. For example, *Oncinocalyx*, on the inland borders of the MMO, is sister to *Teucridium* (Lamiaceae) of New Zealand (North and South Islands) (Fig. 4.12; Cantino *et al.*, 1999).

In a clade of *Lilaeopsis* (Apiaceae), Bone *et al.* (2011) found that a New Zealand clade (*L. novae-zelandiae* p.p. and *L. ruthiana*) is sister to *L. brisbanica*, locally endemic around Brisbane (Bostock and Holland, 2010). This MMO–New Zealand clade is sister to *L. polyantha* of south-eastern Australia (south of Brisbane to Eyre Peninsula) and Tasmania.

Euphrasia cuneata (Orobanchaceae) of northern New Zealand is sister to *E. ramulosa* of the MMO, while these two are closest to *E. phragmostoma* of Tasmania (morphological studies by Barker, 1982).

The mistletoe *Muellerina* (Loranthaceae) comprises five species of south-eastern Australia, all present at the MMO and two (*M. flexialabastra* and *M. myrtifolia*) locally endemic there (*Australia's Virtual Herbarium*, 2011). The sister of *Muellerina* is *Ileostylus* of New Zealand (Vidal-Russell and Nickrent, 2008).

Corokia (Argophyllaceae) has one species endemic to the MMO (at the eastern end of the McPherson Range), with the others on Lord Howe Island, New Zealand, the Chatham Islands and Rapa Island.

Dysoxylum fraseranum (Meliaceae) is a local endemic at the McPherson Range and is closely related to *D. bijugum* of New Caledonia, Vanuatu and Norfolk Island (morphological studies by Mabberley, 1998).

Vesselowskya (Cunoniaceae) is endemic to the MMO (Rozefelds *et al.*, 2001) and has its closest relatives in New Caledonia (*Cunonia*, *Pancheria* and *Weinmannia*).

The liverwort *Dinckleria* (Plagiochilaceae) includes two species (D. Glenny, personal communication, 12 May 2008):

1. A northern Tasman species: MMO and New Zealand (Stewart I. to Northland) (*D. fruticella*).
2. A southern Tasman species: Tasmania and New Zealand (Auckland, Campbell and Chatham Islands to Northland) (*D. pleurata*).

The two species show significant allopatry between their MMO–New Zealand and Tasmania–New Zealand areas; within New Zealand only the second species is present on the islands of the Campbell Plateau and Chatham Rise.

In related patterns, the puriri *Vitex lucens* (Lamiaceae) is one of the most characteristic trees of northern New Zealand. It forms a group (*Neorapinia*) with *V. collina* s.lat. of Norfolk Island, New Caledonia and southern Vanuatu, and *V. lignum-vitae* of the MMO and north-eastern Queensland (Mabberley, 1998; Pratt, 2013).

These groups are all present in mainland New Zealand (but not the Campbell Plateau), the MMO and sometimes further north, but are absent from south-eastern Australia. This distribution coincides with the Whitsunday volcanic province–Median Batholith (Fig. 4.13).

The highlands of Cairns and the MMO are each well known as biogeographic centres, and together form an area of endemism for groups such as *Trimenia moorei* (Trimeniaceae). The area between them also includes important endemism. In the Annonaceae, *Meiogyne hirsuta* is endemic to the Cairns area and its sister-group, *M. stenopetala*, is disjunct at the MMO and in Fiji (Thomas *et al.*, 2012b). The gap between Cairns and MMO is not absolute, but is neatly filled by the clade's sister-group, the two species of 'genus' *Fitzalania* (this proved to be nested in *Meiogyne*). The '*Fitzalania*' species range from the mainland near Fraser Island north to the Northumberland Islands, the Whitsunday Islands and Townsville.

Disjunct MMO–Cairns clades such as *Meiogyne hirsuta–M. stenopetala*, in which the gap is filled by a relative, constitute a typical 'leap-frog' pattern and can be explained by the loss of former connections either inland or out to sea. These would also account for groups such as *Trachymene cussonii* (Apiaceae), native to Vanuatu, New Caledonia and Australia, where it is restricted to the Capricorn-Bunker Islands (Batianoff *et al.*, 2009).

Large areas of subsided continental crust exist off the coast of central eastern Queensland. The main ones are the Queensland Plateau (off Cairns) and Marion Plateau (off a point south of Townsville). Before the opening of the Coral Sea, the Queensland Plateau adjoined south-eastern Papua New Guinea and the Louisiade Plateau. The Queensland plateaus are formed from continental crust that has been extended and thinned, and as a consequence of this they have lost buoyancy and subsided. The extension and rifting developed with the seafloor spreading and rifting of the Coral Sea that took place at 62–52 Ma, together with the opening of the Tasman basin (Gaina *et al.*, 1999). As the Coral Sea opened, there was uplift in eastern Queensland and this produced the present main divide.

The Cairns region in north-eastern Queensland

The area around Cairns includes Mount Bartle Frere and Mount Bellenden Ker, 50 km south of Cairns, and the Daintree area, 100 km north. There are many distinctive endemics, such as the liane family Austrobaileyaceae, basal in the circum-Pacific Austrobaileyales. Nevertheless, unlike Tasmania and the MMO, the Cairns region is not known to have any endemic taxa with global sister-groups.

Apart from being a centre of endemism, the Cairns region is also important as a regional break in Tasman–Coral Sea groups. For example, the basal clade in the global family Micropterigidae (Lepidoptera) splits into eastern and western branches, with distributions intersecting at Cairns (Gibbs, 2010):

> Tasmania to Cairns (*Tasmantrix*).
> New Zealand (northern North Island), New Caledonia and Cairns (*Zealandopterix*, *Aureopterix* and *Nannopterix*).

Oraniopsis and the Ceroxyleae (Arecaceae)

A palm endemic to north-eastern Queensland, *Oraniopsis*, has disjunct relatives to the west in Madagascar, the nearby Comoro Islands and India (fossil), and to the east in the Juan Fernández Islands and the northern Andes (Bolivia–Venezuela). These palms make up the Indo–Pacific tribe Ceroxyleae. Trénel *et al.* (2007) found that the pattern of cladogenesis in the tribe matches the breakup sequence of Gondwanan landmasses, with the phylogeny: (*Ravenea* (*Oraniopsis* (*Juania, Ceroxylon*))) corresponding to the area relationships: (Madagascar (Queensland (Juan Fernández Islands, South America))). Nevertheless, Trénel *et al.* (2007) suggested that divergence time analyses are incompatible with Gondwanan vicariance. They estimated that the tribe evolved at 42–24 Ma and so they rejected vicariance in favour of a 'Lemurian stepping stones' dispersal hypothesis. Their calibrations were based on the assumption that palms endemic on the Mascarenes and others on the Comoros were no older than the islands (Savolainen *et al.*, 2006). Two fossil calibrations were also used. All these will give minimum dates only; for example, while the Comoros formed at 5.7 Ma, the hotspot that produced them was active from ~60 Ma. Trénel *et al.* (2007) acknowledged this and, as an alternative, tried fixing the age of the tribe to 80 Ma. They rejected this, because it resulted in an age of the subfamily Ceroxyloideae (245–172 Ma) that was older than the presumed mid-Cretaceous origin of the palm family. Evolutionary rates are likely to have changed, though, and the idea of a mid-Cretaceous origin of palms is itself based on fossil calibrations.

A sequence of differentiation from Cairns to the MMO in an agamid lizard

The agamid *Diporiphora australis* is distributed in woodland of eastern Queensland, between the Cairns area and the MMO. Distribution and phylogeny follow a sequence with the basal clade in the north, the next basal clade lying further south, and so on (Edwards and Melville, 2010). The authors described this as 'sequentially younger divergences moving from north to south' (p.1550), which is accurate, but they also inferred 'sequential colonization from north to south across known biogeographic barriers' (p.1550), and this is unnecessary if the ancestor was already widespread. The authors found that the phylogeny matches a model of 'sequential dispersal' rather than 'simultaneous vicariance', but they did not consider a model of 'sequential vicariance', which is supported here. The phylogenetic breaks in this woodland taxon correspond to major biogeographic breaks also seen in rainforest groups, with the same biogeographic nodes expressed in the two different vegetation types (Edwards and Melville, 2010). This suggests that a general phylogenetic process, such as vicariance, underlies the current ecological division.

Breaks around Torres Strait and the Coral Sea

At the northern border of Australia there is another 'curious case of great differences', matching the break between south-western and south-eastern Australia that Hooker (1860) described. From east Asia to New Guinea, Good (1957) observed gradual floristic change, but across Torres Strait he recognized an abrupt change, with Australia appearing as a 'different world'. Good concluded that 'Nowhere else in the world is there so fundamental a biological change within a distance of less than 100 miles'. The phylogenetic break is often located somewhat north of Torres Strait itself, within the plains of southern New Guinea. For example, death adders (*Acanthophis*) comprise two clades, one in Australia and Merauke (southern New Guinea) and its sister in

New Guinea and Maluku Islands (Wüster *et al.*, 2005). The most diverse reptile genus in Australia, *Ctenotus* (Scincidae), occurs throughout mainland Australia, parts of Tasmania, and the Fly area of southern New Guinea, but does not occur elsewhere in New Guinea. As Maiden (1914) observed, the Fly River savanna is the only part of New Guinea that shows strong affinities with the flora of Australia, and 'It seems strange that the Australian flora has not taken greater possession of New Guinea'. Similar challenging biogeographic breaks occur between mainland New Guinea and the Bismarck Archipelago, and between mainland Australia and Tasmania.

Despite the great difference between Australia and the main part of New Guinea, some clades occur in both areas and these are often interpreted as the result of dispersal. For example, the snake *Pseudonaja textilis* (Elapidae) has a common distribution pattern: widespread in Australia and New Guinea. In their study of the species, Williams *et al.* (2008) did not sample from throughout the vast range, but found good evidence for three allopatric clades:

> South-eastern Australia.
>
>> Central-northern Australia and *cratonic* New Guinea (south-western New Guinea: Merauke).
>>
>> Northern and eastern Australia, and *accreted terranes* in New Guinea (Papuan Peninsula: Central, Milne Bay and Oro provinces).

A dispersal model would infer that the New Guinea populations are the result of two separate colonization events, but the pattern is also consistent with the ancestor of *Pseudonaja* inhabiting the New Guinea region before the genus or the island existed. A widespread ancestor is consistent with the high level of allopatry and the main phylogenetic break coincides with the margin of the Australian craton in south-central New Guinea.

A Coral Sea break in bandicoots

> A clade in bandicoots (Marsupialia: Peramelidae) is structured as follows (Westerman *et al.*, 2012):

> All Australia, Tasmania, local in New Guinea at the Fly River and SE Papuan Peninsula (*Perameles, Isoodon*).
>
> New Guinea – widespread through central and northern mountains (not in Fly River area), also southern Maluku Islands, New Britain and Cape York (*Peroryctes, Microperoryctes, Rhynchomeles, Echymipera*).

The two clades, one mainly in Australia and one mainly in New Guinea, overlap only at Cape York and the south-eastern Papuan Peninsula – that is, around the Coral Sea. This suggests that the original break between the two clades along with subsequent, minor overlap have occurred around this region. (For a break at Cairns, cf. Fig. 4.10.)

North-eastern Queensland and Coral Sea distribution in a honeyeater

The Coral Sea is an important biogeographic region, not just a barrier. The passerine bird *Meliphaga* (Meliphagidae) has two main clades, one distributed around the Arafura Sea (Fig. 4.9) and the other based around the Coral Sea, in north-eastern Queensland and south-eastern New Guinea (Norman *et al.*, 2007). Three of the four species in the Coral Sea clade are endemic to the region, while the fourth also occurs through New Guinea. A further Coral Sea species (from Tagula Island) has not been sampled.

The family Meliphagidae is most diverse in Australia, but *Meliphaga* itself is most diverse in New Guinea. Norman *et al.* (2007) found it was impossible to distinguish which of the two areas was the true centre of origin. Instead, the group may have evolved *in situ*, by vicariance, in the area now occupied by the Arafura and Coral Seas. The pattern in *Meliphaga* is similar to that of the elapid snake *Pseudonaja textilis*, discussed above; the northern Australia–Merauke clade is based around the Arafura Sea, while the north-eastern Australia–Papuan Peninsula clade is based around the Coral Sea.

In the Meliphagidae as a whole, the following phylogeny has been proposed (Nyári, 2011; NG = New Guinea):

> Australia, NG, Sulawesi, Marianas, Carolines, Solomons, Vanuatu, New Caledonia, Fiji, Samoa (not Tonga) (*Myzomela* and *Certhionyx niger*).
>> Australia, NG, New Zealand (*Acanthorhynchus* etc.).
>>> Australia, NG, New Caledonia, Fiji, Samoa, Tonga (*Philemon*, *Foulehaio*, etc.).
>>> Australia, NG (*Melipotes* etc.).
>>>> Australia, NG (*Lichenostomus* s.str., etc.).
>>>> Australia, NG (*Meliphaga*).

This study included a more intensive sampling of the family than the broader survey of Gardner *et al.* (2010), but only two of the clades had strong support. Groups that were not sampled in either study include *Myza* of Sulawesi, *Gymnomyza* of New Caledonia, Fiji and Samoa, and three monotypes from the Solomon Islands (*Stresemannia bougainvillei* of Bougainville, *Guadalcanaria inexpectata* of Guadalcanal and *Meliarchus sclateri* of San Cristobal). Further work is needed to resolve the phylogeny of the family, but the significance of Australia and New Guinea is evident. Each of the main clades occurs in both areas and three of them only occur there. Despite the extensive overlap, differences in the distributions of the clades suggest traces of the original allopatry. For example, the basal node in meliphagids separates two groups:

1. Australia, Melanesia, Sulawesi, Micronesia (*Myzomela* and *Certhionyx niger*).
2. Australia, Melanesia, New Zealand (the rest).

This points to the common meliphagid ancestor being widespread in the west Pacific before splitting into two groups in, say:

1. Northern Australia, northern Melanesia and Micronesia.
2. Southern Australia, southern Melanesia and New Zealand.

Following this differentiation, overlap has developed in Australia and Melanesia with range expansion in one or both clades. Both groups occur in mangrove forest, where the first group is represented by *Myzomela*, the second by *Philemon* and *Lichmera*. The overlap of the first two groups and subsequent meliphagid clades has been by range expansion in groups that were likely to have included mangrove associates.

5   The Tasman–Coral Sea region: a centre of high biodiversity

Huxley (1868, 1873) proposed four main biogeographic regions for terrestrial verte-brates: New Zealand; Australia plus New Guinea and the Philippines; South America north to Mexico; and the rest of the world. The scheme highlights the areas around the South Pacific and also the break between New Zealand and Australia. This last break coincides with the extinct spreading centres (mid-ocean ridges) that formed the Tasman and Coral Sea basins. The biogeographic and tectonic margin is also a distributional centre, as the lands and islands around it make up one of the most distinctive areas of endemism on Earth. This includes the eastern seaboard of Australia and the islands that fringe it, from New Zealand in the south to New Guinea in the north. The region has long been known for its strange vertebrates and these include the following.

- In amphibians, the New Zealand frog genus *Leiopelma*. The three extant species (Holyoake *et al.*, 2001) and three subfossil species (Worthy, 1987) form a group with *Ascaphus* of the north-western US (two species) and, perhaps, two fossil genera of Patagonia. This Pacific clade is sister to all other extant frogs (Pyron and Wiens, 2011).
- In reptiles, the New Zealand tuatara, *Sphenodon*. Together with its extinct relatives, the other members of Rhynchocephalia, *Sphenodon* is the sister-group of Squamata – lizards and snakes. Rhynchocephalia had a circumglobal range in the Mesozoic, but Cenozoic fossils and extant species are only known from New Zealand.
- In birds, the large, flightless ratites (Fig. 3.21). Three families existed in Australasia in historic times, compared with two in South America and one in Africa. The Palaeog-nathae, including the ratites and some fossil groups, are sister to all other living birds.
- In mammals, the monotremes. These egg-laying animals are sister to all the other mammals. They are known from New Guinea, Australia and, as fossils, in Patagonia. The three extant genera are:
  - *Ornithorhynchus* (platypus): one species in Tasmania and the eastern Australian seaboard.
  - *Tachyglossus* (spiny anteater): one species, the most widespread native mammal in Australia, present in all states as well as in New Guinea.
  - *Zaglossus* (an anteater, larger than *Tachyglossus*): three species endemic to New Guinea, fossils in Australia.

This chapter incorporates material previously published in the *Biological Journal of the Linnean Society* (Heads, 2009a), reprinted here by permission of John Wiley & Sons.

To summarize, there are four extant monotreme species in New Guinea, two in eastern Australia and one elsewhere in Australia. The group extends throughout Australia and is more widespread than the Tasman region endemics discussed next, but has its maximum diversity around the western margins of the Tasman Sea–Coral Sea. In New Guinea it is more diverse in the northern part of the island than in the southern part. Northern New Guinea is formed from accreted terranes – fault-bounded blocks with independent histories – while the southern region is part of the Australian craton. The diversity of monotremes in both sectors suggests a widespread distribution before accretion.

Another indication of mammal evolution in the Tasman region is provided by a Miocene mammal fossil from New Zealand which is neither marsupial nor placental (Worthy *et al.*, 2006). The fossil is 'nothing like' anything in the rich Australian record and there is 'little if anything' to suggest the animal was a good flyer or swimmer (p.19422). This taxon, along with monotremes, Leiopelmatidae, and many other endemics all imply early differentiation between New Zealand and Australia.

The Tasman–Coral sea region as a centre of biodiversity

Interest in the Tasman–Coral Sea area as a hotspot of biodiversity has been stimulated by molecular studies. These are showing that many endemic clades in the region have circumglobal sister-groups. Examples include *Amborella* of New Caledonia, basal in angiosperms, and Acanthisittidae of New Zealand, basal in the passerines. (As discussed in Chapter 1, a 'basal' group is just a small sister-group.) In the sister-group of passerines, the parrots (Hackett *et al.*, 2008), the basal group is again a New Zealand clade (*Nestor* plus *Strigops*). The recognition of these and other basal groups in the region is not a complete surprise, as the vertebrate groups of the region cited above have been recognized as unusual since the nineteenth century. Other areas in which many endemics have global sister-groups include Madagascar and Mexico, but an initial survey (Heads, 2009a) suggested there is a higher concentration around the Tasman–Coral Sea region than elsewhere. Fifty-five examples are now known and these are discussed below.

What is biodiversity?

The many Tasman–Coral Sea endemics with global sister-groups raises the following questions. What do these molecular findings mean for evolution in the region? What do the globally basal endemics represent? Assuming that they are ancestral is just as unwarranted as assuming that their large sister-groups are. But the quality of difference or otherness that the basal groups possess is important and related to that of biodiversity; perhaps *Amborella* and the acanthisittids represent high biodiversity. This in turn raises the challenging question of what biodiversity is and how it can be measured. Many writers have assumed that biodiversity can be measured by counting the number of species in an area, but the nature of the species is controversial and it is likely that not all described species are equivalent.

Species concepts

The species concept used here was proposed by Darwin (1859) – the species is just a point on a continuum of differentiation and has no special status. Species differ only in degree and not in kind from groups with other ranks, and a species in one group may not be equivalent to a species in another. Distinct geographic subspecies might be regarded as separate species by some, while a well-marked species with distinct geographic forms could be seen as a subgenus. The genetic and morphological distance between groups, as well as reproductive isolation, are byproducts of the evolutionary trajectory, not the cause of it.

Darwin (1859) accepted that species were natural groups but also argued that there is 'no fundamental difference between varieties and species' (p. 278), or between species and genera (p. 485). The evolutionary explanation freed one 'from the vain search for the undiscovered and undiscoverable essence of the term species' (p. 485). In particular, Darwin analysed the idea, supported in Buffon's creationist work, that reproductive isolation is the essence of the species (Mallet, 2010b). Darwin devoted a whole chapter ('Hybridism') to dissecting this defnition, later termed the 'biological species concept', and he concluded by rejecting it.

In their 'neo-darwinian' synthesis, Dobzhansky (1935) and Mayr (1942) argued that Darwin (1859) had failed to understand species and speciation. Mayr (1982: 269) suggested that Darwin had treated species 'purely typologically', in an essentialist way. Yet Mayr did not cite Darwin's explicit, detailed critique of the biological species concept and his rejection of this concept as essentialist. In contrast with Darwin's more evolutionary approach, Dobzhansky (1935), Mayr (1942) and their modern supporters (for example, Coyne and Orr, 2004) have argued that species are demarcated from subspecific varieties by sharp boundaries and are the only 'real' clades.

The views of the 'neo-darwinian' synthesis came to dominate the field. For at least 70 years biologists have accepted the claim that Darwin did not define species or have a useful concept of species, and that in the *Origin of Species* he 'failed to solve the problem indicated by the title of his work' (Mayr, 1963: 12, cf. Mayr, 1942; Coyne and Orr, 2004). In contrast, Mayr's approach has been praised: 'Hardly any one, until Ernst Mayr published his *Systematics and the Origin of Species* in 1942, really understood that the species was a unique unit, having quite a different significance in the evolutionary system from that of any other category. . . .' (Hardy, 1965). Philosophers of science have also accepted Mayr's idea, overlooked Darwin's analysis and proposed that species, unlike subspecies and subgenera, are individuals, not just classes. In this approach, Darwin's views on species and speciation 'are pretty wonky' (Coyne, 2009).

In contrast with the Modern Synthesis approach, Mallet (2010a) suggested that Darwin did know what he was talking about, and Mallet questioned why Dobzhansky and Mayr 'felt impelled to develop a new, anti-Darwinian definition' of species. Mallet (2010a) pointed out that it is unfair to ignore Darwin's critique of hybrid sterility as the essence of the species, and that Mayr 'paradoxically seems to have promoted a new kind of essentialism. His theoretical concept of species reintroduced a reproductive isolation essence similar to the one Darwin had tried to banish. . . . Deviations of actual organisms from idealized reproductive isolation thereby became less important than the underlying truth of the theoretical species concept . . .'. Mallet (2008) wrote that 'Generations of evolutionary biologists for decades afterwards were brought up on Mayr's textbooks, and this . . . was to cause a catastrophic delay in the progress of understanding speciation'. As Mallet (1995) concluded, 'the modern species concept is an updated, genetic version of Darwin's', whereas the neo-darwinian species concept 'Owes nothing either to genetics or to darwinism'.

In this book it is the evolutionary trajectory, not a particular point along it, that is of interest. Whether a group is treated as a species, subspecies or genus is not relevant to the discussion; instead, the focus is on the process of differentiation in space and time. Darwin's '*Origin of Species*' was written for a general audience and the title is short-hand for 'origin of varieties, species, genera, and natural groups of any rank'. Darwin's main concern was the process, not the particular byproducts, as evolution itself erased any absolute difference between the ranks, and even between rank and no rank.

Ideas on species and speciation are tied to concepts of the spatial component of evolution. Mayr (2004) thought that the problem of the origin of biodiversity had been solved in the early twentieth century, not by the first geneticists (whose work he rejected), but by the Bavarian geographer and naturalist, Moritz Wagner. Mayr often emphasized that his own ideas on the origin of biodiversity (including founder effect speciation; Mayr, 1954) were based on Wagner's *Migrationstheorie* (e.g. Wagner, 1873). In this model, a small founder population moves to a new, isolated locality where a burst of genetic variation triggered by the new environmental conditions leads to speciation. In contrast with Mayr, Darwin regarded Wagner's work as 'increasingly fanatical and scientifically naïve', and described it in his notes as 'most wretched rubbish' (Sulloway, 1979). In Wagner's theory and in the Modern Synthesis, evolutionary change involved acquisition of variation caused by changed 'conditions' and 'needs' (as in Lamarckism). Darwin, though, was well aware of the natural variation in populations and did not need the *deus ex machina* of changed 'conditions' to provide variation (Mallet, 2010a).

Phylogenetic biodiversity

In practice, a systematist's first survey of an area is often a simple species count, using conventional species delimitations. Yet while species counts give a useful preliminary summary, they will not reveal many fundamental aspects of biodiversity and can be misleading. The idea that biodiversity is 'synonymous with species richness' (Hubbell, 2001) is too narrow. One of the most cited papers in biogeography included south-western Australia and New Zealand in a list of hotspots, but no area from Tasmania, eastern Australia or New Guinea (Myers *et al.*, 2000). This is because the hotspots were defined using numbers of endemic species (and extent of habitat loss), rather than a broader measure of phylogenetic diversity.

The splitting and 'taxonomic inflation' that has occurred over the last few decades means that species described in recent years often represent less biodiversity than those outlined earlier, and so, again, not all 'species' are equivalent. The trend to splitting is continuing despite the obvious dangers (Padial and de la Riva, 2006). It has been suggested that in molecular studies of large genera: 'The overwhelming trend has been documentation of polyphyly (and thus genus-level splitting)' (Voelker *et al.*, 2007). Still, polyphyly can be dealt with just as easily by lumping as by splitting, as in the treatment of *Turdus* by Voelker and colleagues.

Even apart from problems of taxonomic inflation, species are not all equal in phylogenetic terms. If a species is the only one in its family, it can represent more diversity than a species of, say, Asteraceae or Curculionidae. Branches from more basal nodes in a phylogeny each represent higher biodiversity than branches from more distal nodes. In a regional example, Malesia (Malaysia, Indonesia, the Philippines and New Guinea) has less than half the number of plant species (42 000) than are found in the Neotropics (90 000), but has 18 more families (Prance, 1994) and so it could be more diverse overall. Rodrigues and Gaston (2002) cited the *sibiricus* group of bumblebees as having highest species richness in South America, but higher phylogenetic diversity in Asia. In *Dactylorhiza* (Orchidaceae), the main centre of diversity was thought to be the British Isles and Germany, because the highest species numbers occur there, but Pillon *et al.* (2006) found greater phylogenetic diversity around the Mediterranean Basin and the Caucasus. They concluded that species numbers 'were not an appropriate measure of diversity'. The south-western Cape region of South Africa has a huge number of plant species, but many belong to just a few, related genera. Forest *et al.* (2007b) concluded that aiming to conserve phylogenetic diversity in the region is a better strategy than relying on simple species diversity.

There are now many species-based methods used for assessing biodiversity levels and conservation priorities (Isaac *et al.*, 2007). These use threatened species, restricted-range endemic species, 'flagship', 'umbrella', 'keystone', 'landscape', 'focal' and 'indicator' species. The 'key biodiversity areas' approach (Eken *et al.*, 2004) has been championed at many high profile conservation meetings and sounded promising, but, as Knight *et al.* (2007) argued, it has 'significant technical limitations that render it unable to identify the most globally important areas for conservation action . . .'. The main problem with all these methods is their sole focus on species and the treatment of all species as equivalent.

Critical early work on biodiversity assessment recognized that not all species represent equal diversity. This work introduced phylogenetic diversity measures that were characterized as either 'node-based' (Vane-Wright *et al.*, 1991) or 'branch length- or feature-based' (Faith, 1992).

Faith (1992) suggested that the phylogenetic diversity of a clade equals the total length (number of characters) of all its branches. By focusing on 'feature' diversity rather than species, this measure deals in a more direct way with the differentiation underlying the phylogeny. Nevertheless, just as species are not all of equal biodiversity significance, not all characters (genetic or morphological) are equal. Relying on branch length can obscure important aspects of a group and lead to the neglect of isolated, basal clades with few unique features (autapomorphies). In conservation, should we aim to protect individual clades that have many distinctive autapomorphies? Or should we try to preserve the overall phylogenetic and biogeographic structure, and thus treat small basal groups at the same level as their diverse, autapomorphous, sister-groups? A single 'species' such as *Amborella trichopoda* represents much less branch length than its sister-group, the rest of the angiosperms, but as sisters the two are equivalent in phylogenetic terms and so a diversity measure that does not depend on branch length would be desirable.

Vane-Wright *et al.* (1991) argued that simple species number is an inadequate measure of biodiversity. They introduced a measure of 'taxonomic distinctness' and based their approach 'exclusively on cladistic procedures'. Their method has the advantage that it incorporates comparison with the sister-group. Yet, in practice, their measure of phylogenetic diversity still relies on numbers of species (or terminals): a species basal to three others is worth more than a species basal to two others and so on. The basal angiosperm, the New Caledonian *Amborella*, would be worth more than the basal passerines, the New Zealand Acanthisittidae, because there are more angiosperm species than passerine species. Nevertheless, a bird species cannot be assumed to be equivalent to an angiosperm species, and a method that did not rely on species numbers (or rank in general) would be preferable.

Vane-Wright *et al.* (1991) did consider that in 'one possible approach' sister-groups could be given equal weight. Although this appears to be an obvious solution, they argued that it had the 'undesirable' effect of overweighting basal taxa. They gave the example of the two species of tuatara (*Sphenodon*) and wrote that to accord them equal status to that of the 6800 living squamates (lizards and snakes) is 'unreasonable'. Yet in phylogeny and taxonomy these sister-groups are given equal status – why not in biodiversity studies? If the species-centrism of the Vane-Wright *et al.* approach is dropped, the first priority in phylogeny-based conservation of these animals would be to ensure that one tuatara population and one squamate population are protected. This does not seem unreasonable – squamates all share many morphological features not found in tuatara (Gauthier *et al.*, 1988, listed 69). The same principle can be applied to the clades of lizards, snakes and others. These views imply that we should aim to conserve the main phylogenetic structure of a group – that is, approach the biota from the base of the tree up, as well as from the top down, as in species conservation.

Phylogenetic diversity measures have been modified in different ways, for example to incorporate range size and geopolitical considerations (Soutullo *et al.*, 2005), and threat status (Redding and Mooers, 2006). Posadas *et al.* (2001) produced an interesting modified version of Vane-Wright *et al.*'s (1991) measure, in which the size of a group (number of species or terminals) is not relevant. Faith *et al.* (2004) criticized Posadas *et al.* for not using branch length, but this was the whole point of their approach. In any case, Posadas *et al.* (2001) concluded by identifying central Chile (Coquimbo to Ñuble) as the area with highest phylogenetic diversity in southern South America. Faith *et al.* (2004) protested that the South Americans' method 'could be detrimental to biodiversity conservation'. Nevertheless, the same area of central Chile (27°–37°S, especially 33°–34°) has maximum species diversity in groups such as Chilean Asteraceae (Moreira-Muñoz and Muñoz-Schick, 2007) and *Alstroemeria* (Alstroemeriaceae; Muñoz-Schick and Moreira-Muñoz, 2003). The area also includes several groups with global sisters and so the high level of phylogenetic diversity there is well substantiated.

Many different measures of biodiversity, whether simple species counts or measures of phylogenetic diversity, have been used to good effect depending on the details of the application. Different measures sometimes show the same pattern. For example, in evaluations of conservation priority areas for Crustacea in Australia and America, rankings based on species and genus richness agreed to a large extent with rankings based on taxonomic, phylogenetic and genetic diversity (Pérez-Losada and Crandall, 2003; cf. Rodrigues and Gaston, 2002). However, in studies on *Piper* (Piperaceae) in Colombia (the country has about 400 species), different measures of diversity highlighted different areas (Jaramillo, 2006). In the Chocó region of western Colombia the genus shows high total species numbers concentrated near Buenaventura in the central part of the region, high numbers of endemic species there and in the south (Nariño Dept.), but highest phylogenetic diversity from Buenaventura northwards to the Atrato River. Jaramillo (2006) identified the last phenomenon as historical, involving a major biogeographic/phylogenetic break between northern South American and Central American floras.

Phylogenetic diversity and endemism

The discussion presented above suggests that assessing the biodiversity of an area should take into account both the phylogeny and biogeography of the biota. Areas show a range of 'distinctness' or biodiversity:

1. Area includes no endemism (e.g. urban or other anthropogenic environments).
2. Area includes minor population differentiation (most natural areas).
3. Area includes locally endemic species (many areas).

Fewer areas have locally endemic higher taxa. At this point in assessing the local endemics, it is desirable to avoid relying on the absolute 'degree of difference' (branch length) of the endemics and whether an endemic is a genus, family, order, etc. The significance of the first is problematic, as mentioned above, and the second, in large part, reflects academic traditions in particular groups. Instead, the spatial relations of the endemic clade can be used.

4. Area includes a locally endemic clade with a local or regional sister-group.
5. Area includes a locally endemic clade with an intercontinental sister-group.
6. Area includes a locally endemic clade with a global (pantropical, pantemperate, or more or less cosmopolitan) sister-group.

A few regions, including the Tasman–Coral Sea area, Madagascar–South Africa, and western Mexico to northern Colombia, require a category of their own:

7. Area includes many locally endemic clades with more or less cosmopolitan sister-groups.

The significance of groups and their areas for evolution, ecology and conservation is taken here to depend on a combination of phylogeny and biogeography. *Amborella* and the New Zealand wrens are not significant because of their species numbers or their branch length. They both look rather ordinary and combine characters found elsewhere in their groups, rather than having many distinctive characters of their own. They are important because of their phylogenetic and geographic context – their global, diverse sister-group, together with their restricted distribution.

Taking the lead from these two groups, in the approach used here, tree topology (not branch length) and biogeography are used to retrieve taxa and areas of global significance. Locally endemic taxa, regardless of rank, that are sister to cosmopolitan groups represent the highest possible values of a variable φ, where φ of a group = the size of the sister-group's range/size of the group's range. A preliminary survey of locally endemic groups with global sisters suggested high regional levels of φ around the Tasman and Coral Sea basins, with a total regional value possibly even higher than in Madagascar/South Africa or western Mexico (Heads, 2009a).

Amborella of New Caledonia has a value of φ equal to the total area of angiosperms/total area of *Amborella*, or the area of land on Earth (not including Antarctica)/about one-third of New Caledonia, $\approx 14\,000$. The New Zealand wrens occupying, say, 75% of New Zealand, and with a global sister, have a value of $\varphi \approx 4000$. Reliance on the number of species in angiosperms and birds is avoided, as is comparing angiosperm species with bird species. The ranks given to the endemic groups, well-known for their subjectivity, are also irrelevant. The levels of φ for these two groups are approaching the maximum possible value (this would be held by a clade present at only one site and with a global sister). Many taxa will have a value of about 1.

Davis *et al.* (2008) used a biodiversity measure incorporating a species' sequence divergence weighted with the inverse of its range size. This is similar to the approach used here, although Davis *et al.* used the range of the group itself rather than in comparison with its sister. Rosauer *et al.* (2009) developed this theme and proposed a measure, also similar to that used here, termed 'phylogenetic endemism'. This combines phylogenetic diversity (based on branch-length) and weighted endemism (the geographic restriction of a group). It avoids assuming that species in one group are equivalent to species in another and uses the geographic range size of group and the sister-group to rate evolutionary diversity. It differs from the measure used here only in its use of branch length. Mooers and Redding (2009) commented 'to the extent that geographically rare species are at greater risk of extinction . . . and that phylogenetically rare [basal] species contribute disproportionally to overall biodiversity . . . it would seem reasonable to formally integrate the two processes . . . Rosauer *et al.* (2009) do just that . . .'. In a paper titled 'The rarest of the rare', Cadotte and Davies (2010) also stressed the importance of both evolutionary diversity and restricted geographic range, and they discussed possible ways of synthesizing this information.

Globally basal endemics in the Tasman–Coral sea region

The following list (from Heads, 2009a, with additions and corrections) cites 55 endemic clades in the Tasman–Coral Sea region that have circumglobal sisters. In other words, these groups show maximum levels of phylogenetic diversity (φ) in the region and so they have outstanding significance for biogeography, ecology, evolution and conservation. The individual centres of endemism occupied by the groups are shown in Figure 5.1

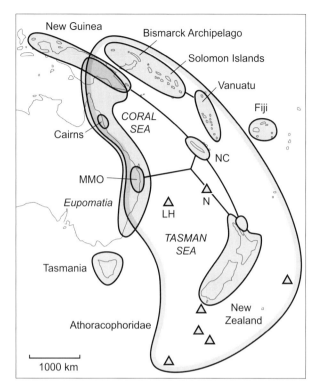

Fig. 5.1 The Tasman–Coral Sea region, with the centres of endemism of the groups with global sisters. LH = Lord Howe Island, MMO = McPherson–Macleay Overlap; N = Norfolk Island, NC = New Caledonia, triangles = other islands. (From Heads, 2009a, reprinted by permission of John Wiley & Sons.)

along with the range of the athoracophorid land snails, which more or less define the region (although the family is not in Tasmania, western New Guinea or Fiji).

Flora

1. In red algae, *Apophlaea* is endemic to New Zealand (main islands, subantarctic islands and Chatham Islands). Its sister is a clade of *Hildenbrandia* that is widespread through north and south temperate zones (Sherwood and Sheath, 1999, 2003).
2. In red algae, *Dione* of the South Island, New Zealand, is sister to the rest of the global order Bangiales (Sutherland *et al.*, 2011).
3. In red algae, *Minerva* of New Zealand is sister to the rest of the Bangiales except *Dione* (Sutherland *et al.*, 2011).
4. In red algae, *Lysithea* of the New Zealand subantarctic islands is basal to a global clade of Bangiales (Sutherland *et al.*, 2011). It is recorded from Macquarie, Auckland and Antipodes Islands (W. Nelson, personal communication, 23 November 2011).
5. In red algae, *Lembergia* Saenger (= *Lenormandia allanii* Lindauer) is known from the Three Kings Islands and Aupouri Peninsula in the far north of New Zealand.

It is sister to *Sonderella* of Victoria, and the two genera form a separate tribe, the Sonderelleae in Rhodomelaceae (Phillips, 2001). Another study found that *Sonderella* is basal in Rhodomelaceae, the largest family of red algae and distributed worldwide (*Lembergia* was not sampled) (Choi *et al.*, 2002).

6. In liverworts, *Goebeliella* occurs in New Zealand (main islands, Auckland and Chatham Islands) and New Caledonia. It is sister to a diverse, cosmopolitan group: Radulaceae, Frullaniaceae, Jubulaceae and Lejeuneaceae (He-Nygrén *et al.*, 2006; D. Glenny, personal communication, 12 November 2007).

7. *Dinckleria* ('*Proskauera*') of New Zealand and south-eastern Australia (cited above) is sister to the cosmopolitan *Plagiochila*, the largest liverwort genus (Heinrichs *et al.*, 2006; Engel and Heinrichs, 2008). In New Zealand, *Dinckleria* is in the main islands, the subantarctic islands and Chatham Islands. In Australia the genus is in Tasmania and at the McPherson–Macleay Overlap (MMO).

8. In the cypress family, Cupressaceae, after *Amentotaxus*, *Taiwania* and *Cunninghamia* branch off, *Athrotaxis* of Tasmania (fossil on mainland Australia) is sister to the rest of the family (Gadek *et al.*, 2000).

9. In angiosperms, *Amborella trichopoda* is a shrub or small tree of New Caledonia and is basal in angiosperms (Burleigh *et al.*, 2011; Soltis *et al.*, 2011). Some studies have proposed instead that *Amborella* is sister to Nymphaeales (a worldwide group). Sequences in the nuclear gene *Xdh* suggested that *Ceratophyllum* (Ceratophyllaceae) is sister to the rest of the angiosperms, followed successively by Amborellaceae, Nymphaeaceae and Austrobaileyales as sisters to the remaining angiosperms (Morton, 2011). Hovever, Stevens (2012) gave a useful discussion and concluded that *Amborella* can still be regarded as sister to the rest of the angiosperms.

10. *Eupomatia* is a liane or scandent shrub. One species, *E. laurina*, occurs in eastern Australia (east of the Great Dividing Range) and Papua New Guinea; the other, *E. bennettii*, is restricted to the MMO. *Eupomatia* is sister to Annonaceae, a pantropical family of lowland rainforest (Sauquet *et al.*, 2003).

11. In monocots, *Maundia* (Alismatales) is endemic to the MMO and sister to a large cosmopolitan clade (Posidoniaceae, Ruppiaceae, Cymodoceaceae, Zosteraceae and Potamogetonaceae) (Stevens, 2012).

12. *Blandfordia* (Asparagales) of south-eastern Australia (MMO to Tasmania) is sister to Lanariaceae of South Africa, Asteliaceae of the Mascarene Islands, Pacific islands and Chile, and Hypoxidaceae, of the seasonal tropics (Graham *et al.*, 2006).

13. *Doryanthes* (Asparagales) has two species of massive, agave-like plants at the MMO, one of which is also in north-eastern Queensland. The genus is sister to a cosmopolitan clade comprising Iridaceae, Xeronemataceae (next) and others (Janssen and Bremer, 2004).

14. *Xeronema* (Asparagales) comprises *X. callistemon*, found on small islets off northern New Zealand (the Poor Knights; Hen and Chicken Islands), and *X. moorei* of New Caledonia. It is basal to a large, cosmopolitan clade of Asparagales (Amaryllidaceae, Agavaceae) (Fay *et al.*, 2000; Janssen and Bremer, 2004).

15. *Isophysis* (Asparagales) from southern and western Tasmania is basal in the more or less cosmopolitan Iridaceae (Goldblatt *et al.*, 2008).

16. In orchids (Asparagales), *Pachyplectron* of New Caledonia is sister to the rest of the subtribe Goodyerinae, an almost cosmopolitan group (Cribb *et al.*, 2003).

17. There are three main clades in Liliales. One is Campynemataceae of Tasmania and New Caledonia (Fay *et al.*, 2006). The others are the trans-Pacific Corsiaceae and a large, cosmopolitan clade with the eight remaining families. The phylogeny of the three clades is still unresolved and in two of the three possibilities Campynemataceae would be sister to a cosmopolitan clade.

18. *Petermannia* is a climber with a woody rhizome found at the MMO and is sister to a cosmopolitan clade of Liliales (Colchicaceae, Alstroemeriaceae and Luzuriagaceae) (Fay *et al.*, 2006).

19. In grasses (Poaceae), the trans-Tasman *Chionochloa* (New Zealand, Lord Howe Island and Mount Kosciuszko in south-eastern Australia) is basal in a global group (most of the subfamily Danthonioideae, see Chapter 1; Linder *et al.*, 2010).

20. In eudicots, *Ranunculus* sect. *Pseudadonis* (Ranunculaceae) is diverse in the New Zealand mountains, and also has one species on Mount Kosciuszko and one in Tasmania. Sect. *Pseudadonis* is sister to *Hecatonia* + *Batrachium*, found worldwide except Antarctica (Hörandl and Emadzade, 2012).

21. *Tetracarpaea* is a subalpine shrub of Tasmania and is sister to the cosmopolitan clade Haloragaceae + Penthoraceae (Saxifragales) (Moody and Les, 2007).

22. Berberidopsidaceae comprise two genera: *Streptothamnus* at the MMO and *Berberidopsis*, with one species at the MMO and one in central Chile (by Concepción) (Moreira-Muñoz, 2007; Stevens, 2012). The family is sister to *Aextoxicon*, also in central Chile (Chiloé Island to about Coquimbo). The two families (forming Berberidopsidales) are basal to a large, cosmopolitan clade, the Caryophyllales + the asterids (Stevens, 2012).

23. *Drosera arcturi* (Droseraceae) of New Zealand, south-eastern Australia and Tasmania is sister to all other *Drosera* species (except *D. regia*), a global clade (Rivadavia *et al.*, 2003).

24. *Canacomyrica* is endemic to New Caledonia and is sister to the rest of the subcosmopolitan family Myricaceae (Herbert *et al.*, 2006). *Canacomyrica* is extant only in New Caledonia, but fossil pollen in New Zealand (Eocene–Miocene) has been attributed to it (Lee *et al.*, 2001). The rest of the family Myricaceae is more or less circumglobal, but absent from Australia, eastern New Guinea, New Caledonia, New Zealand and southern South America.

25. The tree *Balanops*, sole genus of Balanopaceae (Euphorbiales), occurs in New Caledonia (seven species), Vanuatu and Fiji (one species) and north-eastern Queensland (one species). It is sister to a large pantropical complex comprising Chrysobalanaceae, Dichapetalaceae, Euphroniaceae and Trigoniaceae (Stevens, 2012).

26. *Sparattosyce* (New Caledonia) and *Antiaropsis* (New Guinea) make up the basal clade in the pantropical tribe Castilleae (Moraceae) (Zerega *et al.*, 2005).

27. In the pantropical ebony genus *Diospyros* (~500 species, Ebenaceae), a clade of five species in New Caledonia and eastern Australia (Cairns to Sydney) is sister to all species except *D. maingayi* and *D. puncticulosa* (Duangjai *et al.*, 2009).

28. *Teucridium* (Lamiaceae) of central New Zealand (North and South Islands) is sister to *Oncinocalyx* of the MMO (Fig. 4.12; Cantino *et al.*, 1999). This pair is sister to *Teucrium* (~300 species),which is subcosmopolitan but absent in New Zealand.
29. *Pennantia* is in New Zealand (South Island, North Island, Three Kings Islands), Norfolk Island and eastern Australia (eastern New South Wales, disjunct in north-eastern Queensland) (Fig. 3.23; Gardner and de Lange, 2002). The genus is sister to a cosmopolitan clade including Torricelliaceae, Griseliniaceae, Pittosporaceae and Apiaceae/Araliaceae (Stevens, 2012).
30. Myodocarpaceae of New Caledonia and the Coral Sea region (Fig. 3.23) are sister to the cosmopolitan Apiaceae (Tank and Donoghue, 2010; Stevens, 2012).

Fauna

1. In molluscs, Athoracophoridae are a family of terrestrial slugs and are sister to the cosmopolitan Succineidae (Wade *et al.*, 2006). Athoracophorids occur in New Zealand (including the subantarctic and Chatham Islands; Burton, 1963, 1980), New Caledonia (these two countries have most of the species), Melanesian islands, Papua New Guinea mainland ('Schroder Mountains' – probably Schrader Moun-tains) and eastern Australia (Queensland and south to Woolongong, south of Sydney).
2. The centipede *Craterostigmus* has one species in Tasmania and the other in New Zealand (North, South and Stewart Islands; Edgecombe and Giribet, 2008). Some parts of the genome show *Craterostigmus* as basal in centipedes, others have it near the base and still sister to a cosmopolitan group (Giribet and Edgecombe, 2006a).
3. The cicada family Tettigarctidae (Hemiptera) is extant in south-eastern Australia and Tasmania but also includes Mesozoic fossil representatives worldwide. It is sister to the worldwide Cicadidae (Grimaldi and Engel, 2005; Cryan and Urban, 2011).
4. In aphids (Hemiptera: Aphididae), *Aphis coprosmae* of the South Island, New Zealand, has a worldwide sister-group (Kim *et al.*, 2011).
5. Another clade of New Zealand aphids, comprising *Paradoxaphis*, *Aphis healeyi* and *A. cottieri* ('southern hemisphere clade'), also has a worldwide sister-group (Kim *et al.*, 2011).
6. In the littoral beetle *Cafius* (Staphylinidae), a clade of four mainland New Zealand species has a cosmopolitan sister-group (in two of the three phylogenies presented; Jeon *et al.*, 2012; distributions from K.-J. Ahn, personal communication, 26 Decem-ber, 2012).
7. Lepidoptera have the phylogeny (Kristensen *et al.*, 2007):

Micropterigidae: worldwide.

Agathiphaga, one species in north-eastern Queensland and one in the Solomon Islands, Vanuatu and Fiji. Feeding mostly on Araucariaceae.

All other Lepidoptera.

Agathiphaga of the Coral Sea region has a worldwide sister-group.

8. In the worldwide Micropterigidae, the basal clade comprises four genera (*Tasmantrix*, *Aureopterix*, *Zealandopterix* and *Nannopterix*) endemic to Tasmania and east coast Australia (MMO north to Daintree), New Caledonia and northern North Island, New Zealand (Gibbs, 2006; 2010).
9. After the clade in 8 is separated from other Micropterigidae, *Sabatinca* s.str. of New Zealand and New Caledonia is sister to the rest (Gibbs, 2006; 2010).
10. In other Lepidoptera, the family Mnesarchaeidae is endemic to New Zealand (North and South Islands) and is sister to the more or less global group Hepialoidea (ghost moths; not in the Caribbean, West Africa or Madagascar) (Gibbs, 1990; 2006; personal communication, 7 August 2007; J.R. Grehan personal communication, 5 October 2007).
11. In Hymenoptera, Maamingidae is endemic to the North and South Islands of New Zealand (Early *et al.*, 2001) and is sister to the global family Diapriidae, including Mymarommatidae (Munro *et al.*, 2011) (or to the global Diapriidae including Monomachidae; Castro and Dowton, 2006).
12. *Cheimarrichthys* is a fish endemic to New Zealand (North and South Islands). It is diadromous, spending part of its life cycle in freshwater and part in the sea, near the coast. It is sister to a diverse, global clade of marine fishes, Leptoscopidae + Pinguipedidae (Smith and Craig, 2007).
13. In reptiles, the New Zealand tuatara *Sphenodon* (with its widespread, extinct relatives the Rhynchocephalia) is sister to extant lizards and snakes (Squamata).
14. In ducks, the brown teals are a New Zealand group (*Anas chlorotis*, *A. aucklandica* and *A. nesiotis*) that is sister to a worldwide clade including mallards (Johnson and Sorenson, 1999).
15. In Gruiformes s.str., the adzebill (*Aptornis*) was a flightless, turkey-sized bird of the North and South Islands that went extinct in historical times. It forms a clade with Rallidae (cosmopolitan) and Heliornithidae (pantropical). The three groups form a tritomy in analyses that have included a small amount of molecular data from *Aptornis* (Houde, 2009). Whether *Aptornis* is sister to Rallidae, Heliornithidae, or to both, it is sister to a globally widespread group.
16. In other Gruiformes s.str., *Megacrex* (*Amaurornis ineptus*) of New Guinea (lowland rainforest, including mangrove) is sister to a global group of Rallidae (*Rallus* plus *Gallirallus* s.lat.) (Kirchman, 2012).
17. In the seabird *Puffinus* (Procellariidae), a clade of North and South Islands (*P. huttoni*, *P. gavia* and, based on morphology, the fossil *P. spelaeus*) is sister to a group of 23 species and subspecies found widespread globally (J. J. Austin *et al.*, 2004).
18. In parrots, two New Zealand genera form a clade that is sister to all other parrots (Psittaciformes) (Wright *et al.*, 2008; Schweizer *et al.*, 2010). The two genera are *Strigops* (kakapo; South Island, Stewart Island) and *Nestor* (kaka and kea; North, South and Stewart Islands; present on Norfolk and Chatham Islands in historical times, but now extinct there).

19. Aegothelidae (Aegotheliformes; owlet–nightjars) are endemic to the Tasman–Coral Sea region: New Guinea (including the D'Entrecasteaux Archipelago and the Maluku Islands but not the Bismarck Archipelago), New Caledonia, eastern Australia, Tasmania and New Zealand (extinct in the last locality by about 1200 A.D.). The family is sister to a cosmopolitan clade, the Apodiformes (swifts and hummingbirds) (Hackett *et al.*, 2008; Pacheco *et al.*, 2011).

20. In passerine birds, the basal group is the New Zealand wren family Acanthisittidae (Slack *et al.*, 2007; Hackett *et al.*, 2008; Cracraft and Barker, 2009). Four species were extant in historical times and fossil representatives are also known, but only from New Zealand (Tennyson and Martinson, 2006; Worthy *et al.*, 2010).

21–23. Basal Passerida and Corvida. The two, diverse, worldwide clades of passerines are the Passerida and the Corvida (the latter is also known as 'crown Corvida' or 'core Corvoidea'). In the Passerida the four main branches (Irestedt and Ohlson, 2008) have the following distributions:

1. **New Guinea region**: Melanocharitidae (*Melanocharis, Oedistoma, Toxorhamphus*). Widespread on the mainland, islands around the Bird's Head Peninsula (Waigeo, Japen, Misool and Aru) and the D'Entrecasteaux Archipelago.

2. **New Guinea**: Cnemophilidae. Axial ranges, but not on the Bird's Head Peninsula.

3. **New Zealand**: Callaeidae (= Callaeatidae) plus *Notiomystis* (Ewen *et al.*, 2006; Driskell *et al.*, 2007).Widespread in North, South and Stewart Islands until historical times (now rare).

4. Global: all remaining Passerida.

The phylogeny for the groups given by Irestedt and Ohlson (2008) is: 1 (2 (3 + 4)). In other studies, the first three groups appeared instead as a clade, and basal, not in Passerida, but in its sister-group, the worldwide Corvida (Barker *et al.*, 2004). Jønsson *et al.* (2011a) showed groups 1 and 2 as a basal grade in Corvida; 3 was near basal in Passerida. Whether groups 1–3 form a clade or a grade, and whether they are basal in Passerida or in Corvida, these birds highlight the New Zealand–New Guinea sector as a primary centre of differentiation in the passerines. None of the three basal clades occurs on the Bismarck Archipelago, a distinctive absence seen in many groups that are diverse on the New Guinea mainland.

24. *Eulacestoma* (Passeriformes) of the central New Guinea mountains is sister to a global clade (Dicruridae, Corvidae, etc.; Jønsson *et al.*, 2011a).

25. *Ifrita* (Passeriformes) of the central New Guinea mountains is sister to a global clade (Corvidae etc.; Jønsson *et al.*, 2011a).

Other Tasman groups in which traditional morphology suggests a global sister include the glow-worm genus *Arachnocampa* (Diptera: Keroplatidae). This inhabits rainforest gullies and wet caves on both sides of the Tasman. It has two main clades, one in New Zealand (North and South Islands) and one along the eastern seaboard of Australia from Tasmania to northern Queensland (Baker *et al.*, 2008). The genus makes up the Arachnocampinae, one of four subfamilies in Keroplatidae. Of the other subfamilies, the Keroplatinae, Macrocerinae, are each diverse and worldwide, while Sciarokeroplatinae comprises a single species from China (Evenhuis, 2006). It seems likely that the sister-group of *Arachnocampa* will be a worldwide clade.

Interpretation of the Tasman–Coral centre

The localities of the basal angiosperm in New Caledonia and the basal passerine in New Zealand are part of a more general pattern, as the Tasman region is rich in clades with global sisters. What is the reason for this apparent concentration of basal groups in the Tasman and other regions, such as Madagascar and Mexico? These areas could represent: (1) *centres of survival*; (2) *centres of origin* from which their widespread sisters have dispersed; or (3) *centres of differentiation* in widespread groups.

Basal centres as centres of survival

Three of the Tasman endemics with global sisters are relics of groups that were more widespread in the past but have gone extinct in all the other localities. The New Zealand tuatara *Sphenodon* has fossil relatives (the other Rhynchocephalia) that had a circum-global distribution in the Mesozoic, as does the cicada *Tettigarcta*, extant in south-eastern Australia. The plant *Canacomyrica* is extant only in New Caledonia, and in addition there are Cenozoic fossils in New Zealand. There is no fossil evidence from anywhere in the world that any of the 52 other basal Tasman endemics ever occurred outside the region.

Given that the Tasman is a centre of survival for relic groups such as the tuatara and *Tettigarcta*, what is the reason for survival in this particular region? Chance, local ecology and the location of prior diversity have all been suggested as factors determining the location of relictual groups, and the three can be considered in turn.

1. **Chance**. Any distribution can be explained by chance survival in the occupied local-ities and extinction everywhere else. It is impossible to disprove chance survival in a general sense, but it would be incompatible with a strong concentration of globally basal groups in just a few areas.
2. **Local ecology**. It is possible that relic groups have survived in particular regions because of environmental factors. This is unlikely, though, as the main areas of globally basal groups – for example, the Tasman region (tropical and temperate), Madagascar–South Africa (tropical and temperate) and western Mexico (tropical and subtropical) – are so heterogeneous in their range of environments and each shares similar ecological conditions with many other areas.
3. **Locations of prior diversity**. It has been suggested that relics of declining taxa tend to survive in areas where the taxa had prior centres of diversity. This would be the case if, on average, taxa do not go extinct first at the geographic core of

their diversity but at less diverse, peripheral areas. This was described above for the subgenera of *Nothofagus* (Fig. 3.5). The allopatry among the centres of diversity in the *Nothofagus* subgenera suggests that the subgenera differentiated by vicariance, underwent range expansion that led to overlap and later died out in areas of secondary range expansion where they were less diverse to start with. The main centres of diversity have persisted more or less *in situ* since their origin. Following this line of argument, a group with a much poorer fossil record than *Nothofagus*, such as Rhynchocephalia, can be considered. Fossils at least show that the group was diverse and worldwide in the Triassic and Jurassic, persisting in South America until the Late Cretaceous and in New Zealand until modern times. The reason for the group's survival in these particular localities has seldom been addressed, but the pattern in *Nothofagus* suggests it is because of prior centres of diversity in these areas. This is not contradicted by the poverty of *fossil* Rhynchocephalia in New Zealand (and other localities), which reflects the low chances of terrestrial fossilization. (Jones *et al.*, 2009 described the first pre-Pleistocene fossil Rhynchocephalia from New Zealand, from the Miocene.)

Extinction, along with normal, ecological dispersal of individuals, happens everywhere, all the time. Yet this does not justify 'stretching' these everyday phenomena to fit ad hoc hypotheses. Any distribution can be explained by chance dispersal or chance extinction, but this does not represent any real progress in understanding.

Crisp *et al.* (2011) discussed 'Geographically restricted taxa that are species-poor and sister to a species-rich lineage (often referred to as 'relicts')'. They suggested that 'Such lineages have probably been subject to considerable extinction', but there is no reason to assume this as a general explanation. Crisp *et al.* (2011) cited examples from the Tasman region, such as *Amborella* from New Caledonia, with no known fossil record outside the country. The authors concluded that the restriction of these groups to a localized area is 'shrouded in mystery', but this is only because it conflicts with the authors' underlying assumption: 'to a large degree, New Zealand and New Caledonia resemble "oceanic" islands with young, immigrant biota, rather than "continental" islands with relictual "Gondwanan" biota'. Further north in Melanesia, Mayr and Diamond (2001) applied the same dispersal theory and again found mysteries (p. 229) and paradoxes (p. 254), this time in bird distributions. Yet if dispersal theory is rejected and if groups evolve by simple vicariance, these mysteries do not arise. A globally basal group that is endemic in New Zealand or New Caledonia and has no extralimital fossil record is only a mystery because it conflicts with dispersal theory; there is no simple way it can be derived from elsewhere and additional, ad hoc hypotheses have to be proposed to protect the model.

Basal nodes as centres of origin

As discussed in Chapter 1, the basal taxon in a group is often interpreted as the ancestor of the group, rather than just being a small sister-group of the rest. In a group with a phylogeny: (*a* (*b* (*c*, *d*))) with the clades found in areas A, B, C, D, the usual theory suggests the group has attained its distribution by a series of dispersal events from A to B to C and to D. This concept of the evolutionary process, in which groups always establish their range by spreading out, rather than fracturing, has been adopted by many systematists, but also ecologists (Levin, 2000) and paleontologists (Eldredge *et al.*, 2005). Yet the simple allopatry among *a–d* is also compatible with *in-situ*, sequential differentiation of a widespread ancestor, rather than dispersal. The idea that basal groups are ancestral is not accepted here and neither is the idea that their location represents a centre of origin. Basal groups are simply small (less clade-rich) sister-groups and their distribution limits represent early centres of differentiation in already widespread groups, not centres of origin.

Basal nodes as centres of initial differentiation in widespread ancestors

In the example given in the last section, the taxa *a–d* evolved in a geographic sequence of vicariance rather than a sequence of dispersal events. Brown *et al.* (2006) accepted that this process led to the west-to-east phylogeny and distribution of Malesian *Rhododendron*. Using the same approach, Doan (2003) suggested a south-to-north sequence of speciation, not dispersal, in Andean lizards.

In the same way, the evidence suggests that the Tasman Sea region hosts so many globally basal groups because a pool of diversity, or at least genetic propensity for diversity, has existed there since before the modernization of the main groups. The region does not represent a 'centre of origin' for the cosmopolitan groups in which Tasman clades are basal, but rather the break, or one of the breaks, at which differentiation of their modern groups began. The original centre of genetic diversity on the Tasman region terranes has been exposed to high levels of tectonic activity in the Mesozoic, including subduction, obduction (in which crust is ramped up onto land), rifting, orogeny, massive igneous emplacements and regional metamorphism. Similar processes have occurred in other areas with many globally basal groups, including South Africa/Madagascar and western Mexico.

Are the globally basal groups the result of long isolation?

Many studies have attributed endemism in New Zealand and New Caledonia to long isolation on the islands, which rifted from Gondwana (as Zealandia) in the Late Cretaceous. Still, this process cannot explain the globally basal groups endemic in areas such as south-eastern Australia + New Zealand, and it is unlikely to explain globally basal endemics in New Zealand. Before the rifting that opened the Tasman and Coral Sea basins, other phases of isolation occurred in the region, including those caused by the Jurassic–Cretaceous Rangitata orogeny, the mid-Cretaceous marine transgressions and the igneous belts. In any case, the east Australian nodes associated with the Tasman Sea centre all occur on mainland, not islands, and have not been caused by the opening of the Tasman. Likewise, at the south-west Indian Ocean node, the South African and Tanzanian parts of the node centred on Madagascar occur on the mainland. Biogeographers have known about the great diversity of South Africa and Madagascar – mainland

and island – for centuries. In most accounts the two are explained as separate phenomena. Instead, many distributions show that they are related, and the last intercontinental differentiation around South Africa, the Mozambique Channel and Madagascar can be attributed to the Jurassic–Early Cretaceous folding and rifting that took place there. The Tasman region, with its mainland (Australia) and islands formed from rifted continental terranes (New Zealand, New Caledonia), has had a similar history.

Islands have caught the imagination of many biogeographers, and island taxa in the Tasman region and around Madagascar are among the most distinctive. Yet mainland centres of global differentiation and endemism in eastern Australia, South Africa and western Mexico are just as significant. This implies that processes of evolution are regional in scale and that whether a group now occupies an island or a mainland has little significance.

Are the Tasman–Coral Sea groups Gondwanan?

The Tasman region includes part of the margin between Gondwana and the central Pacific. Yet despite the large numbers of endemics in the central Pacific (on the Pacific plate) with relatives in New Zealand, the Tasman–Coral Sea groups are often seen simply as Gondwanan. For example, Barker et al. (2002) accepted the New Zealand Acanthisittidae as sister to all other passerines and also noted that the oldest passerine fossils are from the Early Eocene of Australia. On this basis they supported a Gondwanan origin for the passerines. Ericson et al. (2002) agreed that the Acanthisittidae were sister of all the other passerines, 'supporting a Gondwanan origin and early radiation of passerines'. Edwards and Boles (2002) also argued that 'the New Zealand wrens fall at the base of the passerine radiation, implying an origin of this clade in Gondwana'. Yet the recognition of basal clades in New Zealand need not imply a Gondwanan origin. The 'basal' New Zealand wrens are not the ancestor of the passerines, just a small sister-group. Pre-passerines and proto-passerines could have already been widespread or even circumglobal when the differentiation between Acanthisittidae and the others occurred. This probably happened somewhere around the terranes that now flank the Tasman–Coral Sea basins. The only movement required is the local range expansion needed to explain the overlap of the Acanthisittidae and other passerines within New Zealand, and the overlap of *Amborella* and other angiosperms within New Caledonia.

In another example, various 'wattlebirds' of New Zealand and New Guinea (Callaeidae, Cnemophilidae and others; see p. 194) are basal in a cosmopolitan clade, either Passerida or Corvida. Despite the absence of living or fossil wattlebirds from Australia (there is plenty of suitable habitat there), Cracraft et al. (2004) wrote that 'The basal position of these groups relative to the remaining Corvida provides persuasive evidence that the group as a whole had its origin in Australia (and perhaps adjacent Antarctica)...'. Instead, the phylogeny is good evidence that the group did *not* have a centre of origin in Australia, as it suggests that differentiation in Corvida began around the New Zealand–New Guinea region. Again, this does not mean that the Corvida originated there. Other dispersal-oriented interpretations of the New Zealand–New Guinea wattlebirds have assumed that these birds (Shepherd and Lambert, 2007) or their last common ancestor (Ewen *et al.*, 2006) migrated to New Zealand, but no real evidence or discussion supporting this idea has been presented.

In geology, as in biogeography, New Zealand, New Caledonia and New Guinea are more than just parts of Gondwana. Some of their terranes are fragments of the margin of Gondwana, but some of these had only been accreted soon before rifting (the Ontong Java–Manihiki–Hikurangi plateau), and others were never part of Gondwana (the Northland–East Coast allochthon in New Zealand and similar terranes in New Caledonia and New Guinea; the Loyalty–Three Kings Ridge).

It is sometimes suggested that New Zealand terranes include deep sea sediments and so they are not relevant to terrestrial biogeography. Nevertheless, all the oceanic terranes include at least small amounts of material derived from island arcs or seamounts. This material and the regional biogeographic patterns are the only traces of a former geography. In some cases, the history of a terrane after its formation – for example, its ultimate accretion – is more relevant for biogeography than its origin.

Until the 1970s, New Zealand was seen as the vertical development of a single landscape feature, a great mobile belt termed the New Zealand geosyncline. Following the acceptance of plate tectonics, though, the New Zealand plateau was re-interpreted as a zone of lateral accretion, in which arc after arc has arrived from the east (Pacific) and been added to the pile. This process has continued through the Paleozoic, Mesozoic and Cenozoic, but for much of this time the history of the central Pacific is obscure. No pre-Jurassic rocks exist there, as the formation and growth of the Pacific plate has almost completely destroyed the prior geography of the whole Pacific region. The only fragments left are pieces of seafloor, arcs and seamounts preserved in terranes that have accreted around the Pacific margins. These show that arcs were significant in the Pacific region even before the formation of the modern Pacific plate. Even the lower Paleozoic Western Province terranes of New Zealand include arc-derived material. After their formation somewhere in the 'pre-Pacific' Ocean (Panthalassa), these arc terranes were fused to Gondwana (eastern Australia/Antarctica) and a series of arcs, plateaus and blocks of ocean floor were added through the Paleozoic and Mesozoic. The whole sequence rifted away from Gondwana in the Late Cretaceous (New Guinea has not completely separated), but further collisions with terranes accreting from the Pacific occurred in the Cenozoic.

The Tasman region as a tectonic margin and a biogeographic edge

The Tasman region is more likely to be a centre of differentiation and juxtaposition than a centre of origin or radiation. What were once central Pacific biotas have been piled up in and around the south-west Pacific. The arcs, plateaus and intraplate islands (now seamounts) that they arrived on formed over a wide area of the Pacific and the pre-Pacific (cf. Kerr and Tarney, 2005). Tectonic accretion often involves the docking and suturing of a terrane with a craton. The craton margin is buckled and faulted, with accompanying extension, shortening, volcanism, metamorphism and orogeny. The accreting terrane is also deformed and can undergo tectonic erosion. During accretion, many terranes in New Zealand and New Guinea have been translated horizontally, reduced to thin slivers or removed altogether by subduction, although some of their taxa could have survived by being 'scraped off' or obducted. In a similar way, proto-Japan has experienced large-scale tectonic erosion in multiple stages, with large blocks of continental crust being subducted (Isozaki *et al.*, 2010).

Biologists often regard New Zealand as 'part of Gondwana', but it is more complicated than that and geologists describe it as 'part of the Gondwana margin'. This edge itself was a composite entity; some of the component terranes formed *in situ* while others accreted from the Pacific. Some terranes of New Zealand and New Caledonia were only part of the edge of Gondwana for some time in the Cretaceous before they rifted away again into the Pacific with the new islands. The Tasman region is part of a geological 'edge' with large-scale, structural significance and the biota there also represents a biological 'edge' or 'split', as indicated by the globally basal endemics and many other groups. It is likely that the two phenomena observed in the Tasman region – the geological edge and the biological edge – are related. The Mesozoic geological revolution represented in the Tasman region by the pre-breakup orogeny, magmatism, metamorphism and rifting occurred at the same time as the last major phase of modernization in the Earth's plants and animals. This resulted in the old biota being metamorphosed into the new, Cenozoic biota seen today, and it also established the main phylogenetic and geographic breaks.

The 'invasion from the north' theory of biogeography denied the flora and fauna of the south-west Pacific and the region itself any global significance in the evolution of the main groups. In contrast, a synthesis of biogeography and molecular biology that concentrates on clade distribution and tectonics (rather than fossil-calibrated clock dates, branch lengths and species numbers) indicates that the Tasman region is one of the primary evolutionary centres of modern life. In modern angiosperms, passerines and large numbers of other groups, differentiation of the global ancestors occurred first around the Tasman region, or at least the terranes now accumulated there, and then at other breaks in areas such as Madagascar–South Africa and Mexico–north-western South America.

Through the late Paleozoic and the Mesozoic the Australasian part of the Gondwana margin was dominated by convergence, with terrane accretion, magmatism, orogeny and metamorphism, until about 100 Ma in the mid-Cretaceous. Following a tectonic switch at that time, extension developed in the margin as the subduction zone retreated into the Pacific. Many biogeographic patterns discussed below suggest that the earlier, convergence phase, culminating in the magmatism of the Whitsunday–Median Batholith belt, was more important for the global breaks than the later phase.

The ecological context of the basal Tasman groups: arapod forests

The current ecological associations of the basal groups in the Tasman region provide interesting clues concerning the evolutionary history of the region. Most of the basal groups, at least the terrestrial ones, occur in forest dominated by trees in two conifer families, Araucariaceae (41 species) and Podocarpaceae (156 species). Some members of the genus *Araucaria* were widespread in the northern hemisphere in the Jurassic, and extant Podocarpaceae range north to Mexico and Japan, but the great majority of diversity in both families is in the southern hemisphere. The araucarias have a typical female cone (as in a pine) while the podocarps have a 'fruit' (something like that of a yew), and both groups were long regarded as unrelated offshoots of different northern clades. Nevertheless, the two families are now thought to be sisters (Qiu *et al.*, 2007). This gives a large, southern clade with a concentration of diversity around the Tasman and Coral Seas.

Podocarpaceae

Podocarpaceae are characteristic of many southern hemisphere forests, especially in cooler climates, but also occur north of Gondwana in China, Japan and Mexico. While the group is most diverse around the South Pacific and is fairly diverse in Madagascar and East Africa, there is a pronounced drop in diversity in western Africa and Brazil. The extant distribution suggests the family is mainly a Pacific + Indian Ocean group.

Morley (2011) wrote that the Podocarpaceae first appeared in the Triassic of Gondwana, and although some northern hemisphere Jurassic records are known, 'it has essentially remained a southern or southern-derived family until the present day'. Many former identifications of northern hemisphere fossils as podocarps are now questioned and the ambiguity of many Mesozoic wood morphogenera attributed to extant families has been emphasized. Nevertheless, Morley (2011) cited a convincing fossil pollen record from the north Pacific (Alaska, Miocene). The family is not simply southern or Gondwanan, but mainly distributed around the Pacific and Indian Oceans, with additional records north of Tethys. These represent secondary expansions into the range of northern conifer families such as Pinaceae.

Araucariaceae

The family is extant in Southeast Asia, Australasia and South America, with three genera. Well-preserved Mesozoic fossil cones and foliage of possible crown group Araucariaceae occur in the southern hemisphere. They are also known in the northern hemisphere (United States, England, Japan), but only with two sections of one genus (*Araucaria* sects. *Bunya* and *Eutacta*) and both these are also extant in the south. Fossils of *Araucaria* sections *Araucaria* and *Intermedia* have only been found in the southern hemisphere (Setoguchi *et al.*, 1998). The other two extant genera in the family, *Agathis* and *Wollemia*, are southern hemisphere groups. *Agathis* extends north across the equator in Sumatra, Malaysia and the Philippines. It is possible that the northern members represent secondary range expansions, but a full geographic analysis of the conifer families is needed.

Many of the early fossils of 'araucarian' foliage, wood and pollen lack araucarian apomorphies and the affinities are uncertain (Gilmore and Hill, 1997; Dettman and Clifford, 2005). *Araucarioxylon* fossil wood is consistent with the Araucariaceae but also with other disparate groups, including Cordaitales. *Agathoxylon* wood is ubiquitous across Gondwana and is widespread in the northern hemisphere. Among extant conifers only the Araucariaceae have wood of the *Agathoxylon* type, but in the Jurassic–Early Cretaceous, this could have occurred in other plant groups, such as Pteridospermales and Cheirolepidiaceae (Philippe *et al.*, 2004).

Arapods and ecology

The new concept of Araucariaceae and Podocarpaceae as a clade, the arapods, has biogeographic significance as the group has such a strong concentration of diversity in the southern hemisphere, around the Indian and Pacific Oceans. The concept is also of ecological interest as the arapods are the largest and the tallest trees in the forests of New Zealand, New Caledonia and New Guinea. The arapods include true forest giants as well as the world's smallest conifers (*Lepidothamnus laxifolium* of New Zealand, 4–5 cm high) and the only parasitic conifer (*Parasitaxus* of New Caledonia). Arapods extend from the tropical shores of Borneo (the rhizomatous *Podocarpus micropedunculatus*) to the alpine zone in New Zealand (the subshrub *Lepidothamnus laxifolium*). They form the largest trees in a widespread 'warm' type of rainforest that occurs between the hot forests of the tropical lowlands and the cold *Nothofagus* forests of the higher altitudes and latitudes. In the tropical parts of the Tasman–Coral Sea area, arapod forest is found at mid-montane elevations. These sites include centres of maximum diversity for some of the highest-diversity groups in Australasia (e.g. Ericaceae, birds of paradise; Heads, 2001a, 2003). This tropical montane 'arapod zone' is also the habitat of globally basal endemics such as *Eupomatia* (Fig. 5.1).

There is a parallel case to the arapods in eudicots, with the families Cunoniaceae and Elaeocarpaceae. These are each distributed through the pantropical zone and the southern hemisphere. They were thought to be unrelated but are now interpreted as close relatives in the order Oxalidales. Both families have their main diversity around the Tasman–Coral Seas. They are among the most abundant large trees there, in the lowlands and the mountains, and are often co-dominant with the arapods.

The Tasman arapods support an interesting insect fauna. For example, the genus *Agathis* (Araucariaceae) is the host of *Agathiphaga*, the near-basal lepidopteran (see p. 192). The Araucariaceae of Australasia and South America are also the food plants of the tribe Tomicini, at or near the base of another worldwide group, the bark beetles (Curculionidae subfamily Scolytinae; Sequeira *et al.*, 2000).

The plant-eating Phytophaga comprise about a third of all beetles and include, among others, two large groups, the Curculionoidea (45 000 species) and Chrysomeloidea (35 000 species) (Farrell, 1998). In the Curculionoidea, one of the basal groups in the basal clade (Nemonychidae + Anthribidae) is the subfamily Rhinorhynchinae. This is made up of 19 genera recorded from the Tasman–Coral Sea region (north-eastern Australia, New Guinea, New Zealand and New Caledonia) and America (Chile to Colorado). There is also an Early Cretaceous fossil genus from Brazil. The Australasian members feed on arapod pollen, while the American members are associated with arapods, *Nothofagus* and Pinaceae (Kuschel and Leschen, 2010).

In the Chrysomeloidea, the basal group in the basal clade (Megalopodinae etc.) is the subfamily Palophaginae (Farrell, 1998). This group caters for just three species, which all feed on arapod cones in Australia, New Zealand, Chile and Argentina. In a similar pattern, the curculionoid tribe Pachyurini (Belidae) is made up of 13 genera associated with Araucariaceae in Australia and New Zealand, and with Podocarpaceae/Cupressaceae in Brazil.

Farrell (1998) concluded that all these feeding relationships probably already existed in the Pacific forests of the Mesozoic, as fossil Rhinorhynchinae, Palophaginae and Belidae are all known from Mesozoic strata around the Pacific and Brazil in which *Araucaria* fossils are also prominent. Overall, the evidence indicates that the arapods in the Tasman–Coral Sea region have an autochthonous Mesozoic history there, together with their insects. The intense tectonism in the region, involving accretion, orogeny, back-arc basin formation and many other processes, is reflected in the basal differentiation of Lepidoptera, Coleoptera and other groups.

In another globally basal Tasman group, the New Zealand bat fly, *Mystacinobia* (North Island), plus the undescribed 'McAlpine's fly' of Victoria, A.C.T. and New South Wales, form a clade that is sister to the worldwide dipteran family Sarcophagidae (Kutty *et al.*, 2010; distributions from T. Pape, personal communication, 30 July 2013).

6   Distribution in and around the Tasman region

In Australasian ornithology, the most influential research tradition has been the evolutionary synthesis of Matthew, Simpson and Mayr (Chapter 1). In this approach, authors studied 'the colonization of Australia by birds', and 'Evidently, it was untenable and not worthy of discussion to think that birds could have been in Australia all along' (Joseph, 2008: 265). The situation is changing now, though. One example concerns passerines, an order comprising more than half of all bird species. Molecular work has shown that the Australasian passerines do not comprise unrelated minor offshoots of northern groups, as was assumed, but large groups that are basal in the order and in its subclades. Joseph (2008) concluded that the Australian passerines are 'as old as the hills'. This is an important result, but the temptation is simply to reverse the Matthew model and propose a new centre of origin in Australasia, with dispersal northwards rather than southwards. Instead, Australasia, especially the Tasman–Coral Sea area, represents neither a sink nor a centre of origin, but a global centre of differentiation. One possible explanation for this is related to the distinctive geological history.

Tectonic context of the Tasman region and the evolution of island biotas there

Since the Paleozoic, plate convergence has continued along the south-west Pacific subduction zone. The tectonic history in the Tasman region has been dominated by convergence and accretion of terranes at the plate margin, with accompanying magmatism and orogeny. The convergence led to the immediate precursors of New Zealand, New Caledonia and New Guinea. These were produced when different oceanic terranes, formed in the pre-Pacific, fused with each other and Gondwana. Most of the oceanic terranes included arc-related material or seamounts. Among the last structures to have accreted have been large igneous plateaus. For example, convergence led to the accretion of the Hikurangi Plateau at New Zealand in the Late Cretaceous. This choked subduction and rearranged the subduction geometry in the region.

It is sometimes assumed that the oceanic terranes were devoid of terrestrial life when they collided with Gondwana, and that they emerged so close to the Gondwana coast that they were colonized solely by the mainland biota. This model conforms to the process outlined in the theory of 'island biogeography' (MacArthur and Wilson, 1967); proto-New Zealand was formed by uplift along the margin of Gondwana, and 'the origins of the earliest colonists seem straightforward' (Winkworth et al., 2002b: 514). In this model, at the time of later rifting from Gondwana, the biota of the Zealandia was 'identical' to that of Gondwana (Campbell and Hutching, 2007). Lee *et al.* (2010: 29) wrote that: 'early Zealandia ought to have contained a mammalian fauna similar to that of eastern Gondwana at the time of complete separation at c. 60 million years ago'.

Despite these suggestions, there is no real need to think that the arc and seamount terranes that were integrated into New Zealand were populated only from the Gondwana mainland, or that they had a biota like that of the mainland. Nor is it necessary to assume that the biotas of Gondwana or its rifted daughter fragments, including Zealandia, were ever homogeneous. Many Tasman–Polynesian distributions indicate that the plants and animals of the arcs and intraplate volcanics on the Pacific and pre-Pacific plates were just as significant for Tasman region biogeography as those of Gondwana. It is unlikely that Gondwana was full of terrestrial life and the entire Pacific and pre-Pacific regions were empty. There were always many islands in the Pacific region, even before the formation of the current Pacific basin, and even a small tropical island, let alone a large archipelago, can maintain a diverse biota. Understanding the biogeography of the Tasman region requires understanding the history of the life of the central Pacific, together with that of Gondwana and Tethys.

Two of the most significant events in the history of the Pacific were the formation and growth of the Pacific plate itself, as demarcated by its active margins, and the emplacement of large igneous plateaus soon after the origin of the plate. This dynamic geological history supports a metapopulation model, which differs from the 'island biogeography' model in not requiring a mainland source.

The Pacific plate

The Pacific plate began to form in the mid-Jurassic at the Tongareva triple junction, a meeting point of three spreading ridges that developed near the site of the modern Cook Islands (Larson *et al.*, 2002). From here, the Pacific plate expanded and its dramatic growth took place at the expense of the Phoenix, Farallon and Izanagi plates that surrounded it. Figure 6.1 shows the Pacific plate half-way through its growth in the mid-Cretaceous. It now occupies most of the Pacific basin.

Most of the Phoenix plate was eventually destroyed by subduction, but during subduction some of the old biota preserved on its islands would have been transferred to islands in the Tasman region. Similar biogeographic redeposition and piling-up of diversity probably took place at the east Pacific/western Americas margin as the Farallon plate was being subducted.

Large igneous plateaus in the central Pacific: Ontong Java, Manihiki and Hikurangi

In the Early Cretaceous, when the new Pacific plate was still only a fraction of its eventual size, several large igneous plateaus formed in the central Pacific (Fig. 6.1). The Ontong Java, Manihiki and Hikurangi plateaus were emplaced at the same time and the three probably formed a single, very large plateau (Taylor, 2006; Ingle *et al.*, 2007; Timm *et al.*, 2011). Hoernle *et al.* (2010) wrote that 'Hikurangi basement lavas (118–96 Ma)

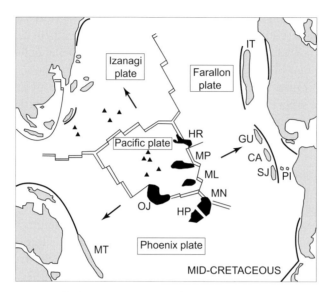

Fig. 6.1 Mid-Cretaceous reconstruction of the central Pacific plateaus and the three spreading ridges that produced the Pacific plate. The active ridges are depicted moving outwards from the centre of the Pacific plate. Spreading ridges as double lines joined by single lines (transform faults). Subduction zones as thicker lines around the Pacific margin. CA, Central America arc; GU, Guerrero terrane; HP, Hikurangi Plateau (separated from the Manihiki Plateau, MN, by the Osbourn Trough); HR, Hess Rise; IT, Insular terrane; ML, Magellan Plateau; MN, Manihiki Plateau; MP, Mid-Pacific Mountains; MT, Median Tectonic arc; OJ, Ontong Java Plateau; PI, Piñón plateau; SJ, San Juan Plateau. (From A.D. Smith, 2007. Reprinted by permission of the Geological Society of America.)

have surprisingly similar major and trace element and isotopic characteristics to the Ontong Java Plateau lavas (ca. 120 and 90 Ma)'. They referred to the Greater Ontong Java Event, 'during which ~1% of the Earth's surface was covered with volcanism . . .'. The eruptions covered an area of some 2 million km^2, the size of Alaska or western Europe, in just a few million years. This was the largest known magmatic event in Earth history (Fitton *et al.*, 2004; Hoernle *et al.*, 2004).

   After their formation in the central Pacific, the plateaus broke up and drifted apart (Fig. 6.2). The Ontong Java plateau was translated westward until it collided with the Solomon Islands (at ~12 Ma). The Hikurangi and Manihiki plateaus were separated by spreading at the Osbourn trough (Worthington *et al.*, 2006); this only ceased when the Hikurangi plateau arrived at the subduction zone (the Chatham Rise) off eastern New Zealand in the Cretaceous, partially subducted, and then choked the system (Whattam *et al.*, 2005). This probably occurred in the Late Cretaceous at ~86 Ma. Soon after, Zealandia rifted from Gondwana (eastern Australia) at 85–80 Ma and so the Hikurangi plateau was part of Gondwana for only 1–6 m.y. of its 120 m.y. history, nearly all of which has taken place in the Pacific. In its tectonics and biogeography, the plateau is a large component of New Zealand that is related to the Pacific, rather than Gondwana. The exposed portion of the Hikurangi plateau (now entirely submerged) covers an area

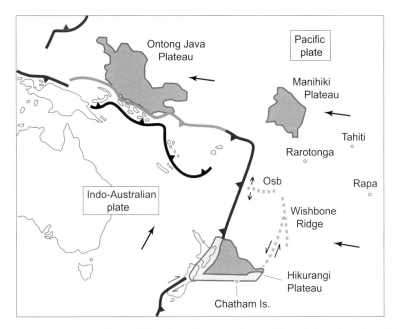

Fig. 6.2 Present position of three large igneous plateaus (the subducted portion of the Hikurangi plateau in light grey). Osb = Osbourn Trough. Significant localities for New Zealand biogeography are indicated: Rarotonga (Cook Islands), Tahiti (Society Islands) and Rapa Island (Austral Islands). (Simplified from IFM-GEOMAR, 2007.) (From Heads, 2009a, reprinted by permission of John Wiley & Sons.)

the size of modern New Zealand and other parts are subducted beneath the North Island and the Chatham Rise.

The Ontong Java Plateau is a large, mainly submerged plateau that abuts the eastern side of the Solomon Islands. It is the world's largest 'large igneous province'. Geological investigations of the immense structure of the plateau have been described as 'pinpricking the elephant' and 'no simple model appears to account satisfactorily for all of the observed first-order features' (Tejada *et al.*, 2004: 133). One of the most interesting discoveries for biogeography was a thick succession of basaltic volcaniclastic rocks on the eastern salient of the plateau which erupted in a subaerial setting (Thordarson, 2004). Fossilized or carbonized wood fragments have been found near the bottom of four of the eruptive members (Fitton *et al.*, 2004). On the Manihiki plateau, the basement and the Early Cretaceous volcaniclastic layers were formed by subaerial and shallow-water eruption (Ai *et al.*, 2008).

Some geologists have suggested that the Ontong Java plateau formed in the central Pacific near the Gorgona plateau, now accreted in western Ecuador, western Colombia and the Caribbean. As the Ontong Java plateau was carried west to the Solomon Islands, the Gorgona plateau moved east and collided with north-western South America (Kerr and Tarney, 2005; Chicangana, 2005). Chicangana discussed the possibility of a single, great igneous province forming in the Cretaceous central Pacific and then breaking up. In any case, while these authors relied on geochemical, geophysical and microfossil evidence, their ideas would also account for the Melanesia–tropical America connections of many molecular clades. These indicate that the history of the central Pacific plateaus, including their drift to the Pacific margins, has had profound effects on evolution in the Pacific. The tectonic history has been proposed as an explanation for trans-tropical Pacific distributions in terrestrial groups (Heads, 2012a), and also near-shore marine groups such as amphipods (Myers and Lowry, 2009) and barnacles (O'Riordan *et al.*, 2010). The Early Cretaceous Ontong Java, Hikurangi and Manihiki plateaus were emplaced at the same time (Aptian, ∼120 Ma; Bryan and Ernst, 2007) as another large igneous province, the Whitsunday volcanic province/Median Batholith (Figs. 4.12, 4.13). This extended from north-eastern Queensland to Antarctica and, as with the central Pacific eruptions, had a profound impact on biological communities.

Fossil evidence indicates that the Ontong Java event affected marine life, although it is not associated with a mass extinction (Hoernle *et al.*, 2004). Some populations and taxa will have been extirpated, but the new volcanic strata, submarine and terrestrial, would have soon been occupied by plants and animals. Many types of vegetation, including rainforest on the islands off northern New Guinea and alpine vegetation in New Zealand, thrive around active volcanism. Pyroclastic deposits of ash and lapilli, and even new lava flows, are soon colonized by the weedy taxa of the region.

3D projections of the Hikurangi Plateau show a spectacular, rifted north-eastern margin and, in the interior of the plateau, massive seamounts up to 24 km across (Hoernle *et al.*, 2004). The seamounts are flat-topped guyots, indicating that they formed as volcanic islands that were later eroded to sea level by wave action and then subsided to 2000 m below sea level. The central Pacific plateaus all have numerous seamounts. Fossils of montane, moss-forest groups, including land snail taxa, are known from several low atolls in the Pacific, and this is further evidence that the islands were mountainous earlier in their history.

The accretion of the igneous plateaus around the Tasman region is part of a long history of terrane accretion there. Later, in the Cenozoic, allochthonous ophiolite terranes (sections of seafloor with their seamounts) were emplaced in New Guinea, New Caledonia and northern New Zealand, after arriving from the north-east. An island arc terrane associated with the ophiolites, the Loyalty Ridge–Three Kings Ridge–Northland Plateau seamount chain, also accreted at some stage to New Caledonia and New Zealand. The arc was active in the Eocene and possibly the Cretaceous. Its history is important for biogeography as the biotas of the Loyalty Islands and the islands off Northland are so distinct from those of the adjacent mainlands of New Caledonia and New Zealand.

New models of intraplate magmatism and the biogeography of Polynesia

In traditional, dispersal interpretations of oceanic island biotas, a volcanic island just 'pops up' at a random location and its distance from the mainland is the only aspect of its location that is considered relevant. Islands are assumed to be populated from the nearest mainland and endemism is explained by isolation from the mainland. But if the location of volcanism is not random and instead recurs in particular areas, volcanic islands would instead 'inherit' their flora and fauna from prior volcanic islands in the same region. This probably occurs at subduction zones, where new islands are produced as long as subduction continues. Biologists might accept that old populations can survive on young strata at plate *margins*, as in Melanesia, but many would argue that populations on *intraplate* volcanics, as in Polynesia and Micronesia, must be the result of long-distance dispersal. Nevertheless, if there were prior islands in these regions, long-distance dispersal in the modern groups would not be necessary. For example, the conveyor-belt, hotspot model for intraplate groups such as the Hawaiian Islands provides one possible mechanism for metapopulation survival in Polynesia, although the whole topic of intraplate volcanism is controversial.

The plate tectonics model was proposed in the mid-1960s and accounted for the distribution of most of the volcanism on Earth's surface, as this occurs at plate boundaries. Yet the model did not account for areas of profuse volcanism within plates, for example at Hawaii and other Pacific islands. Hotspots and a mechanism for producing them – mantle plumes – were proposed as a second mechanism, independent of plate tectonics, to explain this type of volcanism. The mantle plume model became popular, but more recent explanations for 'anomalous' volcanism, based instead on effects of plate tectonics, have developed alternatives (Foulger and Natland, 2003). In these new models, stress fields produce great fissures and systems of fissures through the lithosphere, along which seamount provinces and linear chains form (see papers in Foulger *et al.*, 2005 and Foulger and Jurdy, 2007). As with the mantle plume/hotspot theory, these mechanisms would enable the long-term survival of metapopulations.

The islands of Polynesia and Micronesia are some of the most isolated in the world. But despite their isolation and small size, as a group they maintain many, well-marked taxa that are widespread in the central Pacific and endemic there (Heads, 2012a). Many of these taxa would have colonized the reef and subaerial parts of the igneous plateaus (the islands are now seamounts) in the Cretaceous from other arcs and islands in the region. Many islands would have been transported to Zealandia on the Hikurangi plateau and accreted there, soon before Zealandia rifted from Gondwana.

Central Pacific endemism in western and eastern Polynesia, or the area covered by the Ontong Java, Manihiki and Hikurangi plateaus, has direct importance for the biogeography of the Tasman Sea region. Many of the patterns coincide with the igneous plateaus and other centres of Cretaceous volcanism in south-eastern Polynesia/Line Islands, and with the Pacific plate margin itself. The islands, plateaus, ridges and seamounts of Polynesia have a long history of volcanism, and the patterns of allopatry and endemism suggest that dynamic metapopulations have survived on old systems of young, individually ephemeral islands. The Polynesian biota has been just as important for the Tasman–Coral Sea region as the more famous South Pacific and Gondwana biota of Fagales, Proteaceae, ratites and others, in the south and west.

Pre-breakup geology of Zealandia

Zealandia (including New Caledonia) was assembled from its component terranes before being rifted away, as a new unit, from Gondwana. Before the component terranes of Zealandia assembled, different models suggest that they were spread out along an arc

6000 km long (Adams *et al.*, 2007), or even longer in the model of Wandres and Bradshaw (2005). This belt of terranes extended from the New Guinea region to somewhere south-east of Tasmania. Interaction between this belt and the terranes of coastal Australia has been critical for evolution in the region, while the telescoping of the belt to form New Zealand has probably distorted many distributions and formed others.

In New Zealand prior to Gondwana breakup, the immediate pre-breakup phase was characterized by orogeny, large-scale volcanism and crustal extension. When back-arc basin formation began with seafloor spreading in the Tasman basin, a block composed of older and younger accreted terranes was separated from Gondwana as Zealandia. The rifting cut across some of the earlier features, such as the Whitsunday Volcanic Province–Median Batholith (Fig. 4.13) and biogeographic patterns suggest that breakup itself – the seafloor spreading – was of less phylogenetic importance than pre-breakup events. The pre-breakup magmatism formed a single, vast belt through Queensland, New Zealand and Antarctica, and the importance of the pre-breakup Rangitata orogeny has also been mentioned. By the time rifting in the Late Cretaceous separated New Zealand and New Caledonia from Gondwana, the biota of the whole New Zealand–New Guinea sector had already been exposed to repeated phases of biological and geological accretion, extension, uplift and magmatism.

The Tasman region as a centre of biodiversity later disrupted by slab rollback

The Tasman region represents a biological edge situated on the Gondwana/Pacific margin. The biogeographic interest of the region is matched by the dynamism of its tectonic history. This has involved continuous subduction with associated terrane accretion, obduction, metamorphism, volcanism, orogeny and rifting.

Although it seems counter-intuitive, the Tasman region is a zone of overall convergence (between the Pacific and Indo–Australian plates), but it is also the most striking example of episodic basin formation in the world (Schellart *et al.*, 2006). Since the mid-Cretaceous, a whole series of basins have opened in the disintegrating continental margin of Gondwana, behind (west of) the subduction zone as this has migrated into the Pacific. The first basin, the Tasman, opened from 82 Ma to 52 Ma (Late Cretaceous to Early Eocene), beginning in the south. The spreading centre propagated northwards and the Coral Sea basin opened from 62 Ma (Early Paleocene) to 52 Ma. Later basins opened around New Caledonia, the Loyalty Islands, Vanuatu and Fiji, and the latest basin is opening behind the easternmost arc, the Tonga–Kermadec Ridge.

The basins have formed in the back-arc (the region behind a volcanic arc, on the side away from the trench) by extension or stretching in the crust. In several cases extension has caused the crust to break completely, and seafloor spreading has developed. This back-arc extension has equal significance for geology and biology. The opening of the rifted basins would have caused much vicariance and the subsequent closing of several basins would have caused overlap.

Back-arc basins are built behind, or by, migrating island arcs, which lengthen and increase their curvatures through time (Hamilton, 1988). The critical feature for back-arc basin formation is the migration of the arc. Soon after the introduction of plate tectonics theory, geologists realized that subduction trenches and the hinge of the descending slab, at the top of the trench, are not stable features, but can move (Stegman *et al.*, 2006). A significant number of trenches migrate in a direction opposite to the motion of the subducting plate (i.e. in a retrograde or oceanward direction) and this has been termed rollback or retreat of the trench, slab or hinge. The negative buoyancy of the descending slab means that it is likely to sink vertically (rather than subducting forwards and downwards) and so will migrate away from the upper plate (Heuret and Lallemand, 2005). The motion of the upper plate is also important in determining this slab rollback. Suggested rates of rollback range from 2 to 15 cm/year and the rollback of the Australian/ Pacific plate margin into the Pacific has been continuous since rifting began at ~100 Ma. The current zone, along the boundary of the Pacific plate and the Australian plate, is marked by a strip of island arc crust running between New Zealand, Tonga, Fiji, Vanuatu, Solomon Islands and New Guinea (van Ufford and Cloos, 2005; Schellart *et al.*, 2006).

Retreat of the plate margin into the Pacific would have had several implications for biogeography. First, it caused extension of the continental crust, with intracontinental thinning and rifting, and seafloor spreading. The rift basins would have fractured the biota. Second, since the Cretaceous, a major subduction zone, with an active volcanic arc and its biota, has migrated from the Australian coast eastward into the Pacific, as far as Tonga.

Distribution within the Tasman region

Many Tasman groups with intercontinental connections around the Indian, Tethys and Pacific Ocean basins were discussed in Chapter 3. This chapter concentrates on distribution patterns within the Tasman–Coral Sea region itself.

As mentioned already for passerine birds, molecular phylogenies have led to reassessments of biogeographic models. Several have suggested that the Tasman region is not a dispersal 'sink', but a centre of origin and radiation. Instead, the region is interpreted here as a centre of differentiation in widespread ancestors (caused by pre-breakup tectonics or extensional basin formation) and also a zone of biotic juxtaposition (caused by convergence and accretion). The Pacific biotas are now 'piled up' in and around the south-west Pacific. Their original distributions may have been greatly reduced during this process, as the arcs and plateaus that the groups arrived on originated over a wide area of the Pacific and pre-Pacific Oceans. Following continued terrane accretion and tectonic erosion, many terranes in New Zealand, New Caledonia and New Guinea have been reduced to thin slivers. It is likely that others have been completely subducted, although some of their taxa may have survived by being 'scraped off' and obducted onto adjacent terranes.

Traditional, fossil-based biogeographic narratives have been based on the twin concepts of centre of origin and chance dispersal. These explained all patterns but attention was deflected away from underlying tectonics and onto the details and vagaries of stratigraphy. Island biogeography has focused on the ages of strata on individual islands and atolls, but it should also consider the subduction zones, spreading centres, back-arc basins and accreting terranes that have generated the landscape and determined aspects of metapopulation dynamics and evolution.

Important new geological discoveries include fossil wood on the vast Ontong Java Plateau, evidence for a large, long-lived Paleogene island between New Zealand and New Caledonia (Meffre *et al.*, 2007), and many large, drowned seamounts on the Hikurangi, Manihiki and Northland Plateaus. These are all compatible with the initial differentiation of passerines and angiosperms having taken place around the terranes now located in the south-west Pacific.

The current islands and island chains in the Tasman region (for example, the Solomons and Vanuatu chains, the Lord Howe seamounts and the Norfolk Ridge islands) are often interpreted as landbridges or 'stepping-stones'. Despite their high levels of endemism, they are seen as important routes of dispersal into the Tasman region or from one part of it to another, not as evolutionary centres in their own right. Nevertheless, the island endemics with global sisters, such as the New Zealand wrens and the New Caledonian *Amborella*, cannot easily be explained by dispersal from elsewhere. Old endemic taxa have probably persisted in the region they evolved in by surviving as metapopulations, and in this way the different centres of endemism around the Tasman and Coral Seas can be much older than the islands and reefs of the present geography.

Australia–New Zealand connections

In the Tasman region, the many trans-Tasman disjuncts have often been discussed and possible explanations debated. For trans-Tasman affinities in Hyriidae, freshwater mussels, Graf and Ó Foighil (2000) concluded in favour of vicariance caused by seafloor spreading and wrote 'Among the most damning evidence against long-distance avian dispersal of freshwater mussels is that it has never been observed, it is purely hypothetical'. As the authors observed, while vicariance is a single event with many consequences, a dispersal explanation for a general pattern 'refers to an essentially infinite series of improbable events . . .'. This debate is ongoing.

Australia–New Zealand clades and their incongruence with Gondwana breakup

Some authors have argued that a vicariance/Gondwana breakup model is contradicted if an Australia–New Zealand endemic has a sister-group in South America. This is because New Zealand broke away from Gondwana in the Late Cretaceous (~84 Ma), while Australia and South America remained connected, via Antarctica, until the Paleogene. So authors often argue that a phylogeny determined by vicariance should be: (New Zealand (Australia + South America)). In liverworts, Vanderpoorten *et al.* (2010) found that the Australian and New Zealand floras are sisters and this is supported by the genera endemic to these two areas. The authors described the Australia–New Zealand relationship as a discrepancy from a continental drift scenario and 'a vicariance hypothesis'. Nevertheless, the breakup of Gondwana was not the only cause of vicariance in the region. As discussed above for chironomid midges, three related groups in Australia and New Zealand, New Zealand and South America, and South America and Australia probably diverged at nodes *within* what became New Zealand, South America and Australia, *before* Gondwana breakup. This suggests that pre-breakup tectonics and evolution have produced a phylogeny that is incongruent with the breakup sequence and has produced the many sister-group relationships between Australia and New Zealand. Pre-breakup differentiation in the Gondwana biota would contradict the usual assumption that the biota was uniform throughout the supercontinent, and this was probably unrealistic anyway. Biogeographers sometimes assume that the breakup itself was a simple process and was the only process generating vicariance, but the rifting process that led to breakup was complex and lasted at least 20 m.y.

The idea can be illustrated with a typical distribution from a group of felt scales (Hemiptera: Eriococcidae) that feed on *Nothofagus* and have the phylogeny (Hardy *et al.*, 2008):

> Australia and New Zealand.
>> Australia and New Zealand.
>> South America (basal grade), Australia and New Zealand.

Because this is incompatible with the breakup sequence, the authors assumed that chance dispersal has played an important role in the evolution. Instead, orogeny and large-scale magmatism around the Tasman region and South America *before* rifting could have caused the episodic breakup of a widespread ancestor in that region, resulting in the observed phylogeny.

Kunzea (Myrtaceae)

Trans-Tasman affinities are often involved with the standard disjunction between southeastern and south-western Australia (Chapter 4). *Kunzea* is a southern Australasian group, with three main clades as follows (de Lange *et al.*, 2010):

> Western Australia (subgen. *Salisia*).
>> Eastern Australia (subgen. *Angasomyrtus* and subgen. *Kunzea*).
>> Eastern Australia plus New Zealand (subgen. *Niviferae*).

The main break separates western Australia from eastern Australia + New Zealand. The trans-Tasman disjunction in the third clade formed after this break, although perhaps only soon after. The phylogenetic sequence and the biogeography are compatible with the first event being caused by mid-Cretaceous marine transgressions, and the second event resulting from Late Cretaceous opening of the Tasman basin.

Although opening of the Tasman has been of secondary importance in the phylogeny of *Kunzea*, it is still significant below subgenus level. The event and its possible mode of action on evolution was discussed by McDowall (2008: 203), who wrote: 'Once New Zealand disconnected from Gondwana . . . taxa would immediately have begun dispersing to New Zealand'. Yet taxa would have been migrating into and out of the area that became New Zealand (as part of their normal ecology) before it separated, and opening of the rift probably caused a reduction, not an increase, in this. In many groups, differentiation would have developed as the rifting continued, although regular migrations across the Tasman and Coral Sea have continued in groups such as whales, large sharks, seabirds, cuckoos and others.

Trans-Tasman Onychophora

The phylum Onychophora (*Peripatus* etc.) is thought to be ancient because of its phylogenetic position and also its Paleozoic fossil record (Poinar, 2000). Fortey and Thomas (1993) wrote that 'Like other evolutionary enigmas, they are probably more diverse in the southern hemisphere'. New Zealand has two genera of Onychophora, *Ooperipatellus* and *Peripatoides*. Both are endemic and they form a monophyletic group together with certain Australian genera (Allwood *et al.*, 2010). This Australia–New Zealand clade is sister to genera from central Chile and South Africa, implying breaks across the South Pacific and the southern Indian Ocean.

Clock analyses led Allwood *et al.* (2010) to conclude that the age of *Peripatoides* is consistent with a vicariant origin of the genus following the rifting of New Zealand from Gondwana. The dates calculated for *Ooperipatellus* are less informative. Allwood *et al.* (2010) calibrated the clock analyses by using the Cretaceous opening of the Atlantic to estimate the divergence date of two sister clades in South America (*Metaperipatus*) and Africa (*Peripatopsis*).

Trans-Tasman marine groups

Marine groups found around reefs and shores often show the same biogeographic patterns as taxa of dry land, probably because the tectonic history of a region is shared by both. For example, as with terrestrial groups, many marine clades are endemic to south-eastern Australia and New Zealand. In the past, this restricted distribution has been explained by dispersal in ocean currents from Australia to New Zealand or vice versa. Yet molecular studies of several taxa show substantial genetic divergence between sister clades in south-eastern Australian and New Zealand waters. Waters *et al.* (2007) documented examples in the gastropods *Scutus* and *Austrolittorina*. Both have planktonic larvae and so, in theory, their means of dispersal could allow trans-Tasman movement, but in fact there is no evidence of recent or ongoing population connectivity.

New Zealand–Tasmania connections: barnacles and crickets

One important trans-Tasman area of endemism is Tasmania + New Zealand. This area is a globally basal centre of endemism, defined by local endemics with global sisters such as the centipede *Craterostigmus* (Chapter 5). The same region appears in marine groups. In morphological work, the Ibliformes are an order of barnacles (Buckeridge and Newman, 2006), comprising genera in:

Tasmania and New Zealand (*Idioibla*).
Tasmania and New Zealand (*Chaetolepas*).
New Zealand (*Chitinolepas*).
Tasmania, Australia and Melanesia to Japan, Madagascar and the Red Sea (*Ibla*).
West Africa (*Neoibla*).

In addition, there is a possible fossil from the Carboniferous of Illinois. Four of the five or six genera overlap in the Tasmania/New Zealand region and at least *Idioibla* and *Chaetolepas* probably occurred there before Tasmania and New Zealand rifted apart.

In Tasmania–New Zealand groups, whether marine or terrestrial, the disjunction itself is dramatic but it is only a secondary aspect of the pattern. The main question is: how did the overall distribution, Tasmania + New Zealand, come to exist to begin with, before it was rifted apart? Why are the many groups with this range absent from mainland Australia, New Caledonia and New Guinea, where there is often suitable habitat and related forms? Again, this probably implies tectonic activity around Tasmania + New Zealand as a single area, before it broke up in the Late Cretaceous.

A terrestrial example of the Tasmania–New Zealand area occurs in the Rhaphidophoridae (Orthoptera). These are the cave crickets, known in New Zealand as cave wetas. One member of the group, subfamily Macropathinae, has a phylogeny (Allegrucci *et al.*, 2010):

South Africa.
  New Zealand.
  New Zealand + Tasmania.
  South America.

As the authors noted, the pattern is consistent with a simple vicariance history. Subsequent, local dispersal within New Zealand would explain any overlap between the two clades there. The Macropathinae are sister to the Asian subfamilies of Rhaphidophoridae, and this is also consistent with plate tectonic vicariance (Allegrucci *et al.*, 2010). Allegrucci *et al.* (2010) suggested there is a dilemma, though, as calibrating a split of the New Zealand and Tasmania clade with Tasman Sea rifting would push back the origin of the family to 600 Ma. Yet this would only be true if large changes in evolutionary rates were ruled out and this cannot be justified. As another alternative, the authors suggested that long-distance dispersal might solve the problem, although they admitted that 'this is difficult to reconcile with much that is known about [the group's] ecology and population genetics' (p. 128). Instead, the group's phylogeny and distribution are regarded here as good evidence for extreme changes in evolutionary rate.

New Zealand–Kosciuszko connections: *Chionochloa* (Poaceae), a group with a circumglobal sister

This trans-Tasman area of endemism lies to the north of the Tasmania–New Zealand groups just discussed. The tussock grass *Chionochloa* dominates many grasslands in

New Zealand, including the subantarctic Auckland and Campbell Islands. It also occurs on Lord Howe Island (one endemic species) and in Australia on Mount Kosciuszko, near the New South Wales/Victoria border and 200 km inland (one endemic species). Evolution at and around this northern sector (New Zealand–Kosciuszko) is likely to be just as old as evolution at the southern one (New Zealand–Tasmania), as *Chionochloa* is sister to a global group and is difficult to derive from anywhere else (see Chapter 1; Linder *et al.*, 2010).

Biogeographic discussion about *Chionochloa* has concentrated on the disjunction in the distribution, rather than the area of endemism itself. Pirie *et al.* (2009b: 386) wrote 'It is clear from the topological position and age of the single Australian species, *C. frigida*, which is nested within *Chionochloa*, that New Zealand formed the source area ...'. Yet the position of the Australian species, nested among species from New Zealand, does not necessarily mean it originated there, or that there was any trans-Tasman dispersal in the genus (cf. Fig. 1.6).

Chionochloa itself is sister to a large global complex, and so, assuming allopatric differentiation between the two, there has been subsequent overlap within the south-eastern Australia–New Zealand region. Yet there are significant differences between the distributions of the two. Both New Zealand members of the global sister-group, *Austroderia* (*Cortaderia* p.p.) and *Rytidosperma*, are absent south of Stewart Island on the subantarctic islands, where *Chionochloa* is present. Conversely, both are present on the Chatham Islands, where *Chionochloa* is absent (Edgar and Connor, 2000). The simplest explanation of this allopatry is that it is a trace of the original vicariance between a south-western *Chionochloa* and a north-eastern and global sister-group. The distribution patterns within each group suggest that the overlap between them took place before differentiation of the modern species and probably before rifting in the Tasman basin.

New Zealand–McPherson–Macleay Overlap connections: *Paralamyctes* centipedes (Henicopidae)

New Zealand–Tasmania groups, such as the basal centipede *Craterostigmus*, often occur in western and southern New Zealand localities, in particular the subantarctic islands Fiordland and Nelson. In *Chionochloa* the connection lies to the north, in New Zealand, Lord Howe Island and Mount Kosciuszko. In some groups both tracks are seen; in the *Hebe* complex (Plantaginaceae) there is a Tasmania–New Zealand clade and also a Kosciuszko–New Zealand clade (Meudt and Bayly, 2008). In the fern *Asplenium hookerianum*, a haplotype in Tasmanian populations is shared with populations in south-western New Zealand, while the single haplotype found in Victorian populations was unique but most similar to one found in populations from central and eastern North Island (Perrie *et al.*, 2010). Other trans-Tasman links link Tasmania with north-west Nelson in the northern South Island (e.g. in *Poranthera*; Phyllanthaceae, cited on p. 165).

Groups located further north and east in New Zealand show additional trans-Tasman affinities with the McPherson–Macleay Overlap (MMO) and north-eastern Queensland.

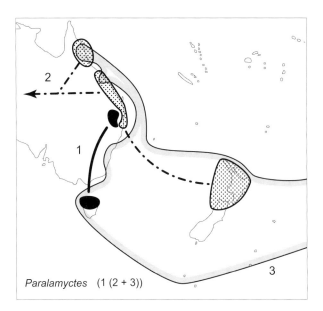

Fig. 6.3 The centipede genus *Paralamyctes*, showing the three main clades and their distribution in the Tasman region. Clade 2 also occurs in southern India, Madagascar, Mozambique and South Africa. (Data from Edgecombe, 2001; Giribet and Edgecombe, 2006a, b). (From Heads, 2009b, reprinted by permission of John Wiley & Sons.)

Differentiation between this region and the Kosciuszko/Tasmania–New Zealand region coincides with the pre-breakup Whitsunday–Median Batholith belt of magmatism. The New Zealand–MMO connection is exemplified by the Indian Ocean component of *Paralamyctes*, a centipede with three main clades (Fig. 6.3; Edgecombe, 2001; Giribet and Edgecombe, 2006b):

> Tasmania and the MMO.
>> Indian Ocean basin: south-eastern Africa, Madagascar, eastern Queensland, including MMO, northern New Zealand.
>> South Pacific basin: Queensland (not MMO), New South Wales, Tasmania, New Zealand (including subantarctic islands), Chile.

Here the Tasman region represents both a centre of overlap (for Indian and Pacific Ocean clades) and a centre of endemism, features that are repeated in the Liliales of the region (Campynemataceae and allies). The clade boundaries in *Paralamyctes* do not correspond with the main geographic boundaries, such as the Tasman Sea. Instead, as Giribet and Edgecombe (2006b) emphasized, the different areas of endemism within Australia have separate affinities with different areas outside Australia. The overall division is into three groups, with one each in eastern, central and western areas. The Tasman Sea has opened after the origin of the main groups, and it has fractured the range of two of them.

New Zealand–New Guinea connections

New Zealand–New Guinea distribution is a distinctive biogeographic pattern that skirts Australia and is not predicted in dispersal models. Nevertheless, it has been reported in many morphological studies. For example, *Oreomyrrhis buwaldiana* (Apiaceae) of New Guinea is closest to *O. colensoi* of New Zealand (Mathias and Constance, 1955), *Epilobium hooglandii* and *E. detzneranum* (Onagraceae) of New Guinea, and *E. prostratum* of New Guinea, the Maluku Islands and Sulawesi are closely related to each other and to the New Zealand species *E. gracilipes* (Raven and Raven, 1976), *Euphrasia* sect. *Pauciflorae* (Orobanchaceae) is in New Zealand, New Guinea and Sulawesi (Barker, 1982), and in *Kelleria* (Thymelaeaceae) similar, pedicellate species occur in New Zealand (*K. multiflora*), New Guinea and Borneo (*K. ericoides, K. patula*) (Heads, 1990a).

In ferns, *Thelypteris* is represented in the southern hemisphere by *T. confluens*, recorded from New Zealand, New Guinea and localities further west (Sumatra, Thailand, India and Africa; Holttum, 1973). In lichens, *Pseudocyphellaria carpoloma* is widespread in New Zealand, Chatham Islands and Norfolk Island, and is also recorded from the Huon Peninsula, Papua New Guinea (Galloway, 1994).

In fauna, the syrphid fly *Anu* of New Zealand is probably sister to *Giluwea* from New Guinea (Thompson, 2008), and the New Zealand clade of the butterfly *Lycaena* (Lycaenidae) is sister to *Melanolycaena* of New Guinea (de Jong and van Dorp, 2006). In Ichneumonidae (Hymenoptera), morphological studies indicated that *Certonotus fractinervis* of New Zealand is sister to *C. vestigator* of New Guinea. The pair is sister to *C. monticola* of New Guinea and north-eastern Queensland, and this group in turn has a sister in eastern Australia, from Tasmania to north-eastern Queensland (*C. avitus, C. geniculatus*) (Gauld and Wahl, 2000). The authors noted that *Aphanistes* is another ichneumonid with a New Zealand–New Guinea relationship.

Rousseaceae (Asterales) and other groups

Carpodetus (Rousseaceae) is widespread through the main islands of New Zealand and also in montane forests of the Solomon Islands and New Guinea. Only the New Zealand species has been analysed in sequencing studies, but although the different species show superficial differences they share a similar architecture (the trunk is an unstable orthotropic monopodium with plagiotropic lateral branches; personal observations).

The other members of the Rousseaceae (Fig. 6.4) are widespread around the Tasman region and in Mauritius and make up the sister-group of a global family, Campanulaceae (Stevens, 2012). Members of Rousseaceae have the phylogeny (Lundberg, 2001):

Mauritius (*Roussea*).

Eastern Australia (*Cuttsia, Abrophyllum*).

New Zealand, New Guinea and the Solomon Islands (*Carpodetus*).

Overall, the three clades are based around the Indian Ocean group. The last two clades occupy western and eastern margins of the Tasman and Coral Sea basins and so the

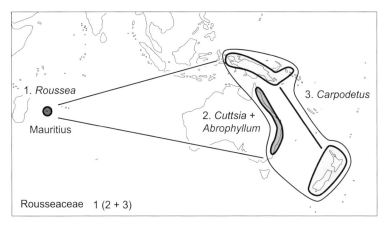

Fig. 6.4 Distribution of Rousseaceae (Lundberg, 2001).

break coincides with the seafloor spreading that opened the basins. *Roussea* is described as a 'woody liane' or 'climbing shrub'. The others are mainly shrubs and small trees.

Molecular studies confirm the New Zealand–New Guinea connection

Molecular studies of different plant and animal groups have confirmed the existence of a direct biogeographic connection between New Zealand and New Guinea. The pattern is recorded in Ericaceae (Fig. 4.15), Plantaginaceae (*Parahebe* s.lat.; Albach and Meudt, 2010), Asteraceae (*Olearia* clade; Cross *et al.*, 2002; a clade in *Abrotanella*; Swenson *et al.*, 2012) and Boraginaceae (clade in *Myosotis*; Winkworth *et al.*, 2002a). In *Coprosma* (Rubiaceae), one of the five main clades is endemic to New Zealand (*C. crenulata, C. serrulata*), New Guinea (*C. discoloris*) and Borneo (*C. crassicaulis*) (Markey, 2005). In ferns, the genus *Cyathea* includes the following group (Papadopulos *et al.*, 2011):

New Zealand, south-eastern Australia (*C. cunninghamii, C. colensoi*).
New Zealand (including Auckland Islands) (*C. smithii*).
Lord Howe Island (*C. brevipinna*).
New Guinea (*C. nigrolineata, C. coactilis, C. foersteri, C. pachyrrhachis*).

The New Zealand–New Guinea connections are related to a more general connection between New Zealand and the Coral Sea region, including Cairns, Papua New Guinea and the Solomon Islands (but not the MMO). Groups with this pattern include *Passiflora* subgen. *Tetrapathaea* (Krosnick *et al.*, 2009), a clade in Winteraceae (*Pseudowintera* and *Zygogynum*; Heads, 2012a, Fig. 9.5) and a marine polychaete endemic to New Zealand and the Great Barrier Reef (Watson, 2009). The plant group *Korthalsella* sect. *Heterixia* occurs in New Zealand (two species), New Guinea and the Cairns region (*K. papuana*), also Borneo and Sumatra (*K. geminata*; position based on morphology) (Papadopulos *et al.*, 2011).

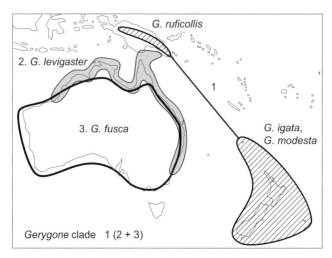

Fig. 6.5 Distribution of a clade in *Gerygone* (Acanthizidae) (Nyári, 2011).

In several cases the New Zealand–New Guinea clade has a sister in Australia, as in the ichneumonids cited above, and this is consistent with vicariance around breaks in the Tasman and Coral Sea basins. Examples include the following animals and plants.

Passerines: *Gerygone* (Acanthizidae) and meliphagids

A New Zealand–New Guinea connection is evident in the passerine bird *Gerygone* (Acanthizidae), which includes the following clade of allopatric species (Fig. 6.5; Nyári and Joseph, 2012):

New Zealand (*G. modesta*) and Norfolk Island (*G. igata*), + New Guinea (central axial mountains) (*G. ruficollis*).
Australia (*G. fusca*), + northern and eastern Australia, southernmost New Guinea (*G. levigaster*).

In the honeyeaters, the passerine family Meliphagidae, Nyári (2011) proposed the following clade (Fig. 6.6):

SW and E Australia, Tasmania (*Acanthorhynchus*).
Widespread though drier Australia (*Certhionyx variegatus*).
New Guinea (*Pycnopygius*).
New Zealand (*Prosthemadera*, the tui).

Nyári (2011) did not sample the New Zealand bellbird, *Anthornis*, but other studies found the genus to be the sister of *Prosthemadera* (Driskell *et al.*, 2007; Gardner *et al.*, 2010).

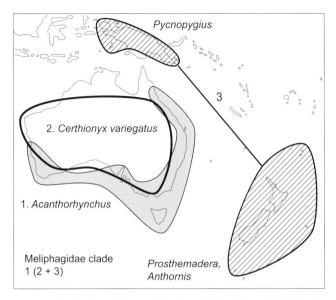

Fig. 6.6 Distribution of three clades in Meliphagidae (Nyári, 2011).

The distribution of the New Zealand–New Guinea clades does not reflect any obvious ecological factors (New Zealand is more similar to south-eastern Australia than to New Guinea) and is probably not due to extinction in Australia (where the sister-groups occur). Instead, it can be explained by simple vicariance with their mainland Australian sister-groups.

In other passerines, the New Zealand Callaeidae (with *Notiomystis*) have appeared as sister to the New Guinea Cnemopholidae (Driskell *et al.*, 2007; Shepherd and Lambert, 2007), although this result has not been reproduced (Irestedt and Ohlson, 2008; Jønsson *et al.*, 2011a). The New Zealand–New Guinea connection seen in *Gerygone* and meliphagids also occurs in the following plants.

Raoulia s.lat. and *Craspedia* (Asteraceae: Gnaphalieae)

Molecular studies have shown that a clade of Asteraceae, referred to here as *Raoulia* s.lat., is endemic to New Zealand and New Guinea (Fig. 6.7) (Breitwieser *et al.*, 1999; see Glenny, 1997, for the New Guinea members). (The group includes *Raoulia* and several small genera: *Leucogenes*, *Ewartiothamnus*, *Rachelia*, *Anaphalioides* and New Zealand '*Helichrysum*'. The internal branches of this clade have only weak support and intergeneric F1 hybrids show high fertility; Smissen *et al.*, 2007.) Breitwieser *et al.* (1999) saw the New Zealand–New Guinea distribution of *Raoulia* s.lat. as unusual and they suggested long-distance dispersal across the gap. Nevertheless, as indicated above, New Zealand–New Guinea connections have been reported in many morphological and molecular studies.

Raoulia s.lat. is a member of the cosmopolitan tribe Gnaphalieae. This group has its basal members in southern Africa, and its maximum species diversity there and in

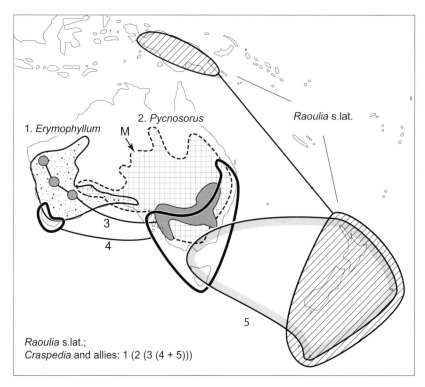

Fig. 6.7 Distribution of two clades (not sisters) in Gnaphalieae: *Raoulia* s.lat. (diagonal lines) and *Craspedia* and allies: 1 = *Erymophyllum*; 2 = *Pycnosorus*; 3 = *Craspedia haplorrhiza*; 4 = *Craspedia* 'AUS 1 clade'; 5 = *Craspedia* 'AUS 2 clade' + 'NZ clade'. M = MacDonnell Ranges. (From Breitwieser *et al.*, 1999 and Ford *et al.*, 2007.)

Australia. In addition to occupying montane areas in southern Africa and Australia, Gnaphalieae are also diverse in arid, central Australia. The groups in the region show high levels of allopatry at all ranks. For example, the main clade of Gnaphalieae in New Zealand after *Raoulia* s.lat. is the *Craspedia* group (Fig. 6.7; Ford *et al.*, 2007). The two are not immediate sisters, but show interesting partial allopatry; *Raoulia* s.lat. is in New Guinea but not Australia, while the *Craspedia* group is in Australia but not New Guinea.

Craspedia itself occurs in southern Australia, Tasmania and New Zealand, while its closest relatives, *Pycnosorus* and *Erymophyllum*, occur in more inland areas of Australia (Fig. 6.7). The first break in the group is between inland Western Australia (*Erymophyllum*) and the rest; the second is between inland eastern Australia (*Pycnosorus*) and the rest. From central Australia the sequence of differentiation then proceeds to southern Australia (clade 3 versus the rest) and then south-eastern Australia (between clades 4 and 5). Only at that stage does the Late Cretaceous opening of the Tasman rift apart clade 5, suggesting that prior differentiation coincided with the mid-Cretaceous marine transgressions of central Australia.

Craspedia includes several Tasmania–mainland Australia clades. For example, *C. glauca* of coastal shrubland and forest in Tasmania is sister to the alpine *C. lamicola* of Mount Kosciuszko. Ford *et al.* (2007) inferred multiple colonizations from mainland Australia to Tasmania, but the pattern is also compatible with the sundering of multiple former connections between the two. The opening of Cretaceous basins in the Bass Strait region would provide a possible mechanism.

In *Craspedia* clade 5 (Fig. 6.7), the New Zealand species are sister to a group in southeastern Australia and Tasmania. There are no dated fossils of the genus or its sister but 'very low' divergences between New Zealand and Australian/Tasmanian samples led Ford *et al.* (2007) to suggest late Tertiary or Quaternary dispersal. The mode of dispersal is 'a matter of conjecture' as the pappus is deciduous and long-distance wind dispersal is unlikely. Ford *et al.* (2007) suggested that carriage on muddy bird feet is 'plausible'. This has also been suggested for other trans-Tasman plant groups with no obvious means of long-distance dispersal, but it does not explain why successful dispersal in *Craspedia* only occurred once (the New Zealand species form a single clade). In addition, there are no obvious bird vectors. Ad hoc hypotheses could be proposed, such as extinction of suitable bird vectors, but then further ad hoc hypotheses would also be needed for all the other groups that share the trans-Tasman distribution of *Craspedia* while, as in that genus, having no obvious means of trans-oceanic dispersal. Finally, the high degree of endemism and vicariance in the tribe Gnaphalieae, the Australian members and the genus *Craspedia* (e.g. its absence from New Guinea) all argue against chance dispersal as the overall mode of spatial evolution. It seems more likely that traditional ideas on the time course of evolution in Asteraceae (based on a literal reading of the fossil record) are incorrect.

The *Hebe* complex (Plantaginacae)

Veronica s.lat. is a plant group with a global distribution and a paraphyletic, basal grade of herbs in the northern hemisphere. This has been interpreted to mean that the woody, South Pacific clade – the *Hebe* complex of Australasia, Polynesia (Rapa Island) and Patagonia – was derived from herbaceous, northern hemisphere ancestors (Albach and Chase, 2001; Wagstaff *et al.*, 2002; Albach and Meudt, 2010). Instead, the biogeography suggests that the ancestral complex was already circumglobal and included both woody and herbaceous forms before the modern clades differentiated.

Within the *Hebe* complex, four of the clades are trans-oceanic disjuncts; two span the Tasman basin, one spans the Tasman and Coral Sea basins (present in New Zealand and New Guinea), and one is centred on the central Pacific (Fig. 6.8). (The phylogeny is from Albach and Meudt, 2010, based on combined ITS and cp DNA data, and Meudt and Bayly, 2008.) The traditional names for the clades, at generic rank, are cited with the figure. The *Hebe* complex was treated as *Veronica* subgen. *Pseudoveronica* in Albach and Meudt (2010), but names for the separate clades are not available in *Veronica*. Other authors have retained a multiple genus classification for the group (Bayly and Kellow, 2006; Meudt and Bayly, 2008).

Of the clades mapped in Figure 6.8, only the trans-central Pacific *Hebe* can form a tree with a well-developed trunk; the other, western groups are always shrubs or subshrubs. (This character geography is repeated in other groups.) In traditional theory, the distributions of clades 2–6 would be attributed to dispersal from a centre of origin in

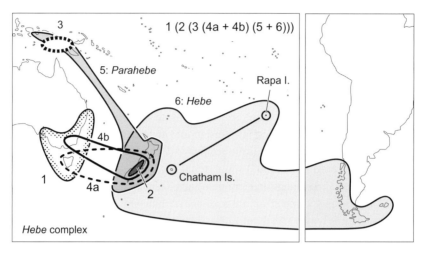

Fig. 6.8 Distribution of the *Hebe* complex (Plantaginaceae). 1 = *Derwentia*; 2 = *Leonohebe*; 3 = *Detzneria*; 4a = *Chionohebe*; 4b = *Hebejeebie*; 5 = *Parahebe*; 6 = *Hebe*. (Phylogeny from Albach and Meudt, 2010.)

New Zealand, but this does not explain the allopatry of the groups outside New Zealand. The distributions are interpreted here as the results of a widespread South Pacific ancestor broken into allopatric clades at nodes around terranes that now comprise New Zealand. Secondary overlap of the clades developed later in New Zealand.

Meudt and Bayly (2008) discussed the two trans-Tasman connections in the *Hebe* complex. *Chionohebe* is endemic in the South Island and Tasmania (cf. *Craterostigmus* etc.) and *Hebejeebie* is endemic to the South Island and Mount Kosciuszko (cf. *Chionochloa* etc.). *Chionohebe* and *Hebejeebie* are each represented in Australia by a single species that also occurs in New Zealand. Meudt and Bayly (2008) noted that seed dispersal in these small alpine plants is probably effective only over short distances – metres, at most – and they concluded that the mechanisms of trans-Tasman dispersal are 'unknown' and 'hard to envisage'. They considered the possibility of human transport, but were sceptical of this idea, as the plants are restricted to remote, alpine areas. In the other clades of the *Hebe* complex, a dispersal model does not explain the New Zealand–New Guinea disjunction, or why *Hebe* s.str. would disperse to Rapa and South America but not Australia, when *Hebejeebie* and *Chionohebe* both did.

Olearia (Asteraceae: Astereae)

A clade in *Olearia* ('clade 2' in Cross *et al.*, 2002) has a trans-South Pacific range (Fig. 6.9) similar to that of the *Hebe* complex. Further parallels include a New Zealand–New Guinea affinity (the *Olearia arborescens* group and *O. velutina*) and a trans-Tasman group (*Celmisia* and relatives). *Celmisia* is the plagiotropic, south-western version of its orthotropic, north-eastern relatives (*Olearia* spp.) in New Zealand, Australia and New Guinea. Plants of the alpine *Celmisia hectorii*, for example, can form extensive, woody, non-rooting mats over bare rock.

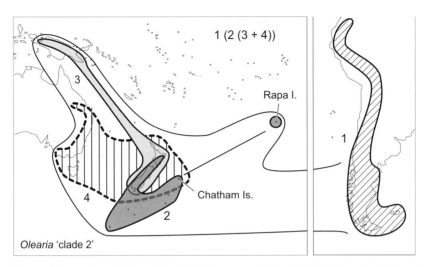

Fig. 6.9 Distribution of *Olearia* 'clade 2' (Asteraceae). 1 = *Chiliotrichum*; 2 = *Pleurophyllum* and *Olearia* 'macrocephalous clade'; 3 = *O. arborescens* etc.; 4 = *Celmisia*, *Pachystegia* and *Olearia* p.p. (Cross *et al.*, 2002.)

Olearia 'clade 2' also includes *Pachystegia* of Marlborough, *Olearia* 'macrocephalous clade' of New Zealand, *Damnamenia* and *Pleurophyllum* on the New Zealand sub-antarctic islands, and, as a basal member, the Patagonian–Andean *Chiliotrichum* group (Bonifacino and Funk, 2012). *Pacifigeron* from Rapa Island is thought to be related to *Celmisia* and *Pleurophyllum* but has not been sequenced.

Despite the obvious similarities between the *Hebe* complex and *Olearia* 'clade 2', the spatial sequence of phylogeny in the two is not the same; in the first it begins at breaks around the Tasman Sea and New Guinea, while in the second it begins at breaks between Australasia and Patagonia. In some approaches (vicariance cladistics) this 'incongruence' would be seen as evidence that the two did not share a single history of exposure to the same sequence of geological events. Yet the distinctive similarities in other aspects of the patterns suggest that chance is not the explanation. One possibility is that the geological events involved took several million years to complete, and so they would have caused episodic, pre-breakup evolution within *Olearia*, *Hebe* and the others, at nodes in both South America and Australasia over the same time period.

Euphrasia (Orobanchaceae)

Euphrasia is a plant genus with a circumglobal distribution. In a morphological classifi-cation (Barker, 1982), the New Zealand members were allocated to four separate clades and each occurs outside New Zealand in different localities, namely, south-eastern Aus-tralia, New Guinea and Sulawesi, Juan Fernández Islands and Patagonia (Fig. 6.10). This arrangement shows many parallels with the patterns in the *Hebe* complex (Fig. 6.8) and *Olearia* (Fig. 6.9), proposed in molecular studies. All three groups are absent from New Caledonia, indicating that their ancestors were not present on or around any of the New

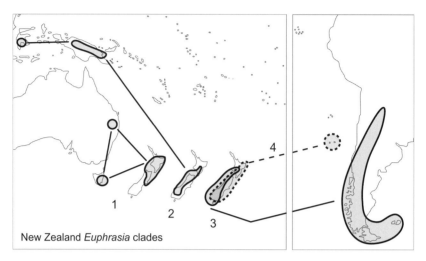

Fig. 6.10 Distribution of New Zealand *Euphrasia* clades (Orobanchaceae) (distribution and morphological phylogeny from Barker, 1982; New Zealand distributions from Heads, 1994b.)

Caledonia precursor terranes. The four clades of *Euphrasia* show considerable overlap in New Zealand, but even within the country they maintain significant allopatry and all four overlap only in north-west Nelson province (Heads, 1994b). This allopatry is interpreted as a trace of the original vicariance and indicates a main node in Nelson.

A note on the Juan Fernández and Desventuradas Islands

The Juan Fernández Islands and their northern neighbours, the Desventuradas Islands, have a distinctive biota (Moreira-Muñoz, 2011) that shows significant connections to New Zealand. In *Euphrasia*, Barker's (1982) morphological studies suggested that a group from one part of New Zealand forms a clade with plants from Juan Fernández Islands, while another New Zealand group extends further south in New Zealand (to Stewart Island) and forms a clade with plants from mainland South America (Fig. 6.10). The New Zealand–Juan Fernández affinity was also cited above in *Mida* (Santalaceae) (Chapter 3). In other angiosperms, a well-supported clade, *Coprosma* s.str. (Rubiaceae) ranges from Borneo to Australasia, the Pacific islands and Juan Fernández, but is absent from South America (Heads, 1996; Markey, 2005). *Haloragis erecta* (Haloragaceae) of New Zealand is sister to *H. masatierrana* of Juan Fernández (Moody and Les, 2007). *Apium* (Apiaceae) includes a New Zealand–Juan Fernández clade (Spalik *et al.*, 2010).

In marine littorinid gastropods, Waters *et al.* (2007) found a phylogenetic affinity between Australasian and Juan Fernández taxa that excludes mainland South America. The authors described this as 'intriguing' and they cited additional examples of the pattern in seaweeds. In other groups, the toothed whale *Mesoplodon traversii* (Ziphiidae) is known only from New Zealand and Juan Fernández Islands.

The fish family Chironemidae (Perciformes) occurs in coastal waters around southern Australia, New Zealand, Juan Fernández and the Desventuradas Islands (Meléndez and Dyer, 2010). Molecular estimates of divergence time between south-east Pacific chironemids and their western relatives pre-date the emergence of the current Juan Fernández islands (Burridge *et al.*, 2006). This is consistent with successive biotic transfer from older, now submerged islands to younger ones (cf. Moreira-Muñoz, 2011).

In addition to these seaweeds, seed plants, fishes and whales, the following mosses are known from Australasia and Juan Fernández, but not South America (from Muñoz *et al.*, 2004, Supporting Online Material).

Aus = Australia, NZ = New Zealand, JF = Juan Fernández Islands, S Af = South Africa.

> *Dicranoloma menziesii*: Aus, NZ, Lord Howe, Norfolk, Macquarie, Auckland, Campbell, Chathams, JF.
> *Distichophyllum assimile*: Aus, JF.
> *Grimmia pulvinata*: S Af, Aus, NZ, Lord Howe, Chathams, JF.
> *Leptodon smithii*: S Af, Aus, NZ, JF.
> *Macromitrium microstomum*: Aus, NZ, JF.
> *Racopilum tomentosum*: Aus, Norfolk, JF.
> *Rhizogonium novaehollandiae*: Aus, NZ, Auckland, Campbell, Chathams, JF.
> *Rhynchostegium tenuifolium*: Aus, NZ, Lord Howe, Kermadecs, Macquarie, Auckland, Chathams, JF.
> *Sematophyllum subpinnatum*: S Af, Aus, JF.
> *Trichostomum brachydontium*: S Af, Aus, NZ, Kermadecs, JF.
> *Weissia controversa*: S Af, Aus, NZ, Lord Howe, Norfolk, Kermadecs, JF.
> *Zygodon menziesii*: Aus, NZ, Macquarie, Auckland, Campbell, Chathams, JF.

Muñoz *et al.* (2004) also recorded similar patterns in the liverworts *Fossombronia pusilla*, *Marchantia foliacea*, *Monoclea forsteri*, *Plagiochasma rupestre*, and the lichens *Chondropsis sorediata*, *Cladonia subradiata*, *Parmeliella nigrocincta*, *Parmelinopsis spumosa*, *Pseudocyphellaria dissimilis*, *P. mooreana* and *P. physciospora*.

The question is not 'How did the groups get from Australasia to Juan Fernández (or vice versa)?' but 'Why is there such an important break between Juan Fernández and the South American mainland, one that affects both terrestrial and marine groups?' The islands maintain the south-eastern margin of a large central Pacific biota absent from South America and this anomaly needs to be explained.

The distribution pattern: Juan Fernández Islands + Desventuradas Islands is documented in marine groups such as fishes (*Chironemus bicornis*; Meléndez and Dyer, 2010) and seabirds (Heads, 2012a, Fig. 8.10). It also occurs in terrestrial groups; for example, *Dendroseris* (Asteraceae: Lactuceae) of Juan Fernández is sister to *Thamnoseris* of the Desventuradas (B. Baldwin, personal communication, in Kim *et al.*, 2007). The two genera are stout shrubs or small trees and this is unique in the tribe Cichorieae (= Lactuceae), which are otherwise herbs, rarely subshrubs, small shrubs or scandent vines (Bremer, 1994). (Anomalous woodiness in Pacific plants is discussed in Heads, 2012a.) The Desventuradas biota shows the same break with the mainland seen in Juan Fernández groups. For example, the shorebird *Procelsterna* (Sternidae) is distributed from the West Pacific (Lord Howe Island and others) to the Desventuradas. The genus

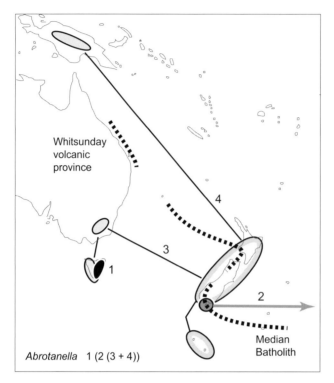

Fig. 6.11 Distribution of *Abrotanella* (Asteraceae: Senecioneae), showing the four clades (Wagstaff *et al.*, 2006). The second clade also occurs in southern South America. (From Heads, 2012c, reprinted by permission of John Wiley & Sons.)

is absent from the American mainland but is replaced there by the Inca tern *Larosterna*, endemic to the coasts of Peru and Chile. (The phylogeny is not yet resolved.)

Abrotanella (Asteraceae)

The distribution of many terrestrial groups is based around ocean basins and so is incongruent with the traditional biogeographic regions, the continents. Biogeographic patterns also deconstruct land areas at a smaller scale and this occurs in many Tasman taxa. A good example is *Abrotanella*, a genus of alpine cushion plants found in Australasia and Patagonia (Heads, 1999; Wagstaff *et al.*, 2006). (The chronology of its evolution was discussed in Chapter 2.) The four main clades are all present in Australasia and one of these is shared with South America (Fig. 6.11). The distributions are:

> *Eastern* Tasmania.
>
> > A trans-Pacific clade: Stewart Island (southern New Zealand), Juan Fernández Is. and Patagonia.
> >
> > > A trans-Tasman clade: *Western* Tasmania, Australia, New Zealand.
> > >
> > > A trans-Coral Sea clade: New Zealand, New Guinea.

There is overall vicariance among the four main clades and significant overlap only in New Zealand. The clades' regions do not correspond with the geographic areas, such as Tasmania, Stewart Island and New Zealand. The breaks separating the clades do not occur in the Tasman or Pacific ocean basins, but between western and eastern Tasmania, in Stewart Island, and between south-eastern Australia and New Guinea. On Stewart Island, the trans-Tasman clade is represented by *A. linearis* at lower elevations and *A. pusilla* on Mount Anglem, while the trans-Pacific clade is represented by *A. muscosa*, a subalpine species. The specimens listed by Swenson (1995) indicate that the two clades overlap at just one locality, the summit ridge of Mount Anglem where *A. pusilla* and *A. muscosa* both occur.

The main phylogenetic breaks occurred in Tasmania and Stewart Island, before rifting in the Pacific split clade 2 and rifting in the Tasman split clades 3 and 4. One possible mechanism for the initial breaks would be the Early Cretaceous pre-breakup extension and plutonism that developed in and around Tasmania (e.g. at Port Cygnet) and Stewart Island (Allibone and Tulloch, 2004). The Escarpment fault traverses Stewart Island, separating the northern half of the island (including Mount Anglem) from the southern half, and was active at 110–100 Ma. The break between clades 3 and 4 is compatible with differentiation at the Whitsunday Volcanic Province–Median Batholith, followed by local overlap in New Zealand.

Wagstaff *et al*. (2006) concluded that distributions in *Abrotanella* 'undoubtedly' reflect a 'convoluted history of dispersal', but their analyses were based on a literal interpretation of the fossil record, and chance processes do not explain the precise biogeographic patterns – allopatry occurs at intercontinental, regional and local scales. The spatial evolution of *Abrotanella* could hardly be less convoluted than it is and the first three main clades show almost complete allopatry. Despite the large trans-Pacific range, overlap is local and occurs on a single mountain in Stewart Island and perhaps three localities in central Tasmania. Much of the lower-level differentiation within the genus is also allopatric (see Heads, 1999).

Abrotanella is probably an old group. Its New Guinea–New Zealand–south-eastern Australia connections suggest the group originated when terranes that were spread along the Australia margin from New Guinea to Tasmania were translated along the margin, then juxtaposed and sutured to form proto-New Zealand.

Abrotanella itself, a typical South Pacific disjunct, is the basal member of the global tribe Senecioneae. This repeats the pattern seen in *Nothofagus*, basal in the Fagales. Senecioneae is a large, cosmopolitan clade that includes *Senecio* (1500 species), the trans-tropical Pacific Blennospermatinae (see above) and many others (Wagstaff *et al*., 2006; Pelser *et al*., 2010). The phylogeny of the tribe is as follows (Pelser *et al*., 2010, combined data set):

South Pacific (*Abrotanella*).
    South Africa: SW Cape Province (*Capelio*).
        South America: NW Argentina to northern Peru (*Chersodema*).
        Cosmopolitan (all other Senecioneae).

In a dispersal model, this pattern implies trans-oceanic dispersal from a South Pacific centre of origin, across the Indian Ocean to South Africa, then across the Atlantic to South America, and then the world. Instead, in a vicariance model, differentiation of a global ancestor has occurred around the South Pacific, then, at the second node, around the south-west Indian Ocean basin. The third break has involved differentiation around the region that began, in the Early Cretaceous, to form the Andes. Subsequent overlap has developed between the cosmopolitan clade and each of the first three clades. The last event, the overlap, probably occurred prior to the Late Cretaceous rifting that broke up groups of Senecioneae such as the South Pacific *Brachyglottis* and relatives (see p. 92), along with *Abrotanella*. Thus the overlap coincides with the last peak in sea levels, in the mid-Cretaceous, and this provides a mechanism for the overlap.

Groups centred on the Tasman and Coral seas

Coronanthereae (Gesneriaceae): distribution between New Zealand and the Coral Sea

The small South Pacific tribe Coronanthereae (Gesneriaceae) is endemic to the Tasman region and central Chile (Fig. 6.12; from Woo *et al.*, 2011; cf. Fig. 3.15). The three clades show clear-cut allopatry among three sectors: south-western, central and eastern.

The allopatry suggests that the distribution of the central clade 2 is not just the result of extinction elsewhere. The simple allopatry can be explained by vicariance in the Tasman region before the separation of Australia and South America. This also accounts for the allopatry among Coronanthereae and its relatives (Fig. 3.15). Woo *et al.* (2011) instead assumed that a basal grade indicated a centre of origin and they supported a centre of origin for Coronanthereae in South America, followed by long-distance dispersal to Australasia. Woo *et al.* (2011) based their dispersal model on this and on an estimated age for the clade (27 Ma). This was calculated using the age of the oldest seamounts on the Lord Howe Rise, dated to 23 Ma, to give a maximum age for *Negria*, endemic to Lord Howe Island (the island itself is dated at 7 Ma). But if *Negria* can be so much older than its island now, this could have also been the case in the past, when *Negria* occupied the 23-Ma island. In general, the age of the strata that a clade inhabits is not related to the age of the clade.

The now-submerged seamounts in the Lord Howe chain (Chapter 2) were not the only prior land in the region. The entire Lord Howe Rise itself is part of Zealandia, and is formed of buoyant continental crust. Before, during and after its rifting from Gondwana, Zealandia was emergent, but it was being stretched (extended and thinned), and so became less buoyant. Much of the plateau subsided through the Cenozoic. Many islands in the region were destroyed by the extension and this is documented, along with phases of uplift, on the Norfolk Ridge, a strip of continental crust adjacent to the Lord Howe Rise.

For the South Pacific Coronanthereae, Woo *et al.* (2011) admitted that trans-Pacific dispersal involves long distances and is against current prevailing winds, but noted that Motley *et al.* (2005) postulated long-distance dispersal from the Caribbean to the

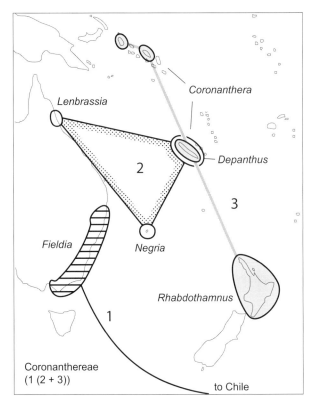

Fig. 6.12 Tribe Coronanthereae (Gesneriaceae). Coronanthereae also include *Mitraria*, *Asteranthera* and *Sarmienta* of central Chile. (From Heads, 2010a, reprinted by permission of John Wiley & Sons. Based on phylogeny from Woo *et al.*, 2011, and distributions from van Balgooy, 1984.)

western Pacific among *Bikkia* and its relatives (Rubiaceae). Woo *et al.* concluded: 'As unlikely as trans-Pacific dispersal may seem, migration via Antarctica seems less likely for Coronanthereae'. (Antarctic migration would have been possible until ~35 Ma, but this is older than the authors' estimates for the maximum age of the tribe.) With respect to means of dispersal, Woo *et al.* (2011: 450) wrote that:

Several different mechanisms have been suggested for Gesneriaceae as a whole. However, there are no dispersal data within this family, so the proposed mechanisms, all of which may be applied to Coronanthereae, are *speculative*.... Dispersal across vast distances may be *difficult to envisage*... Many... dispersal cases involve angiosperms not known to be particularly good dispersers, and they have *challenged the notion of how species disperse*... [Italics added.]

These are crucial points. Woo *et al.* (2011) added that 'Additional ecological studies will be essential to determining how fruits and seeds of Coronanthereae have managed transoceanic dispersal'. Yet ecological studies are unlikely to shed much light on processes that have occurred only once in the millions of years of the group's history, and comparative biogeographic studies might be more useful.

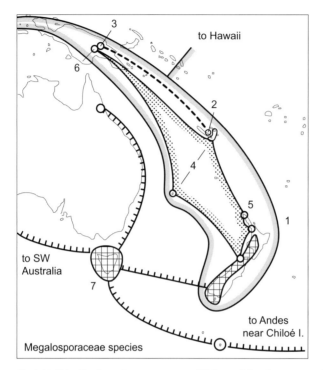

Fig. 6.13 Distribution of some groups of lichens (Megalosporaceae). 1 = *Megalospora atrorubicans*; 2 = *M. hillii*; 3 = *M. weberi* (2 and 3 share a distinctive combination of characters); 4 = *M. sulphureorufa*; 5 = *M. bartlettii*; 6 = *M. granulans* (4, 5 and 6 share a distinctive combination of characters); 7 = *Austroblastenia, Megalospora lopadioides, M. queenslandica, M. kalbii* (the last three share a distinctive spore type and are linked with *Austroblastenia* in molecular studies). (From Sipman, 1983; Galloway, 2007; Kantvilas and Lumbsch, 2012.)

Weber (2004b: 93) concluded that a Mesozoic ('Gondwana') origin for the family Gesneriaceae 'appears very plausible from a geographical view, [but] it is hard to conceive that the family is old enough to match the geological events of Gondwana breakup'. How old could Gesneriaceae be? The family has no fossil record (Roalson *et al.*, 2008), but, as mentioned in Chapter 3, the geographic break between the two widespread clades coincides with the Romeral fault zone in Colombia (the Cretaceous boundary between the South American craton and the accreted terranes in the west) and the South Loyalty Basin (which opened in the Cretaceous and later almost closed again).

Megalosporaceae lichens

Similar patterns to those of Coronanthereae have been described in studies of the lichen family Megalosporaceae (Fig. 6.13; Sipman, 1983; Galloway, 2007; Kantvilas and Lumbsch, 2012). Again, there are affinities in Lord Howe Island–New Caledonia and in northern New Zealand–Melanesia. Between New Caledonia and New Guinea there are two parallel tracks that show precise local allopatry within both islands. All these groups lie oceanward of another set of affinities in Australia/southern New Zealand–Chile.

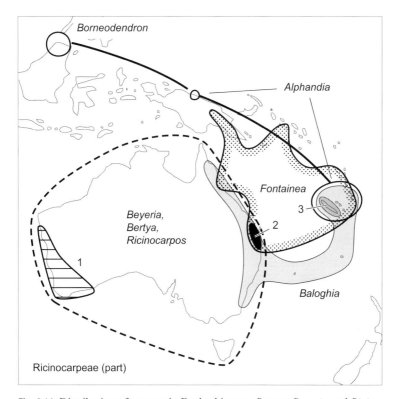

Fig. 6.14 Distribution of a group in Euphorbiaceae. *Bertya*, *Beyeria* and *Ricinocarpos* (Australia) form a clade sister to *Baloghia*, *Fontainea* (sisters) and *Borneodendron* (Euphorbiaceae; Wurdack *et al.*, 2005; K.J. Wurdack, personal communication, 15 May 2008). *Alphandia* (New Caledonia–New Guinea) and the New Caledonian endemics *Myricanthe* and *Cocconerion* have not been sampled but are probably related to the first six genera. The centres of diversity for *Beyeria* (south-western Australia; line fill), *Fontainea* (MMO; black) and *Baloghia* (New Caledonia; grey) are indicated. (From Heads, 2010a, reprinted by permission of John Wiley & Sons.)

Ricinocarpeae (Euphorbiaceae)

The tree *Fontainea* (Euphorbiaceae) is endemic around the margins of the Coral Sea (Fig. 6.14) and has a centre of diversity at the MMO (black in the figure; Heads, 2010a). *Fontainea* forms a clade with *Baloghia*, a trans-Tasman Sea group most diverse in New Caledonia (grey in the figure) and *Borneodendron* of Borneo. *Alphandia* (New Caledonia–New Guinea) and the New Caledonian endemics *Myricanthe* and *Cocconer-ion* were not sampled but probably also belong here (Wurdack *et al.*, 2005; K.J. Wurdack, personal communication, 15 May 2008).

This group, centred around the Tasman and Coral Seas, is sister to a widespread Australian clade, *Bertya*, *Beyeria* and *Ricinocarpos*, in which *Beyeria* has a species concentration in south-western Australia (line fill in Fig. 6.14). Thus the main breaks in a widespread Australasian ancestor have developed along the western and eastern margins of the Tasman–Coral Sea basin.

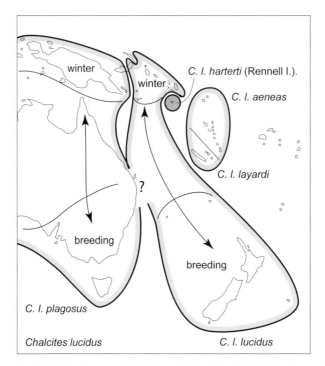

Fig. 6.15 Distribution of *Chalcites lucidus* (Cuculidae) (also in south-western Australia and Lesser Sunda Islands). Annual migrations occur in two of the subspecies, as indicated (Mayr, 1932a; Rand and Gilliard, 1967; Gill, 1998).

The sister of the whole group shown in Figure 6.14 is *Reutealis*: Philippines, plus *Aleurites*: India to Polynesia, including Hawaii (this is often cultivated and the original distribution is obscure). All these genera make up the south-west Pacific tribe Ricinocarpeae.

Trans-Tasman migrations in landbirds – the cuckoos

The distribution of the cuckoo *Chalcites* (or *Chrysococcyx*) *lucidus* is shown in Figure 6.15. Each year populations migrate between southern Australia (breeding) and New Guinea (*C. l. plagosus*), and between New Zealand (breeding) and the Coral Sea islands (*C. l. lucidus*) (Mayr, 1932a; Rand and Gilliard, 1967; Gill, 1998). Whether or not the two groups of populations warrant separate subspecies is controversial. In addition, the details of the migration routes around central eastern Australia are unclear, and banding or molecular studies are needed. But the birds have allopatric, sedentary relatives on Rennell Island (south-western Solomon Islands), Vanuatu and New Caledonia, consistent with evolution of the whole complex by simple vicariance.

The migrations of New Zealand *Chalcites lucidus* are among the longest trans-oceanic migrations of any landbird. Hutton (1872) suggested that the birds follow routes that represent former coastlines. These have changed over time while the birds' annual migrations have remained the same, following old shores of the Lord Howe Rise,

Norfolk Ridge and other features that are now submerged. Repeated strandings of migrating marine mammals at certain localities suggest that the same process applies here. McCann (1964) mapped what he termed the 'coincidental distributional pattern' of strandings in New Zealand and this would be explicable if whales and dolphins were following old features of the continental shelf that have been uplifted.

In addition to *Chalcites*, there is one other cuckoo breeding in New Zealand, *Urodynamis* (or *Eudynamys* s.lat.) *taitensis*. In winter this species migrates to islands in the south-west and south-central Pacific: Palau and New Guinea to the Marquesas, Pitcairn and Line Islands (Ellis *et al.*, 1990). This defines a vast Pacific range, but does not include Australia or islands west of New Guinea. In fact, no birds breeding in Australasia migrate to Asia beyond Wallace's Line (Dingle, 2008), and this is another aspect of migration that probably reflects tectonics.

Phylogenetic breaks around New Caledonia

The archipelago of New Caledonia is well known as an area of endemism, but is also important as a rare zone of overlap in widespread clades that are otherwise allopatric. The overall patterns suggest that the local overlap occurs near the original phylogenetic break.

Melicytus and allies (Violaceae), with a note on tree architecture

Melicytus species are some of the most common trees in New Zealand. The genus belongs to a pantropical group of trees and lianes, the *Hybanthus* complex, which comprises an Indian Ocean clade ('*Hybanthus*' p.p., *Pigea* etc.) and a trans-tropical Pacific clade (*Melicytus* etc.) (Fig. 6.16; data from Feng, 2005 and Tokuoka, 2008). The breaks between the two groups are located at the Atlantic Ocean and the Tasman basin. At the Tasman break there is secondary overlap in New Caledonia and at the MMO. A clade of *Scaevola* (Goodeniaceae) shows a similar distribution, with Indian Ocean and Pacific Ocean sister-groups having significant overlap only in New Caledonia (Heads, 2010a, Fig. 11). These patterns indicate Indo–Pacific groups with initial breaks around New Caledonia or its precursor terranes.

Within the Indian and Pacific groups of Violaceae, many of the subclades also show a high level of allopatry. One example is the clade *Melicytus* + *Hymenanthera*, comprising two sister genera in the Tasman region (Fig. 6.17A). Through most of south-eastern Australia, New Zealand, Samoa and Vanuatu, this is the sole representative of the pantropical *Hybanthus* complex and the simplest explanation is that it evolved over this region by allopatric differentiation, probably during pre-breakup extension. Overlap between *Melicytus* + *Hymenanthera*, and any other member of the *Hybanthus* complex, is restricted to the MMO and Fiji, and so the only dispersal required is within these two small areas. There is no need to invoke trans-Tasman chance dispersal of the *Melicytus* + *Hymenanthera* group in either direction, and chance dispersal would be incompatible with the overall allopatry between the group and its relatives.

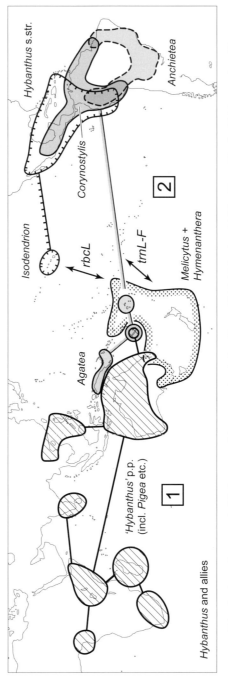

Fig. 6.16 Distribution of a group in Violaceae and its two main clades (Feng, 2005; Tokuoka, 2008). (From Heads, 2010a, reprinted by permission of John Wiley & Sons.)

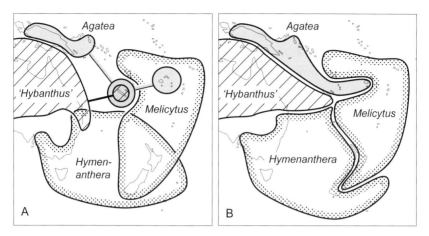

Fig. 6.17 (A) *Melicytus* and allies (Violaceae) (part of the clade shown in Figure 6.16). (From Mitchell *et al.*, 2009.) (B) Reconstruction of the group shown in (A).

The *Melicytus + Hymenanthera* clade can be analysed further. Both members of the pair overlap in the three main islands of New Zealand and in Norfolk Island, but elsewhere they are allopatric (Mitchell *et al.*, 2009). (The authors referred to the two as clades A and B, but the earlier, formal names are well known and available at generic rank.) Their distributions are:

Hymenanthera: New Zealand (three main islands), Chatham, Norfolk and Lord Howe Is., SE Australia and Tasmania.
Melicytus: New Zealand (three main islands), Norfolk I., Vanuatu–Fiji (the species is sister to the New Zealand *M. micranthus*), Kermadec Is. (the plants are related to the New Zealand *M. ramiflorus*), and Samoa–Tonga (one unsampled species, possibly related to the Kermadec Is. form).

Mitchell *et al.* (2009) calibrated a molecular clock using a fossil-calibrated age for the split between *Melicytus* and *Hybanthus*, and the ages of Lord Howe and Norfolk for the endemics there. They proposed a centre of origin in New Zealand and trans-oceanic, long-distance dispersal to Australia and elsewhere. Yet this does not account for the great intercontinental break in the *Hybanthus* complex into Indian and Pacific Ocean groups (clades 1 and 2 in Figure 6.16), or the considerable allopatry between *Melicytus* and *Hymenanthera*. The only dispersal that is required in modern *Melicytus* and modern *Hymenanthera* is within the New Zealand mainland and Norfolk Island. This is needed to explain the geographic overlap there between the western *Hymenanthera* and the northern *Melicytus*. One possible reconstruction involves a break along the central New Zealand/Norfolk Ridge part of the Zealandia (Fig. 6.17B, shown on the modern topography to allow comparison). The main break in the whole *Hybanthus* complex lies near New Caledonia.

The affinities of *Melicytus + Hymenanthera*, as shown by variation in *trnL-F*, are with the Melanesian *Agatea* and a trans-Pacific track at equatorial latitudes, while *rbcL* sequences link the pair instead with the Hawaiian *Isodendrion* and its northern Pacific connections. The biogeographic patterns of variation seen in characters and genes underlie those of clades. The geographic distributions of different genes and characters that show incongruent phylogenies within a group are of special interest. They can indicate ancestral polymorphism that has been overwritten by modern differentiation, and so different phases of vicariance are recorded. The *rbcL* connection crosses the north-eastern Pacific, the *trnL-F* connection is trans-tropical Pacific, and both connections are well documented in many groups.

Within their area of overlap, *Hymenanthera* and *Melicytus* show ecological vicariance, as most forms of *Hymenanthera* are shrubs of open habitats (coastal and alpine shrubland and rocky sites), while *Melicytus* is a tree in closed forest. The ecology is probably determined by the morphology, as *Melicytus* has larger leaves than *Hymenanthera*. *Hymenanthera* is not just a miniaturized version of *Melicytus*, though, and has a very different overall architecture that determines the leaf size. What determines the architecture?

Melicytus and *Hymenanthera* conform to a pattern in which trees of the central Pacific with orthotropic (vertical) axes are related to plants with a strong plagiotropic (horizontal) component on the western margin of the Pacific, in Australasia. The western plants are divaricate, ericoid or lianescent. The overall pattern seen in the western, plagiotropic *Hymenanthera* and the eastern, orthotropic *Melicytus*, also occurs in *Mearnsia* (western Pacific lianes) and *Metrosideros* (central Pacific trees) (Myrtaceae), in the *Hebe* complex (Plantaginaceae) and in *Coprosma* (Rubiaceae). The overlap of the orthotropic and plagiotropic groups in New Zealand creates major centres of diversity there. The overall phylogeny indicates that the overlap of *Hymenanthera* and *Melicytus* in the three main islands of New Zealand developed after the primary 'Hybanthus complex' break in New Caledonia, but before the origin of the modern trans-ocean basin clades, and this would be coeval with the landscape revolutions of the mid-Cretaceous.

Agathis (Araucariaceae)

The conifer *Agathis* has most of its diversity in the Tasman region where it includes New Zealand's largest tree, the kauri. Molecular studies show that the New Zealand species is sister to the rest (Fig. 6.18), with the other two groups showing a break between New Caledonia and Vanuatu, at the current plate boundary (Stöckler *et al.*, 2002; Knapp *et al.*, 2007; Biffin *et al.*, 2010). The three clades have differentiated at a single node located somewhere south-east of New Caledonia, around the old plate margin and the node seen in the *Melicytus* group (Fig. 6.17). Biffin *et al.* (2010) supported a young age (49 Ma) for *Agathis*, but this was based solely on the fossil record.

A case-study: Ericaceae in the Tasman region

The main clades in Ericaceae, the heath family, were introduced above (p. 112). Many members have small, leathery leaves and tolerate harsher conditions than most woody

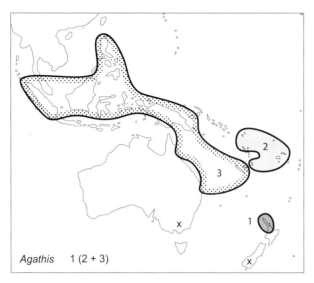

Fig. 6.18 Distribution of *Agathis* (Araucariaceae) (Stöckler *et al.*, 2002; Knapp *et al.*, 2007). X = fossils.

plants. They occupy habitats such as cool mountains, volcanogenic areas and poorly drained, swampy land; many are epiphytes. Because of this ecology and their ability to persist in marginal habitat, many have conserved biogeographic patterns over long periods. The diverse genus *Gaultheria* (Ericaceae: Vaccinioideae) has trans-Pacific tracks at northern, equatorial and southern latititudes, along with trans-Tasman connections (Fig. 3.16).

Other members of Ericaceae are diverse in the southern Tasman region and these belong to the subfamily Styphelioideae (formerly Epacridaceae). One example of this group that has already been mapped is the pair comprising *Agiortia* at the MMO and its trans-Tasman–Coral Sea sister *Acrothamnus* (Fig. 4.15). The basal structure of subfamily Styphelioideae as a whole is as follows (Fig. 6.19; K.A. Johnson *et al.*, 2012):

> Tasmania (*Prionotes*), + southern Chile/Argentina (*Lebetanthus*).
>> Tasmania and New Zealand (*Archeria*).
>>> South-western Australia (*Oligarrhena*, *Dielsiodoxa* and *Needhamiella*).
>>> Malesia and Australasia to Rapa, the Marquesas Islands and Hawaii (all other genera).

The first three clades form a paraphyletic basal grade in the south. This indicates either a centre of origin in the south, with the tropics colonized later, or a sequence of breaks in a widespread Pacific group that began in the southern areas (south-western Australia, Tasmania, New Zealand and Patagonia). This would account for the high level of allopatry among the three southern, basal clades, with overlap restricted to Tasmania. The last clade of Styphelioideae (4 in Figure 6.19) has overlapped the basal grade in

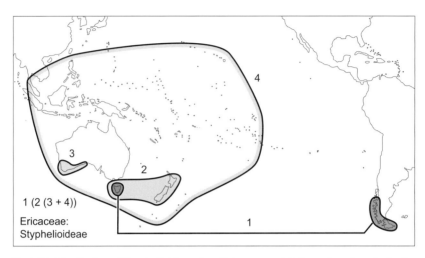

Fig. 6.19 Distribution of Styphelioideae. 1 = *Prionotes* and *Lebetanthus*; 2 = *Archeria*; 3 = *Oligarrhena*, *Needhamiella* and *Dielsiodoxa*; 4 = all other genera. (Phylogeny from K.A. Johnson *et al.*, 2012.)

south-western Australia, Tasmania and New Zealand, but not in America, where the group's absence probably reflects the original allopatry.

Styphelioideae are the only Ericaceae present in a great arc through Western Australia, the New Zealand subantarctic islands and the Chatham Islands. Yet through most of their range they occur together with their sister-group, Vaccinioideae (including *Gaultheria* and others). The overlap between the two subfamilies probably occurred before the Late Cretaceous, as trans-Tasman and trans-Pacific groups occur in both subfamilies.

Within the widespread clade 4 of Styphelioideae (Fig. 6.19), *Monotoca* is a western group, in western and eastern Australia, including Tasmania. It is sister to an eastern group, *Montitega* (formerly known as *Cyathodes dealbata* and *C. pumila*), in Tasmania and New Zealand (Fig. 6.20; Quinn *et al.*, 2003; Albrecht *et al.*, 2010; Puente-Lelièvre *et al.*, 2013). The main break does not occur at the Tasman Sea or between western and eastern Australia. Instead, it is located in, or around, Tasmania (cf. Fig. 1.10), with subsequent overlap there. The break would have formed before the other breaks developed in the Tasman basin and between eastern and western Australia in the mid to Late Cretaceous. A similar break occurs in another member of Styphelioideae clade 4 (Fig. 6.19) and is discussed next.

Dracophyllum and *Sphenotoma* (Ericaceae: Styphelioideae)

Dracophyllum (including *Richea*) grows in many communities around the Tasman basin, from the subantarctic islands to north-eastern Queensland and from the lowlands to the alpine zone (Fig. 6.21). Although the plants are dicots, the leaf blade is linear and forms a broad sheath at the base, as in many monocots, and because of this they have been termed 'grass trees' and 'pineapple shrubs'. The sister-group of *Dracophyllum* is *Sphenotoma* of south-west Western Australia and together they make up tribe Richeeae.

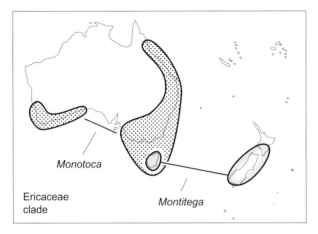

Fig. 6.20 Distribution of two sister genera, *Monotoca* and *Montitega* (Ericaceae subfam. Styphelioideae) (Quinn *et al.*, 2003; Albrecht *et al.*, 2010).

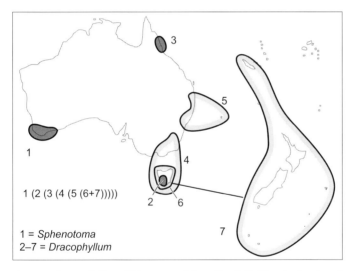

Fig. 6.21 *Dracophyllum* (Ericaceae subfam. Styphelioideae), showing the six main clades (Wagstaff *et al.*, 2010a).

The distribution of the main clades, elucidated by Wagstaff *et al.* (2010a), is shown in Figure 6.21.

   The three basal clades are in the west, and the fourth is in the east. Traditional theory would interpret this as the result of dispersal from south-western Australia eastward, ending with long-distance dispersal from Tasmania to New Zealand. In an alternative, vicariance model, a group widespread through Australia and what became Zealandia began to differentiate at a break between Western Australia and localities further east. One possible mechanism for this would be the last major geological event in the region, the marine transgression of the mid-Cretaceous. Following this, differentiation began in eastern Australia, including a break between Tasmania and the rest of the Tasman

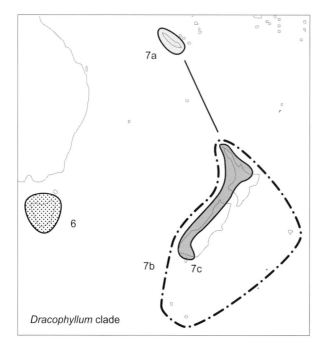

Fig. 6.22 *Dracophyllum* clades 6 and 7 of Figure 4.16 (Wagstaff *et al.*, 2010a).

region that coincides in time and space with Cretaceous basin formation in Bass Strait. Differentiation in the Tasman region culminated with a split between a Tasmanian group (6) and one found throughout Zealandia (7). This would coincide with the Late Cretaceous opening of the Tasman basin.

The distribution of the tribe Richeeae (*Dracophyllum* and *Sphenotoma*) might suggest that the group has died out in central Australia with increasing aridity through the Miocene. There is another possibility, though, as the sister-group of Richeeae is present in central Australia (with groups such as *Leucopogon* s.lat. in the Pilbara, MacDonnell Ranges and the Top End). The sister-group is also in New Guinea–Polynesia, another area where Richeeae are absent (K.A. Johnson *et al.*, 2012). In south-western and eastern Australia and in New Zealand the two groups overlap. This suggests that the absence of Richeeae as such from central Australia and New Guinea is due to phylogeny, rather than extinction.

The *Dracophyllum* clade in New Zealand and New Caledonia (7, shown on Fig. 6.22 with its sister-group in Tasmania) has three main subgroups. One is in New Caledonia, one is in the western parts of central New Zealand and one is widespread in the New Zealand region, with a centre of diversity in the eastern South Island. Again, the three groups show a high level of allopatry, with subsequent, local overlap along the west of New Zealand. In this area there is ecological vicariance, as members of the western New Zealand clade are forest trees, while those of the widespread New Zealand group are smaller shrubs, prostrate mats and cushion-plants of open habitat, such as swamps and alpine vegetation.

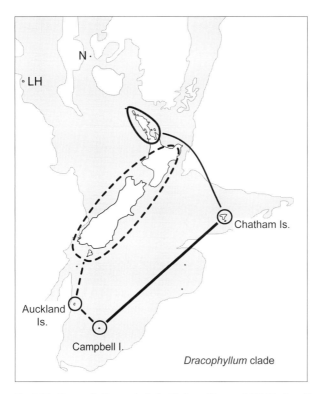

Fig. 6.23 *Dracophyllum* subclade 7b from Figure 4.17. Broken line = *D. longifolium* and allies. *D. scoparium* on Campbell Island is closely related to *D. paludosum* of the Chatham Islands (thick line). The thinner line indicates the affinity between *D. arboreum* of the Chatham Islands and *D. adamsii, D. sinclairii* and relatives of the northern North Island. Grey area = Zealandia (3000 m isobath). LH = Lord Howe Island on Lord Howe Rise, N = Norfolk Island on the Norfolk Ridge. (From Heads, 2009b, reprinted by permission of John Wiley & Sons.)

The widespread New Zealand group of *Dracophyllum* (Fig. 6.23) needs further molecular study, but three possible clades based on morphology occupy allopatric sectors of the New Zealand plateau. The first of these clades includes most of the species (*D. longifolium* etc.) and occurs on mainland New Zealand, Auckland Islands and Campbell Island. The second putative clade is on Campbell and Chatham Islands (*D. scoparium* and *D. paludosum*). The third clade shows an interesting affinity between the other Chatham Islands species (*D. arboreum*) and species of the northern North Island (*D. viride* and *D. sinclairii*).

Whether or not these last connections are confirmed in *Dracophyllum*, a similar Chatham Islands–Northland connection has been found in molecular studies of the fern genus *Asplenium*. *Asplenium flabellifolium* is widespread from south-western Australia to North Island, New Zealand (about 5000 km) (Fig. 6.24). Despite this wide range it does not occur on the Poor Knights (20 km off north-eastern North Island) or Chatham Islands, and these are the only known localities of its rare sister-species, *A. pauperequitum* (Perrie and Brownsey, 2005; Cameron *et al.*, 2006; Papadopulos *et al.*, 2011). The allopatry

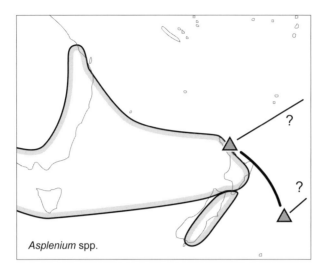

Fig. 6.24 Two sister-species in the fern genus *Asplenium*: *A. flabellifolium* (line with grey, also in south-western Australia) and *A. pauperequitum* (triangles and thick line). Possible vicariance between southern (octoploid) and northern (hexaploid) forms of *A. flabellifolium* is shown (Brownsey, 1977; Dawson *et al.*, 2000). *A. pauperequitum* may be related to Pacific island species that have not yet been sequenced (lines with queries). (From Heads, 2009b, reprinted by permission of John Wiley & Sons.)

suggests the two species are the derivatives of a widespread Australia–New Zealand ancestor. As Perrie and Brownsey (2005) suggested, *A. pauperequitum* could be related to Pacific island species that have yet to be sequenced. In traditional theory the northern New Zealand–Chatham Islands group would be explained by dispersal from New Zealand or extinction on the mainland. Instead, the distributions suggest they are relics of a biota that was once widespread around the islands (now seamounts) of the Hikurangi Plateau and other terranes (Northland–East Coast allochthon) now accreted to northern New Zealand.

Discussions about vicariance in biogeography have often concentrated on the effects of plate tectonic rifting. Yet the phylogenetic break in *Dracophyllum* at the Tasman Sea basin is a late development in the history of the group and its relatives, which are widespread in Australia. The distributions and phylogeny suggest that the differentiation between *Sphenotoma* and *Dracophyllum* took place prior to the Late Cretaceous opening of the Tasman basin, and so the epicontinental seas that bisected Australia in mid-Cretaceous times provide a possible mechanism for the split. As discussed in Chapter 4, these marine transgressions were the result of worldwide rises in sea level and also broad, shallow subsidence in central Australia, caused by plate tectonics.

Plant architecture in *Dracophyllum*

One of the morphological extremes in *Dracophyllum* is represented by trees, such as the New Caledonian *D. involucratum*, with massive leaves and inflorescences, and small floral bracts that are quite different from the leaves. At the other morphological extreme in the genus are prostrate, woody mats and cushions just a few centimetres high with terminal, solitary flowers, as in *D. muscoides* (southern South Island) and *D. minimum* (Tasmania). In these last plants, the leaves are small (2–3 mm long in *D. muscoides*, compared with up to 70 cm in the tree forms) and very similar to the bracts.

For *D. involucratum*, Wagstaff *et al.* (2010a) wrote that the homology of its inflorescence structure 'needs to be investigated in greater detail . . . Its unique inflorescence structure is undoubtedly an autapomorphy'. The homologies are not clear and the cushion plants are far from being just miniaturized forms of the trees. As Oliver (1928) observed: 'Each flowering peduncle [in *D. involucratum*] might be compared with a separate branch of such a species as *D. minimum*'. In this interpretation, the leaves in the *Dracophyllum* cushions are homologous with the bracts, not the leaves, of the tree forms, and the whole cushion plant is homologous with the inflorescence, not the whole plant, of the tree forms. As in the tremands (Elaeocarpaceae), the divaricates and the *Chamaesyce* group of *Euphorbia* (see above), many of the styphelioids and other ericoid, heath-like plants may represent 'bract plants'. In these, the entire shoot system represents an inflorescence in which most of the flowering has been suppressed.

7 Biogeography of New Zealand

New Zealand is polyphyletic in terms of its geology as it is made up of several terranes that each have separate histories and affinities. For example, three intraoceanic arc terranes, now accreted as the Brook Street terrane of New Zealand, the Téremba terrane of New Caledonia and the Gympie terrane of the McPherson–Macleay Overlap, are correlated. In one sense the taxa associated with the terranes can be said to have 'dispersed' to New Caledonia, New Zealand and the McPherson–Macleay Overlap with the accreting terranes and their associated volcanic islands. Unlike the chance events of long-distance dispersal, though, geological dispersal involves entire communities.

New Zealand geology

New Zealand was uplifted along the eastern edge of Gondwana during the Late Jurassic–Early Cretaceous phase of the Rangitata orogeny. The orogeny produced major mountain ranges where there had been sea, and this belt of land formed the basis of modern New Zealand. During this orogeny and the subsequent rifting, the topography of the New Zealand region was turned inside out and there was a similar effect on the biota. The orogeny was not the 'origin' of New Zealand in a strict sense, but it was the last, major modernization of the region and its biota. Before the orogeny, the New Zealand 'basement' terranes (Western Province, Eastern Province and Median Batholith terranes) were spread out along the east coast of Gondwana (east Australia and Antarctica) in an arc at least 6000 km long (see Chapter 6). Later, they were translated along the margin and, during the Rangitata orogeny, amalgamated with the margin itself (Western Province terranes) and with each other (Wandres and Bradshaw, 2005; Adams et al., 2007). This phase would have seen a juxtaposition and fusion of biotas, along with biotic differentiation caused by uplift, subsidence, strike-slip movement and magmatism.

After the Rangitata orogeny, rifting in the Late Cretaceous (84 Ma) separated the whole New Zealand region of continental crust (including New Caledonia) from Australia and Antarctica. The large numbers of clades disjunct between these regions indicate that much of the biota had already differentiated by this time, implicating pre-breakup events such as the Median Batholith magmatism and Rangitata orogeny.

Three major phases of uplift have been recorded in New Zealand since the Paleozoic (Kamp et al., 1989). In Jurassic–Early Cretaceous time the Otago schist belt was uplifted in association with the Rangitata orogeny. During the Late Cretaceous, western parts of the Otago schists were uplifted. In the late Cenozoic there was widespread uplift in the Kaikoura orogeny. This was under way by 7 Ma and the greatest rates of uplift have occurred over the last 1 m.y.

The crust in the New Zealand region shows signs of thickening and shortening – effects of accretion and orogeny – but there has also been great extension. Much of the crust is thinner than that expected for an unextended orogen and the inferred Mesozoic arc–trench gap is greater (Mortimer *et al.*, 2002). Pre-breakup extension developed in the mid-Cretaceous and was focused in areas such as the Great South Basin off south-eastern South Island, and in grabens and metamorphic core complexes on the mainland.

Geologists sometimes argue that a large land region with many endemics was 'completely covered' by a Cretaceous or Cenozoic feature, whether this is the sea, an ultramafic nappe or an ice sheet. But paleogeography around plate margins is often more complicated than this. Some geologists have suggested that New Zealand was completely covered by Oligocene seas, but many biologists and paleontologists have rejected the idea (Tennyson *et al.*, 2010). Instead, they have suggested that some land must have been present, even if it was composed of small, short-lived islands.

North-eastern New Zealand

Many distinctive endemics occur along the island chain off north-eastern New Zealand. These are often allied with groups further south in the East Cape–Cook Strait area, but also show close relationships with groups outside the New Zealand mainland, for example on the Chatham Islands, Kermadec Islands, Norfolk Island and New Caledonia. The geological history of north-eastern New Zealand has been dominated by the arrival of the Hikurangi Plateau (Fig. 7.1), the Northland–East Coast allochthon (Fig. 7.2) and the Loyalty–Three Kings Seamount Ridge (Fig. 7.2). These structures have travelled for some distance from the north-east before docking with the New Zealand basement.

As the Australia/Pacific plate margin has migrated eastward into the Pacific (large arrow in Fig. 7.1), rifts have opened behind it in a west to east sequence: Tasman Sea, Coral Sea, then the New Caledonia, Norfolk–South Loyalty, North Loyalty, South Fiji, North Fiji and Lau-Havre Basins (Fig. 7.1). In addition to the eastward retreating primary subduction zone (1 in the figure), the model discussed here (Schellart *et al.*, 2006) accepted two other subduction zones in the region (2 and 3). Parts of 1 and all of 2 are now inactive. Subduction zone 2 and then subduction zone 3 have developed behind the first subduction zone by splitting lengthwise from it (and inheriting its biota). The critical factor is that the two new subduction zones had a different polarity from their parent. Subduction beneath them has been directed eastward, while the subduction zones themselves have migrated by slab rollback *westward*. Subduction zone 2, with the Loyalty–Three Kings–Northland Plateau arc (lying east of Three Kings Islands), migrated westward until it collided with the Zealandia continental crust. The trench has passed westward beneath the northern North Island and now lies buried there, inactive. Its arc (the Northland Plateau seamount chain), as usual, lies about 200 km behind the

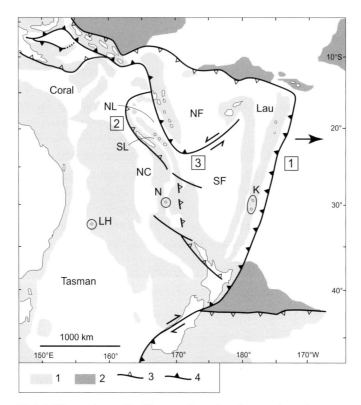

Fig. 7.1 The south-west Pacific tectonic setting: 1 = continental or arc crust; 2 = oceanic plateau; 3 = inactive subduction zone; 4 = active subduction zone. Numbers 1–3 in boxes indicate the subduction zones. Large arrow indicates eastward retreat (rollback) of the primary subduction zone. K = Kermadec Islands; LH = Lord Howe Island; N = Norfolk Islands. Back-arc basins between the Tasman–Coral and Lau basins as follows: NC = New Caledonia; NF = North Fiji; NL = North Loyalty; SF = South Fiji; SL = South Loyalty. (From Schellart *et al.*, 2006. Reprinted by permission of Elsevier.)

trench, east of the mainland. Ophiolite terranes that developed in front of the arc have been obducted onto the North Island at East Cape and Northland (Fig. 7.2).

The Poor Knights, Hen and Chickens, and other islands off north-eastern New Zealand

These islets off north-eastern North Island make up a centre of endemism and many of the groups have their immediate affinities outside New Zealand. For example, one species of *Xeronema* (Asparagales) occurs on the Poor Knights and Hen and Chickens, while the only other species occurs in New Caledonia (Fig. 7.2). The pair make up the Xeronemataceae, basal in a global complex of Asparagales (Chapter 5). In a similar pattern, the tree *Hymenanthera novae-zelandiae* (Violaceae) from northern New Zealand islands (Three Kings, Poor Knights etc.) is sister to the Norfolk Island endemic *H. latifolius* (Mitchell *et al.*, 2009).

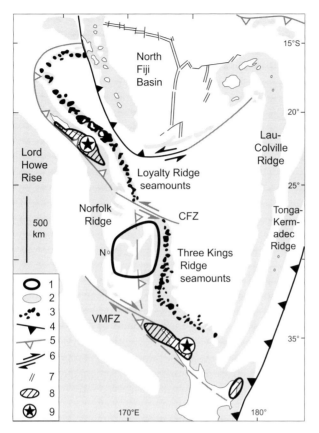

Fig. 7.2 New Caledonia and northern New Zealand: CFZ = Cook Fracture Zone; N = Norfolk Island; VMFZ = Vening Meinesz Fracture Zone. 1 = area emergent 38–21 Ma, destroyed by subsequent extension (Meffre *et al.*, 2007); 2 = continental/arc crust; 3 = seamounts (subduction-induced arc volcanics); 4 = subduction zone; 5 = New Caledonia fossil subduction zone; 6 = strike-slip fault; 7 = spreading ridge; 8 = obducted mafic allochthons; 9 = the monocot *Xeronema* (Xeronemataceae). (From Schellart *et al.*, 2009. Reprinted by permission of Elsevier.)

The Poor Knights and the Hen and Chickens belong to a chain of islands that runs from the Three Kings Islands in the north, south-east to the Bay of Plenty. Along with the many endemics on the chain, many taxa on the islands also occur on some headlands and peninsulas of the adjacent mainland. In many cases an endemic on the islands has an obvious vicariant on the mainland. Endemism on the north-eastern islands conforms to a common type of distribution that has been termed 'horstian'. In these distributions, taxa are endemic to belts of islands (or peninsulas) fringing a mainland and are absent on the mainland, despite the islands being closer to the mainland than they are to each other. Other examples of 'horstian' distribution are seen in plants and animals on the fringing islands of New Guinea and Venezuela. This sort of pattern can develop in simple horst–graben systems, with vertical movement of fault blocks, but in practice, the tectonics are often more complex and also involve lateral (strike-slip) displacement.

Apart from their terrestrial endemism, the islands of north-eastern New Zealand are well known for their endemic marine life. The hagfishes are the jawless sister-group of all other vertebrates and are characterized by their interesting biogeographic patterns (Cavalcanti and Gallo, 2007). The largest hagfish known is a giant, fluorescent pink species 1.2 m long, caught at Hauraki Canyon east of the Poor Knights (Mincarone and Stewart, 2006). As with terrestrial groups, marine clades in the area often show affinities among the offshore islands that do not involve populations on the adjacent mainland. For example, the fish *Chironemus microlepis* (Chironemidae) is only known from the Poor Knights, Kermadec, Norfolk and Lord Howe Islands (Meléndez and Dyer, 2010).

As with *Chironemus microlepis*, many of the groups on the north-eastern islands, marine and terrestrial, are endemic to more than one of the island groups. In a dry land example, the monocot *Xeronema callistemon* is endemic to the Poor Knights and also the Hen and Chickens, as is the tree *Hoheria equitum* (Malvaceae) (Heads, 2000). In a simple dispersal model, these species would be assumed to be nested in a local clade of northern New Zealand. Instead, the *Xeronema* is sister to a New Caledonian clade, while the *Hoheria* is sister to a clade of five species distributed through most of the North and South Islands (Wagstaff and Tate, 2011). Both patterns are compatible with early vicariance.

Moore (1957) cited the similar open, rocky habitats of *Xeronema* (i.e. Xeronemataceae) in its three known localities: the Poor Knights, the Hen and Chickens (all at coastal elevations) and in New Caledonia (at 1500 m). She wrote that 'no morphological feature suggests an explanation for the peculiar distribution of the genus' [i.e. as a means of dispersal]. Instead, she compared the distribution with those of other taxa, such as *Meryta* (Araliaceae): Hen and Chickens, Three Kings Islands, New Caledonia and other Pacific islands. In *Meryta* and *Xeronema* the north-eastern New Zealand centre of endemism is related to New Caledonia by 'horstian' groups that are not on the New Zealand mainland. Other Poor Knights groups not on the mainland connect the group with the Chathams to the south-east. The fern *Asplenium pauperequitum* is endemic to the Poor Knights and the Chatham Islands (Cameron *et al.*, 2006) and is allopatric and sister to, not nested in, a widespread mainland group, *A. flabellifolium* of Australia and New Zealand (Fig. 6.24). The simplest explanation for this reciprocal monophyly and the peculiar absence of the widespread, mainland group on both the Poor Knights and the Chatham Islands (Brownsey and Smith-Dodsworth, 1989; Perrie and Brownsey, 2005) is vicariance.

Northland–East Coast tracks and allochthons

Evolution around the north-eastern 'horstian' tracks can be related to the history of the Three Kings–Northland Plateau seamounts (Schellart, 2007). Some of the seamounts have flat tops and limestone caps indicating wave erosion to sea level (Herzer *et al.*, 2009). At least one large, long-lived Tertiary island between New Zealand and New Caledonia on the Norfolk Ridge (see bold line in Fig. 7.2) has been obliterated by extension, and Meffre *et al.* (2007) illustrated a well-preserved leaf fossil from seafloor rocks in the area.

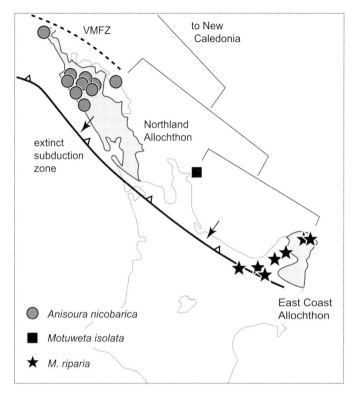

Fig. 7.3 The New Zealand tusked wetas, *Motuweta* and *Anisoura* (Orthoptera: Anostostomatidae; Gibbs, 2001; Pratt *et al.*, 2008) and related allochthonous terranes (Cluzel *et al.*, 2010b). VMFZ = Vening Meinesz fracture zone. Arrows indicate the former south-eastward retreat (rollback) of the now-buried, extinct subduction zone.

Another major tectonic feature in the region is the Vening Meinesz fracture zone (Fig. 7.2; Heads, 1990b). This is a transform, strike-slip fault zone with dextral displacement that runs parallel with the line of endemism. (In dextral or right-lateral displacement, the land on the other side of the fault from the observer moves to the right.)

Mainland geological structures possibly relevant to the north-eastern endemism include the related Northland and East Coast allochthons (Fig. 7.3). Cluzel *et al.* (2010b) concluded that submarine volcanic rocks of both allochthons formed in a single Late Cretaceous to Eocene basin, in association with a subduction zone and its arc. The allochthons (including an ophiolite sequence) were obducted as thrust complexes onto the Northland/East Coast basement and early Cenozoic cover in the earliest Miocene. During this process the allochthons are thought to have been displaced by ∼300 km.

Distributions related to the 'horstian' patterns are seen in groups found in Northland and East Cape regions, with intermediate records on the islands. A typical example is a clade in the orthopteran family Anostostomatidae, the king crickets and true wetas (Fig. 7.3). The clade comprises the genera *Motuweta* and *Anisoura*, New Zealand endemics that occur in Northland, East Cape and the Mercury Islands (Fig. 7.3). These genera are the New Zealand representatives of 'tusked wetas', one of the three main weta clades in New Zealand. *Motuweta* and *Anisoura* form a clade with *Aistus* and *Carcinopsis* of New Caledonia, the only wetas in that country (Pratt *et al.*, 2008).

The Anostostomatidae as a whole has a typical Gondwanan distribution: South Africa, Madagascar, India, Australasia, central and South America (Pratt *et al.*, 2008). Liebherr *et al.* (2011) accepted the weta assemblage as Jurassic, based on an Upper Jurassic fossil. This is a left wing fragment from Port Waikato, New Zealand, described as *Notohagla mauii* (Grant-Mackie *et al.*, 1996). *Notohagla* belongs to the Prophalangopsidae, in a group with the phylogeny: Anostostomatidae (Prophalangopsidae (Haglidae + Gryllidae)). Liebherr *et al.* (2011) did not accept the much younger ages for the weta lineage calculated by Trewick and Morgan-Richards (2005) and based on a questionable evolutionary rate (Brower, 1994; cf. Heads, 2012a).

The New Caledonia–Northland–East Cape clade of tusked wetas shown in Figure 7.3 makes sense in a geological context. Cluzel *et al.* (2010b) discussed geological similarities and differences between the East Coast/Northland allochthons and the Poya terrane of New Caledonia (Fig. 7.2), and suggested that these have been separated by a dextral transform fault, parallel with the Alpine fault. Other New Caledonia–New Zealand similarities are seen in the faunas of the New Zealand Murihiku terrane (Permian–Jurassic) and the New Caledonia Téremba terrane. The similarities are also seen in the 'modern' biota of New Zealand and New Caledonia, including *Xeronema*, *Meryta* and the tusked wetas. Other New Zealand–New Caledonia connections in the extant biotas are well known (see under track 7 in Heads, 2010a).

Northland and East Cape to Cook Strait

Many groups in Northland, the north-eastern islands and East Cape also include localities further south along the eastern side of the North Island, from East Cape to Cook Strait. This region lies next to the Hikurangi Plateau (Fig. 7.1) and has been strongly deformed along the margin with the plateau. A typical example of disjunction across this zone is the basal clade in the diverse, New Zealand-wide skink *Oligosoma* (Hare *et al.*, 2008; Chapple *et al.*, 2009). There are two named species:

1. Northland and its offshore islands, usually within 20 m of the high tide mark (*O. smithi*).
2. East-central North Island (southern Urewera, Taihape, etc.) (*O. microlepis*).

Other groups disjunct between the north-east islands and the eastern North Island include *Hebe parviflora* (Plantaginaceae; Bayly and Kellow, 2006).

Oligocene–Miocene marine transgressions from the east coast (before the rise of the axial Ruahine Mountains) reached Taihape and would have led to the range expansion of coastal elements inland. Following marine regression and the uplift of the mountains, some populations would have been left stranded inland, west of the mountains. This process would explain certain land-locked plant records near Taihape, such as outliers of the eastern *Olearia gardneri* and anomalous inland populations of *O. solandri* (Asteraceae; Heads, 1998a).

Several of the New Zealand offshore island endemics have fossil records on the mainland and so it is sometimes thought that all the island groups were once widespread throughout the country. This cannot be assumed, though, and in island groups with mainland fossil records, the fossils are often restricted to certain areas. For example, *Sphenodon* (tuatara) occurs on the north-eastern islands and also islets further south, around Cook Strait. According to Colenso (1885) 'The old Maoris always said that the tuatara formerly inhabited the headlands of the New Zealand coast (as well as the islets lying off it)'. In addition to sandy coastal habitat, fossils show that tuatara also occupied sandy or gravelly banks of large, braided rivers. Fossils indicate that populations lived in coastal places around the east of the country as far south as Bluff, extending inland in a few areas (Central Otago and the King country), but with west coast records restricted to Nelson (Hay *et al.*, 2003). The group seems to have had a distribution mainly in the east of New Zealand, and the abundance and diversity along the north-eastern islands and headlands reflects this.

Southern New Zealand: Campbell Plateau and the New Zealand subantarctic islands

Zealandia includes the emergent mainland of New Zealand, along with the Earth's largest area of submerged continental crust, made up of the Campbell Plateau, Chatham Rise, Lord Howe Rise and Norfolk Ridge. On the Campbell Plateau, south of New Zealand, the only areas of land are the small, isolated subantarctic islands, but these include distinctive plants and animals. Among the Tasman groups that have global sisters (Chapter 5), *Chionochloa* (Poaceae) is represented on the Campbell Plateau by *C. antarctica*, endemic on Auckland and Campbell Islands. The athoracophorid slugs are represented by endemic species on Macquarie, Auckland, Campbell, Snares and Chatham Islands (Fig. 5.1) (Burton, 1963, 1980).

Endemism in the subantarctic islands reaches at least generic level. In the beetle fauna of the Auckland Islands, the presence of four endemic, monotypic genera and the diversity shown within several other genera is 'odd', and 'not consistent with the present land area of the Auckland Islands' (Michaux and Leschen, 2005: 101). In Diptera, seven genera are restricted to the Auckland Islands, one to Campbell Island, three to both Auckland and Campbell, and one to the Bounty Islands. Michaux and Leschen (2005) cited endemic genera of Hymenoptera, Lepidoptera and spiders on the subantarctic islands, as well as a distinctive absence of some groups that are common on mainland New Zealand.

The islands on the Campbell Plateau are mostly composed of young volcanic strata and so the islands themselves are sometimes regarded as young (e.g. McDowall, 2010: 197). But the continental basement is exposed on several of the islands and Campbell Island has the oldest known rock in New Zealand (Precambrian schist; Campbell and Hutching, 2007). Mesozoic granite is exposed on the Snares and the Bounty Islands, and there is Mesozoic schist on the Chatham Islands. The whole plateau was once above water, before extension and thinning led to reduced buoyancy. Granite has a density of

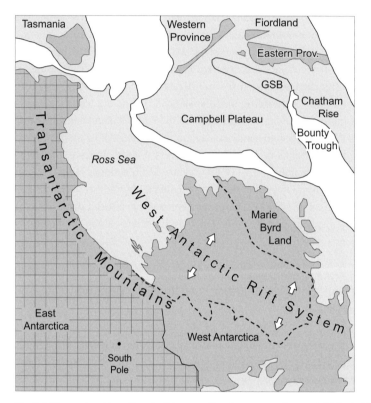

Fig. 7.4 Reconstruction of part of the east Gondwana margin at ~90 Ma (mid-Cretaceous), soon before complete rifting developed between the Campbell Plateau and Antarctica. Grey = continental crust, bounded by the 1500-m isobath, the rifted margin. Dark grey = emergent continental crust. Grid pattern = East Antarctica craton. Broken line = the Cretaceous West Antarctic Rift System, with extension indicated by arrows. GSB = Great South Basin. (From Siddoway, 2008 and Veevers, 2012.)

2.7 g/cm^3, but the same amount of iron (density 7.9 g/cm^3) weighs more than twice as much. Mafic rocks such as basalt contain magnesium and iron (hence 'ma-fic') and are more dense (>3 g/cm^3) than silicic or felsic (feldspar + silica) rocks such as granite. Silicic rocks form light 'continental crust' that is buoyant and usually emergent, while mafic rocks form 'oceanic crust' that is more dense and usually lies below sea level.

If continental crust is extended and thinned, it becomes less buoyant, just as a small iceberg does not project as far above sea level as a large one. The continental crust of the Campbell Plateau is much thinner than normal continental crust, but the nature and timing of the tectonic processes that stretched the plateau crust are not well understood (Grobys *et al.*, 2009). Different authors have related the thinning to Early Cretaceous extension or to the final breakup of New Zealand from Antarctica in the Late Cretaceous. Whether the Campbell Plateau extension took place at the same time as the extension of the Bounty Trough and Great South Basin (Fig. 7.4) is also debated.

Grobys *et al*. (2009) suggested that an early phase of Campbell Plateau extension (Early Cretaceous at ~135–110 Ma) predated the opening of the Great South Basin. Apart from causing the subsidence, extension is relevant for biogeography as a means of vicariance. For example, mainland and Campbell Plateau populations have been separated by extension in the Great South Basin and Bounty Trough (Fig. 7.4).

In Early Cretaceous time, prior to Gondwana breakup, West Antarctica, the Campbell Plateau and the rest of Zealandia formed part of the Gondwana convergent margin (Fig. 7.4). Oceanic crust in the Pacific encroached on Gondwana and subducted beneath it. At a tectonic switch in the mid-Cretaceous (~100 Ma), extension developed in the back arc region, perhaps as the result of slab rollback. This generated the intracontinental West Antarctic Rift System (WARS), the Great South Basin–Campbell Plateau extensional province (Fig. 7.4) and many other rift features in mainland New Zealand. The WARS is the product of multiple stages of intracontinental deformation from Jurassic to Present, with the main extension occurring in mid–late Cretaceous time, between 105 Ma and 90 Ma (Siddoway, 2008). This resulted in >100% extension across the Ross Sea and central West Antarctica. West Antarctica–New Zealand breakup is distinguished as a separate event at 83–70 Ma, developing at the same time as the Tasman basin. Bialas *et al*. (2007) proposed that the whole WARS–Transantarctic Mountains region was a high-elevation plateau with thicker than normal crust before the onset of continental extension. Another model suggests that uplift of the Transantarctic Mountains has occurred in an overall extensional environment during formation of the WARS (van Wijk *et al*., 2008). These events in Antarctica are relevant for the biogeography of clades in Australia–East Antarctica–South America, and their differentiation from sister-groups in New Zealand–West Antarctica–South America (Fig. 3.3).

The Alpine fault: disjunctions and other breaks at a plate boundary

In the South Island, higher levels of species diversity occur in the north (Nelson) and south (Otago and Southland) than in central areas, where the highest mountains are found. Traditional theory attributed this unusual pattern to glaciation in the central part of the island wiping out taxa there. Neiman and Lively (2004) suggested that Pleistocene glaciations produced a gap in the central areas, while movement on the Alpine fault and the rise of the Southern Alps led to an east/west split in taxa. In fact, fault movement and glaciations can, in theory, both lead to similar patterns. This is because there has been 480 km of horizontal, strike-slip displacement along the Alpine fault in addition to the vertical uplift. The horizontal displacement is dextral.

Figure 7.5 (from Musgrave, 2003) shows the horizontal deformation of continental Zealandia that took place in the Neogene. (The blocks show changing relationships between localities; there is no implication of coastlines.) The Alpine fault, part of the Australia–Pacific plate margin (Fig. 7.1), is shown in the figure, along with a marker terrane, the Dun Mountain ophiolite belt and its associated magnetic anomaly (heavy black in the figure). The Dun Mountain terrane has been distorted and eventually ruptured by strike-slip movement along the fault. This probably began in the Miocene.

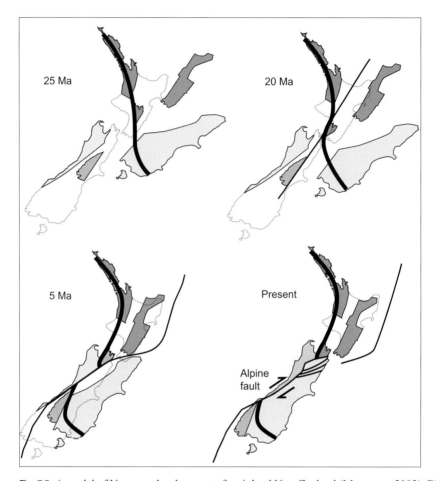

Fig. 7.5 A model of Neogene development of mainland New Zealand (Musgrave, 2003). Black line = Dun Mountain–Maitai terrane. (Reprinted by permission of the Geological Society of America.)

Biogeographic patterns suggest that the strike-slip displacement caused distributional offsets as well as complete disjunctions in the South Island biota. Disjunct patterns that are explicable in this way are now documented in more than 100 taxa (Heads, 1998b; Heads and Craw, 2004). The outcome following fault movement depends on the prior distribution of the group. Groups that were widespread before fault movement and groups that were only on one side of the fault will both be the same after fault movement. Groups with restricted distributions on both sides of the fault to start with (as with the Dun Mountain terrane) can be expected to show signs of disruption.

Harvestmen (Arachnida: Opiliones)

Within the harvestman *Aoraki denticulata* s.lat. (Pettalidae), at least two different clades show disjunction along the Alpine fault (Fig. 7.6; Boyer *et al.*, 2007a). The distributions

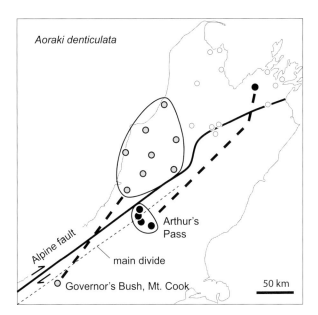

Fig. 7.6 *Aoraki denticulata* (Opiliones: Pettalidae) showing total distribution, two separate clades (black dots, grey dots) and other populations (some not sequenced) (open circles) (from Boyer *et al.*, 2007a). The populations at Arthur's Pass were earlier treated as *A. denticulata major*; the one at Governor's Bush, Mount Cook, was treated as a separate species, *A. longitarsus*.

show parallel disjunctions and this suggests that differentiation within the species had already occurred before transcurrent movement on the fault. The populations in the Arthur's Pass region are separated by the Alpine fault, not the main geographic divide (fine dashed line in Figure 7.6), indicating that uplift along the fault has probably elevated populations but had little phylogenetic effect. The break is instead compatible with strike-slip displacement.

Skinks

Aoraki denticulata shows double affinities disjunct across the Alpine fault and this also occurs in a clade of skinks, *Oligosoma infrapunctatum* and its allies (Fig. 7.7; Chapple and Patterson, 2007; Bell and Patterson, 2008; Patterson and Bell, 2009; Chapple *et al.*, 2009). Strike-slip movement on the fault would explain the dextral off-sets in both the harvestmen and the skinks. In the same group of skinks, another important geological break, the Waihemo fault zone, separates *O. otagense* and *O. waimatense*.

In another skink group, the widespread *Oligosoma nigriplantare polychroma*, Liggins *et al.* (2008b: 3677) found that 'most genetic variation is partitioned across the Alpine Fault'. They concluded that the Alpine Fault initiated allopatric divergence in *O. n. polychroma* and cited 'the importance of the Alpine Fault in the evolution of New

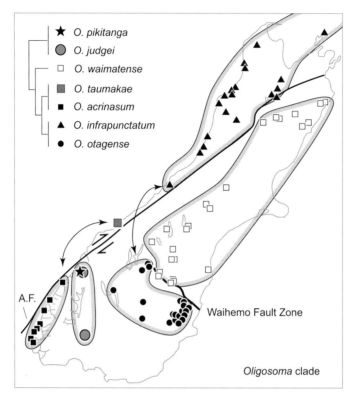

Fig. 7.7 *Oligosoma infrapunctatum* and allies (Scincidae). (Phylogeny from Patterson and Bell, 2009.)

Zealand's biota' (p. 3668). Their clock estimates indicated that genetic divergence at the break predated Pleistocene glaciations.

Kiwis (Palaeognathae: *Apteryx*)

In New Zealand ratite birds, the kiwi genus (*Apteryx*) comprises two groups, brown kiwi and spotted kiwi. The brown kiwi clade has two allopatric groups with a mutual boundary that lies along the Alpine fault, rather than the main divide (Fig. 7.8; Shepherd *et al.*, 2012). Baker *et al.* (1995) interpreted the clade located south-east of the fault (*A. australis* etc.) as a remnant of the 'original ancestral population'. They proposed that this population moved north from Fiordland to Haast and then Okarito, where the population diverged. It then expanded into the northern South Island and all of the North Island. This does not account for the break, though, and this can be explained most simply if the ancestral brown kiwi population was already widespread before the split. The main phylogenetic break occurs between populations of the *A. australis* clade near Haast (Haast Range, Arawata Valley; Holzapfel *et al.*, 2008) and members of the northern group (*A. mantelli* etc.) at Okarito. As with the harvestmen and the skinks, the break

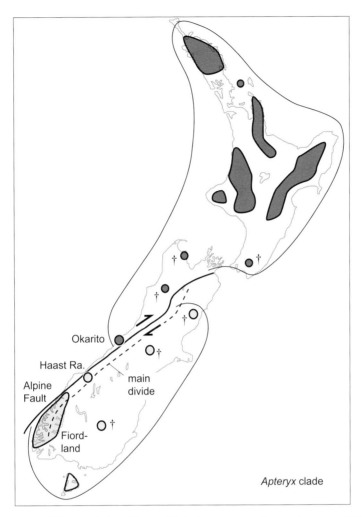

Fig. 7.8 The two clades of brown kiwi: *Apteryx australis* species group (light grey) south-east of the Alpine fault, and *A. mantelli* species group (dark grey) north-west of the fault. Dagger symbols = sequenced, subfossil samples from populations now extinct (Shepherd *et al.*, 2012).

corresponds with the plate boundary, the Alpine fault. Many populations of the southern *Apteryx australis* group lie south-east of the main geographic divide, the Southern Alps. The population at Haast and many in Fiordland lie *west* of the main divide but *east* of the Alpine fault (Fig. 7.8), and so the phylogenetic break corresponds with the fault, not the main divide. This repeats the pattern seen in *Aoraki denticulata*. Wallis and Trewick (2009: 3562) regarded the basal split in the brown kiwi group as Pliocene and suggested it is more likely to reflect earlier uplift than Pleistocene glaciations. Yet if the break were due to uplift it would be located at the main geographic divide (the watershed) rather than the fault itself.

The New Zealand wrens, Acanthisittidae

The basal passerine clade, the Acanthisittidae, has two extant species, *Acanthisitta chloris* and *Xenicus gilviventris*. In both species the Fiordland populations are distinctive and they have been named as *A. chloris citrina* and *X. gilviventris rineyi*. Morphology and behaviour suggest that both the Fiordland forms have their closest relatives in Nelson, not in adjacent Westland (see notes in Heads, 1998b), and a molecular study is needed.

8 Biogeography of New Caledonia

New Caledonia is an archipelago and includes the large island of Grande Terre along with the three smaller Loyalty Islands (see Figure 7.2 for the tectonic setting). The land covers an area of ~19 000 km², about the same as Massachusetts or Wales, and less than a tenth of the area of New Zealand. The biota is distinguished by its high diversity, endemism, peculiar absences and far-flung biogeographic affinities. There is probably no other region of comparable area with such a rich and peculiar flora, one that 'has good claim to be considered the most remarkable in the world' (Thorne, 1965: 1; cf. Good, 1974). In seed plants, apart from the three endemic families (Heads, 2010b), 105 of the 711 genera (15%) are endemic, as are 2324 of the 3004 indigenous species (77%) (Jaffré *et al.*, 2001). It must also be one of the smallest areas in the world with an endemic bird family, the Rhynochetidae (Gruiformes).

Several endemic New Caledonian plant taxa are sister to global groups (*Amborella*) or to widespread northern hemisphere groups (*Canacomyrica*, *Austrotaxus*). Other endemics belong to trans-Indian Ocean groups (*Strasburgeria*), trans-tropical Pacific groups (*Oncotheca*) or Tasman Sea/Coral Sea groups (*Phelline*, *Paracryphia*). Each of these trans-ocean basin groups is sister to widespread northern hemisphere or global groups. Some groups endemic to New Caledonia plus New Guinea (*Sparattosyce* and *Antiaropsis*: Moraceae), and to New Caledonia plus New Zealand (*Xeronema*: Xeronemataceae) are basal in global clades.

What is the explanation for the great diversity in New Caledonia? The biota has often been accounted for in terms of vicariance, with New Caledonia and the rest of Zealandia breaking away, as a unit, from Gondwana in the Cretaceous. Now, with the increasing use of fossil-calibrated clock models, that idea is being replaced with a model of chance, long-distance dispersal from mainland centres of origin. A study of caddisflies (Trichoptera), for example, used a fossil calibration (from Baltic amber) to date nodes on a phylogeny, and by treating these as maximum ages, ruled out earlier vicariance (Espeland and Johanson, 2010).

Biogeographic patterns in the Muscidae (Diptera) have been explained both by vicariance and by trans-oceanic dispersal. Based on its fossil record, the family has been thought to date to the Eocene or perhaps Late Cretaceous, and Couri *et al.* (2010: 56)

This chapter incorporates material previously published in the *Journal of Biogeography* (Heads, 2008b, 2010b) and reprinted here by permission of John Wiley & Sons.

wrote that 'Considering this relatively young age of the Muscidae, the distribution patterns of the New Caledonia species could be explained by the dispersal hypothesis'. In an alternative approach, the dates for the family can be treated as minimum estimates. This has been suggested for one Muscidae subfamily, the Reinwardtiinae. In this group, Couri and Carvalho (2003) suggested that Neotropics–Afrotropics distribution patterns are consistent with Gondwanan vicariance and they accepted an earlier age for the family than that suggested by the fossils. The authors explained Neotropics–Afrotropics distributions in the Muscidae subfamily Dichaetomyiinae in the same way. Likewise, vicariance in the Diptera of the Tasman region can be accounted for by the opening of the basins around New Caledonia, terrane accretion, orogeny and strike-slip movement, as in New Zealand (Chapter 7). There has been less molecular work carried out in New Caledonia than in Australia or New Zealand, but some interesting biogeographic patterns have already been documented and include the following examples (see also Heads, 2008a,b; 2010a,b).

Absences in the New Caledonian biota

Plants such as *Coprosma* s.str. (Rubiaceae) surround New Caledonia on all sides but are absent from it and this distinctive pattern also ocurs in animal life. One interesting example occurs in the passerine family Petroicidae. The family has three main clades (Christidis *et al.*, 2011):

> *Petroica*, etc.: mainland Australia (widespread), Tasmania, New Guinea, Bismarck Archipelago, Solomon Islands, Vanuatu, Fiji and Samoa **(not New Caledonia)**, New Zealand (including Auckland, Snares and Chatham Islands), Norfolk Island.
>
> > *Microeca*, etc.: mainland Australia (widespread), Tasmania, New Guinea, New Britain and **New Caledonia**.
> >
> > *Eopsaltria*, etc.: mainland Australia (widespread), New Guinea.

The *Petroica* clade surrounds New Caledonia (in New Guinea–Vanuatu–Norfolk–Australia), but does not occur in the archipelago itself. This strange absence might be explained by the vagaries of chance dispersal to the other areas, or to chance extinction in New Caledonia. Yet the clade is present on many islands that are much smaller (such as Norfolk and Tanna) and the second petroicid clade is present on New Caledonia. This suggests that the absence of the *Petroica* clade from New Caledonia is due to phylogeny, rather than chance or ecology.

The petroicid that is present on New Caledonia was formerly treated as an outlier in the Australian genus *Eopsaltria*. In fact, it has turned out to be sister to a group that is widespread throughout Australia, New Guinea and New Britain (*Microeca* plus *Monachella*). This means the New Caledonia group is more likely to be derived by phylogenetic vicariance than by derivation from any one part of Australasia or Melanesia. (In a dispersal model, dispersal would need to have occurred before any mainland differentiation, as in simple vicariance.)

All three petroicid clades occur through Australia and New Guinea, but the distribution of the first two in other localities and the absence of the third from Tasmania suggest the original allopatry of the groups. The overlap in Australia and New Guinea would then be a function of subsequent, local range expansion.

Biogeographic affinities of the New Caledonian biota

Indian Ocean connections

Acridocarpus (Malpighiaceae) has a typical disjunct Indian Ocean distribution. There are about 30 species in Africa and Madagascar, and one endemic to New Caledonia (sister to two Madagascan species). Davis *et al.* (2002) used an Oligocene fossil of Malpighiaceae to calibrate a clock for the genus and they dated the origin of the Madagascar–New Caledonia affinity at 15–8 Ma. Treating this as a maximum age, they inferred a long-distance dispersal event from Madagascar to New Caledonia (~10 000 km). They described *Acridocarpus* as a 'discordant' element in the flora of New Caledonia, as all the other Malpighiaceae there have closest relatives in Australia. Nevertheless, a connection between the Tasman basin and the south-west Indian Ocean is common enough, for example, in *Lordhowea* (Fig. 1.14). Likewise, *Cunonia* (Cunoniaceae) is known from New Caledonia and the south-western Cape region of South Africa (Heads, 2012a, Fig. 9.6).

The main node in *Acridocarpus* is located at the Mozambique Channel and separates a clade of Madagascar and New Caledonia from one of East and West Africa. A break at the same locality occurs in many groups, such as the strepsirrhine primates and the owl genus *Otus*. In the latter, a clade of Madagascar, the Seychelles and Southeast Asia is sister to one from Africa and Eurasia (Fuchs *et al.*, 2008). If the fossil-calibrated dates in *Acridocarpus* are treated as minimum ages (and not converted into maximum ages) a simple vicariance model is possible, with the primary break in a widespread Africa–Australasia ancestor occurring at the Mozambique Channel. The absence from Australia could be due to extinction there following Miocene aridification, but the group is diverse in drier parts of Africa. This suggests that before rifting, the New Caledonia–Madagascar–Africa groups were connected via East Antarctica (cf. Fig. 3.3) and were never widespread in Australia.

Different morphological characters often vary in different ways through a group and often have different geographic distributions. All of these are of biogeographic interest, but not all of the variation can be represented in a hierarchical classification or phylogeny of the group. In the same way, different parts of the genome vary in different ways, as seen in *Lordhowea* (Fig. 1.14). The different character geographies can be obscured in phylogenies that combine the data. In some cases, incongruent results that do not fit with biogeographic preconceptions are left out of the analysis. In *Phyllanthus* (Phyllanthaceae), *matK* sequences linked New Caledonian members of subgen. *Gomphidium* with another New Caledonian clade of *Phyllanthus* (Kathriarachchi *et al.*, 2006), giving

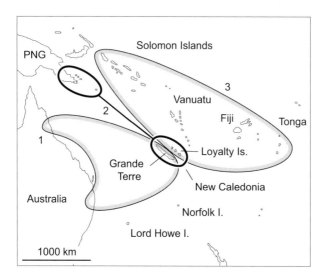

Fig. 8.1 The south-west Pacific region, with distributions of: 1 = the orchid *Acianthus amplexicaulis*; 2 = the tree *Hunga* (Chrysobalanaceae); 3 = the boid snake *Candoia bibroni*. (From Heads, 2008b, reprinted by permission of John Wiley & Sons.)

'geographic coherence'. An ITS tree instead placed the New Caledonian *Gomphidium* with the Madagascan *Phyllanthus betsileanus* (with 92% bootstrap support) and this distribution pattern has many precedents; for example, *Acridocarpus*. Nevertheless, Kathriarachchi *et al.* (2006) concluded that the New Caledonia–New Caledonia tie in *matK* is 'biogeographically more plausible' than the New Caledonian–Madagascar disjunction in ITS. They removed the ITS data of the Madagascan species from their combined (ITS and *matK*) matrix, suggesting that it is 'most likely spurious because it does not fit the general patterns of geographical relationships observed' (p. 643). Determining, on the basis of biogeography, which molecular data should be ignored and which highlighted requires a broad comparison of regional and intercontinental biogeographic patterns. The patterns in *Acridocarpus, Cunonia, Dietes* (pp. 74, 122), *Phyllanthus* 'New Caledonian *Gomphidium*' (ITS sequences), *Lordhowea* (chloroplast DNA sequences) and Crossosomatales (Fig. 3.1) all conform to the standard Tasman region–south-west Indian Ocean connection.

Tasman Sea–Coral Sea groups

New Caledonia shows biogeographic affinities with many other parts of the Tasman region. Three of the northern connections are with Queensland, the Papuan Peninsula and the Vitiaz arc (Solomon Islands to Tonga); these are illustrated in Figure 8.1 with an orchid (*Acianthus amplexicaulis*), a tree (*Hunga*: Chrysobalanaceae) and a snake (*Candoia bibroni*: Boidae).

Connections between New Caledonia and Australia can be broken down into southern (Tasman Sea) patterns and northern (Coral Sea) patterns (Fig. 8.2). The first is

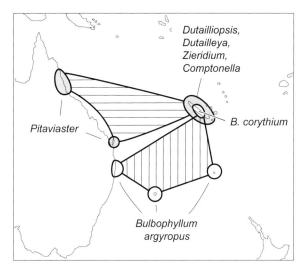

Fig. 8.2 Distribution of *Bulbophyllum argyropus* and the related *B. corythium* (Orchidaceae) (Green, 1994); distribution of a group of five genera in Rutaceae (Hartley, 2001).

illustrated here with the orchid *Bulbophyllum argyropus* of the McPherson–Macleay Overlap, Lord Howe Island and Norfolk Island. In a morphological study this was related to *B. corythium* of New Caledonia (Green, 1994). Patterns further north are seen in a possible group of Rutaceae. This comprises five genera that were keyed together in a morphological revision. The genera occur in New Caledonia (*Dutailliopsis*, *Dutailleya*, *Zieridium*, *Comptonella*) and in eastern Australia (*Pitaviaster*). The latter genus is disjunct at Fraser Island (by the McPherson–Macleay Overlap) and in northeastern Queensland (Hartley, 2001; T.G. Hartley, personal communication, 15 April 2005). The pattern in the five genera is similar to that of *Acianthus amplexicaulis* (Fig. 8.1).

Other Tasman basin affinities in New Caledonian groups are seen in the only parasitic gymnosperm, *Parasitaxus* (Podocarpaceae). This is endemic to New Caledonia and is sister to *Lagarostrobos* (including *Manoao*) of New Zealand and Tasmania (Biffin *et al.*, 2011).

Coral Sea groups in New Caledonia include *Hunga* (Fig. 8.1) and *Spiraeanthemum* s.lat. (Fig. 8.3; Pillon *et al.*, 2009). The latter is the basal clade in the tree family Cunoniaceae, widespread in the southern hemisphere and in the tropics. The two main clades in *Spiraeanthemum* are allopatric over most of their range, with one clade on the inner Melanesian arc (New Caledonia–New Guinea) and the other mainly on the outer Melanesian or Vitiaz arc (Bismarck Archipelago, Solomon Islands, Vanuatu, Fiji). The two clades overlap only in New Caledonia, while the two subclades in the outer arc group overlap only in Fiji. New Caledonia and Fiji are both tectonic composites formed by the fusion of several island arcs and this provides a mechanism for the secondary overlap there in *Spiraeanthemum*, following the original allopatry among the three clades. The conspicuous distributional break between the New Guinea mainland and the Bismarck

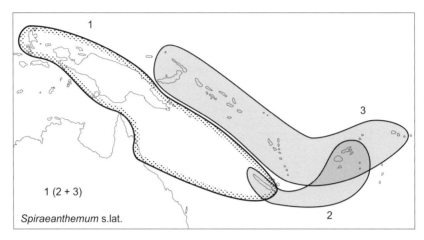

Fig. 8.3 Distribution of *Spiraeanthemum* (Cunoniaceae), showing the three main clades and a phylogeny (Pillon *et al.*, 2009). (From Heads, 2010a, reprinted by permission of John Wiley & Sons.)

Archipelago, as seen in *Spiraeanthemum*, is a classic problem of biogeography. It occurs in many groups and is discussed further in the next chapter.

The bird family Aegothelidae (owlet–nightjars) is of special interest as a Tasman group with a global sister (the swifts and hummingbirds; Chapter 5). Aegothelidae have one clade in mainland Australia, Tasmania and New Guinea, and one in New Caledonia (extant) and New Zealand (Holocene fossil material) (Dumbacher *et al.*, 2003). Rather than emphasizing the simple east/west vicariance around the Tasman/Coral Sea spreading centre, Dumbacher *et al.* suggested that the aegothelids dispersed to New Zealand and New Caledonia from Australia or New Guinea. In support of this they noted that the earliest known fossil is from Australia, the family is unknown outside Australasia and most of the species occur in New Guinea. Yet none of these facts requires an Australian or New Guinean centre of origin, and the phylogeny is compatible with a simple break between a western Tasman clade and an eastern Tasman clade. The opening of the Tasman rift provides a geological mechanism for this.

The New Caledonia–New Zealand connection: is it real?

An account of southern hemisphere biogeography by Sanmartín and Ronquist (2004) is often cited and has been described as 'a comprehensive meta-analysis of austral biogeography' (Waters and Craw, 2006: 354). Nevertheless, the sample size in the study was too small to permit some of the general conclusions that were drawn. For example, none of the plant or animal taxa sampled showed New Zealand and New Caledonia as sister areas, and the authors inferred that there is 'minimal evidence of biotic links between present day New Caledonia and New Zealand' (Sanmartín and Ronquist, 2004: 225; cf. Wallis and Trewick, 2009). The authors concluded (p. 240): 'our results . . . do not support the notion of a common relict late Cretaceous Gondwanan biota in the two land masses', and they explained any affinities between the two areas as due to dispersal between them. Nevertheless, the conclusion has little statistical significance because of the small number of taxa that was sampled.

The standard connection between New Zealand and New Caledonia was well documented in the nineteenth century by authors such as Hutton (1872). In recent morphological studies, a caddisfly of New Caledonia (*Xanthochorema*) is sister to a pair from New Zealand (*Neurochorema* and *Psilochorema*; Ward *et al.*, 2004). Molecular work shows that one New Zealand–New Caledonia group (the plant *Xeronema*) is basal in a worldwide clade, not a recent, secondary derivative from Australia or anywhere else. A New Caledonia–New Zealand clade is also supported in *Dracophyllum* (Fig. 6.22). In other molecular phylogenies the link is evident in *Hachettea* (Balanophoraceae) of New Caledonia, sister to *Dactylanthus* of New Zealand (Nickrent, 2012). This pair is sister to *Mystropetalon* of South Africa, giving a standard Indian Ocean basin track as in *Ixerba* + *Strasburgeria* and their allies (Fig. 3.1). In Monimiaceae, *Hedycarya arborea* of New Zealand is sister to *Kibaropsis* of New Caledonia (Renner *et al.*, 2010). In Myrtaceae, *Metrosideros parkinsonii* of New Zealand is sister to five species of New Caledonia (Papadopulos *et al.*, 2011). In earthworms, most of the many New Zealand species sampled by Buckley *et al.* (2011) made up a single clade and this is sister to a group of *Megascolex* from New Caledonia. In freshwater fishes, *Galaxias brevipinnis* of New Zealand is sister to *Nesogalaxias* of New Caledonia (Burridge *et al.*, 2012). Migrations of humpback whales occur between New Caledonia and the Northland Plateau of New Zealand and are 'as straight as an arrow' (Horton *et al.*, 2011).

New Caledonia–outer Melanesian arc vicariants

The standard break between taxa on the inner Melanesian arc (this includes New Zealand) and the outer Melanesian arc was illustrated in *Spiraeanthemum*, above. The blackfly genus *Simulium* (Diptera: Simuliidae) provides another example, with the following clade (Craig *et al.*, 2001):

Asia, Australia and New Guinea east to Norfolk Island and New Caledonia (subgen. *Nevermannia*). (Mainland and inner Melanesian arc.)

Vanuatu and Fiji (subgen. *Hebridosimulium*). (Outer Melanesian arc.)

Micronesia (Mariana and Caroline Islands) and Polynesia (subgen. *Inseliellum*).

Craig *et al.* (2001) explained the overall distribution as the result of 'routes of colonisation' into the Pacific from a Southeast Asian centre of origin, but they did not mention the well-defined vicariance among the subgenera. The break between New Caledonia and Vanuatu corresponds to the present plate boundary and the North Loyalty basin (Fig. 7.1).

The same break between New Caledonia and Vanuatu is the main node in the caddisfly *Apsilochorema* (Trichoptera). Overall the genus is in Australia, the Pacific islands (not New Zealand) and north to Japan, Russia and Iran. Its sister-group is *Isochorema* in

Chile (Strandberg and Johanson, 2011) and so the pair forms a standard Pacific + Tethys pattern.

Apsilochorema occurs through the south-west Pacific. At several of its localities there – the Solomon Islands, Vanuatu and Fiji – the current islands were never part of Gondwana, and one study of the group concluded: 'it was therefore not considered appropriate to apply vicariance models . . . ' (Strandberg and Johanson, 2011). A vicariance model is still possible, though, as the biota of the Bismarck Archipelago, the Solomon Islands, Vanuatu, Fiji and Tonga inhabited a continuous Vitiaz arc, before it was torn apart. This rifting occurred with slab rollback westward and the opening of the North Fiji Basin (Fig. 7.1). (Note the spreading centres between Vanuatu and Fiji; Fig. 7.2.) In one model (Schellart *et al.*, 2006) the current subduction zone 3 and the associated Vitiaz arc islands (Fig. 7.1; New Hebrides trench and Hunter fracture zone) have been derived from the original subduction zone 1 (Vitiaz trench), which itself has migrated from the Gondwanan margin. The different archipelagos and their endemic metapopulations have been rifted apart, providing a mechanism for vicariance. The metapopulations have survived *in situ*, over long periods of time, persisting on groups of individually ephemeral islands by means of ordinary, local dispersal. Some margins have become inactive, and arcs have been submerged, but volcanism has continued along new arcs nearby (as in subduction zone 3). Within arcs, older islands have subsided while they are replaced with new strata. Recent volcanism has occurred at New Britain, Bougainville, Tanna and Taveuni. If survival by metapopulations and normal dispersal is acknowledged, there is no need to invoke one-off events of extraordinary long-distance dispersal for which there is no supporting evidence except transmogrified clade ages.

Apsilochorema is distributed from Australasia (not New Zealand) north to Japan, Russia and Iran. Strandberg and Johanson (2011) focused on the genus in the Southeast Asia–Pacific region and showed that it consists of three main clades:

Australia (*A. obliquum*).

Sulawesi, eastern Australia and New Caledonia.

Southeast Asia (Burma, Vietnam and Taiwan), the Solomon Islands, Vanuatu and Fiji.

The last two clades show precise allopatry along two great arcs, with part of the break running between New Caledonia and Vanuatu. A dispersal model requires a convoluted migration from Australia to Burma and from there back to Vanuatu and Fiji (Strandberg and Johanson, 2011), suggesting that the pair evolved instead by simple vicariance. *Apsilochorema* itself could have evolved in the same way, by vicariance with its sister-group *Isochorema* in Chile.

New Caledonia–outer Melanesian arc clades

Other groups show a direct connection between New Caledonia and the outer Melanesian arc (the Vitiaz arc). The sylvioid passerine *Megalurulus* (including *Cichlornis*, *Trichocichla* and *Ortygocichla*; ?Locustellidae) shows this pattern (Fig. 8.4). *Cettia* (including

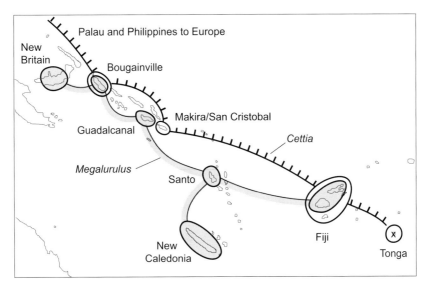

Fig. 8.4 Distribution of the passerine birds *Megalurulus* (?Locustellidae) and *Cettia* in part (Cettiidae). X = fossil. (From Heads, 2010a, reprinted by permission of John Wiley & Sons.)

Vitia; Cettiidae) is another sylvioid bird found along the arc, but it has a different distribution there; for example, it is not in New Caledonia, Vanuatu or the central Solomon Islands (Fig. 8.4; Dickinson, 2003; Lecroy and Barker, 2006). The inner distribution of *Megalurulus* and the outer distribution of *Cettia* would both usually be attributed to island-hopping along the island chains, but this does not explain the allopatry between them. For *Cettia*, at least, Orenstein and Pratt (1983) concluded that the gaps in the distribution 'cannot be entirely explained by accidents of dispersal or unsuitability of habitat on the unoccupied islands'.

In *Megalurulus*, New Caledonia connects with northern Vanuatu (Espiritu Santo)/southern Solomon Islands, rather than with southern Vanuatu (the usual connection with New Caledonia). This northern connection could involve the d'Entrecasteaux Ridge, an extension of the Norfolk Ridge north of New Caledonia that curves eastward to Santo, where it is being destroyed by subduction at the New Hebrides trench (Fig. 7.2). As an alternative, some models propose that the Pacific subduction zone that produced Fiji and Tonga was originally located off New Caledonia and the Norfolk Ridge, and this would provide a southern connection: New Caledonia–Fiji.

The Fiji–Santo connection in *Megalurulus*, and the direct Solomon Islands–Fiji connection in *Cettia* skirt most or all of Vanuatu to the north and may be related to the extinct subduction zone at the Vitiaz trench (Fig. 7.1). This pattern skirting Vanuatu is also seen in plants such as *Schefflera* (*Plerandra*) clade 'Dictyophlebes' (Araliaceae), known only from the Solomon Islands and Fiji (Frodin *et al.*, 2010). Similar patterns skirt Fiji to the north and connect the Solomon Islands with Samoa and northern Tonga. For example, in cicadas, the genus *Moana* is in the Bismarck Archipelago, the Solomon Islands, Wallis and Futuna, and Samoa, while the *Baeturia bloetei* group is in the Maluku Islands, New

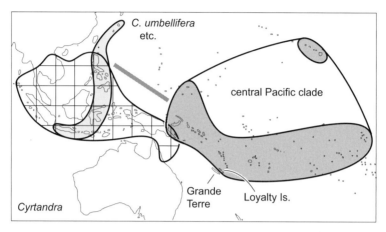

Fig. 8.5 Distribution of *Cyrtandra* (Gesneriaceae): the central Pacific clade, its sister-group (*C. umbellifera* etc.), and all other species (grid). (From Heads, 2010a, reprinted by permission of John Wiley & Sons.)

Guinea, Bismarck Archipelago, Solomon Islands, Vanuatu, Rotuma, Samoa and Tonga (morphological studies by de Boer and Duffels, 1996a).

This dynamic history of slab rollback and migrating arcs has been thought to explain patterns in both terrestrial and marine groups. For example, Bitner (2010) wrote that 'The biogeographical pattern of brachiopod distribution in the South-West Pacific is more readily explained by vicariance . . . , when geological processes such as plate tectonics led to separation of regions and their faunas, than by simple dispersal . . . '. Bitner (2009) noted that the brachiopod fauna of the Norfolk Ridge south of New Caledonia has closer relationships with Fiji than with Australasia, and this follows the geological sequence, with the South Fiji Basin opening after the Tasman Basin (Fig. 7.1).

New Caledonia–Pacific islands groups

The close relationships of many New Caledonian groups with sister-groups in mainland Australia and Southeast Asia are well known. Other New Caledonian taxa are instead involved with groups of the central Pacific islands, resulting in broad, central Pacific patterns of endemism. One example is the shrub genus *Cyrtandra* (Gesneriaceae, ∼600 species) (Fig. 8.5). In this genus, Cronk *et al.* (2005: 1017) suggested that 'Because all oceanic Pacific island species form a well-supported clade, these species apparently result from a single initial colonization into the Pacific, possibly by a species from the eastern rim of SE Asia . . . '. The sister of the widespread Pacific island clade is endemic to the strip: Ryukyu Islands–Philippines–Sulawesi–Java (Clark *et al.*, 2009), but there is no real evidence that the Pacific clade of *Cyrtandra* colonized the Pacific by physical movement. It is a simple vicariant and a dispersal model would not explain why it only colonized the entire Pacific region once, why it has never dispersed back to Asia, or why there is so much local endemism in the species. These are all accounted for in a vicariance model, which also explains the striking distribution in New Caledonia.

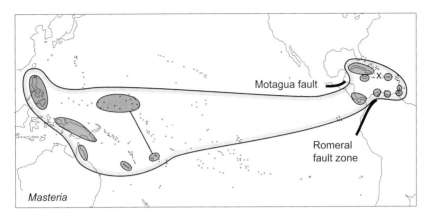

Fig. 8.6 Distribution of the spider genus *Masteria* (Dipluridae). X = fossil record from the Dominican Republic (Platnick, 2011).

In this archipelago, *Cyrtandra* is restricted to the low, flat Loyalty Islands and is absent from Grande Terre, despite the steep, montane gullies there that provide much more typical habitat for the genus. In this and many other groups a surprising phylogenetic and ecological difference occurs between Grande Terre and the Loyalty Islands. This break can be attributed to the independent history of the Loyalty Ridge with respect to Grande Terre and the Norfolk Ridge (Fig. 7.2), rather than to chance dispersal.

New Caledonia–tropical America

Many groups are endemic to Australia, New Caledonia and America. Several of these trans-tropical Pacific groups have greater diversity in New Caledonia than in Australia. In other groups Australia is not involved at all and in these, New Caledonia/Melanesia has a direct connection with South America. The New Caledonia/New Zealand–Chile track involving more or less subantarctic latitudes has been known for over a century, but trans-tropical Pacific connections are often overlooked. Nevertheless, they are also common and the Bataceae and others were cited above (Fig. 3.19). In animals, trans-tropical Pacific tracks are documented in the spider *Masteria* (Dipluridae; Fig. 8.6; Platnick, 2011). The trans-tropical Pacific connection is also seen in the areas of maximum diversity for angiosperm families. These occur in southern Yunnan/Cambodia and southern Mexico (Gaston *et al.*, 1995), or Malesia (most diverse for trees) and north-western Colombia/central America (most diverse for herbs) (Hawkins *et al.*, 2011).

Unlike the subantarctic tracks between Australasia and Chile, trans-tropical Pacific distributions are not explained by Gondwana breakup. They could be explained on an ad hoc basis by extinction in other areas or as the result of long-distance dispersal, but neither of these two ideas accounts for many aspects of the patterns. Instead, trans-tropical Pacific distributions can be related to central Pacific tectonics (Chapter 3). The emplacement of the large igneous plateaus in the central Pacific during the Cretaceous was the largest volcanic event in Earth history. At least parts of the Ontong Java plateau and most of the Manihiki plateau were subaerial during eruption. Their formation was followed

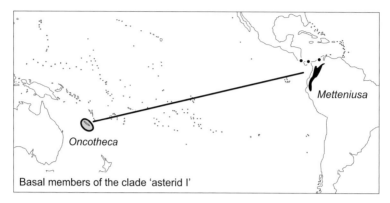

Fig. 8.7 Distribution of Oncothecaceae and Metteniusaceae. These form either a basal clade or the first and second basal branches in the cosmopolitan 'asterid I' clade (Lamiales, Solanales and Gentianales) (González and Rudall, 2007; González *et al.*, 2007). (From Heads, 2010b, reprinted by permission of John Wiley & Sons.)

by their rifting and dispersal with plate movement. This took place westward to New Zealand (Hikurangi plateau), the Solomon Islands (Ontong Java plateau) and elsewhere, and eastward to Colombia and the Caribbean (Gorgona and Caribbean plateaus; Heads, 2012a; Fig. 6.1). This history is compatible with the biogeographic patterns, which include the following.

The angiosperm *Oncotheca*, monotypic in New Caledonia, and *Metteniusa*, known from Costa Rica to Ecuador (Fig. 8.7), form either a basal clade or the first and second basal branches in a cosmopolitan clade of Lamiales, Solanales and allies (asterid I clade in Stevens, 2012) (González and Rudall, 2007; González *et al.*, 2007). The second topology indicates a trans-tropical Pacific sequence of initial differentiation.

In birds, the family Rhynochetidae, endemic to New Caledonia, is related to Eurypygidae of tropical America (Guatemala to Brazil), and also to *Aptornis*, a New Zealand fossil (Houde, 2009; Trewick and Gibb, 2010). In molecular studies, the trans-tropical Pacific Eurypygidae + Rhynochetidae is sister to a diverse, cosmopolitan clade comprising Caprimulgiformes, Aegotheliformes and Apodiformes (Hackett *et al.*, 2008).

Molecular studies are revealing more and more trans-tropical Pacific connections. A clade in Opiliones suborder Cyphophthalmi comprises Troglosironidae of New Caledonia plus Neogoveidae of tropical America/West Africa (Boyer *et al.*, 2007b; Sharma and Giribet, 2009), and Boyer and Giribet (2007) described the connection as 'intriguing'. A similar trans-Pacific plus trans-Atlantic pattern occurs in plants: *Montrouziera* (Clusiaceae) is endemic to New Caledonia (the spectacular flowers of *M. gabriellae* are the largest in the flora) and forms a clade with *Platonia*, *Lorostemon* and *Moronobea* of tropical South America (Fig. 8.8; Ruhfel *et al.*, 2011).

A similar trans-Pacific pattern occurs in Polygalaceae tribe Moutabeeae, which is represented in New Caledonia (*Balgoya*), New Guinea and the Solomon Islands (*Eriandra*), and northern South America (*Barnhartia*, *Diclidanthera* and *Moutabea*) (Fig. 8.9; van Balgooy and van der Meijden, 1993). The tribe is basal in the worldwide Polygalaceae

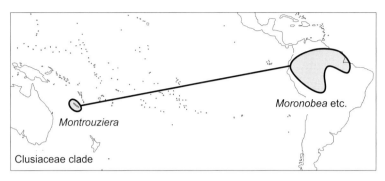

Fig. 8.8 Distribution of *Montrouziera* and allies (Clusiaceae) (Ruhfel *et al.*, 2011).

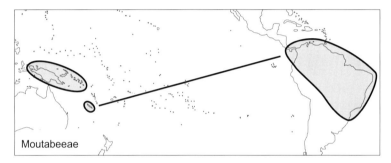

Fig. 8.9 Distribution of the tribe Moutabeeae (Polygalaceae) (van Balgooy and van der Meijden, 1993; Eriksen, 1993; Forest *et al.*, 2007a). (From Heads, 2010a, reprinted by permission of John Wiley & Sons.)

(minus *Xanthophyllum*) (Eriksen, 1993; Forest *et al.*, 2007a), indicating that the trans-tropical Pacific region is an important centre in its own right, not just a sink for dispersal from elsewhere.

Trans-tropical Pacific groups with global sisters also occur in shorebirds, Charadri-iformes. The snipe *Coenocorypha* (Scolopacidae) has extant species only on the New Zealand subantarctic islands (Auckland, Campbell, Snares, Antipodes) and Chatham Islands. Extinct populations are recorded on the New Zealand mainland, Norfolk Island, New Caledonia and Fiji (Fig. 8.10). Together, the present and past localities encompass most of Zealandia plus Fiji. *Coenocorypha* is sister to '*Gallinago*' *imperialis*, known from the Andes of Peru, Ecuador and Colombia. This trans-Pacific pair is sister to the remaining species of *Gallinago*, widespread on all continents including Australia, but absent, except as vagrants, from Zealandia (Gibson and Baker, 2012).

Trans-Pacific distribution is also evident in orchids. The New Caledonian genera *Clematepistephium* and *Eriaxis* are sister taxa, and the pair is sister to *Epistephium*, widespread in tropical South America (Cameron, 2003). This trans-tropical Pacific group belongs to the tribe Vanilleae, a group that 'exhibits one of the most intriguing continental disjunctions in the Orchidaceae' (Cameron, 2003); *Vanilla* is pantropical while other genera occur around the Pacific in Southeast Asia, Malesia, Australia,

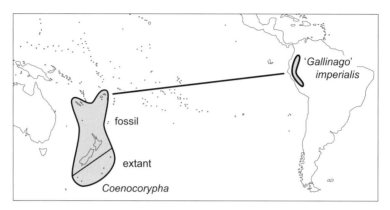

Fig. 8.10 Distribution of *Coenocorypha* and its sister-group '*Gallinago*' *imperialis* (Scolopacidae) (Gibson and Baker, 2012).

Oceania and South America. Cameron (2000) attributed this to the Mesozoic evolution of the group and a series of vicariance events. Given the distributions, it would be surprising if the large igneous plateaus of the central Pacific were not involved.

The shrubs in the genus *Augusta* (including *Lindenia*; Rubiaceae) have a trans-tropical Pacific distribution in New Caledonia, Fiji and tropical America, from Mexico to Brazil (Motley *et al.*, 2005). Another trans-tropical Pacific group of Rubiaceae, the Catesbaeeae–Chiococceae complex, occurs in the western Pacific (Philippines to New Caledonia and Tonga; not Australia), the Galapagos and throughout tropical America (Motley *et al.*, 2005). Within the Catesbaeeae–Chiococceae complex, a New Caledonian clade (*Bikkia* species plus *Morierina*) is sister to *Siemensia* of western Cuba. Much of the Caribbean crust was formed in the Pacific and the biogeography is consistent with this history. Motley *et al.* (2005: 327) suggested a New World origin for the Catesbaeeae–Chiococceae complex because of a paraphyletic basal grade there (but cf. Fig. 1.6). The most species-rich area for the Catesbaeeae–Chiococceae complex is the Greater Antilles region (cf. *Masteria*, Fig. 8.6) and Motley *et al.* (2005) proposed these islands as the ultimate centre of origin, inferring long-distance dispersal from there to the West Pacific. Nevertheless, Cuba is about 12 000 km from New Caledonia and, in addition, similar patterns occur in groups such as *Masteria* (Fig. 8.6), suggesting that these distributions could have arisen instead by *in-situ* evolution and vicariance. The same history, mediated by seafloor spreading and magmatism in the central Pacific since the mid-Cretaceous, would account for the patterns shown in Figures 8.6 to 8.10.

In the harvestmen, the group Zalmoxoidea (Opiliones suborder Laniatores) has its seven basal clades in the Neotropics, with the seventh being sister to an Old World genus, *Zalmoxis* (Fig. 8.11; Sharma and Giribet, 2012). *Zalmoxis* is distributed from the Seychelles and Mauritius to Fiji and the Marshall Islands. Because of the paraphyletic basal grade in tropical America, Sharma and Giribet (2012) proposed a centre of origin in America, with a single dispersal event across the tropical Pacific. (Dispersal via Beringia or the southern Pacific was discussed and ruled out.) Sharma and Giribet (2012) saw the trans-Pacific pattern as a 'biogeographic conundrum' and the mechanism as a 'matter

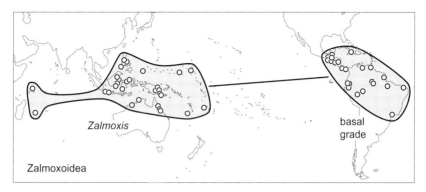

Fig. 8.11 Distribution of Zalmoxoidea (Opiliones: Laniatores) (Sharma and Giribet, 2012).

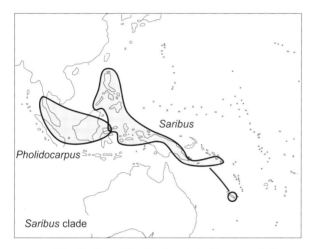

Fig. 8.12 Distribution of the *Saribus* clade (Arecaceae) (Bacon *et al.*, 2012).

of speculation'. They dated the origin of *Zalmoxis* to mid-Cretaceous (92 Ma, a fossil-calibrated, minimum date) and suggested that Indo–Pacific colonization was achieved by rafting on floating vegetation. But that does not explain why it only occurred once, why it happened at 92 Ma, or why the same pattern is replicated in many other groups. A model incorporating a widespread Indo–Pacific–American ancestor and the breakup of the large Cretaceous central Pacific plateaus would explain all three.

A case-study: the New Caledonian palms and their relatives

Palms are a conspicuous, well-studied group, and are represented in New Caledonia by 10 genera. The affinities of these illustrate the main biogeographic connections of the area (Figs. 8.12–8.15). The molecular phylogeny presented by Baker *et al.* (2009, Fig. 2) is followed here, along with the classification of Pintaud and Baker (2008).

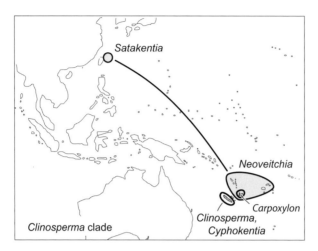

Fig. 8.13 Distribution of the *Clinosperma* clade (Arecaceae). (From Heads, 2010b, reprinted by permission of John Wiley & Sons.)

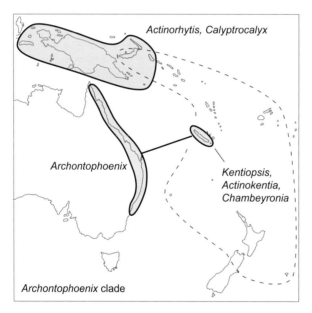

Fig. 8.14 Distribution of the *Archontophoenix* clade (Arecaceae). (Distribution of *Basselinia* clade as fine broken line). (From Heads, 2010b, reprinted by permission of John Wiley & Sons.)

The New Caledonian fan palm, *Saribus jeanneneyi* (subfamily Coryphoideae; Fig. 8.12), is one of the rarest palms in the world and is known from just a few individuals in southern Grande Terre (the three other lineages of New Caledonian palms, discussed below, are all widespread in Grande Terre). *Saribus* ranges from New Caledonia to the Philippines and its sister-group, *Pholidocarpus*, is in western Malesia and Thailand (Bacon *et al.*, 2012).

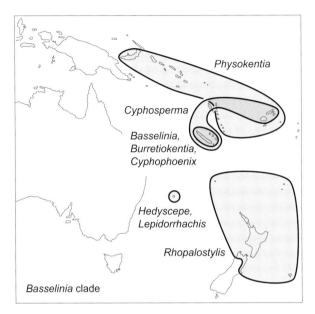

Fig. 8.15 Distribution of the *Basselinia* clade (Arecaceae). (From Heads, 2010b, reprinted by permission of John Wiley & Sons.)

The other New Caledonian palms make up a remarkable assemblage of nine genera (eight endemic) in the tribe Areceae (subfamily Arecoideae). The nine genera in New Caledonia belong to three groups of the 'Western Pacific' clade (Baker *et al.*, 2009):

1. The *Clinosperma* clade in New Caledonia, Vanuatu/Fiji and the southern Ryukyu Islands near Taiwan (Fig. 8.13).
2. The *Archontophoenix* clade in New Caledonia, eastern Australia and New Guinea, including neighbouring islands (Fig. 8.14).
3. The *Basselinia* clade, lying east of the last clade and overlapping with it only in New Caledonia and New Britain (Fig. 8.15).

The three clades have their relatives in quite different places. The simplest explanation for this is an origin of the groups by vicariance, followed by overlap in New Britain, New Caledonia, Vanuatu and Fiji caused by tectonic juxtaposition of terranes and arcs, local range expansion, or both. (The three groups plus two others that are not in New Caledonia – *Ptychosperma* and allies, and *Laccospadix* and allies – form the 'Pacific clade' of Baker and Couvreur, 2013a). (For notes on Bayesian dating of palms by Baker and Couvreur 2013a, see p. 39.)

In the New Caledonian palms, the distribution of the *Clinosperma* clade (Fig. 8.13) is notable for its large disjunction to the north-west, across Micronesia. Another palm, *Clinostigma*, has a similar distribution: Samoa/Fiji, Vanuatu, Solomon Islands, New Ireland, Caroline and Bonin islands, skirting mainland New Guinea to the north. Uhl and Dransfield (1987: 437) described this as a 'great arc . . . a most interesting distribution'. In a similar pattern, the gecko genus *Perochirus* has what Heinicke *et al.* (2011) called a 'bizarre' distribution, as the species are endemic to Micronesia and Vanuatu, with fossils on the Marianas and Tonga. In dispersal interpretations, these western Pacific arcs would be regarded as routes of migration, along 'stepping-stone' islands. Instead, they are interpreted here as allopatric sectors and fragments of widespread central Pacific groups. The groups have survived around centres of volcanism at plate margins and also intraplate localities.

Distribution of the Trachycarpeae and their allies, the date palms (Phoeniceae)

The New Caledonian–Malesian genus *Saribus* (Fig. 8.12) is a member of the tribe Trachycarpeae (Fig. 8.16). *Saribus* is represented in neighbouring Vanuatu by *Licuala*, in eastern Australia by *Livistona* and in the central Pacific (Fiji to Hawaii) by *Pritchardia*. The sister-group of *Pritchardia* is *Copernicia*, found in Cuba (20 species), Hispaniola (two species) and South America (three species). The tribe Trachycarpeae (18 genera) is most diverse in and around the Pacific, and is more or less allopatric with its sister-group, the Phoeniceae (date palms), based around the Atlantic and Indian Oceans. The two tribes have separated through the breakup of a circumglobal ancestor, with subsequent overlap of the two descendants along the Tethys sector.

Within the largely allopatric tribes, Trachycarpeae and Phoeniceae, differentiation of several of the genera has also been allopatric. The central Pacific representative, *Pritchardia*, probably indicates evolution on and around the large igneous plateaus that were emplaced in the Cretaceous. *Pritchardia* or its ancestors would have also occupied other Cretaceous volcanics in its region, such as the Line Islands. This archipelago extends for 4000 km between south-eastern Polynesia and Hawaii, and would have been composed of high islands in the Cretaceous although these have now all been reduced to atolls and seamounts. The central Pacific *Pritchardia* is sister to the American *Copernicia*, but deriving *Pritchardia* by dispersal into the Pacific from America (Baldwin and Wagner, 2010) does not explain the neat allopatry between *Pritchardia* and its other allies at the plate margin in the west Pacific, or why the vast Pacific region was only invaded once.

One genus of Trachycarpeae, the fan palm *Livistona*, ranges from the Horn of Africa to Australia and the Solomon Islands. Crisp *et al.* (2010) investigated whether the genus represented a Gondwanan relict or a Miocene immigrant in Australia, and their study relied on fossil-calibrated dates. These minimum dates were transmogrified into maximum ages, used to rule out early vicariance and cited as support for a centre of origin/dispersal model. If the ages are instead treated as minimum ages, they are consistent with a vicariance model. Around the present-day coasts of Australia several species of *Livistona* inhabit the landward edge of mangrove forest (*L. benthamii*, *L. concinna*, *L. drudei*) (Dowe, 2009). In addition, the genus is represented in central Australia, at the MacDonnell Ranges, by the locally endemic *L. mariae* (Fig. 8.16). These observations can be explained if the ancestral complex occupied mangrove edge and was stranded around the MacDonnell Ranges following the recession of the Cretaceous seas.

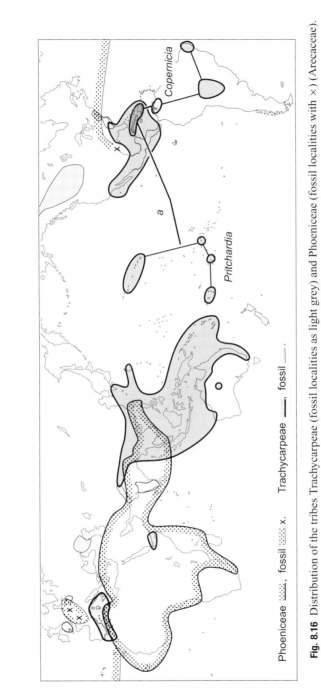

Phoeniceae ⬚, fossil ⬚ x.   Trachycarpeae ▬, fossil ▬.

Fig. 8.16 Distribution of the tribes Trachycarpeae (fossil localities as light grey) and Phoeniceae (fossil localities with ×) (Arecaceae). Line *a* indicates the affinity between *Pritchardia* and *Copernicia*. (Uhl and Dransfield, 1987; Bacon *et al.*, 2012).

The pattern seen in New Caledonian palms, with different clades having sister-groups in different areas, also occurs in other plants. The pantropical genus *Diospyros* (ebony, Ebenaceae) is a good example. There are four clades in New Caledonia (Duangjai *et al.*, 2009):

> Clade 1 (*Diospyros balansae*, etc.) has its sister in eastern Australia (Cairns to Sydney), indicating a Cretaceous split with the opening of the Tasman Sea. This Tasman clade is basal to a global group (the entire genus except for *D. maingayi* and *D. puncticulosa*).
>
> Clade 2 (*D. calciphila*, etc.) has its sister in the Hawaiian Islands, indicating evolution around central Pacific volcanic centres.
>
> Clade 3 (*D. fasciculosa*) has its sister *D. ehretioides* in Southeast Asia, Australia, Vanuatu and Fiji.
>
> Clade 4 (*D. olen*) has its sister in the Solomon Islands, Vanuatu, Fiji, Tonga and Samoa. This is compatible with terrane accretion in New Caledonia from the north-east (see below).

(Clades 3 and 4 are sisters; the others are interspersed with clades from other areas.) As in palms, the high level of allopatry among the four ebony clades can be explained by original vicariance in a widespread ancestor, followed by overlap of all four clades in New Caledonia with terrane accretion there. The only evidence for local range expansion is within New Caledonia itself (all clades), between clades 1 and 3 in Australia, and between clades 3 and 4 in Vanuatu and Fiji. The overall process of diversification in New Caledonia is one of juxtaposition rather than radiation.

A similar pattern occurs in the polycentropodid Trichoptera of the Australasian region (Johanson *et al.*, 2012). The family is made up of 90 species in seven genera, and the highest species diversity is recorded from New Caledonia. The New Caledonian species are not a monophyletic assemblage, but belong to three distinct clades that have the following distributions (the three clades form a monophyletic group):

> New Caledonia–Australia (Clade D).
> ((New Caledonia–New Zealand) (Australia–Chile)) (Clade E1).
> (New Caledonia–Chile) northern Neotropical region (Clade E2).

From this the authors inferred that New Caledonia was colonized three times, once from Australia, once from New Zealand and once from Chile. Yet this does not explain why dispersal only occurred once from each source, and the subclade allopatry suggests vicariance (overlap is restricted to within New Caledonia). In this model, a widespread southern and eastern Pacific ancestor has divided into three widespread groups: clade D in pre-New Caledonia and the west, clade E1 in New Caledonia and the south, and clade E2 in New Caledonia and the east. Modern New Caledonia represents the original break that separated all the three clades, with local overlap occurring there later. The pattern suggests that the original break *among* the clades probably occurred in the pre-New Caledonia region of Gondwana prior to breakup (perhaps during terrane accretion), with breakup then producing the disjunctions *within* each clade. The polycentropodid clade comprising D, E1 and E2, termed clade A, is widespread around the southern and eastern

Pacific, but is absent in Vanuatu and Fiji. Here it is replaced by its sister, clade B, also found in other localities around the world; part of the break between these two large clades occurs between New Caledonia, and Vanuatu + Fiji.

Evolutionary ecology of the Trachycarpeae and Phoeniceae

In traditional analyses, geographic centres of origin are located using centres of diversity, 'basal' clades and grades, the oldest fossil, and so on. In the same way, an ecological centre of origin or ancestral niche is often proposed for a group (Chapter 1). The group is inferred to have spread from there into the habitats and the general ecological space that it now occupies. In contrast, in a vicariance process the ancestor of a group was already widespread spatially and occupied a wide ecological range before the modern group existed.

In the palm tribe Trachycarpeae, the New Caledonian member, a species of *Saribus*, occurs on soils derived from ultramafic rock. Other members of the tribe include the Central American *Brahea* and the Asian *Maxburretia*, *Guihaia*, *Rhapis* and *Trachycarpus*; all of these occur on limestone, and most are on karst (ecological data from Uhl and Dransfield, 1987, and Dowe, 2009). The affinity reflects the close physiological, ecological and tectonic relationship between limestone and ultramafic floras. Both rock types characterize belts of subduction and accretion, where taxa of uplifted coral reefs are pre-adapted for life on the basic soils of obducted ultramafic terranes. The ecological ranges of the other genera of Trachycarpeae are as follows:

> *Pritchardia*: rainforest (Hawaii), uplifted coral reef near the sea (other Pacific Islands). In Fiji, plants occur only on limestone islets and limestone lagoon cliffs, where they grow right to the edge of the sea and are subject to salt spray (Heads, 2006a).
> *Johannesteijsmannia*: undisturbed, primary rainforest; kerangas heath forest (*J. altifrons*).
> *Pholidocarpus*: freshwater and peat-swamp forest.
> *Saribus*: mangrove margins, limestone near seashore, ultramafic rocks, swamp forest, lowland rainforest.
> *Licuala*: landward fringe of mangrove (*L. spinosa*), forest on limestone (*L. calciphila*), peat-swamp forest (*L. paludosa*), lowland rainforest.
> *Livistona*: mangrove (*L. saribus* etc.), limestone (*L. halongensis* is only known from limestone), freshwater and peat-swamp forest, heath forest, lowland and montane rainforest, woodland, savanna, by streams and permanent water bodies in desert.
> *Acoelorrhaphe*: coastal brackish swamps.
> *Serenoa*: (south-eastern USA) pinelands, prairies, coastal sand-dunes, often forming dense swards. The architecture is unusual in the tribe as the stem is subterranean and prostrate or surface-creeping. Similar plagiotropic axes occur in many mangroves, including the mangrove palm *Nypa*.
> *Rhapidophyllum*: low-lying moist to wet areas with rich humus, calcareous clay or sandy soils in woods or swamps. In grottos, limestone sinks and shaded pinelands, but usually associated with limestone.

Copernicia: grassland, savanna, woodland; forms vast stands in South America in
areas subject to annual flooding.
Colpothrinax: open, sandy country, open forest on ridges.
Washingtonia: by springs and streams in deserts.
Chamaerops: (western Mediterranean) sandy or rocky ground, usually near the sea.

The sister tribe of Trachycarpeae is Phoeniceae, with only one genus, *Phoenix* (including
the date palm). Species occur around oases and watercourses in semiarid areas, with a
few in tropical monsoon areas. *Phoenix paludosa* occurs in perhumid parts of Asia where
it is confined to the landward fringe of the mangrove. Trachycarpeae and Phoeniceae
are more or less allopatric.

To summarize, the Trachycarpeae and Phoeniceae occupy a wide series of habitats
that is typical of more diverse tropical groups: back-mangrove, limestone and ultramafic
sites, monsoon forest, savanna, swamp forest, heath forest, lowland and montane forest.
This ecological series in the palm tribes can be derived from a weedy complex of back-
mangrove and associated habitats that spread through inland areas with the advance
of Cretaceous seas. Later they were stranded inland with the retreat of the seas and
onset of Cenozoic orogeny. It is often suggested that the biogeography of a group is
determined by its ecology, but in the model proposed here a group's current ecology
is determined in large part by its location. Populations derived from a widespread
Trachycarpeae/Phoeniceae ancestor have ended up stranded in areas that became desert
in Arabia, rainforests in Southeast Asia, atolls in the Pacific and flood plains in South
America. The current taxa have not *invaded* these different, extreme habitats, adapting
as they went. They have *evolved* there and their ancestors were in the areas before the
modern habitats existed. Because of their pre-adapted morphology and physiology these
groups survived in the regions they inherited, while other groups without pre-adaptations
have died out.

New Caledonian terranes and tectonic history

New Caledonia is not just a fragment of Gondwana but is a complex mosaic of
eight allochthonous terranes – fault-bounded blocks of crust with independent histories
(Fig. 8.17; Heads, 2008b). The four New Caledonian basement terranes formed from
island arc-derived and arc-associated material which accumulated in the pre-Pacific
Ocean, not in Gondwana. The basement terranes amalgamated and were accreted to
Gondwana (eastern Australia) in the Late Jurassic/Early Cretaceous, but in the Late
Cretaceous they separated from Australia with the opening of the Tasman Sea and the
breakup of Gondwana. In the Eocene the basement collided with an island arc, leading
to further terrane accretion. The arc – possibly the Loyalty Ridge – arrived from the
north-east and so the collision would have resulted in New Caledonia–central Pacific distributions. In this way New Caledonia has developed as a geological composite formed
from Gondwanan and Pacific components. New Guinea and New Zealand also have this

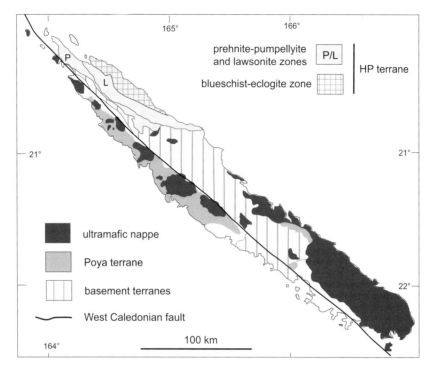

Fig. 8.17 Grande Terre terranes (based on Baldwin *et al.*, 2007). (From Heads, 2008b, reprinted by permission of John Wiley & Sons.)

structure and the three countries have similar histories, although New Guinea has only been partly rifted from Australia.

Many authors have attributed New Caledonian endemism to the long isolation of the main island and the variety of habitat types present. Nevertheless, the isolation from New Zealand, caused by the subsidence of the Norfolk Ridge, is only Cenozoic, not old enough to account for the globally basal endemics in New Caledonia (the orchid *Pachyplectron*, the shrub *Amborella* and others; see Chapter 5). The range of habitats in New Caledonia accounts for the survival of an archaic biota but does not, on its own, explain its origin. This is more likely to be related to the different histories of the terranes in the region. For example, New Caledonia has an exceptional number of marine molluscs (Marshall, 2001; Bouchet *et al.*, 2002) and is the world centre of diversity for families such as the Volutomitridae (Bouchet and Kantor, 2003). Marshall (2001: 41) proposed that 'Since the Melanesian arc [including New Caledonia] is situated at current or former [convergent] boundaries of the Australian and Pacific lithospheric plates, species richness there is probably due at least partly to progressive accumulation of taxa transported on the plates'. This process, with taxa being 'scraped off' at a subduction zone, would account for some of the most distinctive biogeographic phenomena in and around New Caledonia. It would also imply that large sections of crust, where the current endemics occupied islands in earlier times, no longer exist. Large structures in the region, such as the South Loyalty Basin between Grande Terre and the Loyalty Islands (Fig. 7.2), have been more or less destroyed by subduction but may have included prior areas of endemism that have left traces in the extant flora and fauna.

Paleogene marine transgression in New Caledonia: was it complete?

Deep-water sedimentary rocks dated as Paleogene occur in parts of New Caledonia. Some geologists (Pelletier, 2007; Cluzel *et al.*, 2012) and biologists (Pole, 1994; Espeland and Murienne, 2011) have inferred from this that the entire land mass of New Caledonia was submerged from the start of the Paleocene until the mid-Eocene (65–45 Ma). These authors have supported an 'entirely long-distance dispersal' explanation for the island's biota. Nevertheless, without a stratum of a single age covering the island, total submersion is difficult to prove from geological data, and in such a dynamic region tectonic uplift and subsidence are often quite localized.

Ladiges *et al.* (2003) discussed the total submersion theory and cited the large, mainly Australian, clade formed by the eucalypts (*Eucalyptus* and allies, Myrtaceae). This is represented in New Caledonia only by the endemic *Arillastrum* – no *Eucalyptus* is present. Ladiges *et al.* (2003) noted that long-distance dispersal is unlikely, owing to the limited dispersal capacity of the seeds. They concluded that patterns such as this, together with high levels of endemism in New Caledonia and the presence of ancient angiosperms such as *Amborella* and Winteraceae, all imply that 'emergent land existed in the vicinity of New Caledonia throughout its rifted Late Cretaceous and Cainozoic history'. Other authors have also agreed that complete submergence is unlikely, based on biogeographic evidence (Morat *et al.*, 1984; de Laubenfels, 1996; Jolivet, 2008; Jolivet and Verma, 2008a, b; 2010; Sharma and Giribet, 2009). Very small tropical islets often preserve diverse biotas and they could have provided effective microrefugia in or around New Caledonia. Lowry (1998: 181) summed up the situation:

Geologists have contended that during at least some of [the Paleogene] all of the land area comprising New Caledonia must have been submerged. Inference from the modern flora, however, strongly suggests that at least a portion of the land must have remained exposed throughout this process, serving as a refugium – although these sites may have been situated to the south and/or west of the present day Grande Terre in areas that are now submerged. Many attributes of New Caledonia's flora, such as its high generic and familial diversity, and the presence of numerous primitive groups, would be particularly difficult to explain by invoking long-distance dispersal . . .

In the fauna, Bauer *et al.* (2006) dated New Caledonian gecko lineages back to the Late Cretaceous and ruled out the possibility that the whole island was submerged in the Paleogene.

Murienne (2009a) also accepted the possibility of refugium islands. In another study, Murienne (2009b: 1433) concluded that the different hypotheses – island refugia during flooding, versus long-distance dispersal after flooding – cannot be separated on the basis of the phylogenies alone and he approached the problem using molecular dating studies. These indicate that most New Caledonian groups are of Cenozoic age (cf. Grandcolas *et al.*, 2008, Fig. 2) and so Murienne (2009b) proposed that the 'only valid interpretation' for these groups is long-distance dispersal. Nevertheless, the two methods employed in the dating studies (treating the oldest fossils and clock dates derived from these – minimum ages – as maximum ages; using the age of the young islands now emergent on the old Loyalty Ridge to date the taxa there) will both give underestimates of clade age (Chapter 1; Heads, 2008b). In the same way, a review of the molecular studies on New Caledonian plants found that all lineages were regarded as very young (<37 Ma) (Pillon, 2012), but the dates were all calculated using the same method – treating fossil-calibrated youngest possible clade dates as oldest clade dates.

Murienne (2009b: 1434) concluded that 'the presence of numerous relict groups in New Caledonia remains puzzling', but this is only because the clades were assumed to be so young. Likewise, Grandcolas *et al.* (2008) accepted that the New Caledonian biota was wiped out by 'total submersion' during Paleogene flooding and so the current biota is the result of subsequent long-distance dispersal. But they concluded that the 'Phylogenetic relicts remain puzzlingly enigmatic...' (p. 3312). Neither Murienne (2009b) nor Grandcolas *et al.* (2008) mentioned the New Caledonian terranes or their history, although this provides a simple explanation for what are otherwise enigmatic aspects of the biota.

Terrane tectonics in New Caledonia

The main islands of New Caledonia – Grande Terre and the Loyalty Islands – represent emergent parts of two ridges, each more than 2000 km long (Fig. 7.2). Grande Terre itself is made up of seven distinct terranes (Fig. 8.17). These show evidence of two metamorphic–tectonic events. One is dated as latest Jurassic–Early Cretaceous (150 Ma; Cluzel and Meffre, 2002). It is associated with plate convergence in a subduction zone and amalgamation/accretion of the composite basement. This orogeny and metamorphism was followed by a phase of rifting associated with the breakup of Gondwana that formed the Tasman, New Caledonia and South Loyalty Basins. The second phase of metamorphism in New Caledonia occurred in the Eocene (~44 Ma; Spandler *et al.*, 2005b) and represents collision of the basement (by then part of the Norfolk Ridge) with an island arc, possibly the Loyalty Arc.

Biogeographers are familiar with rifting in the Tasman Sea Basin causing disjunction between eastern Australia and New Caledonia (Ladiges and Cantrill, 2007). The two phases of metamorphism in New Caledonia, with associated terrane accretion and orogeny, were probably also important for New Caledonian biogeography and would have involved deformation and accretion of biological distribution patterns in the region.

Cluzel *et al.* (2012) cited three phases of development in New Caledonia:

1. Permian–Early Cretaceous: opening and closure of basins, ending with assembly of 'basement' terranes.
2. Late Cretaceous–Eocene: rifting, followed by convergence and terrane obduction.
3. Oligocene–Holocene: convergence between New Caledonia and the subduction zone at Vanuatu.

The two phases of terrane assembly and metamorphism are both recognized in New Zealand. The first culmination is equivalent to the Rangitata orogeny, the last major reorganization of New Zealand geography and biogeography. The second phase is equivalent to the obduction of the East Coast and Northland allochthons. The terranes of New Caledonia include the following (Cluzel *et al.*, 1994, 2001, 2005; Aitchison *et al.*, 1995, 1998; Meffre *et al.*, 1996; Cluzel and Meffre, 2002; Cluzel *et al.*, 2012).

1. Koh terrane; 2. Central Chain terrane; 3. Téremba terrane; 4. Boghen terrane

Together these make up the New Caledonia basement (Fig. 8.17). All are pre-Cretaceous and were juxtaposed, folded and metamorphosed in a Late Jurassic–Early Cretaceous orogeny. The New Caledonia basement terranes formed in the ocean at an unknown distance off the eastern Gondwana coast and are dated as Carboniferous to Jurassic. The basement terranes comprise arc-derived or arc-associated material, including some terrigenous sediments, and ophiolites (sequences of ocean floor crust and mantle). They can be compared with several New Zealand Eastern Province terranes.

The basement terranes show that precursor arcs, probably with a diverse biota, already existed in the pre-Pacific Ocean before it was invaded by the growing Pacific plate, with its own arcs and biotas, from the Jurassic onwards. The ophiolite and arc terranes amalgamated to form the composite New Caledonia basement and were accreted to the Lord Howe Rise/East Australia part of Gondwana in the Late Jurassic/Early Cretaceous. For most of their history the New Caledonian basement terranes were not part of Gondwana and this is reflected in the many Pacific affinities in the New Caledonian biota. These are quite distinct from Australian–Indian–African (Gondwanan) affinities in other clades there. The basement terranes were accreted to each other and Gondwana by the Early Cretaceous, but by the Late Cretaceous were separating from Gondwana, as part of a large block of continental crust (Zealandia) that also included Lord Howe Rise and New Zealand. The history of the basement terranes has been played out mainly in the pre-Pacific and Pacific; they only formed part of Gondwana for one episode in their evolution, although this was an important one.

After rifting began in the Tasman Sea and the basins off New Caledonia, widespread subsidence also occurred within New Caledonia. Rocks formed at this time include deltaic sandstone, coal shale and conglomerates with blocks up to 40 cm in diameter, indicating high relief. Through the Paleogene there was progressively less terrigenous sediment, and deposition of deeper-water marine sediments began. As discussed above, several biologists have accepted the idea, proposed by some geologists, that all of New Caledonia was submerged at some time in the Paleogene. However, other biologists have preferred to stress biological evidence and this does not support total submergence. Instead, as in New Zealand, the Paleogene would have been a period of changing coastlines and enhanced speciation.

Seafloor spreading in the Tasman Basin ended at the start of the Eocene and a period of convergence began. An Eocene–Oligocene collision zone can be traced in New Zealand, New Caledonia, Rennell Island (south-western Solomon Islands) and south-eastern Papua New Guinea (Aitchison *et al.*, 1995). In New Caledonia a mid-Eocene collision of the basement with an intraoceanic island-arc system to the north-east disturbed the basement subsidence and led to the renewed elevation of New Caledonia. During the collision the following four terranes were accreted to the New Caledonian basement, all from the north-east.

5. Poya terrane

This terrane (Fig. 8.17) is thought to have formed a unit with the New Zealand Northland–East Coast allochthon, before being separated from it by strike-slip faulting (Cluzel *et al.*, 2010a). It is a basaltic mélange of oceanic crust formed as part of the South Loyalty Basin in the Late Cretaceous–Paleocene. The terrane is allochthonous and was formed perhaps 200–300 km north-east of its present location. Following its translation to the south-west it was obducted onto the New Caledonia basement in the Eocene and was then itself overthrust by the ophiolitic nappe. (During obduction, seafloor crust is ramped up onto land at a convergent margin, instead of being subducted as is usually the case.) In the Poya terrane, local alkali basalts accompanied by Paleocene pelagic foraminifera in carbonate sediments represent remnants of intraplate seamounts or islands.

6. The ultramafic nappe

This is a peridotite nappe 3500 m thick that is assumed to be the base of an obducted ophiolite, a seafloor sequence. The nappe is the dominant geological feature of New Caledonia and is famous for its nickel deposits, soils derived from serpentinite (peridotite that has been hydrated after obduction), and endemic plants. The great massif in southern Grande Terre is the largest single unit and smaller massifs occur along the northern half of the west coast. The nappe shows no direct, genetic relationship with the Poya terrane or the Pouébo terrane (next), but dips north-east and is more or less continuous with the oceanic crust of the South Loyalty Basin, between Grande Terre and the Loyalty Islands. As with the Poya terrane, the ultramafic terrane represents part of the basin that was obducted onto the Norfolk Ridge in the Eocene.

Many ophiolites are now seen as remnants of oceanic forearc basins that have been stranded on continental margins in the course of arc–continent collisions (Milsom, 2003). This interpretation implies that before their obduction, ophiolites were associated with a volcanic arc 100–200 km away. In New Caledonia there is good evidence for such a relationship, as the Loyalty Island chain lies 100 km north-east of the mainland ophiolite and is probably an old island arc. This suggests that the ophiolite represents the forearc basin of the Loyalty Ridge. The tectonic and spatial relationship between ophiolites and arcs is reflected in the many plant groups that are each endemic to ultramafic rock plus limestone.

Around the same time that the Poya terrane and the ophiolite were colliding with New Caledonia basement, other ophiolite terranes were also obducted in Paleogene New Zealand and New Guinea. In New Zealand, Late Cretaceous to Paleocene ophiolites and sediments were obducted as the Northland–East Cape allochthon (Mortimer, 2004). In the model discussed above, these represent the forearc of the Northland Plateau seamount chain. North of New Caledonia, the Santa Cruz Basin (between the Solomon Islands and Vanuatu) and the Pocklington Basin (off south-eastern Papua New Guinea) opened at the same time as the South Loyalty Basin. Remnants of the Pocklington Basin appear as ophiolites obducted in the Owen Stanley Mountains of Papua New Guinea.

7. High-pressure (HP) metamorphic terrane

In north-eastern Grande Terre, allochthonous metamorphic rocks of blueschist–eclogite facies make up the Pouébo and Diahot terranes. These are exposed in a north-west–south-east-trending anticlinal range, ~175 km long and 35 km wide (Cluzel *et al.*, 2001). This range includes the highest mountain in the country, Mont Panié (1650 m). The rocks here represent part of a sediment-filled basin that was buried by subduction and underwent high-pressure metamorphism at depths of up to 60 km in the Eocene. Baldwin *et al.* (2007) interpreted the blueschist–eclogite zones (the Pouébo and Diahot terranes), along with bordering lawsonite and prehnite–pumpellyite zones as a single high-pressure (HP) terrane. The three fault-bounded zones have resulted from metamorphism of Late Cretaceous to ?Eocene volcanics similar to those of the Poya terrane, and, in particular, the Pouébo (HP) and Poya terranes seem to be related. Protoliths of the Pouébo terrane formed between Late Cretaceous and Eocene (85 and 55 Ma) (Spandler *et al.*, 2005b). The age matches that of the Poya terrane and is cited as evidence, together with shared aspects of their geochemistry, for a direct link between the two. However, the Pouébo terrane is not simply a metamorphosed equivalent of the Poya terrane and includes a diversity of rocks which indicate a mixed derivation from both oceanic and continental terranes. Associated with the metamorphosed, arc-related basalts there are metamorphosed terrigenous sediments, including remnants of cliff conglomerates, that are absent in the Poya terrane.

After burial and metamorphism of the HP terrane during arc–continent collision there was a phase of extension and the terrane was exhumed at a rapid rate. This resulted in a narrow orogen, <100 km across. The HP terrane has been interpreted as a metamorphic core complex (Aitchison *et al.*, 1995). Peak metamorphism (44 Ma) predates by 10 m.y. the obduction of both the Poya terrane and the ophiolite, which marked a renewed phase of compression (Spandler *et al.*, 2005b). One model proposes multiple episodes of compression and extension during the Eocene, in a belt of 'oscillating orogenesis' (Rawling, 1998; Rawling and Lister, 1997, 1999). In most cases, orogenesis is accompanied by evolution and so the repeated orogeny suggested here would probably have had substantial effects on the biota.

The deep burial of the Pouébo and Diahot terranes and their rapid exhumation represent major tectonic events. It is obvious that the endemic taxa of the modern Mont Panié, for example, did not always survive on the rock strata they occupy now, as these have been buried tens of kilometres under the earth and then uplifted. Yet if the groups were tolerant of disturbance they would have survived in the region of the HP terrane, on upper parts of the crust that have been eroded. Any land in the region – whether continental crust, accreting seafloor, volcanic island, or low atoll – would have had a biota and this would have been affected by the extreme tectonism. The biotas of the HP terrane and the Loyalty Islands are distinctive, with many endemics. The Eocene collision also accounts for other aspects of biogeography; for example, the interdigitation of Loyalty Islands and Grande Terre biota evident in north-eastern Grande Terre and southernmost Grande Terre. The original Grande Terre/Loyalty Islands vicariance probably goes back to the earlier opening of the South Loyalty Basin in the Late Cretaceous.

Belts of orogenesis, metamorphism, intrusion and obduction all represent phases of physiographic dynamism, and the multiple compression/extension events and oscillating orogenesis suggested for the HP terrane are an extreme form of this. The compression/extension has induced breaks in widespread and local metapopulations, changes in local boundaries and elevations of populations, hybridism and other evolution. After the tectonism ended, many of the new distribution patterns, including endemism and disjunction, were left more or less 'frozen' in place.

Emplacement of the HP terrane, ophiolite obduction and the West Caledonian fault

Complex field relationships in the HP terrane have led to controversy over its structural evolution (cf. Rawling and Lister, 2002). Baldwin *et al.* (2007) supported oblique collision between the Norfolk Ridge and the Loyalty Arc, with the latter (and its subduction zone 2 in Fig. 7.1) converging on the former from the north-east from the earliest Eocene. Following peak metamorphism (44 Ma), the entire HP terrane was exhumed as a large, more or less coherent, block during a phase of extension (44–34 Ma). Baldwin *et al.* (2007) inferred that the HP terrane and the ultramafic rocks were spatially separated prior to exhumation of the HP terrane. Obduction of the Poya terrane and the ultramafic nappe, and exhumation of the HP terrane coincide in time but could have occurred in different along-strike regions of the plate boundary. Final movement of the HP terrane took place along the plate margin and led to the juxtaposition of the terrane against the other New Caledonia basement terranes. This occurred after the Oligocene and was possibly caused by strike-slip movement on the West Caledonian fault (Fig. 8.17).

The data just discussed suggest that the HP terrane, the Poya terrane, the ultramafic nappe and the South Loyalty Basin all represent parts of a single, sediment-covered basin – a giant South Loyalty Basin. This opened as a back-arc basin to the north of the New Caledonia basement and later closed (as a forearc basin) as the North Loyalty Basin opened. The age range of the South Loyalty Basin (85–55 Ma) is almost identical to the period of opening of the Tasman Basin. The opening and closing of basins, caused by seafloor spreading and migrating subduction zones, has continued since the Late Cretaceous and provides a mechanism for many biogeographic disjunctions and overlaps in the region.

8. The Loyalty Ridge

The three Loyalty Islands are formed from coral reef material uplifted at 2 Ma. The islands represent the emergent part of the Loyalty Ridge and they form discrete bathymetric highs with distinctive volcanic morphology. The Loyalty Ridge is a mainly submarine feature that extends between Vanuatu and New Zealand, parallel to the Norfolk Ridge (Fig. 7.2; Cluzel *et al.*, 2001; Schellart *et al.*, 2006). Its geology is not well understood owing to the lack of basement outcrops and the thick carbonate cover, but the ridge is probably the remains of an ancient island arc. Eocene andesite has been recovered from the northernmost seamount on the ridge (Bougainville Guyot, west of Vanuatu; Collot *et al.*, 1992).

The Loyalty Islands: the Loyalty arc and the Vitiaz arc

The early Cenozoic location of the south-west Pacific plate boundary (and its associated island arcs) is of special biogeographic interest. The arc developed along the plate margin and from the Eocene onwards there is ample evidence in the volcanic rocks for south-west-dipping subduction. In contrast, the nature and location of the plate boundary east of the Norfolk Ridge from the Late Cretaceous to the Eocene is uncertain.

In Kroenke's (1996) model, the Vitiaz arc formed as an intraoceanic arc in the central Pacific, over 1000 km east of New Caledonia, and converged on New Caledonia through the Cenozoic. (Kroenke suggested the 'Eua Ridge of Tonga' was attached to the eastern side of the Norfolk Ridge until it was detached by back-arc basin formation at 40 Ma.)

In other reconstructions (Crawford *et al.*, 2003; Sdrolias *et al.*, 2003; Schellart *et al.*, 2006), the south-west Pacific subduction zone was originally located next to the eastern Norfolk Ridge. Schellart *et al.* (2006, Fig. 3) proposed that this subduction zone (1 in Figure 7.1) split lengthwise to form the Loyalty and then the Vitiaz arcs (zones 2 and 3 in Figure 7.1). The original precursor arc stretched along the plate boundary from New Zealand to New Guinea or the central Pacific. Crawford *et al.* (2003, Fig. 3) showed the belt as a Cretaceous subduction zone/island arc, but with question marks; Cluzel *et al.* (2006) portrayed it as a 'Late Cretaceous-Palaeocene extinct and/or subducted arc'. Picard *et al.* (2002) and Sdrolias *et al.* (2003) also concluded that the Loyalty–Three Kings Ridge was active back to at least the Cretaceous.

The south-west Pacific is the classic example of episodic back-arc basin formation (see Figure 7.1 for the Late Cretaceous–Cenozoic basins; from Schellart *et al.*, 2006). Back-arc basin formation began with the opening of the Tasman basin and South Loyalty Basin (Late Cretaceous), and basins continued to open behind the subduction zone as it retreated into the Pacific.

South Loyalty Basin

By the Late Cretaceous, rifting was under way in the Tasman basin (84 Ma). About the same time, west of Grande Terre, opening of the New Caledonian basin was separating Grande Terre/Norfolk Ridge from the Lord Howe Rise. Soon after (~75 Ma), east of Grande Terre, the South Loyalty basin began to open, separating Grande Terre/Norfolk Ridge from the original Pacific subduction zone and its arc (zone 1 in Fig. 7.1). The South Loyalty basin reached a maximum width of at least 750 km and its formation would explain the vicariance seen between many Norfolk Ridge/New Caledonia taxa and their sister-groups in the Loyalty Islands, Vanuatu and the Solomon Islands.

At 55 Ma (Paleocene–Eocene boundary), 20 m.y. after it began to open, the South Loyalty basin began to close again. This occurred as the basin was consumed at an east-dipping New Caledonia subduction zone that originated between the Grande Terre/Norfolk Ridge and the Loyalty arc (zone 2 in Fig. 7.1). Grande Terre/Norfolk Ridge and the Loyalty forearc continued to converge until their collision at 44–35 Ma. This would have led to the secondary juxtaposition of biotas on the continental ridge and along the arc. Following the collision, subduction began along the western side of the Norfolk Ridge and lasted until ~25 Ma. Seismic tomography indicates that an extinct subduction zone is buried 80 km beneath the west coast of Grande Terre and Oligocene granitoid intrusions in the ultramafic nappe display features of volcanic arc magmas (Cluzel *et al.*, 2005). After subduction in zone 2 ceased, subduction began in zone 3, off Vanuatu.

North Loyalty Basin

When the South Loyalty Basin had reached its maximum width (earliest Eocene), it began to be subducted and the North Loyalty Basin began to open (Cluzel *et al.*, 2001; Schellart *et al.*, 2006). Later, the North Loyalty Basin in turn was closed as the North Fiji Basin opened. In this model, the South and North Loyalty Basins were both major features that have opened and then closed as other basins opened behind the main subduction zone.

From ~50 Ma the original Pacific arc was split along its more or less north–south axis by formation of the North Loyalty Basin, dividing the arc into the Loyalty–Three Kings Arc in the west and the Vitiaz arc in the the east. (The Vitiaz arc was itself later rifted apart.) This accounts for the standard biogeographic connection between the Loyalty Islands and the Vitiaz arc (e.g. the snake *Candoia bibroni*; Fig. 8.1). Much of the North Loyalty Basin has been destroyed by subduction beneath Vanuatu, where the process continues today and causes volcanism. The current, active arc runs from New Britain and the Solomon Islands to Vanuatu and Fiji/Tonga (subduction zone 3), and has been separated from the original Pacific arc by the opening of the North Fiji Basin. This has also separated Vanuatu from Fiji.

Tectonic evolution of New Caledonia: a summary

In its tectonics, the south-west Pacific is one of the most complex regions on Earth. Geological evidence of its history is limited as most pre-Cretaceous oceanic crust has been subducted and much on-land geology consists of younger volcanics and limestone. It is not surprising that there are many different models for the region's evolution over the past 100 m.y. There is even less geological information available for earlier periods, such as the Jurassic/Early Cretaceous, that have been critical in the early evolutionary and spatial development of modern groups.

New Caledonia evolved within the Australia–Pacific plate boundary zone and its history has involved compressional and extensional tectonism, strike-slip displacement, terrane accretion, orogeny, volcanism and metamorphism. The following chronology, for the Cretaceous onwards, is summarized from several papers (Crawford *et al.*, 2003; Sdrolias *et al.*, 2003; Cluzel *et al.*, 2005; 2006; Spandler *et al.*, 2005a; 2005b; Schellart *et al.*, 2006; Baldwin *et al.*, 2007):

- **150–100 Ma (Latest Jurassic–Early Cretaceous)**. Convergence along what is now the eastern side of the Norfolk Ridge (perhaps the Gondwana–Pacific plate boundary), with accretion of the basement terranes, metamorphism and orogeny (equivalent to the New Zealand Rangitata orogeny).

- **100–80 Ma (Late Cretaceous)**. Extensional regime culminating in the Cenozoic phase of back-arc basin formation (breakup of eastern Gondwana). The basins began to open in the eastward sequence: Tasman basin (Late Cretaceous: 84 Ma), the New Caledonia basin (west of Grande Terre) and the South Loyalty basin (east of Grande Terre) (Late Cretaceous: 75 Ma), and the Coral Sea basin (Paleocene: 61 Ma). Later the South Fiji, North Fiji and Lau basins opened.

- **55 Ma (Paleocene–Eocene boundary)**. Major change in plate boundary processes. Cessation of seafloor spreading in Tasman, Coral, New Caledonia and South Loyalty basins. In the South Loyalty Basin convergence (subduction) replaced divergence (seafloor spreading). The basin itself began to be subducted and, instead of opening, began to close.

- **44 Ma (Middle Eocene)**. Renewed convergence along the Vitiaz arc, active from at least the Eocene. The separate archipelagos (Solomon Islands, Vanuatu, Fiji, Tonga) were originally continuous but were later rifted apart.

- **44–35 Ma (Middle to Late Eocene)**. Opening of the North Loyalty Basin led to convergence in the South Loyalty Basin and collision of the Loyalty arc with Norfolk Ridge. 44 Ma: Peak metamorphism in the HP terrane. By this time, most of the South Loyalty basin had been subducted. Arrival of the Norfolk Ridge at the subduction zone jammed the subduction system, and subduction switched to west of the Norfolk Ridge.

- **40–34 Ma (Middle to Late Eocene)**. Exhumation of the HP terrane. 34 Ma: Juxtaposition of the terrane against the other basement terranes. Obduction of the Poya terrane during this same period or soon after. 38–34 Ma: Ultramafic nappe obducted.

- Later subduction and back-arc basin formation have occurred further east. The New Britain–New Hebrides trench began forming at 27 Ma, the North Fiji Basin at 10 Ma and the Lau Basin at 5 Ma. These are all currently active.

Areas of endemism in New Caledonia

The following areas of endemism (Heads, 2008b) are based mainly on plant distributions (Aubréville *et al.*, 1967–present). Several of the patterns have already been reported in molecular studies of different groups.

1. Loyalty Islands

The islands are formed from raised carbonate platforms dated at 2 Ma. They are much smaller, lower (138 m) and flatter than Grande Terre. Nevertheless, they host a rich rainforest biota with many endemics. The biota of the Loyalty Islands, while not as rich as that of Grande Terre, is distinguished by many taxonomic differences. Virot (1956) observed that the problem of the geological origin of the Loyalty Islands is reflected in a biogeographic problem: their biota shows surprising diversity for such low islands and it is not just an attenuated subset of the Grande Terre biota. As already stressed, the geological history of the Loyalty Ridge is distinct from that of the Grande Terre basement and includes volcanic arc rocks dated as Eocene. Tectonic models of the south-west Pacific propose that this arc was split off lengthwise from the original Cretaceous arc in the region and would have inherited its biota.

The biogeographic enigma of the Loyalty Islands lies in their unexpected differences from Grande Terre. Loyalty Islands endemics include groups of palms, parrots and many others. Apart from the endemism, many Loyalty Islands taxa have closer affinities with groups in southern Vanuatu and Fiji than to any in Grande Terre. This disjunction can be explained by the opening of the North Loyalty Basin that took place as the Loyalty arc approached Grande Terre and the South Loyalty Basin closed. On the other hand, Grande Terre plus the Loyalty Islands together form a centre with many endemics. One example is the parrot genus *Eunymphicus*, with one species on Grande Terre and one on the Loyalty Islands, and this is explicable by the earlier opening of the South Loyalty Basin. *Eunymphicus* is sister to *Cyanoramphus* (New Zealand, Lord Howe, Norfolk, New Caledonia and French Polynesia). The two genera show a great disparity in the size of their current geographic ranges, but the distribution area of *Eunymphicus* probably underwent a significant reduction as the giant South Loyalty basin was almost entirely subducted.

Groups that are found on the Loyalty Islands and elsewhere, but not on Grande Terre, are of special interest. For example, there are no indigenous snakes on Grande Terre but two families, Boidae and Typhlopidae, are represented on the Loyalty Islands. As mentioned already, the boid *Candoia bibroni* is widespread from the Solomon Islands to Tonga, but in New Caledonia it occurs only on the Loyalty Islands (Fig. 8.1). The skink *Emoia* is endemic to the tropical Pacific islands and is widespread there with 74 species, but in New Caledonia it is present only on the Loyalty Islands. The gecko *Gehyra* extends from Madagascar to the eastern Pacific (cf. Fig. 3.18), but in New Caledonia is only on the Loyalty Islands; the species there is also in Vanuatu (Flecks *et al.*, 2012).

The rainforest shrub *Cyrtandra* (Gesneriaceae), mentioned above (Fig. 8.5), is diverse in places such as New Guinea, the Solomon Islands and even in the smaller archipelagos of Vanuatu (11 species) and Fiji (37 species). There is an endemic species on the Loyalty Islands and this is unexpected, as *Cyrtandra* usually occurs in montane or submontane rainforest, often around steep, shaded gullies. Also unexpected is the *absence* of the genus from Grande Terre, where typical habitat is abundant. Green (1979) suggested that *C. mareënsis* of the Loyalty Islands arrived by dispersal from Vanuatu, but there is no real evidence for this. It is unlikely that the genus would be able to disperse to the Loyalty Islands but not to the nearby Grande Terre. *Cyrtandra* is interpreted here as part of an old, central Pacific biota (including boid snakes and many others) that survives as a relic on the Loyalty Islands and the Vitiaz arc, but has never invaded Grande Terre. The most likely reason is that *Cyrtandra* and its intercontinental clade are represented there by another main clade of Gesneriaceae (Coronanthereae; Fig. 3.15) and the break is phylogenetic.

Alyxia stellata (Apocynaceae) repeats the 'widespread central Pacific clade' pattern seen in *Cyrtandra* (Fig. 8.5). It is a widespread Pacific species, found in diverse habitats on many islands east from Micronesia (Palau) and Queensland to Hawaii and south-eastern Polynesia (Middleton, 2002). The only substantial islands in this vast region that it is not recorded from are New Guinea, the Bismarck Archipelago and Grande Terre, but it does occur on the Loyalty Islands.

In stick insects, several genera occur on Grande Terre and the Loyalty Islands (*Leosthenes*, *Carlius* and *Asprenas*), but the Loyalty Islands also have genera, such as *Cladomimus*, *Graeffea* and *Gigantophasma*, that are more closely associated with other Pacific islands, including New Guinea and the Solomon Islands (Buckley *et al.*, 2010). *Gigantophasma* is endemic to the Loyalty Islands and morphological studies suggested that it is closest to the Monandropterini of Réunion and Mauritius. The only other clade in the subfamily Tropidoderinae is Tropidoderini, widespread in mainland Australia (Hennemann and Conle, 2008).

Modern debate on the Loyalty Islands biota

Tronchet *et al.* (2005) argued that the entire flora of the Loyalty Islands was 'almost certainly derived from elements that reached the archipelago by long-distance dispersal, either from the New Caledonian mainland or other more distant islands such as Vanuatu'. This was based on the idea that there have never been islands on the Loyalty Ridge prior to the current ones. This is unlikely, though, and tectonic and biogeographic evidence both suggest that there were prior islands. These islands inherited an arc biota from the early Pacific arc that had separated from New Caledonia + Australia in the Cretaceous, at about the same time that seafloor spreading was beginning *west* of Grande Terre (Norfolk Ridge).

The reefs and islands in the Loyalty group are built on subsided volcanic basement that has been uplifted in the last 2 m.y. The presence of thick reefs indicates significant prior subsidence and it cannot be assumed that the current islands on the Loyalty Ridge are the only ones to have ever existed there. This would also be unlikely, given the endemism and biogeography of the biota. For example, in the New Caledonian cockroach genus *Angustonicus*, Pellens (2004) stressed the 'extreme endemism' of the species in both Grande Terre and the Loyalty Islands, and emphasized the 'nearly complete lack of sympatric distributions' among the species. Nevertheless, Murienne *et al.* (2005) adopted a similar approach to that of Tronchet *et al.* (2005) and equated the age of *Angustonicus* species endemic to the Loyalty Islands with the age of the carbonate rocks there (2 Ma).

Murienne *et al.* (2005: 5) were confident that the diversification of *Angustonicus* in New Caledonia 'cannot be dated to earlier than the emergence of the Loyalty Islands', that the paleogeographic date is 'convincing . . . clear and unambiguous', and that *Angustonicus* 'first colonized the Loyalty Islands a maximum of 2 Myr ago from the New Caledonian mainland'. Yet the proposed dispersal model does not account for the central biogeographic enigma of the Loyalty Islands – the profound difference between their biota and that of Grande Terre. The Loyalty Ridge is a major tectonic and magmatic feature, and while there is no direct geological evidence for prior islands along it, these are likely, given that the ridge is a subsided, poorly known arc structure buried under thick carbonate and located in a region of large-scale vertical and horizontal deformation. In any case, many Loyalty Ridge groups show a closer phylogenetic bond with islands to the north-east rather than with Grande Terre, and this needs to be explained.

A clock study of crickets in Eneopteridae subfamily Eneopterinae found that New Caledonian members were too young for vicariance, even 'under the most conservative dating scenario' (Nattier *et al.*, 2011). Yet the clock calibrations relied on island age to date island endemics (on Fiji and Samoa) and the formation of the Panama Isthmus to date Pacific/Atlantic differentiation (Heads, 2005a, b), and so the actual clade ages are likely to be much older than was suggested. Nattier *et al.* (2011) tried 'To make the biogeographical test even more conservative by pushing back the dating as far as possible', aiming at a maximum possible age, but to do this they relied on the age of a fossil, a minimum date (they used the only known fossil attributed to Eneopterinae). The authors concluded that the group is too young for vicariance and that a model of *in-situ* survival 'is clearly refuted', but their argument was flawed. The distributions of the clades are especially interesting, though, and their correlation with tectonics indicates they are more valuable than island age, fossil age or branch length for dating purposes. One clade has a well-supported phylogeny:

> Grande Terre (*Agnotecous*).
>> Loyalty Islands (*Lebinthus lifouensis*).
>> Vanuatu (*Lebinthus santoensis*).

The primary break is between Grande Terre and the Loyalty Islands (not between the Loyalty Islands and Vanuatu, as might be expected from geography) and it corresponds with the standard break at the South Loyalty Basin (which began to open by seafloor spreading at 75 Ma).

2. Loyalty Islands, Ile des Pins and southernmost Grande Terre

Some groups with this common distribution are also in Tonga (for example, *Nicotiana fragrans*, *Xylosma orbiculatum*). The pattern suggests that these groups were widespread around the South Loyalty Basin before this was reduced to a fraction of its former width.

3. Loyalty Islands–(southernmost Grande Terre)–north-eastern Grande Terre

This distribution pattern is described and illustrated elsewhere (Heads, 2010a).

4. North-western Grande Terre (the West Coast peridotite belt)

The north-western area is dominated by a series of ultramafic massifs. There are considerable biotic differences among these, and the four largest all have endemic plant taxa (Jaffré, 1980). The region includes some of New Caledonia's most distinctive plant endemics (Fig. 8.18). Many taxa endemic to the north-western region are widespread there, e.g. the monotypic genus *Myricanthe* (Euphorbiaceae s.str.) (Fig. 8.18A) and the

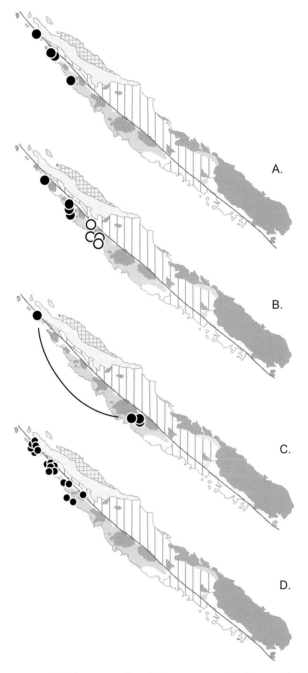

Fig. 8.18 North-western Grande Terre groups. (A) *Myricanthe* (Euphorbiaceae). (B) The two varieties of *Corchorus neocaledonicus* (Malvaceae), var. *neocaledonicus* (circles) and var. *estellatus* (dots). (C) *Alstonia deplanchei* var. *ndokoaensis* (Apocynaceae). (D) *A. deplanchei* var. *deplanchei*. (From Heads, 2008b, reprinted by permission of John Wiley & Sons.)

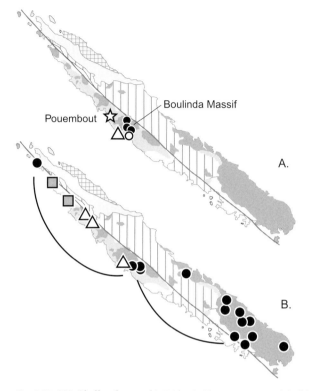

Fig. 8.19 (A) *Phyllanthus nothisii* (dots), *P. avanguiensis* (circle) and *P. pindaiensis* (triangle) (Phyllanthaceae), *Oryza neocaledonica* (Poaceae; star). (B) *Melaleuca gnidioides* var. *gnidioides* (dots), *M. gnidioides* var. *microphyllus* (triangles) and *M. brevisepalus* (squares) (Myrtaceae). (From Heads, 2008b, reprinted by permission of John Wiley & Sons.)

unusual *Corchorus neocaledonicus* (Malvaceae; Fig. 8.18B). Other groups show disjunctions within the region, e.g. *Alstonia deplanchei* var. *ndokoaensis* (Apocynaceae) is disjunct between northern and southern parts of the region (Fig. 8.18C) and surrounds *A. deplanchei* var. *deplanchei* (Fig. 8.18D).

Local endemics at the southern node in this disjunction, around Pindaï and the Boulinda Massif, include *Phyllanthus pindaiensis*, *P. nothisii*, *P. avanguiensis* (Phyllanthaceae; triangle, dots and circle in Figure 8.19A) and *Pittosporum aliferum* (Pittosporaceae). The only Pacific island species of *Oryza* (rice; Poaceae) is *Oryza neocaledonica*, a local endemic at Pouembout in north-western Grande Terre (star in Figure 8.19A). In its morphology, the species appears to be closest to *Oryza meyeriana* of Malesia (Morat *et al.*, 1994). Other species endemic here include *Solanum hugonis* (Pouembout and Poya) and the distinctive pachycaul *Captaincookia* (Rubiaceae), known from Pouembout and Pindaï. *Melaleuca gnidioides* var. *gnidioides* (Myrtaceae) occurs around the same locations and also disjunct in the north, surrounding *M. gnidioides* var. *microphyllus* and *M. brevisepalus* (Fig. 8.19B), as in *Alstonia deplancheii* (Fig. 8.18). *Melodinus guillauminii* (Apocynaceae) has a similar disjunction, with the gap filled by *M. scandens*.

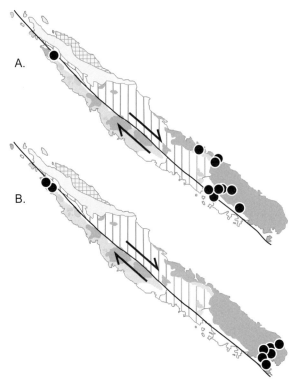

Fig. 8.20 Southern Grande Terre–north-western Grande Terre disjunction. (A) *Xylosma nervosum* (Salicaceae). (B) *Phyllanthus guillauminii* (north) and its putative sister species *P. pronyensis* (south) (Phyllanthaceae). (From Heads, 2008b, reprinted by permission of John Wiley & Sons.)

5. North-western Grande Terre–southern Grande Terre: dextral disjunction along the West Caledonian fault

Many taxa endemic on the Grande Terre ultramafics are widespread there, but other ultramafic taxa have a more restricted range. There is often a major biogeographic break between taxa of the north-western ultramafic belt and relatives on the southern massif, separated by the West Caledonian fault. Some taxa show a boundary at the fault and others show a remarkable dextral disjunction along it, with populations in the north-west separated by more than 100 km from those in the south. In all, 86 examples of the disjunction in morphological taxa of plants, and moths and lizards have been mapped (Heads, 2008a), and two examples are shown in Figure 8.20. Cases documented in recent studies include *Taeniophyllum muelleri* (Orchidaceae), also present in north-eastern Australia and New Guinea) (Pignal and Munzinger, 2011). Another example, *Acianthus amplexicaulis* (Orchidaceae), is disjunct along the West Caledonian fault and is also disjunct in north-eastern Australia (Fig. 8.1).

The West Caledonian fault is controversial; some geologists deny that it exists (Cluzel *et al.*, 2001) while others interpret it as a major structure that has undergone 150–200 km of horizontal displacement since the Miocene (Brothers and Blake, 1973; Rawling and Lister, 1997; 1999). Biogeographic data also support the idea that the fault is important and that it has undergone large-scale strike-slip movement. The fault or fault zone has caused the biological disjunction by pulling populations apart, as discussed above for the New Zealand Alpine fault (Chapter 7).

Baldwin *et al.* (2007) described the West Caledonian fault as a 'major structural break'. They concluded that 'Movement along this fault is a subject of continued controversy, but it may provide a critical constraint on the tectonic evolution of this part of the SW Pacific.' (p. 7). They noted that different authors have interpreted the fault as a downbuckle, a median fault line, a thrust, the trace of a subduction surface, and, of special relevance for biogeography, a dextral transcurrent fault zone. This last model suggests the feature is a long-lived fracture zone with some 150 km of dextral offset and a major vertical component of offset. The high-pressure (HP) metamorphic terrane of north-eastern Grande Terre lies east of the West Caledonian fault. A genetic relationship between the fault and the metamorphism/exhumation of the HP terrane is suspected, but is difficult to examine because of poor exposure. The HP terrane and the ultramafic rocks were spatially separated prior to exhumation of the HP terrane, but they were juxtaposed in post-Oligocene time and this could have been caused by movement on the West Caledonian fault.

Titus *et al.* (2011) also accepted the West Caledonian fault zone (WCFZ) as a major feature and proposed a complex history for it. They suggested that it may have been the original structure that allowed ophiolite obduction. Following obduction, there was dextral reactivation of pre-existing structure in the Belep shear zone (i.e. the northern part of WCFZ). In this process, the Belep Islands region was displaced away from New Caledonia. Later, during Neogene extension, the Belep shear zone was reactivated as the WCFZ, a large detachment fault that was active in the unroofing of the Pouebo terrane in the north-eastern HP province and displacement of the north-west coast ophiolite massifs (klippes).

Recent molecular work has corroborated disjunction along the fault in animals and plants. Sadlier *et al.* (2004) wrote that the skink *Kanakysaurus*, endemic to north-western Grande Terre and the Belep Islands, is most similar 'in overall appearance and biology' to the forest-dwelling species of the endemic New Caledonian genus *Marmorosphax*. Yet in phylogeny, *Kanakysaurus* was strongly supported as sister to *Lacertoides*, endemic to the southern tip of Grande Terre (Fig. 8.21A; Sadlier *et al.*, 2004). This affinity represents a disjunction along the West Caledonian fault. A similar example occurs in geckos, with the genus *Correlophus* (Diplodactylidae) disjunct along the West Caledonian fault (Fig. 8.21B; Bauer *et al.*, 2012).

A study of the pitcher plant *Nepenthes* (Nepenthaceae) in New Caledonia found a disjunct pattern consistent with that of the skinks and geckos. Populations west of the fault form disjunct clades with plants from east of the fault (Fig. 8.21C; Kurata *et al.*, 2008). As the authors noted, the phylogenetic relationships do not reflect the geographic distribution, but they would be congruent with fault disjunction. In its biogeography, the disjunct West Caledonia fault pattern matches the dextral disjunctions in New Zealand along the Alpine fault.

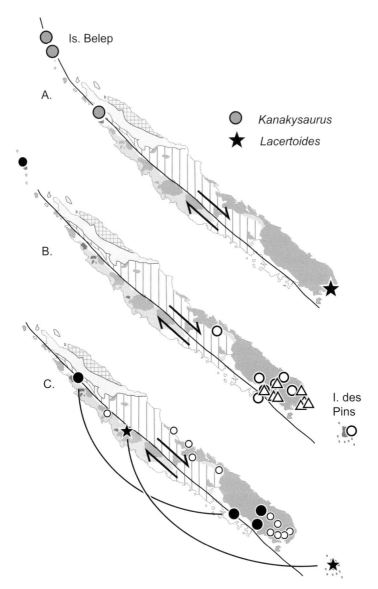

Fig. 8.21 (A) Distribution of two sister genera of skinks, *Kanakysaurus* and *Lacertoides* (Sadlier *et al.*, 2004). (B) Distribution of *Correlophus* (Gekkonidae s.lat.: Diplodactylidae). Black dot = *C. belepensis*, open circle = *C. ciliatus*, triangle = *C. sarasinorum* (Bauer *et al.*, 2012). (C) Distribution of *Nepenthes* (Nepenthaceae) in New Caledonia, showing two clades (stars and black circles) and other populations (open circles) (Kurata *et al.*, 2008).

6. Southern Grande Terre (the southern ultramafic massif)

The main ultramafic massif in southern Grande Terre is well known as a centre of endemism. (This and the next three patterns are documented in different groups of plants; Heads, 2008b.)

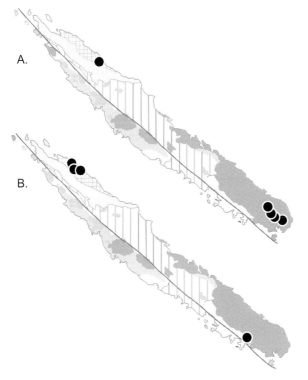

Fig. 8.22 Southern Grande Terre–north-eastern Grande Terre disjunction. (A) *Podocarpus lucienii* (Podocarpaceae). (B) *Symplocos gracilis* (Symplocaceae). (From Heads, 2008b, reprinted by permission of John Wiley & Sons.)

7. North-eastern Grande Terre

This important centre of endemism is equivalent to the HP terrane. Most of the endemism is in the eastern core of the metamorphic belt, where the highest mountains of Grande Terre are located.

8. Disjunction between southern Grande Terre and north-eastern Grande Terre

The disjunction between the south and north-east is illustrated here with morphology-based groups in *Podocarpus* and *Symplocos* (Fig. 8.22A and B). The tectonic relationship between the two different terranes is still being resolved, but the biogeographic disjunction is a common one. These groups are absent from the basement terranes in central Grande Terre.

9. North-eastern Grande Terre–central Grande Terre–south-western Grande Terre

This pattern, along the axial mountains of Grande Terre, is discussed in Heads (2008a).

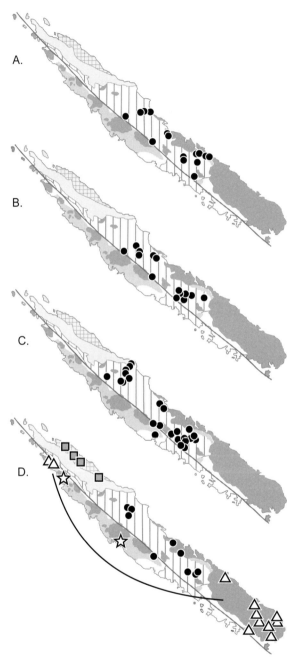

Fig. 8.23 New Caledonia basement endemics. (A) *Pittosporum malaxanii* (Pittosporaceae). (B) *Pittosporum morierei*. (C) *Amborella trichopoda* (Amborellaceae). (D) *Bocquillonia lucidula* (dots, Euphorbiaceae) and the related group *B. nervosa* (squares), *B. longipes* (stars) and *B. spicata* (triangles, disjunct). (From Heads, 2008b, reprinted by permission of John Wiley & Sons.)

10. Central Grande Terre (basement terranes)

Many biologists have explained the high floristic diversity and endemism in New Caledonia as the result of adaptation to the ultramafic rocks there. These rocks are high in
magnesium and iron. Yet although most genera in the New Caledonian flora are present
on the ultramafic massifs, many distinctive New Caledonian plants are not. Among the
other terranes, the HP terrane is one important centre of endemism (pattern 7, above).
The basement terranes comprise another, illustrated here with two *Pittosporum* species
(Pittosporaceae; Fig. 8.23A and B). *Amborella* is the 'basal' angiosperm and is of special
phylogenetic interest. One feature of its biogeography (Fig. 8.23C) has been neglected:
not only is *Amborella* absent from the ultramafics (Jaffré *et al.*, 1987; Lowry, 1998), it
is almost restricted to the basement terranes. The map in the Flora (Aubréville *et al.*,
1967–present) includes one queried record in the north-east (not shown on Fig. 8.23C),
and one anomalous record west of the West Caledonian fault (shown here but not on the
map at http://www.endemia.nc/). Nevertheless, the main distribution is on the basement
terranes.

   Bocquillonia lucidula (Euphorbiaceae; Fig. 8.23D) is another basement endemic, with
the exception of some marginal populations on ultramafics. It is keyed out in Aubréville
et al. (1967–present) next to the *Bocquillonia nervosa–B. longipes–B. spicata* complex,
widespread in Grande Terre but absent from the basement. The biogeography of the
group as a whole is typical and illustrates several of the main patterns listed here,
including the north-eastern centre (pattern 7), the north-western belt (pattern 4) and a
West Caledonian fault disjunction (pattern 5) in *B. spicata*.

   The basement terranes form a centre of endemism but this involves at least two separate
subcentres. Both are occupied by *Amborella* (Fig. 8.23C). The north-western subcentre is
illustrated by *Baloghia balansae* (Euphorbiaceae; Fig. 8.24A), surrounded by *Baloghia
buchholzii* which is in the north and south but absent from the basement. The south-
eastern subcentre is exemplified by the orchid *Chamaeanthus aymardii* (Fig. 8.24A). The
disjunct sister-pair *Pittosporum mackeei–P. bernardii* (Fig. 8.24B) occupies the same
two areas.

   In the north-western subcentre of the basement region, Mount Aoupinié (Fig. 8.25A;
Central Chain terrane) is well documented as an area of endemism for lizards and plants.
Phyllanthus aoupiniensis is endemic to Mount Aoupinié and is related to *Phyllanthus
cherrierei* of Mount Arago (also Central Chain terrane), to the east (Fig. 8.25A). Other
Phyllanthus species (Fig. 8.25B) illustrate independent, parallel connections of Mount
Aoupinié and Mount Arago with the HP terrane centre of endemism in north-eastern
Grande Terre (pattern 7, above). These indicate early biogeographic connections between
the basement and precursors of the HP terrane.

   The examples of endemism cited here for the basement terranes are from morphology-
based accounts, but molecular studies on a well-sampled group have confirmed the
significance of Mount Aoupinié. The cockroach *Lauraesilpha* (Tryonicidae) is a Grande
Terre endemic and its basal clade is endemic on and near Mount Aoupinié (Fig. 8.25C;
Murienne *et al.*, 2008). Its sister-group surrounds it, with one component in the south and
the other disjunct in the north-east. Murienne *et al.* (2008) emphasized that the pattern
is not correlated with rainfall or with soil type. Instead, vicariance between groups on

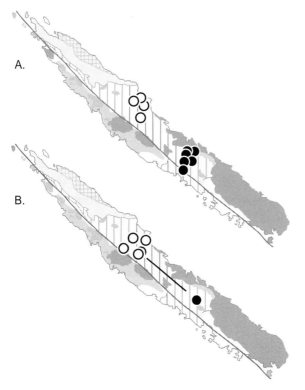

Fig. 8.24 Basement distribution. (A) *Baloghia balansae* (Euphorbiaceae; black circles) and *Chamaeanthus aymardii* (Orchidaceae; open circles). (B) *Pittosporum mackeei* and *P. bernardii* (Pittosporaceae). (From Heads, 2008b, reprinted by permission of John Wiley & Sons.)

the Central Chain terrane and their sister clades in the HP and southern terranes could date to the terranes' initial separation by the opening of the South Loyalty Basin.

Judging from the distribution map in the flora, *Amborella* (Fig. 8.23C) has a small number of populations growing on ultramafic terrane in the southern part of its range where the basement terrane meets the ultramafic rocks. *Bocquillonia lucidula* (Fig. 8.23D) is similar. Several other taxa (*Syzygium brachycalyx*, *S. propinquum*, *Pleurocalyptus austrocaledonicus*: Myrtaceae; *Salaciopsis megaphylla*: Celastraceae; *Pittosporum letocartiorum*: Pittosporaceae) are restricted to the central third or so of the island and are found mainly, but not exclusively, on the basement. Some of these groups have putative sisters found only *off* the basement. Again, this is compatible with the basement area of endemism reflecting geographic aspects of phylogeny rather than edaphic or climatic factors.

Recent studies on New Caledonian groups include further examples of the patterns described in Heads (2008a, b; 2010a, b). For example, in *Polyplectropus* (Trichoptera), disjunctions are evident between the south and the north-east (*P. aberrus*) and between the central terranes and the north-east (*P. nathalae*) (Johanson and Ward, 2009). The Mount Aoupinié endemic *P. aoupiniensis* is closest in its morphology to the north-eastern *P. hovmoelleri*, repeating the disjunction shown in *Phyllanthus* (Fig. 8.25B).

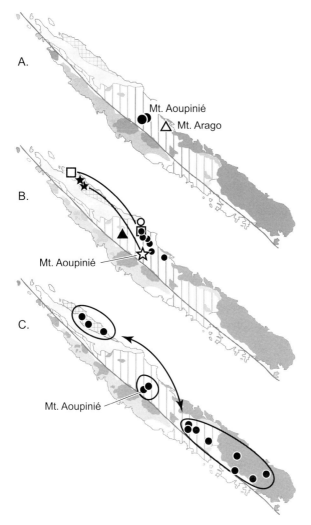

Fig. 8.25 Basement–HP terrane distribution. (A) *Phyllanthus aoupiniensis* (Phyllanthaceae) (Mount Aoupinié) and *P. cherrieri* (Mount Arago). (B) *Phyllanthus moratii* (triangle), *P. margaretae* (white star) related to *P. mandjeliaensis* (black stars), *P. vespertilio* (squares), *P. pseudotrichopodus* (dots) and *P. trichopodus* (circle). (From Heads, 2008b, reprinted by permission of John Wiley & Sons.) (C) A clade in the cockroach genus *Lauraesilpha* with the Mount Aoupinié clade sister to a disjunct group (Murienne *et al.*, 2008).

Terrane tectonics and biogeography in New Caledonia

Most aspects of terrane tectonics, apart from the final breakup of Gondwana by seafloor spreading, are overlooked in biogeographic studies. The 'units' created by seafloor spreading are treated as if they were pure elements and had no history apart from this event. A typical study compared the phylogenies of southern hemisphere taxa with the breakup sequence of Gondwana and found much incongruence, at least for plants (Sanmartín and Ronquist, 2004). The authors suggested the plant taxa were too young to have been affected by breakup (the possibility of their being too old was not considered) and that distribution patterns were caused by chance, long-distance dispersal after Gondwana breakup. Nevertheless, the terminal areas that the authors used in their area cladograms, including 'New Zealand' and 'New Caledonia', are all biogeographic and geological composites, and this means that the analyses were compromised. In particular, groups linking *different* parts of the different geographic areas can appear to be incongruent with each other if the geographic areas are accepted as units (cf. Heads, 1999; Ladiges and Cantrill, 2007).

Sanmartín and Ronquist (2004) treated New Zealand as a single area because it was 'one unit' when it rifted from Gondwana (in fact it was connected to New Caledonia), but the New Zealand biota was already diverse at this time. The biogeographic evidence suggests that precursors of connections such as South Island–Tasmania; North Island–McPherson–Macleay Overlap; and Poor Knights–New Caledonia, already existed then. In addition, large terranes that are critical for New Zealand biodiversity, the Northland and East Coast allochthons, accreted in the Oligocene, long after the breakup of Gondwana (Whattam *et al.*, 2004). Sanmartín and Ronquist (2004) also accepted New Caledonia as an area, but, as with New Zealand, it is the product of terrane amalgamation (and associated metamorphism and orogeny) that took place both before the breakup of Gondwana, in the Jurassic/Early Cretaceous, and after Gondwana breakup, in the Eocene. The amalgamation of New Caledonia terranes in the two orogenies involved the metamorphism of rocks, landscapes and living communities. Because of uplift, strike-slip displacement and other processes, the episodes of accretion would have been just as significant for regional biogeography as classic vicariance by seafloor spreading.

Phases of modernization for geography and biogeography occurred about the same time in other areas, such as the western Americas and the Caribbean. The Greater Antilles show many parallels with New Caledonia and New Zealand. A synthesis of ecology and evolution in Caribbean *Anolis* lizards stressed the importance of plate tectonics and terrane accretion (Roughgarden, 1995). Fossil *Anolis* material dated at 20 Ma and possibly even 40 Ma on Hispaniola is 'indistinguishable' from extant species there, and so the *Anolis* lizard populations represent 'living strata'. Roughgarden (1995) argued that the assemblage of large communities, such as those on Cuba and Hispaniola, has resulted from packages of species combining when tectonic blocks fused to form a single island, rather than from the addition of single species, one by one, as in chance dispersal. Roughgarden (1995: 185) concluded: 'An overall implication of plate tectonics for terrestrial ecology is that relatively fast-acting ecological interactions such as competition and predation are far from sufficient to explain the structure and composition of ecological communities. Instead, ecological communities are fashioned as much by relatively slow geologic processes as by fast species interactions'.

One correspondent agreed that groups on archipelagos may comprise metapopulations connected by local dispersal, and that these may be much older than the individual islands they inhabit. Given this dynamic ecology, though, he questioned the idea that once the archipelagos were accreted to a continent, the organisms would remain on their terranes.

The problem that needs to be explained is this: why are so many groups (such as *Amborella*) restricted to certain terranes and not others when there is no obvious soil or climatic factor causing the boundary? When an arc is accreted to a continent, the arc biota is brought face to face with another biota on its own terranes. A group on the accreting archipelago may be a 'weedy metapopulation' *in its own area*, but if that area is compacted and crushed up against another, there is no reason to assume that all components of both biotas will merge. Wallace's line is a classic example of this sort of boundary, but there are many other terrane boundaries (former plate margins) within land masses such as New Guinea and New Caledonia that coincide with biogeographic boundaries.

Terranes, the age of taxa and Mediterranean biogeography

In the traditional evolutionary chronology based on the fossil record, the Mesozoic–Paleogene tectonics discussed here for New Caledonia would be too old to be relevant for modern species. Species distributions have been assumed to reflect present-day ecology, even if the species themselves are accepted as somewhat older. Instead, the precise clade geography indicated by molecular phylogenies, such as those cited in this book, indicates that taxa and their distributions have evolved together, and that both can be much older than was thought.

Several studies have rejected the evolutionary clock and fossil-based chronology and instead examined spatial correlation of molecular clades with tectonics. Some are cited in Chapter 2. Another study used this approach in the Mediterranean region and found that haplotype distributions in the oak *Quercus suber* (Fagaceae) show 'remarkable conformity' with the regional tectonic history (Magri *et al*., 2007). This history was dominated in the Oligocene by terrane rifting and dispersion, and later by terrane accretion around the margins of the western Mediterranean. Magri *et al*. (2007) inferred an early Cenozoic origin for *Q. suber*, with subsequent displacement on the terranes. For at least 15 m.y. the populations have persisted on each terrane without detectable modifications of chloroplast DNA; Magri *et al*. (2007) cited this as an example of 'longterm permanence *in situ* and prolonged evolutionary standstill'. They also compared the distribution pattern of the oak with a similar one in *Pinus pinaster*. Hampe and Petit (2007) discussed this work and noted that examples of great antiquity of lineages in this region are starting to accumulate in molecular studies. Hampe and Petit (2007) emphasized that the interpretation proposed by Magri *et al*. (2007) rejected long-distance, chance dispersal and instead concluded that genetic patterns reflect tectonic history. Magri *et al*. (2007) were able to reach these conclusions because of the distributions themselves, which were 'extremely clear-cut', and because the authors introduced a simple but critical methodological improvement – oldest fossils (in this case Miocene) were used to set a minimum limit for age, not a maximum limit as in many clock studies.

The Mediterranean oaks and pines cited here have evolved in the early Cenozoic. Similar timing has been proposed in New Caledonian palms; Pintaud *et al*. (2001: 453) argued 'It is unlikely that the local endemism of the New Caledonia at specific and generic level in putative refuge zones can be explained by Pleistocene allopatric speciation'. Pintaud *et al*. suggested instead that the taxa and their distributions, including several north-eastern Grande Terre–southern Grande Terre disjunctions, are the result of earlier

Cenozoic events. Groups such as *Amborella*, the basal angiosperm endemic to the New Caledonian basement terrane, are much older.

Ultramafic terranes: ophiolite obduction, serpentine soils and biogeography

Ultramafic endemism has fascinated botanists since the first modern taxonomists and biogeographers documented it in sixteenth century Italy (Heads, 2005c). The ultramafic terranes are the lower parts of ophiolite suites – obducted slices of ocean floor crust and upper mantle – and have a distinctive geochemistry and tectonic history. The evolutionary relationship between the ultramafic terranes and the endemism they hold remains controversial and a new approach is suggested here.

Endemism on ultramafic – limestone (base-rich) substrate

The 'calcareous riddle' (Ewald, 2003) is a classic problem in biogeography and asks the question: Why are there so many calciphilic species in the central European flora? This has a counterpart in the 'ultramafic riddle' of places such as New Caledonia. The two problems are related by the taxa that are restricted to ultramafic and limestone sites. In New Caledonia, trees such as *Bocquillonia sessiliflora* (Euphorbiaceae), *Balanops vieillardii* (Balanopaceae) and *Syzygium pseudopinnatum* (Myrtaceae) grow on ultramafic rocks and the soils derived from them on Grande Terre, but on calcareous substrate on the Loyalty Islands and Ile des Pins. Santalaceae in New Caledonia have 'zones de prédilection' on ultramafic and calcareous soils (Aubréville *et al.*, 1967–present). The five *Euphorbia* species in New Caledonia are all on calcareous substrate (Morat *et al.*, 2001), while a close relative of *Euphorbia*, the New Caledonian endemic *Neoguillauminia*, is restricted to ultramafic terranes. This is a common pattern; plant taxa known only from ultramafic rock and limestone occur in Tuscany (Selvi, 2007), the Balkan Peninsula (Papanicolaou *et al.*, 1983; Stevanović *et al.*, 2003), the Greater Antilles (Judd *et al.*, 1988; Graham, 2002; Hong *et al.*, 2004; Barker and Hickey, 2006; Grose and Olmstead, 2007; Vorontsova *et al.*, 2007a), Malesia (Heads, 2003), and New Zealand (Heenan and Molloy, 2006). In a Greater Antilles clade of Euphorbiaceae, Jestrow *et al.* (2012) found a phylogeny:

Eastern Hispaniola; on limestone (*Garciadelia*, four species).

Coastal Cuba, Bahamas, western Hispaniola and Jamaica; on limestone (*Lasiocroton*, seven species).

Inland Cuba; on ultramafic soils (*Leucocroton*, 26 species).

The phylogeny does not necessarily imply an invasion into inland Cuba and onto serpentine soils. Instead the simple allopatry suggests the ancestor was already widespread throughout the Cuba–Hispaniola–Jamaica region on both habitats, but only these habitats. Soils on both ultramafics and limestone are characterized by being drought-prone and showing high base saturation, although in limestone soils the exchange complex is dominated by calcium, in ultramafic habitats by magnesium. *Normandia* (Rubiaceae) is a pioneer plant on young, ultramafic soils of New Caledonia and is also unusual in having high levels of calcium in its leaf tissue (http://www.endemia.nc/). The problem of its history has an ecological and physiological component as well as a phylogenetic and biogeographic one. *Normandia* is related to *Durringtonia*, endemic to the McPherson–Macleay Overlap and a rare plant in swamps with permanent subsurface water. On Moreton Island it occurs behind the beach in brown or black peat-like material overlain by fine coastal sand (Heads, 1996). What is the link between the ultramafic habitats of *Normandia* and the beaches of Queensland?

Taxa that are able to survive on ultramafics and limestone will thrive around zones of subduction and obduction. Their ecology will often allow them to persist there for tens of millions of years as metapopulations, surviving on volcanic islands, beaches, atolls, uplifted reefs and obducted basins with their ophiolites.

Evolution of ultramafic flora

Pole (1994: 629) wrote that:

Much of New Caledonia's unique or otherwise interesting plant life at the specific level is restricted to, and presumably a result of, soils developed on its widespread ultramafic rocks. However, some genera are restricted to this substrate, and since the ultramafics were emplaced as an obducted slice of ocean floor in the Late Eocene . . . , it requires an element of special pleading to argue that these lineages date back to the Cretaceous, 40 million years earlier, as most biogeographers have.

This interpretation assumes that taxa have stayed in place, not just in the region but also on the same substrate they evolved on; taxa endemic to strata first available in the Late Eocene cannot be any older than that date. This is unlikely, though, as the ultramafic terrane would have been colonized by the local flora and fauna as it was emerging from the sea. The ultramafic nappe would have inherited biota from other terranes already in the region, such as the Poya terrane (now covered in large part by the ultramafic nappe), and limestone strata that were later removed by erosion. As indicated, the ultramafic flora in New Caledonia shows many affinities with the limestone flora, and the ophiolite was probably emplaced near island arcs where coral reef limestone would have been abundant. Thus, there is no need for the flora currently preserved on the New Caledonian ultramafics to have originated on these or any other ultramafics. The biota is one of former island arcs that colonized the ophiolite sequence as it was emplaced on land.

Fiedler (1985: 1716) discussed heavy metal tolerance in North American *Calochortus* (Liliaceae) and argued in a similar way, suggesting that the tolerance:

. . . is an exaptation in the sense that it may be a character evolved for another use (*or no other function*), which presently is coopted for its current role for life on ultramafic substrates. [It] may be a plesiomorphic character that perhaps has been repeatedly lost throughout the clade, rather than an apomorphy derived through selection for ultramafic substrates. Thus, tolerance of trace metal accumulation . . . is a feature that enhances plant fitness but not necessarily one that evolved repeatedly and specifically for life on ultramafic substrates. [Italics added.]

De Kok (2002) concluded that the occurrence of New Caledonian species on ultramafic soils is a homoplasious character in the respective genera and the result of either pre-adaptation or frequent shifts. He also observed that 'In the minds of some botanists serpentine soils seem to possess almost magical properties. Not only are they said to preserve in isolation so called "primitive" taxa, they can at the same time act as an evolutionary laboratory [producing "derived" taxa]' (De Kok, 2002: 235). The problem in understanding just how ultramafic endemism evolves is because of the reliance on concepts from soil chemistry and current ecology rather than tectonics and biogeography. As Proctor (2003: 105) wrote, in New Caledonia 'The variation in species richness on the ultramafics is difficult to explain. The degree of endemism varies too; it is probably less dependent on soil characteristics than on historical factors'. Pillon *et al.* (2010) also supported a pre-adaptation model for the ultramafic flora.

In New Guinea, botanists have known for years that the ultramafic terranes were strong foci of endemism and, as usual, it was felt that this was caused by edaphic rather than historical factors. Yet Polhemus (1996) pointed out that many animals show similar patterns and that this weakens the edaphic hypothesis. Instead, Polhemus recognized that the ultramafics are significant for biogeography because they indicate the location of prior island arc terranes that have collided with continental crust. Arc-related ultramafics in ophiolite suites are now embedded in the basement terranes of New Zealand, New Caledonia, New Guinea and the Philippines. The remnants of all the accreted arc systems have been crushed between other arc terranes and continental crust fragments, but have left a biological signature in the disjunct distributions of living taxa.

To summarize, many authors have argued that the diversity of rock types in New Caledonia is a fundamental cause of the high floristic diversity. Lithological diversity has permitted the survival of diverse flora (and fauna) through the Cenozoic, but the diversity of substrates is probably not the original cause of the biodiversity and it does not account for the regional biogeography. Instead, this suggests that the conspicuous absences in the New Caledonia flora and the presence of endemic groups both result from the prior location of the component terranes and biotas, rather than soil chemistry.

It is sometimes proposed that the extensive ultramafic outcrops in New Caledonia have discouraged the establishment of certain groups on the islands, but most of the land in New Caledonia does not consist of ultramafic rock. In addition, even groups that are diverse in other ultramafic areas are depauperate in New Caledonia. For example, grasses are diverse in Cuba, with many endemics there and many species on the ultramafics. Yet in New Caledonia grasses show much lower levels of endemism than other families; Dawson (1981) described this as 'puzzling'. Indigenous grasses are also depauperate in Fiji; *Lepturus*, with two species, is the only genus there with more than one well-defined species (Heads, 2006a). This indicates that the situation in New Caledonian grasses reflects a regional, low-diversity anomaly for the group in the central Pacific, and this is also seen in animal groups such as bees.

Towards a new model of New Caledonian biogeography

The centres of endemism in New Caledonia show a close relationship with the terranes. The basement terranes together constitute a centre of endemism that has been neglected, although it includes taxa such as *Amborella*. The Poya terrane outcrops around the margins of the ultramafic terrane. The two show a close association at a local scale and it has not been possible to distinguish them in terms of their biogeography. The ultramafic terrane is well known as a centre of endemism; part of its basiphile biota has probably been inherited from the mafic Poya terrane and from limestones of earlier arc terranes. The HP terrane in the north-east, as with many orogens, is a major centre of endemism. The Loyalty Ridge has an independent tectonic history from that of Grande Terre and there are also biotic differences between the two, with old taxa endemic to the Loyalty Islands surviving on the young islands. Tectonic history also explains some of the disjunctions between the New Caledonian centres, including some caused by strike-slip displacement.

A similar correlation of accreted terranes and biogeographic patterns is also evident in New Zealand, New Guinea, Borneo (Heads, 1990b; 2001b; 2003), Southeast Asia, parts of western America (Heads, 2012a) and the Mediterranean (Magri *et al.*, 2007). New Caledonia is composed of seven or eight terranes, the larger New Zealand of about nine (Mortimer, 2004), and the much larger New Guinea of 32. For the more diverse New Caledonian groups, Morat *et al.* (1984) cited 'surprisingly active speciation in view of the small surface of the island'. The high number of terranes juxtaposed in New Caledonia could explain this. The distributions indicate that taxa have persisted in and around the New Caledonia region, but not always on the strata they currently occupy. Old endemics have survived at old centres of volcanism by colonizing younger volcanic strata from older ones (for example, along the Loyalty Ridge). In addition, slices of seafloor ramped up onto land will inherit taxa from the limestones of local island arcs and intraplate islands, some of which were later destroyed.

The case of the Loyalty Islands biota, with its history of independence from the Grande Terre biota, has parallels in the Lau Group of eastern Fiji and Rennell Island of the south-western Solomon Islands. All are young, volcanic/limestone islands perched on much older, subsided submarine ridges. Each preserves a biota that differs from that of the older 'mainland' in their island group and instead shows biogeographic affinities with other, distant archipelagos. The distributions reflect the underlying tectonics (migrating subduction zones and their arcs) rather than the exposed stratigraphy. For example, the skink *Eugongylus rufescens* is in northern Australia (Cape York Peninsula), Maluku Islands, New Guinea, Admiralty Islands (north of New Guinea) and the Solomon Islands: Rennell Island only (McCoy, 2006). The duck genus *Anas* (Anatidae) includes a clade, the grey teals, distributed from Madagascar to Australia, New Guinea, Rennell Island and New Zealand, but not in the main Solomon Islands or New Caledonia (Johnson and Sorenson, 1999). Because of these and other similar patterns, the bird fauna of Rennell represents a 'paradox' for dispersal theory (Mayr and Diamond, 2001) while the biota of the Loyalty Islands is an 'enigma' (Virot, 1956). In a vicariance model, the biogeography of these islands, along with the Lau group in Fiji and others, is a simple consequence of Paleogene tectonics and *in-situ* survival of subduction zone weeds.

Biologists were quick to appreciate that rifting at divergent plate margins can lead to vicariance. In a similar way, at convergent margins – subduction zones – terranes and biotas can be juxtaposed. Convergent margins are often sites of orogeny and strike-slip movement, and both of these can have direct biological effects. These processes, suggested here for New Caledonia and New Zealand, are also indicated in New Guinea, the Philippines, Taiwan, Japan and along the west coast of the Americas. New Caledonia and its extraordinary biodiversity have developed in a context of back-arc basin formation, terrane accretion, metamorphism, orogeny and marine transgression. The biogeographic patterns of deformation, differentiation, juxtaposition and overlap reflect this history.

9 Biogeography of New Guinea and neighbouring islands

New Guinea is a mountainous, tropical island more or less covered with rainforest. The geology is complex, with all main rock types represented, and this is discussed in more detail below. Together with its offshore islands, New Guinea is well known for its high biodiversity and endemism (Heads, 2001a, b, c; 2002a, b; 2003). It is often assumed that members of the biota originated in Asia or Australia, and then dispersed to New Guinea. Nevertheless, a phylogeny: (Australia (Australia (Australia (New Guinea)))) is compatible with episodic vicariance of a widespread ancestor at an early node in Australia. In the same way, the principle of parsimony suggests that many phylogenies with the pattern: (Australia (New Guinea (Australia (New Guinea, Australia)))) have begun with a widespread ancestor, not at a restricted centre of origin. The Australia–New Guinea bird family Maluridae was used to illustrate these ideas (pp. 20–21), and the same principles apply to Asia–New Guinea connections.

Biogeography of the New Guinea region

New Guinea and Australia

The Torres Strait region, between Australia and New Guinea, marks an important biogeographic break, but the strait only formed at 7–15 000 B.P. and so it is too young to explain the profound biogeographic difference between the two countries (Chapter 4). It is also too young to explain the significant endemism around the strait itself, or disjunctions between islands in the strait and areas such as Sulawesi and tropical America (as in pachychilid snails; p. 108). Mid-Cretaceous coastlines divided Australia and were more complex in the Torres Strait region than in any other part of the continent (Fig. 4.1). As with Bass Strait, Torres Strait is a recent geographic feature that masks earlier, Mesozoic or Paleogene breaks between mainland Australia and its neighbours.

Many groups show simple allopatry between Australia and New Guinea; divisions are often evident at the Arafura Sea, Torres Strait, southern New Guinea and Cairns (Fig. 4.9). For example, in the elapid snake *Pseudechis australis*, Kuch *et al.* (2005) found evidence for five species, with the phylogeny:

Southern New Guinea.

Northern Australia (Kimberley, Arnhem Land, NW Queensland) (three spp.).

Australia, widespread.

There is no evidence for dispersal between Australia and New Guinea, in either direction, as the two main clades are allopatric (at Torres Strait). The first break is at Torres Strait, the second break is somewhere between northern and southern Australia. The only dispersal required in the groups themselves is in the widespread Australian clade, to explain its overlap with the second clade. Whether the direct ancestor of this Australia–New Guinea clade attained its distribution by vicariance alone or by range expansion can only be determined by comparison with its sister-group and others.

In other groups, the break between mainly Australian groups and mainly New Guinea groups occurs in northern Australia or southern New Guinea. For example, the emus of Australia (*Dromaius*) and the cassowaries of New Guinea and Cairns (*Casuarius*) (Fig. 3.21) are an example of a break at Cairns. A break at Cape York is the primary node seen in Phalangeridae (Meredith *et al.*, 2009), a family of marsupials structured as follows:

Australia, widespread (*Trichosurus*), + Kimberley (*Wyulda*).

Sulawesi (*Ailurops* + *Strigocuscus celebensis*).

New Guinea (including Waigeo, Biak and Manus Islands) and Cape York Peninsula (*Spilocuscus*).

Sulawesi and the Maluku Islands (*Strigocuscus pelengensis*).

Maluku Islands, New Guinea, Cape York Peninsula and the Solomon Islands (*Phalanger*).

The Australian clade *Trichosurus* + *Wyulda* is sister to the Melanesian cuscuses (Sulawesi to the Solomon Islands). The two clades overlap only in Cape York Peninsula, indicating an early break in this region.

Invasion from Asia? *Rhododendron* and the Ericoideae (Ericaceae)

Groups with a phylogeny of the type: Asia (Asia (Asia (New Guinea))) are often cited as evidence for dispersal from Asia to New Guinea. A group in *Rhododendron* is a typical example (Fig. 9.1; Goetsch *et al.*, 2011). The authors wrote: 'In view of the basal positions of subsects. *Pseudovireya* (Asia) and *Malayovireya* (Western Malesia) . . . the phylogeny leads us to conclude that evolutionary divergence of the various clades accompanied a dispersal that was polarized in an eastward direction.' Brown *et al.* (2006) agreed that there was an eastward phylogenetic polarity in the group but suggested that this represented a sequence of eastward differentiation, rather than eastward migration. How can these two different models be assessed?

The *Rhododendron* group (Fig. 9.1) comprises a basal clade on mainland Asia (subsect. *Pseudovireya*), a paraphyletic grade in western Malesia (clades 2–5), and a clade

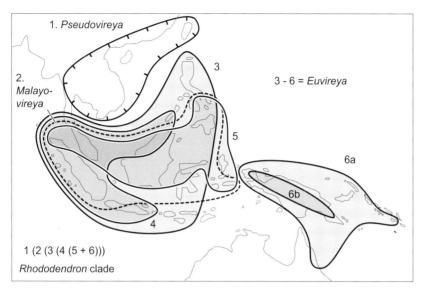

Fig. 9.1 Distribution of a clade in *Rhododendron* (Ericaceae) (Goetsch *et al.*, 2011).

(6a and 6b) in eastern Malesia, the New Guinea region. The breaks between mainland Asia and Malesia, and between eastern and western Malesia, both show strict allopatry and are consistent with an origin of the groups by vicariance. In addition to the allopatry, there is extensive overlap in western Malesia (centred on Aceh, the Malay Peninsula/Singapore and northern Borneo) and in central New Guinea. The distribution of the *Rhododendron* clade mapped in Figure 9.1 repeats the pattern shown in Figure 1.6, with a paraphyletic basal grade of five groups in the west, and an eastern clade sister to one of these. The overlaps only require phases of local range expansion within the regions, not overall eastward migration.

The south-eastern limit of *Rhododendron* occurs in the Solomon Islands and Queensland (with the clade shown in Figure 9.1), where there are two species near Cairns. This also marks the south-eastern limit of *Rhododendron*'s inclusive clade, subfamily Ericoideae. Why does this group have such a restricted distribution in Australia, when it is so abundant in South Africa and elsewhere? Why did *Rhododendron* and the subfamily stop dispersing when they reached Cairns and the Solomon Islands? The pattern reflects the overall phylogeny and distribution of the main clades in the family Ericaceae and so the boundary is probably caused by historical, phylogenetic factors rather than ecological ones. As outlined in Chapter 3, Ericoideae have their main massing of species around the Indian Ocean, whereas Vaccinioideae are based instead around the Pacific, although they are widespread in the Old and New Worlds. In Malesia, the Ericoideae and Vaccinioideae both overlap and both show similar species diversity: Ericoideae have 292 species, Vaccinioideae have 256 species, 88% of the Ericoideae. Further west, the figures are quite different; in China, Ericoideae have 576 species, the Vaccinioideae have 192 species, only 33% of the Ericoideae (*Flora of China*, at http://flora.huh. harvard.edu/china/). In Madagascar, there are 35 Ericoideae, 7 Vaccinioideae (20%).

The Ericoideae and Vaccinioideae have undergone overlap in Malesia, but when? Goetsch *et al.* (2011) proposed that the overlap is reflected in the phylogeny of *Rhododendron*, as it shows the genus invading from west to east. The phylogeny does not necessarily indicate movement, though. Instead, close parallels in the Malesian differentiation and distribution of the two subfamilies indicate they overlapped prior to their modern differentiation.

In eastern Malesia, many distinctive, standard distribution patterns are repeated in both subfamilies and some are cited in the following list. Most of these are large-scale disjunctions that coincide with tectonic features (Heads, 2003). The numbers refer to the figures (distribution maps) in Heads (2003). Map numbers for Ericoideae are in regular type, map numbers for Vaccinioideae are in bold italics:

New Guinea–Mount Kinabalu, Borneo (25, 26; *42, 61*); Philippines–Sulawesi (10, 16, *53, 55*); Philippines–western New Guinea accreted terranes (20, 22; *38, 40, 41, 59, 67*); Philippines–eastern New Guinea accreted terranes (18, 24, 31; *51, 63, 64*); New Guinea, mainly western (11, 19; *49, 50*); New Guinea, mainly eastern (13, 17; *44, 68*); central/southern Sulawesi–New Guinea (9, 12, 13, 22, 23, 25, 29, 32; *37, 40, 55, 56, 67*); Bird's Head Peninsula–eastern New Guinea accreted terranes (9, 14, 30; *38, 47, 64, 69*); western New Guinea craton–eastern New Guinea accreted terranes (14, 16, 17, 18, 26, 27; *48, 74*); western New Guinea accreted terranes–eastern New Guinea accreted terranes (30; *62, 72*); allopatry at or near the craton margin in New Guinea (20, 29; *38, 52, 71*).

Styphelioideae only have 14 species in Malesia – not enough to see any of the diagnostic patterns listed above, except in two groups (*Decatoca* and *Styphelia* subgen. *Cyathodes*) that occur in Malesia only in eastern Papua New Guinea.

The many similarities between genus and species distributions in the Ericoideae and Vaccinioideae in their areas of overlap, such as Malesia, indicate that the overlap of the subfamilies developed early. This means that there is no need for *Rhododendron* itself to have migrated eastward into New Guinea, although the subfamily could have, before *Rhododendron* differentiated. The clades in Figure 9.1 show perfect allopatry among Indochina, western Malesia and eastern Malesia, suggesting that the Indochina and Malesian clades originated by simple vicariance, as did the New Guinea clade of *Rhododendron* (6a + 6b) and its western Malesian sister. At some stage there has been range expansion of the *Rhododendron* clades in western Malesia, causing overlap there.

The clade shown in Figure 9.1 plus its sister-group *Discovireya*, known from Sumatra to New Guinea (not mapped), make up *Rhododendron* section *Schistanthe* (formerly *Vireya*). Webb and Ree (2012) analysed *Schistanthe* using DEC (in LAGRANGE) and SHIBA (a program incorporating ideas on island biogeography). The programs located a centres of origin in mainland Asia/Sundaland, with dispersal eastward to New Guinea and Australia. The authors concluded that the occurrence of the basal taxa in the west, plus the presence of fossil *Rhododendron* primarily in Laurasia, 'intuitively suggest' the same interpretation. The programs have been written to incorporate particular ideas; for example, they will always find a localized centre of origin and reject multiple area versus multiple area vicariance. Yet, as Webb and Ree (2012) admitted, the algorithms 'may misrepresent the biology of speciation for some groups', and the stark allopatry of the three regions in Figure 9.1 (mainland Southeast Asia, western Malesia and New Guinea) indicates that vicariance among large regions has been a main factor in the history of *Rhododendron*.

New Guinea as a global centre of differentiation in several bird groups

Pigeons: the dodo and its relatives

In dispersal theory, New Guinea has been seen as a 'sink', with a biota made up of waifs and strays that arrived from elsewhere in recent, Neogene times. This traditional view predicts that New Guinea clades will not have any special phylogenetic significance. Yet molecular studies show that groups in New Guinea, such as the endemic passerine families Cnemophilidae and Melanocharitidae, are each sister to global groups, not derived within them. With respect to timing, molecular work has also found that New Guinea clades such as birds of paradise (Paradisaeidae) have an 'unexpectedly long history' (Irestedt *et al.*, 2009).

The idea that New Guinea has received its plant and animal lineages from elsewhere can be tested using the phylogeny of the better-known New Guinea groups, such as birds. Pigeons and parrots are two of the most valuable groups for biogeography as their distribution patterns tend to be clear-cut. One interesting group of pigeons includes the dodo, the tooth-billed pigeon of Samoa and distinctive New Guinea forms. It has the following phylogeny (Pereira *et al.*, 2007) (New Guinea is in bold to highlight its importance):

New Guinea (*Otidiphaps*).
    New Guinea (*Trugon*).
      New Guinea (*Goura*).
        Samoa and (fossil) Tonga (*Didunculus*).
          Nicobar Islands (off Sumatra) via Indonesia and **New Guinea** to the Solomon Islands; fossil on New Caledonia and Tonga (*Caloenas*). (The fossil *Bountyphaps*, from Henderson Island in eastern Polynesia, seems closest to *Didunculus* or *Caloenas*; Worthy and Wragg, 2008.)
          Mascarenes (*Raphus*, the dodo, of Mauritius, and *Pezophaps*, the solitaire, of Rodrigues; both extinct).

(*Microgoura* of the Solomon Islands is extinct and was not sampled; most authors have placed it near *Goura*.) In the usual model, New Guinea is colonized by dispersal from Asia or Australia, but it is not possible to explain the pigeon clade in this way without invoking ad hoc processes. If the usual scenario is simply reversed, and New Guinea is proposed as a centre of origin, long-distance trans-oceanic flights are still required in the modern clades (the last three listed above). In a vicariance model, the ancestor of the group as a whole was already widespread around the Indian and south-west Pacific Oceans when its modern clades diverged. This occurred first around the New Guinea terranes (the first three breaks), then the Samoan–Tongan clade broke from the rest (at the Pacific plate margin).

Other widespread groups with New Guinea basal include the stick insect clade Lanceo-cercata, with the following phylogeny (Buckley *et al.*, 2010):

New Guinea, Maluku Islands, Sulawesi (*Dimorphodes*).

Mascarenes (*Rhaphiderus* and *Monandroptera*).

Australia, Lord Howe I., New Guinea, Solomon Is., Fiji, Cook Is.

New Zealand (10 genera: *Clitarchus*, etc.) and New Caledonia (10 genera: *Carlius* etc.). (Records of *Leosthenes* in New Guinea are likely to refer to something else; T. Buckley, personal communication, 7 November 2012.)

The group has a pattern: New Guinea (SW Indian Ocean + Pacific), while other groups such as the freshwater fishes Bedotiidae + Melanotaeniidae (p. 328) have a pattern: south-west Indian Ocean (northern New Guinea (southern New Guinea and Australia)). The incongruence in the branching sequence of the two suggests that the break around New Guinea/Sulawesi and the break around Madagascar/Mascarenes developed over the same period of time, and large-scale rifting is recorded in both areas during the Cretaceous.

The New Guinea endemics with widespread, intercontinental sister-groups show that the high biodiversity in New Guinea involves more than just large numbers of species. Many of the basal groups also have distinctive morphology. For example, *Goura* of New Guinea is the largest extant pigeon and is bright blue. It has a spectacular head-dress, perhaps the last trace of the casque of the cassowaries and similar structures in dinosaurs. Reduction and fusion at the anterior end of the animal is one of the main trends in vertebrate evolution, and other relictual anterior structures are seen in New Guinea groups such as the bird of paradise *Pteridophora*.

The *Goura* lineage is now restricted to New Guinea, but the distribution has probably undergone considerable tectonic shortening. The three allopatric species are in the low-lands of the Bird's Head Peninsula plus the Raja Ampat islands (*G. cristata*), southern New Guinea (*G. scheepmakeri*), and northern New Guinea from Geelvink Bay (includ-ing the islands) to Madang, with a disjunct outlier in Milne Bay (*G. victoria*) (IUCN, 2013). The modern distributions suggest that pre- and proto-*Goura* populations were widespread through the Mesozoic south-west Pacific, on the Australian craton compo-nent of New Guinea and also the island arcs and intraplate volcanics that were later amalgamated to form the mainland. As with many New Guinea groups, *Goura* shows an interesting absence from New Britain and it is also absent from the Huon Peninsula, the Owen Stanley terrane and the Milne Bay islands.

Parrots

The parrots are one of the most conspicuous and diverse bird groups in New Guinea. The phylogeny of the parrot order, Psittaciformes, is as follows (Wright *et al.*, 2008; Schweizer *et al.*, 2010):

New Zealand (the three main islands and Norfolk I.). (Strigopidae: *Strigops* and *Nestor*.)

Australia (throughout), Philippines, Sulawesi, New Guinea and the Solomon Islands. (Cacatuidae; cockatoos.)

Pantropical (and New Zealand). (Psittacidae, including Loriidae.)

The strict allopatry of the first two clades (Strigopidae and Cacatuidae) suggests initial vicariance of all three clades (rather than a centre of origin in Australasia) followed by range expansion of the third group. The overall pattern in the parrots is similar to that of their sister-group, the passerines (Hackett *et al.*, 2008); both are diverse, global groups that have several basal clades endemic in Australasia.

The cockatoos (Cacatuidae) are parrots with a phylogeny (White *et al.*, 2011):

Australia and Tasmania, widespread (*Calyptorhynchus*).

New Guinea lowlands, Cape York Peninsula (*Probosciger*).

Australia and Tasmania (widespread), Philippines, Sulawesi, New Guinea, Bismarck Archipelago and Solomon Islands, fossil on New Caledonia (four genera).

After the basal *Calyptorhynchus* is split off, *Probosciger*, mainly in New Guinea, is sister to the rest, widespread in Australasia. *Probosciger*, the Palm cockatoo, is a distinctive, large, black bird. There is no real evidence or need for a centre of origin outside New Guinea. The main genetic break within *Probosciger* is not at Torres Strait, but at the Weyland Mountains in north-western New Guinea, by the margin of the Australian craton (Murphy *et al.*, 2007). This break separates populations from the Bird's Head Peninsula (accreted terrane) from all populations to the east and corresponds with a tectonic boundary active in the Paleogene.

The parrot family, Psittacidae (including the lories, Loriidae) is a pantropical group with the following area phylogeny (based mainly on Joseph *et al.*, 2011; cf. Wright *et al.*, 2008; Schweizer *et al.*, 2011; Kundu *et al.*, 2012). The importance of the New Guinea region (in bold) is emphasized:

Atlantic group: Africa, but not Madagascar (*Poicephalus* and *Psittacus*) + all American parrots.

Indian Ocean group: **New Guinea mountains** (*Psittrichas*, the Vulturine parrot), sister to Madagascar, Mascarenes, and Seychelles (*Coracopsis*).

Indian Ocean group: **New Guinea, Bismarck Archipelago, Solomon Islands** (*Micropsitta*), sister to Africa (not Madagascar), Mauritius, southern Asia, all Australia, New Guinea and the Solomon Islands (fossil in Vanuatu and Fiji) (*Psittacula, Eclectus, Aprosmictus*, etc.).

Indo-Pacific group: **New Guinea** (*Psittacella*), sister to Africa, Madagascar (*Agapornis*), all Australia, New Guinea, New Zealand and Pacific Islands (*Platycercus, Cyanoramphus, Neophema, Agapornis, Lorius*, etc.).

Analyses of these parrots (Schweizer *et al.*, 2011) and Indian Ocean passerine clades such as Campephagidae (Jønsson *et al.*, 2010c) have located centres of origin in Australasia, using centre of origin programs such as DEC implemented in LAGRANGE, and Bayes-DIVA. The same studies have found young clade ages, but these relied on calibrating phylogenies with island ages and fossils. An alternative approach assumes that Psittacidae and Campephagidae were both already widespread before their modern clades differentiated.

Following the origin and widespread overlap of the last three psittacid groups around the Indo–Pacific, each has differentiated with the basal node separating a New Guinea group from the rest and so this arrangement can be explained with a single New Guinea event. The New Guinea endemics are basal in widespread Indian Ocean or Indo–Pacific clades and cannot be derived from any particular area in Australia or Asia.

The basal break in the parrot order is between New Zealand and the rest, while in the parrot family three of the four clades have a basal group in New Guinea. These patterns indicate, again, the global importance of the Tasman–Coral Sea region and, in particular, New Zealand and New Guinea.

Songbirds (Oscines) in Australasia

The sister-group of parrots is the passerines (perching birds), with the phylogeny:

New Zealand (Acanthisittidae: two extant species, three fossil).

Pantropical, mainly in the Americas (Suboscines: ~1000 species).

Worldwide (Oscines: ~4000 species).

Most passerine species are oscines, or songbirds. Several Australia/New Guinea groups are basal in oscines overall, and in remaining worldwide clades of oscines, New Guinea families are basal. The phylogeny of the oscines (from Jønsson *et al.*, 2011a) is as follows (NG = New Guinea):

Australia (Menuridae and Atrichornithidae).

Australia, NG (Ptilonorhynchidae and Climacteridae).

Australia, NG, Micronesia, Melanesia and Polynesia, east to New Zealand and Samoa, plus one species (*Gerygone sulphurea*, in mangrove etc.) from Borneo to Thailand (Meliphagoidea: Meliphagidae, Acanthizidae, Maluridae – see pp. 20–21, Dasyornithidae).

Australia, NG (Orthonychidae).

Australia, NG (Pomatostomidae).

Worldwide (Corvida and Passerida); basal groups in **NG and New Zealand** (see p. 194).

The phylogeny is not compatible with the traditional idea that Australasian oscines are unrelated, secondary offshoots of northern stock. Apart from the honeyeaters

(Meliphagidae) and the bowerbirds (Ptilonorhynchidae), most of the basal oscine groups are small and little known outside the region. The concentration of the basal oscines in Australasia is obvious; all five groups occur in Australia, and four of these are also in New Guinea. This suggests that the early breaks in oscines occurred at nodes located around Australia and New Guinea, or at least the terranes that became New Guinea. Instead of the Australasian region being a centre of origin for the oscines (Christidis and Norman, 2010), the region can be interpreted as an early centre of differentiation in a global ancestor.

Christidis and Norman (2010) noted that in oscines, several diverse genera are present in Australia and New Guinea, but have much of their diversity outside these areas, either in the Pacific islands or Indonesia (e.g. *Gerygone*, *Myzomela*, *Philemon* and *Lichmera* from the Meliphagoidea, *Pachycephala*, *Coracina* and *Rhipidura* in Corvida, *Zosterops* in Passerida). In these birds, 'It is of interest that *Pachycephala*, *Rhipidura*, *Gerygone* and *Myzomela* all contain species that are mangrove specialists, while *Philemon* and *Lichmera* [and *Zosterops*] also have species that occur in mangroves...'.

The mangrove forests of Australia and New Guinea harbour the world's richest fauna of birds restricted to mangroves (Nyári, 2011). In Australia, 12 species are confined entirely to mangroves, 16 other species occupy mangroves in parts of their range and 80–90 other species visit mangroves. Although an Australasian centre of origin for oscines is not accepted here, the mangrove-associate ecology of several basal oscines and many other oscines is relevant to their history. It would have enabled their survival in oceanic regions and permitted the colonization of new volcanic arcs and islands from others nearby. Weedy, mangrove-associate ancestors of the Pacific birds would have colonized the mid-Cretaceous volcanic plateaus and islands, and their modern descendants have inherited them, as atolls. Other basal oscines have been stranded in New Guinea and Indonesia, where they survive in mangrove and inland rainforest. Many oscine lineages have been stranded in central Australia, where some have persisted and diversified around salt lakes and in other arid habitats.

One of the basal oscines that is well represented in and around mangrove forest is *Gerygone* (Acanthizidae), mentioned already in connection with its New Zealand–New Guinea clade (Fig. 6.5). The genus as a whole reaches its northern and eastern limit along a line: New Guinea mainland, Louisiade Archipelago, south-western Solomon Islands (Rennell Island), northern Vanuatu (Banks etc.), New Caledonia and New Zealand. The genus tolerates high levels of disturbance and is common in mangroves, urban gardens and many types of secondary vegetation. Despite this, it is absent from the Bismarck Archipelago, the Solomon Islands (except for Rennell), Fiji and Tonga – most of the islands that make up the Vitiaz arc. As mentioned already, Rennell Island belongs to a separate tectonic and biogeographic system that involves New Guinea and New Caledonia. Although *Gerygone* is absent from most of the Vitiaz arc, it is represented on Rennell, Vanuatu and New Caledonia by *G. flavolateralis* (cf. *Myiagra caledonica* – Monarchidae; Heads, 2012a, Fig. 6.4). The absence of *Gerygone* and all other Acanthizidae from the Bismarck Archipelago and several other Vitiaz arc islands is difficult to explain in terms of dispersal ability or ecology, but this pattern is repeated in many groups and is discussed below.

Gerygone includes the following clade of mainly allopatric species, with mangrove habitat emphasized (the last five species are mapped in Figure 6.5) (Nyári and Joseph, 2012):

New Guinea (coasts, widespread), northern Australia (**mangrove**) (*G. magnirostris*).

New Caledonia, northern Vanuatu, SW Solomon Islands (Rennell I.) (*G. flavolateralis*).

NW coast Australia (**mangrove**) (*G. tenebrosa*).

Australia. All mainland except the driest deserts and the far north (*G. fusca*), + northern and eastern Australia, southernmost New Guinea (**mangrove**) (*G. levigaster*).

New Guinea: central axial mountains (*G. ruficollis*).

New Zealand (widespread, some populations resident in **mangrove**; Beauchamp and Parrish, 1999) (*G. igata*).

Norfolk Island (*G. modesta*).

An extinct Lord Howe species may also belong in this group. Most of the clades occur in mangrove and there is no reason to assume that this is a secondary development. The ecological series from mangrove to lowland and high montane forest is seen in the New Zealand–New Guinea clade (the last three species), while the series from mangrove to desert is seen in Australia. The first cline can be explained by uplift in the New Guinea orogen, and the second by stranding of inland populations following marine transgressions.

As mentioned already, *Gerygone flavolateralis* is endemic to Melanesian islands (Rennell, Vanuatu and New Caledonia) and is sister to a widespread Australasian group (the group mapped in Figure 6.5, plus *G. tenebrosa* of north-western Australia). This is incompatible with an origin of *G. flavolateralis* by dispersal from either New Guinea, Australia or New Zealand. A vicariance origin with the opening of the South Loyalty Basin (between the proto-Vitiaz arc and continental Gondwana) would account for the group's distribution. The South Loyalty Basin opened at the same time as the Tasman Basin and so the opening of the latter can account for the phylogenetic break in the sister-group, between New Zealand–Norfolk Island–New Guinea on one hand, and Australia on the other (Fig. 6.5).

New Guinea as a centre of differentiation in Corvida and Passerida

After the basal oscines, concentrated in Australia and New Guinea, are split off from other oscines, two large, worldwide clades of oscines remain – the Corvida (sometimes referred to as 'crown' or 'core' Corvida or Corvoidea) and the Passerida.

The Corvida

In Corvida, two small groups in montane New Guinea – the Melanocharitidae and Cnemophilidae – form a basal grade (see p. 194; Jønsson *et al.*, 2011a). The other main groups in Corvida have New Guinea representatives and, again, these are often basal or near-basal in widespread clades. For example, *Eulacestoma* and *Ifrita* are endemics in the central mountains of New Guinea and each has a global sister-group (including Corvidae and others). It is not surprising that a DIVA analysis proposed New Guinea, or the proto-Papuan archipelago, as the centre of origin for the Corvida as a whole (Jønsson *et al.*, 2011a). Nevertheless, the trans-oceanic patterns suggest that the proto-Papuan archipelago was instead a primary centre of vicariance in a moderate number of worldwide ancestors, with repeated differentiation events taking place there.

Despite their differences, the dispersal and vicariance interpretations both agree that the New Guinea clades are not recent immigrants; instead, they evolved early in the history of their groups, more or less *in situ*. Although the topography of the New Guinea region has changed a great deal since this early evolution, with uplift all along the orogen, terane accretion, strike-slip movement and volcanism, many groups have persisted. Even without accepting their centre of origin model, it is possible to agree with Jønsson *et al.* (2011a: 2332) on the precursor islands of New Guinea:

Our hypothesis that islands can be centers of evolution and adaptation *depends on persistence* of island populations and contrasts with the widely held idea that populations on small islands are prone to extinction... highly diversified lineages have survived for millions of years in archipelagoes, such as the Hawaiian and Galapagos Islands, and... estimated ages of many individual island populations of Lesser Antillean birds range into the millions of years... [Italics added.]

The family Oriolidae (Corvida), with notes on the Vitiaz Strait

One group of Corvida, the Old World orioles (Oriolidae), is widespread through the Old World. There are five main clades, distributed as follows (Jønsson *et al.*, 2010b; Johansson *et al.*, 2011):

Northern and eastern Australia, Timor, New Guinea (*Sphecotheres* incl. *Pitohui* p.p.).

New Zealand (*Turnagra*).

Northern and eastern Australia, New Guinea, Maluku and Tanimbar Islands (*Oriolus* 1).

Philippines (most diverse) to Burma (*Oriolus* 2).

Sulawesi (not Philippines) and Indonesia, to Asia, Europe and Africa (*Oriolus* 3).

Turnagra pairs either with *Oriolus*, as shown here (from Johansson *et al.*, 2011) or with *Sphecotheres* (Zuccon and Ericson, 2012). A dispersal model for the family requires a centre of origin in Australia and New Guinea, a leap across Wallacea to mainland Asia, and back-colonization of the Philippines and Sulawesi (Jønsson *et al.*, 2010b). Yet this does not explain the neat break between the Maluku Islands–New Guinea–Australia group (*Oriolus* 1) and its sister in Sulawesi, Philippines, Asia, Europe and Africa (*Oriolus* 2 and 3). This same, localized break at the Molucca Sea occurs in many groups (Heads, 2001a).

The family Oriolidae ranges from Africa to New Guinea, but is absent across the Vitiaz Strait in the Bismarck Archipelago and the Solomon Islands, despite apparently suitable forest habitat there. The break at Vitiaz Strait, only 80 km wide, is not easy to explain in terms of ecology or means of dispersal and Jønsson *et al.* (2010b) wrote: 'there is no obvious reason why orioles have not colonised the Melanesian archipelagos'. They suggested that Oriolidae are poorly adapted to island life, but members of the family are widespread and diverse in the islands of Indonesia and the Philippines.

The birds of paradise are the best known example of the biogeographic break at Vitiaz Strait. They are diverse on the mainland but have never been found in the archipelago, despite suitable habitat there (extensive lowland and montane rainforest) and many expeditions looking for them. Vitiaz Strait also marks an important distributional limit in as many as six other passerine families, namely, Acanthizidae, Ptilonorhynchidae, Grallinidae, Cracticidae, Melanocharatidae and Cnemophilidae. Passerine genera present in Australia and New Guinea, but not in the Bismarck Archipelago, include the melaphagids *Lichmera*, *Meliphaga* and *Melithreptus*. Plant examples include the tree family Dipterocarpaceae. In mosquitoes, *Anopheles faurati* comprises a single clade in the Bismarck Archipelago, Solomon Islands and Vanuatu, with the rest of the species in mainland New Guinea and northern Australia (Ambrose *et al.*, 2012).

The gecko *Cyrtodactylus* has a more or less continuous distribution from Tibet to the Solomon Islands, but it is absent from New Britain and New Ireland, although it surrounds these islands on New Guinea, Manus Island, the Louisiade Islands and the Solomon Islands. Oliver *et al.* (2012) interpreted the absence on New Britain and New Ireland to mean that 'dispersal is a rare stochastic event and/or that the current distribution of these islands is not indicative of their historical position and proximity to New Guinea'.

Fritz *et al.* (2012) compared the biogeography of four passerine families, Oriolidae, Paradisaeidae, Campephagidae and Pachycephalidae. They wrote that 'These families all originated in the same region, the Indo-Pacific archipelagos, with all the most basal species on New Guinea'. There is one obvious difference in the distributions of the families though. Oriolidae and Paradisaeidae occur east only as far as New Guinea and do not cross Vitiaz Strait into the Bismarck Archipelago, the Solomon Islands, New Caledonia or Polynesia. In contrast, Campephagidae and Pachycephalidae both have significant representation in this eastern region (10/93 species, and 6/53 species, respectively).

Fritz *et al.* (2012) attributed differences in the distributions of the four bird families to different dispersal and colonization abilities, and speculated that these might be determined by ecological traits. Nevertheless, it is difficult to suggest an ecological trait that would explain the limit of Oriolidae and Paradisaeidae, along with the other families, at Vitiaz Strait. The break occurs in many groups and also marks an important tectonic boundary (the margin of the orogen), suggesting that it reflects geological and phylogenetic history rather than ecological traits.

The family Dicruridae (Corvida)

The drongos (Dicruridae) range from Africa to the Solomon Islands, and are sister to monarchids, corvids, birds of paradise and rhipidurids (Jønsson *et al.*, 2011a). There are three clades (Pasquet *et al.*, 2007):

India to Borneo.

Africa to Bali.

Himalayas to Australia, New Guinea, Bismarck Archipelago and the Solomon Islands.

All three clades overlap in the region extending from India to Borneo. Pasquet *et al.* (2007) accepted India–Borneo as a centre of origin, with dispersal of the group east across Wallace's line to New Guinea. A simpler explanation for the pattern involves two breaks in a widespread Africa–Solomon Islands group somewhere in the India–Borneo region, followed by overlap there.

The family Pachycephalidae and *Mohoua* (Corvida)

The Pachycephalidae range from India to Samoa, Norfolk Island and south to Tasmania. Jønsson *et al.* (2010a: 253) wrote that 'the presence of several New Guinea humid forest species at the base of the phylogeny suggests a Papuan origin of the family', and they inferred long-distance colonizing flights from there direct to Vanuatu and elsewhere (Jønsson *et al.*, 2010a, Fig. 4). Nevertheless, the Vanuatu species, *P. ornata*, is basal to a widespread clade of Australia and northern Melanesia and so the proposed dispersal routes become very convoluted. Instead, New Guinea can be interpreted as a primary centre of differentiation in a widespread ancestor.

Although Pachycephalidae occur in localities such as Tasmania, Lord Howe Island and Norfolk Island, they are absent from New Zealand. This is not predicted in a dispersal model, but is compatible with vicariance as the Pachycephalidae and its close relatives (Psophodidae, Oriolidae, etc.) are represented in New Zealand by the endemic *Mohoua*. This genus appears to constitute a separate family (cf. Norman *et al.*, 2009; Jønsson *et al.*, 2011a). The absence of *Mohoua* on Lord Howe and Norfolk Islands can then also be explained by simple vicariance.

Some examples from snake life: the elapids

The Elapidae are a pantropical family of venomous snakes. Members include the most poisonous land snakes, *Oxyuranus microlepidotus* (Australia) and *Pseudonaja textilis* (Australia and New Guinea). The Australian snake fauna is unique in the world as it is the only one dominated by the venomous Elapidae (more than 90 species), rather than the primarily non-venomous Colubridae (10 species) (Keogh *et al.*, 2003). This anomaly in Australia provides a clue to the original biogeographic split of elapids and colubrids, the two largest snake families. Pyron *et al.* (2011) gave the phylogeny for the relevant group of snakes as:

Worldwide, in Australia but rare in the south-west quarter (Colubridae etc.; 304 genera).

China and India to northern Australia (Homalopsidae; 10 genera).

Pantropical, more or less throughout Australia (Elapidae: 61 genera).

Africa (Lamprophiidae: 11 genera).

The rarity of colubrids in south-western Australia and the absence of homalopsids in southern Australia is balanced by the diversity and ubiquity of elapids in these areas, suggesting that the groups have differentiated during pre-breakup vicariance at breaks around Australia.

Within Elapidae there are five components (Pyron *et al.*, 2011):

A basal grade with four small groups in Africa, Asia and the Americas (not in Australasia).

A clade of the Indo–Pacific (marine) and Australasia (terrestrial) (Hydrophiinae).

Hydrophiinae is sister to *Elapsoidea*, widespread in sub-Saharan Africa. The simple allopatry indicates that the two have evolved from a widespread southern clade that has split around the margins of Africa into the African *Elapsoidea* and a widespread, marine and terrestrial, Indo–Pacific clade. The sea snakes in this group occur through the warm and temperate Indian and Pacific Oceans, but are absent from the Atlantic.

Although the Hydrophiinae include marine species that are widespread around the coasts of the Indian and Pacific Oceans, most of the diversity in the group is in Australasia. In this region there are sea kraits (*Laticauda*), true sea snakes (*Hydrophis* etc.) and ~100 terrestrial species. The latter occur in all parts of Australia, including the desert, and are widespread in New Guinea at lower elevations.

The phylogeny of the Hydrophiinae, as sampled by Sanders *et al.* (2008), is:

Coastal waters and shores from India to Fiji (*Laticauda*, the sea kraits).

New Guinea, including islands around Bird's Head Peninsula (*Micropechis*).

Terrestrial in Australia and Melanesia, coastal marine through Indo–Pacific (*Furina, Aspidomorphus, Toxicocalamus, Demansia* and a clade of 27 genera).

The Melanesian *Micropechis* is sister to the remaining oxyuranines, in which most other basal forms are exclusively (*Aspidomorphus* and *Toxicocalamus*) or partly (*Furina* and *Demansia*) Melanesian. As Sanders *et al.* (2008: 690) observed, 'This pattern of predominantly mesic, basal Melanesian forms is found in other Australasian squamates (e.g. agamids . . .) and mammals (e.g. murids . . .)'. Similar patterns occur in the pigeons and cockatoos (see above) and are consistent with a widespread Indo–Pacific ancestor undergoing early differentiation around nodes in New Guinea. The Melanesian snake *Micropechis* is sister to a large group found throughout Australasia and also further afield. In cockatoos (after the basal *Calyptorhynchus* is split off), *Probosciger*, mainly in New Guinea, is sister to the rest, widespread in Australasia. The pattern also occurs in the basidiomycete *Lentinula* (shiitake mushrooms), as a New Guinea group has its sister in China, Nepal, New Guinea, Australia, Tasmania and New Zealand (Hibbett *et al.*, 1998). To summarize, a similar pattern, with New Guinea clades basal in widespread west Pacific or even Indo–Pacific groups, occurs in mushrooms, snakes and parrots. This is probably a general phylogenetic/tectonic phenomenon that is not related to ecology, means of dispersal or movement into or out of the archipelagos that became modern New Guinea.

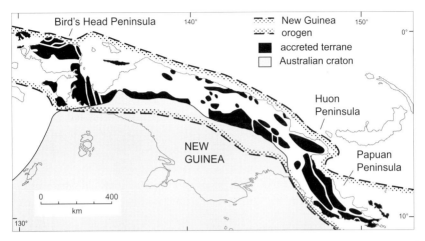

Fig. 9.2 Geological map showing the Australian craton (grey), the New Guinea orogen (between the heavy broken lines) and the accreted New Guinea terranes (black). Blank areas on the map are successor basins (simplified from Pigram and Davies, 1987).

Biogeographic differentiation in mainland New Guinea

Early in the twentieth century, Dutch oil geologists proposed that New Guinea represented a compaction of northern and southern structural elements, and this idea was integrated with local biogeography in early syntheses (Toxopeus, 1950). In this model, there have been plants and animals in the New Guinea region since before New Guinea itself was formed. The model contrasts with traditional dispersal scenarios, in which New Guinea begins as an empty receptacle, a biological vacuum, and is populated after it forms, from Asia and Australia.

Later developments of the vicariance model for New Guinea have proposed ancestral metapopulations that survived in a shifting landscape of many small islands on many, separate terranes (Pigram and Davies, 1987; Michaux, 1994; Polhemus, 1996; Polhemus and Polhemus, 1998; de Boer, 1995; de Boer and Duffels, 1996a, b; Heads, 2001a, b; 2002a, b; 2003). The southern part of New Guinea forms the northern edge of the Australian craton, whereas northern New Guinea is made up of some 32 independent terranes that have encroached from the Pacific side and docked with the craton (a few are displaced parts of the craton margin) (Fig. 9.2). Accretion of the allochthonous terranes began in the Late Cretaceous (Davies, 2012).

The craton margin has been deformed by an orogeny that developed along it, in association with the terrane accretion. A high mountain range follows the central axis of New Guinea from the west to the east, and a main break in New Guinea biogeography and geology – the craton margin – is located within this range. The southern parts of the range are part of the Australian craton, while the northern parts are allochthonous, accreted terranes. The Bird's Head Peninsula in the west of the island, the Huon Peninsula in the central part and the Papuan Peninsula (the 'tail' of the island) in the east, are all formed from accreted terranes, as are several islands off the western and eastern ends of New

Guinea. The Finisterre terrane (Huon Peninsula), for example, is an Oligocene–Miocene arc and one of the last terranes to be accreted. The next terranes to accrete will be the islands of New Britain and New Ireland. All the accreted terranes are important areas of endemism. For example, the sorcerer's tree *Magodendron* (Sapotaceae) is restricted to the Papuan Peninsula terranes (Heads, 2003), although it is sister to diverse, widespread clades of Southeast Asia, Australia and New Caledonia (Bartish *et al.*, 2011).

Many terrane boundaries (former plate margins) in mainland New Guinea are well defined, but the details of paleogeography in the region are poorly known and even the history of the subduction zones is understood only in general terms. Significant questions include: 'How many island arcs existed north of the Australian plate during the Cenozoic, how did they form, what was their polarity and what was their accretion history?' (Baldwin *et al.*, 2012). Hall (2012) wrote: 'For most of the Cenozoic the New Guinea Limestones, which now form the high mountains, were part of a wide and long-lived carbonate shelf north of the Australian land mass'. He cited 'numerous minor unconformities, hard grounds, minor clastic intervals, and signs of karstic alteration suggesting intermittent emergence' and concluded: 'it is impossible at present to map the palaeogeography in detail; it seems improbable that the entire shelf was ever completely emergent, but it is likely that the shelf was an area of numerous low islands during most of the Cenozoic'. This history explains how mangroves have been trapped and uplifted in the orogeny that developed through New Guinea.

The craton margin is not obvious in the current topography of New Guinea, but it is one of the most important tectonic and biogeographic features in the country. Southern groups, such as *Nothofagus* (Nothofagaceae) and *Amylotheca* (Loranthaceae), have species massing in New Guinea on the southern, cratonic part of the mountainous axis, whereas Malesian groups, such as *Archidendron* (Fabaceae), *Aglaia* (Meliaceae) and *Amyema* (Loranthaceae), are most diverse on the accreted terrane portion of the mountains (Heads, 2001c). Boundaries between basement and accreted terranes are also significant in the biogeography of New Caledonia and New Zealand. Molecular studies have begun to clarify relationships between New Guinea and the rest of Australasia, but as yet there are few detailed molecular studies of diversification within New Guinea. Nevertheless, preliminary surveys have already corroborated the biogeographic importance of the craton margin, as in the following group of fishes.

Differentiation around the craton margin in the freshwater rainbowfish (Melanotaeniidae: Atheriniformes)

The basal parrots include the trans-Indian Ocean link: New Guinea (*Psittrichas*)–Madagascar, Mascarenes, Seychelles (*Coracopsis*). Similar Indian Ocean links occur in freshwater fishes, as the rainbowfish (Melanotaeniidae) are endemic to Australia and New Guinea (81 species) and have their sister-group, Bedotiidae, in Madagascar (Fig. 9.3; Sparks and Smith, 2004b; Unmack *et al.*, 2013). There is no close affinity between bedotiids and mainland African atheriniforms. As with the Milyeringidae of north-western Australia and Madagascar (Chapter 4), Sparks and Smith (2004b)

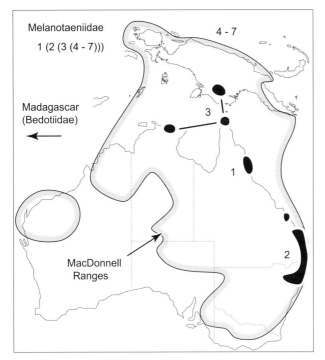

Fig. 9.3 Distribution of Melanotaeniidae clades 1, 2, 3 and 4–7. 1 = *Cairnsichthys*; 2 = *Rhadinocentrus*; 3 = *Iriartherina* (Unmack *et al.*, 2013).

concluded that the Indian Ocean pattern 'is congruent with the break-up of Gondwana, not a scenario reliant on Cenozoic trans-oceanic dispersal' and a molecular clock study gave tentative support for this (Unmack *et al.*, 2013).

The distribution and phylogeny imply that the ancestor was distributed in Madagascar, Australia (absent to the south-west of the MacDonnell Ranges) and New Guinea (absent from much of the Papuan Peninsula and the Bismarck Archipelago). Within melanotaeniids, a basal grade of three genera occurs at the McPherson–Macleay Overlap, north-eastern Australia and southern New Guinea (Fig. 9.3; Unmack *et al.*, 2013). Thus the overall phylogeny shows that differentiation within Bedotiidae + Melanotaeniidae has occurred first around Madagascar and then around the Tasman–Coral Sea centre, as in many groups. There is no ecological reason why the basal grade in melanotaeniids could not have been in north-western Australia (Pilbara–Kimberley) or northern New Guinea, and this indicates an underlying tectonic cause for the distribution.

After the three basal genera of melanotaeniids have split off, the large clade that remains illustrates the way in which Australia has often been 'invaded' by a wave of differentiation arriving from New Guinea (Fig. 9.4). New Guinea itself does not appear as an area and, instead, the two basal breaks in the phylogeny occur around the accreted terranes. The first break is between group 4 (Bird's Head Peninsula) and the rest, and the second is at the craton margin, between group 5 (northern New Guinea) and the rest, on

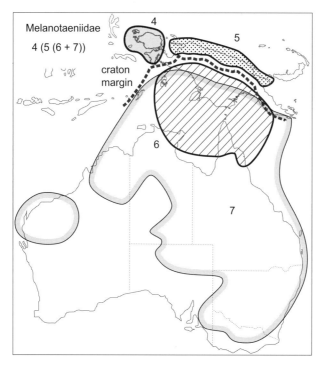

Fig. 9.4 Distribution of Melanotaeniidae clades 4–7. 4 = 'Western' ('E'); 5 = 'northern' ('F'); 6 = 'goldiei group' ('D'); 7 = 'nigrans group' ('C') + 'australis group' ('A') + 'maccullochi group' ('B'). (Unmack *et al.*, 2013; clade names based on letters are from McGuigan *et al.*, 2000.)

the Australian craton. In the clade on the Australian craton, group 6 (Arafura and Coral Sea basins) is sister to the widespread group. In frogs, a craton/accreted terrane split that is well-defined geographically and well-supported statistically occurs in the microhylid *Mantophryne lateralis* s.lat. (between clade A, B, C + D and clade E, F + G; Oliver *et al.*, 2013).

Unmack *et al.* (2013) wrote that it is surprising and puzzling that hybridism in Melanotaeniidae seems very rare in the wild, but species (and even genera) hybridize readily in captivity. The molecular study also showed evidence of prior introgression in the wild populations. This suggests that groups such as the Melanotaeniidae and many plants in New Zealand (*Coprosma*, *Hebe*) represent Cretaceous hybrid swarms that have been left more or less frozen in place, although renewed hybridism and divergence can be provoked in cultivation.

Butcherbirds (Cracticidae): a parallel to the rainbowfish

The trans-Indian Ocean pattern in the melanotaeniid/bedotiid group has a parallel in passerine birds, in the following clades (Kearns *et al.*, 2013):

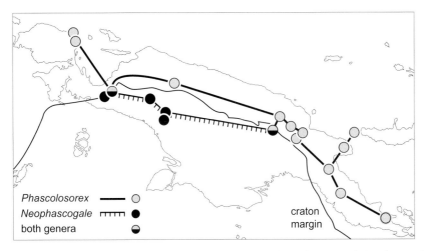

Fig. 9.5 Distribution of two sister genera of marsupials, *Phascolosorex* and *Neophascogale* (Dasyuridae). (Heads, 2001b. Reprinted by permission of John Wiley & Sons.)

Africa, Madagascar, India and Sri Lanka, east to Borneo, Java (*Philentoma*) and Bali (*Hemipus*). (Prionopidae, Malaconotidae, Platysteiridae and Vangidae s.lat.)

New Guinea (including the Raja Ampat and Louisiade Archipelagos west and east of the mainland, but not the Bismarck Archipelago) and Australia (Cracticidae).

The two clades show simple allopatry, with a break between Bali and New Guinea, and so there is no need for any trans-Indian Ocean dispersal in the modern families. This can be compared with the break in the rainbowfish, somewhere between Australia/New Guinea and Madagascar. Cracticidae has the following arrangement:

New Guinea (except Merauke Ridge–Oriomo Plateau in southern New Guinea) (*Peltops*).

All New Guinea, Australia and Tasmania (*Cracticus* incl. *Gymnorhina*).

Southern and eastern Australia, Tasmania (*Strepera*).

The primary node separates a northern New Guinea group from one in all New Guinea and Australia, at a break somewhere around the craton margin.

The phascolosoricine marsupials of New Guinea: a New Guinea orogen clade with a division at the craton margin

The marsupial genera *Phascolosorex* and *Neophascogale* (Dasyuridae) form a clade that is endemic to the New Guinea orogen. Within the orogen, the two genera are allopatric, with one north of and one south of the craton margin (Fig. 9.5; Heads, 2001a; Westerman *et al.*, 2008). The two genera together make up the sister-group of *Dasyurus* in Australia and New Guinea (widespread), and *Sarcophilus* in Australia (Pleistocene fossils) and Tasmania (extant).

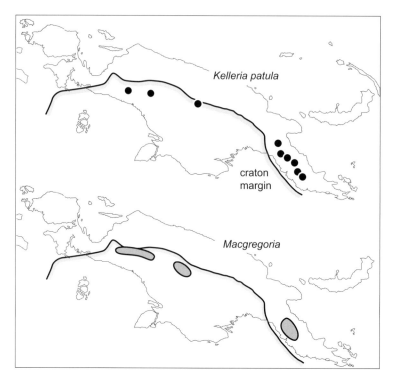

Fig. 9.6 Two examples of disjunction in New Guinea along the craton margin (line) (Heads, 1999). (Top) The alpine shrub *Kelleria patula* (Thymelaeaceae). (Bottom) The subalpine bird *Macgregoria* (Meliphagidae).

Uplift, strike-slip and disjunction in the New Guinea orogen

De Jong (2004: 411) explained the distribution of certain high-elevation New Guinea butterflies by suggesting that the species or an ancestor 'may have simply been carried up by the uplift of the mountains'. This process would explain many aspects of montane biota in general. Terranes have also moved horizontally, as well as vertically, along faults, sometimes for long distances, and many disjunctions suggest that this has also had direct effects on the biogeography.

One of the main tectonic breaks in New Guinea occurs at the craton margin, where phylogenetic breaks occur in groups such as marsupials and melanotaeniid fishes. The main events at the craton margin include the accretion of the northern terranes and this has involved strike-slip movement of the terranes, with both dextral and sinistral offsets. In biogeography, apparent dextral disjunction occurs in groups such as the plant *Kelleria* (Thymelaeaceae) and the bird *Macgregoria* (Meliphagidae) (Fig. 9.6).

Similar disjunctions along the craton margin occur in beetles: a species of *Mecyclothorax* (Carabidae) on the Finisterre Mountans (Huon Peninsula) is closest to one on Mount Trikora (= Wilhelmina Peak) (Liebherr, 2008). Other examples occur in the Ericaceae genera *Rhododendron* and *Vaccinium* (Heads, 2003, Figs. 27 and 48).

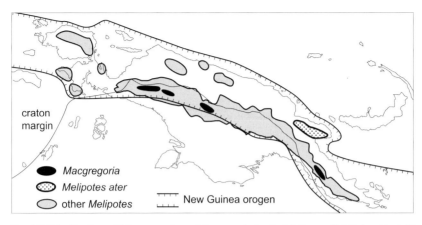

Fig. 9.7 Distribution of *Macgregoria* and the two clades in *Melipotes* (Meliphagidae) (Beehler *et al*., 2007).

New Guinea is much larger than New Zealand and New Caledonia, and has a more diverse biota, but there are many parallels in the accretion tectonics and the biogeography of the three countries. Their history has been dominated by terrane accretion, with associated metamorphism, magmatism, folding, strike-slip movement, obduction and orogeny.

The relationship between the central New Guinea ranges and the coast ranges: an example from honeyeaters (Meliphagidae)

In dispersal theory, taxa on the offshore islands north of New Guinea dispersed there from the mainland. Within the mainland, the Bird's Head Peninsula and the Huon Peninsula also form 'islands', separated from the main axial range of New Guinea by lowland valleys. There are many clades endemic to the peninsulas and, as with the offshore island groups, these are thought to have dispersed across the valleys from the central mountains (Diamond, 1972, 1985).

A typical example concerns two genera of honeyeaters (Meliphagidae). As shown in Figure 9.6, *Macgregoria* has a disjunct distribution in the central axial range of New Guinea. Its sister-group, *Melipotes*, is endemic to the New Guinea orogen and is widespread there with eight species (Fig. 9.7; Beehler *et al*., 2007). The distribution of *Melipotes* includes some of the northern peninsulas and northern coastal ranges that represent accreted terranes. The genus is absent from the high mountains of the southern Maluku Islands and the Bismarck Archipelago, areas that do not form part of the New Guinea orogen.

In the pair *Macgregoria* + *Melipotes*, Beehler *et al*. (2007) interpreted the 'basal' clade, *Macgregoria*, as indicating the centre of origin. Within *Melipotes* itself, the basal group (from morphological studies) is *M. ater* of the Huon Peninsula, and so Beehler *et al*. (2007) inferred dispersal from the cordillera (*Macgregoria*) to the Huon Peninsula (*M. ater*), and then back again, to give rise to the rest of *Melipotes*. However, if *Melipotes* is such an effective long-distance colonizer, it would be expected to occur in the Bismarck Archipelago. In fact, it is absent there, along with mainland New Guinea groups such as birds of paradise, bowerbirds, acanthizids and many others, suggesting distribution mediated by tectonics rather than chance dispersal.

The dispersal model for *Macgregoria* and *Melipotes* explains the pattern by rare events of chance dispersal between the central range and the coast ranges. In a vicariance model, the only dispersal needed is local range expansion of *Melipotes* and/or *Macgregoria* in parts of the central range, to account for their overlap there. The ancestor was already widespread on the terranes that became the cordillera and coast ranges before the modern clades differentiated. The first differentiation took place within the central ranges and separated the local *Macgregoria* from the widespread *Melipotes*. The next break was between *M. ater* on the Huon Peninsula and the rest of *Melipotes*, widespread through the rest of the New Guinea mountains, but absent from the Huon Peninsula. This last distribution can be compared with the well-known tree *Araucaria cunninghamii* (Arau-cariaceae) (de Laubenfels, 1988), which is widespread in the New Guinea mountains but notably absent from the Huon Peninsula and the nearby Adelbert Range. In other words, it is absent from the extensive suitable habitat (montane rainforest) of the Finisterre island arc terrane, as well as the Bismarck Archipelago.

Birds of paradise (Passeriformes: Corvida: Paradisaeidae)

Some of the biogeographic affinities among the New Guinea terranes can be illustrated with one of the best-studied groups in the region, *Paradisaea* (Paradisaeidae). The genus is widespread through New Guinea and on islands off the Bird's Head and Papuan Peninsulas (Fig. 9.8; Irestedt *et al.*, 2009).

Paradisaea rudolphi, the blue bird of paradise, differs from the others in its mor-phology, behaviour and ecology. It occurs only in eastern New Guinea (Papua New Guinea) and at higher elevations than the other *Paradisaea* species (Fig. 9.8; clade 1). The main split within *P. rudolphi* corresponds with the craton boundary. This suggests that *P. rudolphi* and the remaining *Paradisaea* clade already occupied Pacific terranes as well as the Australian craton before mid-Cenozoic accretion and uplift. Differentiation at the craton margin could have occurred before or during these processes; for example, during compression and the shortening of prior clines. In this process, the end members of a group that was earlier distributed across hundreds of kilometres or more into the Pacific would have become juxtaposed in the mountains of New Guinea. Distributions in other groups suggest that oblique convergence and strike-slip have produced significant biological disjunctions at the craton margin, as discussed above (p. 332).

After the differentiation of *P. rudolphi* from the rest, the next break in the genus is between *P. guilielmi* of the Huon Peninsula (Fig. 9.8; clade 2) and the rest. The Huon Peninsula is an accreted terrane (one of the latest to accrete) with a different history from the other New Guinea terranes. If the peninsula had been colonized from the main range, peninsula endemics would be related to a specific central range population, their original source. Instead, *P. guilielmi* is sister to a clade found throughout the rest of New

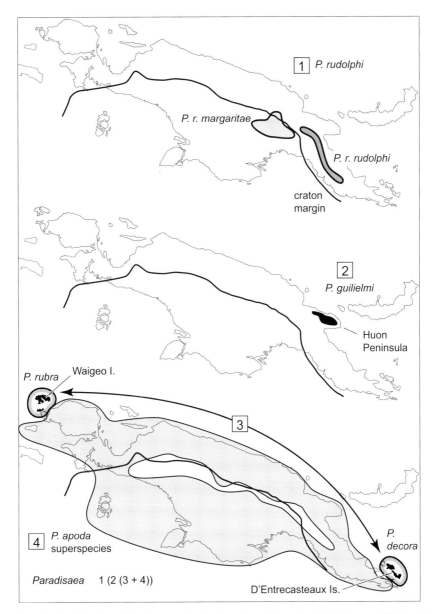

Fig. 9.8 Distribution of the three clades in *Paradisaea* (Paradisaeidae) (Heads, 2002a). (Top) *Paradisaea rudolphi*. (Middle) *P. guilielmi*. (Bottom) *P. rubra*, *P. decora* and the *Paradisaea apoda* superspecies.

Guinea, just as *Melipotes ater* of the Huon Peninsula is sister to the rest of *Melipotes*, found throughout montane New Guinea (Fig. 9.7).

After *Paradisaea guilielmi* of the Huon Peninsula separates from the others, a disjunct clade on offshore islands (Fig. 9.8; clade 3) breaks from the mainland group. The disjunct clade is distributed on:

West Papuan Islands: Waigeo and Batanta (*Paradisaea rubra*).

D'Entrecasteaux Islands: Fergusson and Normanby (*P. decora*).

Both sets of islands represent distinctive tectonic structures. The West Papuan Islands form an ophiolite terrane, 5 km off Salawati Island and the Bird's Head Peninsula, and separated from them by the Sorong fault. The D'Entrecasteaux Islands are a metamorphic core complex (the D'Entrecasteaux terrane), lying 15 km off the Papuan Peninsula.

The affinity between *P. rubra* and *P. decora* is a good example of the disjunct biogeographic connections among the islands north of New Guinea. Instead of the two island species each being derived from adjacent mainland species, they are related to each other, despite being 2000 km apart. The pair as a clade is not derived from any particular mainland population, but is sister to the widespread mainland clade (Fig. 9.8; clade 4).

Paradisaea rubra on islands off the Bird's Head Peninsula shows obvious morphological differences from the only other nearby congener, *P. minor* (part of the *P. apoda* group) on the adjacent mainland. Differences of this kind, between birds on offshore islands and those on the New Guinea mainland, led Mayr (1954) to combine founder dispersal and founder effect speciation, and establish the basis for much modern biogeography. Nevertheless, although the West Papuan Islands provide the classic examples of dispersal, molecular analysis of *Paradisaea* does not show direct phylogenetic connections between the island taxon and mainland forms, as predicted in dispersal theory. Instead, the island taxon *P. rubra* is sister to another island taxon 2000 km away.

Similar 'horstian' connections to those of *Paradisaea rubra* and *P. decora* occur in many New Guinea groups, with disjunctions among the northern peninsulas and offshore islands (Heads, 2002a, 2003). As in New Zealand, the disjunctions could be caused by horizontal, strike-slip movement along the Sorong fault and others, rather than vertical, horst-graben tectonics, and this could have taken place either during or after terrane accretion. The sister of the *P. rubra–P. decora* pair is the *P. apoda* complex, and this is widespread and diverse on the New Guinea mainland. It is often assumed that a local endemic can be due to vicariance, while a widespread distribution must be the result of dispersal (Mayr, 1965), and this has been suggested for New Guinea groups (Craft *et al.*, 2010). Nevertheless, the distributions in the birds of paradise are consistent with an *in-situ* origin of the widespread *P. apoda* complex by differentiation from the local *P. rubra–P. decora* clade.

The Bird's Head Peninsula islands of Waigeo, Batanta and nearby Kofiau (but not Salawati) comprise an obducted ophiolite, the Waigeo terrane. Mayr (1940) suggested that the straits between Batanta and Salawati Islands (Sorong Fault), although less than 5 km wide, 'have prevented the crossing' of 17 species of Salawati birds to Batanta and five (including *Paradisaea rubra*) from Batanta to Salawati. Rather than this major break reflecting the current geography, the distributions indicate that the break has resulted from tectonic change. This could have involved extension between the mainland and the offshore Waigeo–D'Entrecasteaux sector (as in a traditional, horstian model), or convergence, transpression and long-distance translation of terranes along strike-slip faults.

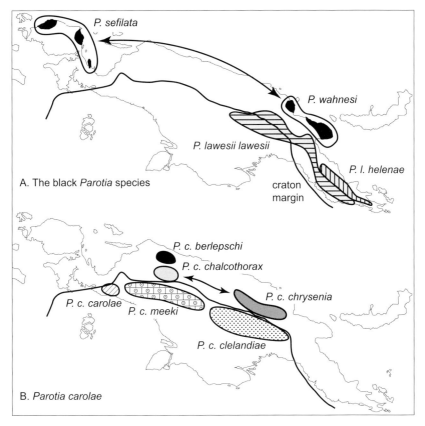

Fig. 9.9 Distribution of *Parotia* (Paradisaeidae). (A) The black *Parotia* species. (B) *P. carolae*. (From Heads, 2002a. Reprinted by permission of John Wiley & Sons.)

The *Paradisaea* species show affinities among the island terranes at each end of New Guinea. In a similar way, groups on the peninsulas are often related to each other or to clades on the islands, rather than to clades of the central range. For example, the genus *Parotia*, another bird of paradise, shows a wide disconnection among the accreted terranes. *Parotia sefilata* of the Bird's Head Peninsula is sister to *P. wahnesi*, endemic to the Huon Peninsula. The sister-group of this pair, *P. lawesi*, is restricted to eastern New Guinea (Fig. 9.9A; Irestedt *et al.*, 2009), recalling *Paradisaea rudolphi*. The remaining gap in *Parotia* in central New Guinea is filled by the last *Parotia* species, *P. carolae* (Fig. 9.9B). *P. carolae* as such is absent from the Bird's Head, Huon and Papuan Peninsulas, but it is represented there by the first three *Parotia* species. Other mainland New Guinea birds absent from all three peninsulas, despite the abundance of suitable habitat there, include *Pteridophora* (Paradisaeidae) and *Loboparadisea* (Cnemophilidae). Within *Parotia carolae*, morphological studies suggest possible connections between the accreted terrane subspecies *P. c. chalcothorax* and *P. c. chrysenia* (indicated on Fig. 9.9B).

The phylogenies and distributions of *Paradisaea*, *Parotia* and others indicate that dispersal from the central mainland to the outlying peninsulas and offshore islands is not a viable explanation for the distribution. Ecological speciation is also unlikely, as Mayr concluded after carrying out transects across parts of New Guinea (Mayr, 1954). Ecological differences between the mainland and the offshore islands are likely to be less than the great differences within the mainland.

One of the main objections raised against tectonic vicariance models is that the groups are not old enough. Birds of paradise, and passerines in general, are often thought to have speciated only in the Pleistocene. In contrast, biogeographic analysis of birds of paradise concluded that the species and the subspecies had originated earlier in the Cenozoic (Heads, 2001a, 2002a). This idea of pre-Pleistocene evolution was based on the geographic correlation of bird distributions in New Guinea with mid-Cenozoic tectonics. In their molecular analysis of birds of paradise, Irestedt *et al.* (2009) also concluded that the family has an 'unexpectedly long history'. They wrote that the family is 'an older clade than previously suggested' (p. 10), with most speciation taking place not in the Pleistocene, but in the Miocene and Pliocene. Irestedt *et al.* (2009) did not calibrate the phylogeny with a fossil, but with a tectonic event, the split of New Zealand and Antarctica (at 85 to perhaps 65 Ma). They took this date as indicating the time of separation between New Zealand wrens (Acanthisittidae) and all other passerines.

Whether or not the split of Acanthisittidae from other passerines occurred together with the New Zealand/Gondwana split, there is no need to rely on this one particular node – far away in space and phylogeny – when studying the birds of paradise. There are many spatial parallels between the distributions of the birds in New Guinea and the terranes, and these parallels mean that estimates of clade age can be made (Heads, 2001a, 2002a).

The breaks in the bird of paradise clades are attributed here to activity on and around the craton margin and the many strike-slip faults that delimit the accreted terranes. These were active through the Cenozoic. An earlier phase of paradisaeid history involved the extensive overlap of many clades – 10 of the 15 genera range from one end of New Guinea to the other – and one possible mechanism for this is provided by the high sea levels of the Cretaceous. Range expansion along tropical coasts of low-lying land areas during extensive flooding suggests a tolerance of freshwater swamp or mangrove forest and implies that the five or six genera of birds of paradise recorded in mangrove, including the widespread mainland clade of *Paradisaea* (Heads, 2002a), retain aspects of an earlier ecology. *Seleucidis*, for example, inhabits a range of lowland forest types, but has a particular affinity with permanently flooded coastal swamp forest dominated by *Pandanus* and sago palm (*Nypa*). The high elevation of the blue bird of paradise, *Paradisaea rudolphi*, in the central mountains probably developed after the clade had already evolved. It can be attributed to Neogene uplift of populations that were located in what became the orogen.

The disjunctions in *Paradisaea* and *Parotia* that skirt the north coast of New Guinea belong to a series of similar disjunctions seen in many groups, including the crow family Corvidae, one of the closest relatives of the Paradisaeidae (Jønsson *et al.*, 2011a). A study of all 40 species in *Corvus* (Jønsson *et al.*, 2012) showed a phylogeny: Eurasia

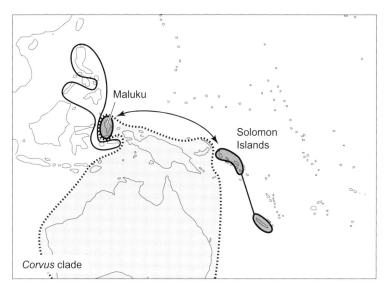

Fig. 9.10 Distribution of a clade in *Corvus* (Corvidae) (Jønsson *et al.*, 2012).

(all Greater Antilles (world)). One clade in the world group comprises the following branches (Fig. 9.10):

> New Caledonia, Solomon Islands, northern Maluku Islands (*Corvus moneduloides*, *C. woodfordi*, *C. meeki*, *C. validus*).
>
> > Philippines, southern Maluku Islands (*C. violaceus*).
> >
> > Australia (widespread), New Guinea and the Bismarck Archipelago, northern Maluku Islands (*C. coronoides* etc.).

The precise phylogeny among the groups (also *C. enca* of Sumatra to Sulawesi, and *C. unicolor* and *C. typicus* of Sulawesi) is still not resolved, but the individual clades are well supported. The first group includes a great disjunction between the Solomon Islands and the Maluku Islands. The gap extends all through New Guinea and the Bismarck Archipelago, and is filled by the sister-group. The disjunction is a common pattern in traditional morphological studies (Heads, 2003, Figs. 49, 81, 87); similar Maluku Islands–Bismarck Archipelago patterns are discussed in Heads (2001b). The disjunctions are compatible with sinistral strike-slip among the terranes of northern New Guinea.

A similar disjunction across northern New Guinea is documented in palms; the *Veitchia* clade + *Balaka* clade, and the *Adonidia* clade form a monophyletic group disjunct between the Solomon Islands and Biak Island (by the Bird's Head Peninsula) (Zona *et al.*, 2011). The gap is filled by the sister-group, the *Drymophloeus* clade (Fig. 9.11), indicating an origin of the pattern by simple vicariance, with subsequent overlap only on Biak.

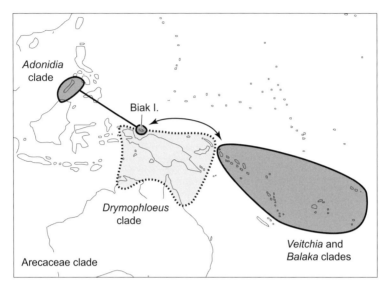

Fig. 9.11 Distribution in a clade of palms (Zona *et al.*, 2011).

The Milne Bay islands: new rifts and old snakes

Paradisaea decora illustrates a centre of endemism in the small islands off the eastern Papua New Guinea mainland, in Milne Bay province (Fig. 9.8). There is a close biogeographic relationship between the Milne Bay islands and the accreted terranes of the mainland. For example, the colubrid snake *Tropidonophis aenigmaticus* has a distribution spanning both regions. The main phylogenetic boundary within this species is not between the islands and the mainland, but at the craton margin, where *T. aenigmaticus* meets craton endemics such as *T. parkeri* (Heads, 2001b, Fig. 38).

Nevertheless, the D'Entrecasteaux Islands also have independent biogeographic connections with the craton, that do not involve other accreted terrane groups. The polymorphic python *Morelia viridis* has three main clades, with the following distributions (Rawlings and Donnellan, 2003):

Northern New Guinea (accreted terranes).

Milne Bay: D'Entrecasteaux Is. (accreted terrane).

Australia and southern New Guinea, to Port Moresby (mainly on the craton).

Dispersal theory would suggest that the D'Entrecasteaux Islands form is nested in a Papuan Peninsula clade, but in fact it is sister to a group that is widespread through southern New Guinea and north-eastern Queensland, suggesting an early break rather than a recent dispersal.

In Milne Bay, the D'Entrecasteaux and Louisiade Archipelagos form a southern chain, while the Trobriand Islands and Woodlark Island lie along a northern ridge (Fig. 9.12). The D'Entrecasteaux Islands are larger and higher, but even Rossel Island in the Louisiade group has an endemic genus of trees (*Rosselia* – Burseraceae;

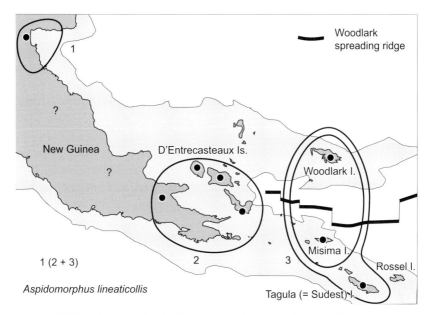

Fig. 9.12 Differentiation in the elapid snake *Aspidomorphus lineaticollis* (Metzger *et al.*, 2010).

Forman *et al.*, 1994). In the microhylid frog *Mantophryne*, a clade endemic to Rossel, Tagula and Misima is sister to a diverse, widespread, mainland clade (Oliver *et al.*, 2013).

   The crustal plate that includes the New Guinea orogen is being subducted beneath the Solomon Islands, and this process is leading to the opening of the Woodlark basin. The Woodlark rift between Woodlark Island and the Louisiade group develops eastward into a divergent spreading centre, a mid-ocean ridge that is generating the Woodlark basin (Ferris *et al.*, 2006). The Woodlark rift is one of the most rapidly extending rift systems known ($\sim$20–40 mm/year; Baldwin *et al.*, 2012). (The other, current seafloor spreading around New Guinea is between Manus and New Britain.) Metamorphic core complexes have been exhumed in the rift, on the D'Entrecasteaux Islands and in the Suckling–Dayman massif on the Papuan Peninsula, during rapid uplift in the Pliocene–Pleistocene (Taylor, 1999). Rift propagation in mainland Papua New Guinea is presently focused on the Gwoira fault zone in Goodenough Bay (Mann and Taylor, 2002; Daczko *et al.*, 2011).

   The snake *Aspidomorphus lineaticollis* (Elapidae) of New Guinea has six lineages (Fig. 9.12) that show deep divergences and have the following phylogeny (Metzger *et al.*, 2010):

   Morobe (mainland).

      Milne Bay mainland + D'Entrecasteaux Is.

         Woodlark I. + Misima I.

         Tagula (= Sudest) I.

The group is restricted to the accreted part of the orogen and does not occur on the craton. Metzger *et al.* (2010) noted that Misima and Woodlark Islands were adjacent until rifting in the Woodlark Basin began at ~8 Ma (Little *et al.*, 2007), and so sister-group relationships between their biotic elements are to be expected. Misima and Sudest Islands lie in an eastward extension of the orogen (possibly related to the Owen Stanley terrane of the Papuan Peninsula) that was once continuous with the mainland. There are obvious signs of submergence around Sudest Island, whereas Woodlark Island has been uplifted (Heads, 2001a).

The opening of the Woodlark basin is a dramatic process. Westward propagation of the rift has led to rapid exhumation of metamorphic core complexes in the D'Entrecasteaux Archipelago and material from at least 90 km down, in the mantle, has been brought to the surface at plate tectonic rates (centimetres per year) (Baldwin *et al.*, 2008).

In another species of *Aspidomorphus*, *A. muelleri*, a sample from an accreted terrane of the north coast of New Guinea (Torricelli Mountains) was closest to one from New Ireland. Metzger *et al.* (2010) attributed the affinities to the history of terrane accretion along northern New Guinea. Thus the species in *Aspidomorphus* have evolved in response to both accretion and extension in the New Guinea orogen.

The passerine *Colluricincla megarhyncha* (Colluricinclidae) is a variable bird in New Guinea, and the eight lineages there show high levels of divergence (Deiner *et al.*, 2011). Five of the lineages form the main clade and have the phylogeny:

> NW New Guinea: Waigeo I.
> > Western New Guinea.
> > > Milne Bay: Sudest I.
> > > > Milne Bay: D'Entrecasteaux and Trobriand Is.
> > > > Papuan Peninsula.

Deiner *et al.* (2011) concluded that the bird's evolution corroborates the geological history of the New Guinea region. Waigeo Island (an obducted ophiolite), the D'Entrecasteaux Islands (a metamorphic core complex) and Sudest Island (an accreted metamorphic terrane) reflect the processes of convergence, extension and uplift that have shaped the orogen and produced the phylogenetic breaks. The phylogeny shows that the birds on islands off the Bird's Head Peninsula and in Milne Bay province off eastern New Guinea are not secondary derivatives of mainland groups, but, as with certain birds of paradise, indicate early breaks in a widespread ancestor.

The Bismarck Archipelago and the Solomon Islands

The Bismarck Archipelago and the Solomon Islands include high islands that are more or less covered in rainforest. The highest mountain is Mount Balbi on Bougainville (2715 m/8907 ft). The region is renowned for its biodiversity; for example, the Solomon Islands have more endemic bird species than any other area of similar size in the world (Filardi and Smith, 2005). In addition, the region is characterized by major tectonic features, including several plate boundaries and a large igneous plateau. (For notes on distribution in the region, see Heads, 2012a, Chapter 6.) The islands are seldom visited by biologists, and their complicated biogeography is not well understood. The taxa have often been interpreted as minor variants, derived by recent, chance dispersal from Australia and New Guinea, and the region is not seen as having any real significance for global evolution.

Mayr and Diamond (2001) interpreted the birds of the Bismarck–Solomon region as the result of recent dispersal from mainland Australia and New Guinea, with the speciation caused by Pleistocene sea level changes. This model leaves many questions unanswered. For example: Why are there so many disjunctions between the Bismarck Archipelago and the Bird's Head Peninsula region (e.g. the pigeon *Ptilinopus solomonensis*; IUCN, 2013)? Why are there no birds of paradise on New Britain, although it is so close to the mainland and there is abundant suitable habitat? There are many other mainland groups not on the islands, but the island biota is not depauperate; the mainland endemics are complemented by large numbers of island endemics. The islands are also the main centres of diversity for many groups; for example, the frog genus *Platymantis* (Ranidae) (including *Batrachylodes*, *Ceratobatrachus* and *Discodeles*; Pyron and Wiens, 2011) includes about 40 species in the Bismarck Archipelago and Solomon Islands, and only seven or eight on mainland New Guinea.

In fruit bats (Pteropodidae), *Melonycteris* and *Pteralopex* of the Bismarck Archipelago and the Solomon Islands form a disjunct clade with *Desmalopex* of the Philippines. This clade's sister-group is widespread from the East African coast to the Pacific (*Pteropus* etc.) (see p. 79). The restriction of *Melonycteris* to the Solomon Islands and Bismarck Archipelago is 'puzzling' (Pulvers and Colgan, 2007). The authors reasoned that if the bat has occupied the region during its evolution, suitable habitat 'must have been permanently available in the Melanesian Arc for many millions of years, either as continuously emergent land or as successively emergent, geographically proximal islands' (p. 721). The latter is observed around subduction zones and this would help explain the survival, if not the origin, of the biological diversity in the region.

Other groups in the Bismarck Archipelago–Solomon Islands region are basal in widespread, diverse Australasian groups. Examples include a *Citrus* species on New Ireland with a widespread sister-group of three species in New Guinea (not New Britain), New Caledonia and Australia (Heads, 2012a, Fig. 9.4). An explanation for the pattern based on dispersal to New Ireland is not straightforward – in a simple dispersal model the species would have a centre of origin on mainland New Guinea, but that is not indicated. In contrast, a simple vicariance model is possible as New Ireland belongs to an accreted plate margin system, the Vitiaz arc, with a history quite distinct from that of New Guinea, New Caledonia and Australia. Interesting clades from the Bismarck–Solomon group similar to the *Citrus* example include the following.

Egernia group (Scincidae)

The world's largest extant skink, *Corucia* of the Solomon Islands, is sister to the rest of the *Egernia* group (*Egernia*, *Tiliqua*, *Cyclodomorphus*), widespread throughout New Guinea, Australia and Tasmania (Gardner *et al.*, 2008).

Zoothera (Turdidae)

In this thrush genus, Voelker and Klicka (2008) proposed a Himalayan origin, as the four basal-most species in the genus all have ranges centred there. 'However,' they noted, 'colonization does not proceed in a simple downstream pattern from continent to islands.' One clade is structured as follows:

> Asia and Southeast Asia, to Java and Lombok (*Zoothera dauma*).
>> Solomon Islands (*Z. margaretae*).
>>> Eastern Australia (*Z. lunulata*).
>>>> Tasmania, eastern Australia and New Guinea (*Z. heinei*).
>>>> New Britain and Solomon Islands (*Z. talaseae*).

As in the *Egernia* group, the Solomon Islands clade is basal in a widespread Australasian complex. A dispersal model of the phylogeny would require an initial leap from Java or Lombok to the Solomon Islands across New Guinea, but this is not required if the ancestor was widespread to begin with. The sequence of differentiation jumped from Java to the Solomon Islands – not the birds themselves.

Rhipidura (Rhipiduridae)

In the passerine *Rhipidura* subgen. *Rhipidura*, Nyári *et al*. (2009a) found a phylogeny:

> New Guinea (*R. albolimbata, R. hyperythra*).
>> New Caledonia, Vanuatu, Fiji (*R. verreauxi*).
>>> SE Solomon Islands (Makira I.) (*R. tenebrosa*).
>>>> SW Solomon Islands (Rennell I.) (*R. rennelliana*).
>>>>> New Caledonia, Vanuatu, Solomon Islands (*R. albiscapa* p.p.).
>>>>> Australia, southern New Guinea (*R. phasiana*).
>>>>> Australia, New Zealand (*R. albiscapa* s.str., *R. fuliginosa*).

Here there is a basal grade in the islands, while the Australian members are distal. Again, this is incompatible with the traditional model of dispersal into the Pacific from the mainland. A similar pattern occurs in other passerines, such as Monarchidae (Filardi and Moyle, 2005; Heads, 2012a, Fig. 6.3) and Pachycephalidae, with *Pachycephala ornata* of Vanuatu basal to a group of Australia and Melanesia (Jønsson *et al*., 2010a). The pattern is consistent with early differentiation around the islands, in an ancestor that was already widespread both there and on the mainland. In the model discussed above (Schellart *et al*., 2006), the Vitiaz arc (zone 3 in Fig. 7.1) developed near the arc (zone 1 in Fig. 7.1) that first developed at the Gondwana margin and migrated eastward. The trans-Tasman disjunctions (Australia/New Zealand) in *Rhipidura* suggest that the early breaks in the genus around the Bismarck and Solomon Islands were Late

Cretaceous. This time coincides with the initial rifting of the Pacific arc – the precursor of the Bismarck–Solomon Islands arc – from Gondwana.

In the southern Solomon Islands, Makira (= San Cristobal) has 13 endemic bird species, the highest number of any island in the Solomon archipelago. A dispersal analysis concluded that bird distribution around Makira and Rennell Islands is a 'mystery' and a 'paradox' (Mayr and Diamond, 2001: 249, 254). The distinctive biogeographic patterns may be problematic because they are much older than has been thought. Instead of being related to Pleistocene sea level change events, they probably reflect Cretaceous–Cenozoic tectonics involving the Pacific arc, its derivatives and the Ontong Java Plateau (cf. Fig. 6.2). The high endemism in many groups, together with the endemic clades that are basal in Old World or Australasian groups, suggest that the Bismarck–Solomon region is important as a primary evolutionary centre in its own right and not just as a sink for dispersal from other regions. This would explain many anomalous aspects of biodiversity there. For example, Corner (1967) described the exceptional diversity of the pantropical tree *Ficus* (Moraceae) in the Solomon Islands rainforest. He wrote that the differentiation of the group in the area 'is too closely-knit with geography to admit the prevalence of random dispersal'. He also commented that 'as the islands are very far away, we are led to think of their problems as marginal and irrelevant to the mainspring of life'. He argued instead that they hold 'important keys to the evolution of flowering forest' (Corner, 1969) and he related the problem of the Solomon Islands flora to the trans-tropical Pacific connection.

Geologists do not agree about many aspects of the Cretaceous–Cenozoic history of the south-west Pacific, let alone earlier periods when the basal groups originated. Many tectonic models for the Cretaceous–Cenozoic have been proposed (see references in Schellart *et al.*, 2006); these stress different sets of data and reach different conclusions. As discussed in Chapter 8, some models have the Australia/Pacific plate boundary – a subduction zone with its attendant island arcs – developing at the coast of Australia and migrating into the Pacific (Schellart *et al.*, 2006), and this would have provided an initial component of the Vitiaz arc biota. Another component would have been supplied by the Ontong Java Plateau.

Ongoing collision of the Ontong Java plateau with the Solomon arc has resulted in uplift of the plateau's southern margin and created on-land exposures of the plateau on the islands of Santa Isabel, Malaita and Makira (San Cristobal). In dispersal theory, the Makira bird fauna is a biogeographic mystery (Mayr and Diamond, 2001) and Pulvers and Colgan (2007) cited the 'unexpected' basal position of the Makira species in the Solomons clade of *Melonycteris* fruit bats. In its geology, Makira is also a special case within the Solomon Islands. The Cretaceous basement of the Solomon Islands is divisible into three distinct terrains: a northern 'Ontong Java Plateau Terrain', a southern 'South Solomon mid ocean ridge basalt terrain', and a 'hybrid' eastern 'Makira Terrain' for Makira Island (Petterson *et al.*, 1999). Makira is also unusual because, while Ontong Java Plateau material dating back to 90 Ma is present, there is little or no pelagic sedimentation, as would be expected. This suggests deep, old erosion of a subaerial edifice (Craig *et al.*, 2006). The geological evidence indicates that the Ontong Java Plateau (Solomon Islands), the Melanesian Border Plateau (Samoa), the Manihiki

Plateau (Cook Islands) and the Hikurangi Plateau (north-eastern New Zealand) formed as a single, largely subaerial plateau that later broke up. This history would explain distinctive affinities among the biotas of these widespread areas.

Vanuatu and Fiji

Biology and geology can be integrated, not at the level of an individual island or stratum, but at the more general level of underlying, tectonic structures, such as plate margins, orogens and other belts of metamorphism and magmatism. The islands in the Vanuatu and Fiji archipelagos are formed from young strata (Cenozoic volcanics and limestone) and so the biota is often assumed to have immigrated from some other region. Nevertheless, tectonic models (Schellart *et al.*, 2006) suggest that the island arc clades would have existed earlier on prior arcs in the region. The Solomon Islands–Vanuatu–Fiji arc developed near the original Pacific arc (zone 1 in Fig. 7.1), at the Vitiaz trench, and would have inherited its biota. In the model of Schellart *et al.* (2006) this original arc developed at the coast of Gondwana and would have been colonized, at least in part, from there. Most islands in this complex history have subsided, subducted or been buried under subsequent volcanism. This tectonic reconstruction suggests that phylogenies for the Vanuatu–Fiji region calibrated using the emergence date of the current islands (Murienne *et al.*, 2011) will give unreliable clade ages. In Vanuatu, for example, there is spectacular, active volcanism on Tanna, while further north in the archipelago a large island has disappeared in historical times (Chapter 2). As the system is so dynamic, with islands appearing and disappearing even over ecological time scales, it is expected that the ages of the individual islands will not relate to the phylogenetic age of the metapopulations. Taxa that can colonize new strata and islands in their vicinity by normal, local, dispersal will persist in the region as long as volcanism continues. The process will entail the extinction of many lineages with unsuitable ecology, but there is no need to invoke long-distance dispersal to explain the biota that does persist.

Johanson and Oláh (2012) discussed the biogeography of freshwater Trichoptera in Fiji and noted that the earliest land there has been dated to Middle Miocene. This is only the exposed rock, though. The biota of the present, young arc has descended, *in situ*, from metapopulations on the Cretaceous arc, on its seaward side, and this arc was itself formed on or near Gondwana. Johanson and Oláh (2012) also noted contradictory dates from molecular studies of Trichoptera. While a Fijian group of *Apsilochorema* (Hydrobiosidae) diverged from its Vanuatu sister species at about 16 Ma (a mean age) (Strandberg and Johanson, 2011), a New Caledonian clade of Hydropsychidae diverged from its Fijian sister in the Middle Oligocene (at 29.5 Ma, a mean age), well before the supposed first existence of Fijian land (Espeland and Johanson, 2010).

Groups that probably occurred around the Vanuatu–Fiji region before the current islands existed include many endemics. Fiji's many endemics include a plant family (Degeneriaceae: Magnoliales; Heads, 2006a). This is usually treated as a sister to Himantandraceae (in Sulawesi, New Guinea and Queensland), although variation in the nuclear gene *Xdh* indicated Degeneriaceae was sister to a worldwide clade comprising

Magnoliaceae, Annonaceae and Eupomatiaceae (Morton, 2011). Vanuatu has endemic genera such as the palm *Carpoxylon* (Fig. 8.13) and the passerine *Neolalage* (Monarchidae; Heads, 2012a, Fig. 6.4). In addition, Fiji and Vanuatu include several Tasman region endemics that have global sisters; for example, the tree *Balanops*, the moth *Agathiphaga* and the athoracophorid slugs (Chapter 5). Fiji or Vanuatu endemics that have widespread sisters are seen in groups such as water beetles (Dytiscidae; Balke *et al.*, 2009). The *Rhantus suturalis* group is widespread in the Old World and has a phylogeny:

> Europe.
>> Europe, via central and southern Asia (Iran, Nepal, etc.) to Polynesia.
>> Fiji.

The fact that the Fijian clade is sister to such a widespread Old World group, not nested in it, indicates that it did not originate by dispersal from New Guinea, Australia or any other restricted centre. A literal reading of the phylogeny might infer dispersal from Europe to Fiji and then back again. Balke *et al.* (2009) proposed an even more complex colonization pattern (from a centre of origin in New Guinea) and suggested that one widespread member of the widespread clade (*R. suturalis*: New Zealand to Portugal) is a 'supertramp'. Vicariance of a widespread Tethyan ancestor provides an alternative, simpler explanation for the pattern. At the first node, European forms split off from the rest (with subsequent local overlap) and at the second node the Fijian forms differentiated from the remainder. In this model there is no need for complicated dispersal and back-dispersal in the modern groups half-way around the globe. The question is instead: What caused the breaks around Europe and Fiji? The latter probably involved the early history of the Pacific arc and its migration into the Pacific from the Gondwana margin.

The fig wasp genus *Dolichoris* (Agaonidae) has a similar pattern to that of the *Rhantus suturalis* group, with a local Tasman region group having a widespread Tethyan sister-group (Cruaud *et al.*, 2012). The phylogeny is:

> Africa.
>> Mediterranean, Arabo-Sindic region, India to New Guinea.
>> New Caledonia and Vanuatu (diverse clade).

The authors ruled out vicariance between the last two clades owing to Bayesian clock dates. But the 46-Ma date for invasion of New Caledonia is only a mean; the 95% credibility interval (HPD) range for the date extends from Miocene to Cretaceous, so even with the very narrow Bayesian priors that were used, Cretaceous vicariance is possible.

Similar patterns in other groups suggest a deeper significance for Vanuatu and Fiji than the current islands' size and age might suggest. For example, the geometrid moth *Scotocyma* is widespread from northern Sumatra via eastern Australia to Samoa. In a dispersal model the Fijian species would be nested among Australian or New Guinea clades, but in a morphological phylogeny the Fijian species was sister to the rest of the genus (Schmidt, 2005).

In marine groups, *Paramunida* (Munididae) are deep sea lobsters found from Madagascar to French Polynesia. Most species are restricted to single islands or archipelagos and the few widespread species are likely to be complexes of species with more restricted distributions. A DEC analysis of the Pacific species deduced a centre of origin in the Vanuatu–Fiji–Tonga region, with chance dispersal west and east (Cabezas *et al.*, 2012). Instead, the Vanuatu–Fiji–Tonga region could represent a centre of differentiation in a widespread ancestor.

Analyses of endemism in the Pacific islands region have often been based on the numbers of endemics per individual archipelago, but this overlooks the importance of interarchipelago areas of endemism. Compared with New Caledonia, for example, Vanuatu has few endemics, but there are many restricted to Vanuatu and Fiji, such as the tree *Balanops pedicellata* (Balanopaceae). Dozens of other plants with this distribution are documented (Smith, 1979–1996). The damselfly genera endemic in the region include *Nesobasis* of Fiji (21 species), *Vanuatubasis* of Vanuatu (three species) and *Melanesobasis* of Vanuatu and Fiji (three species). A related area of endemism, Vanuatu–Fiji–Tonga, is defined by groups such as the common rainforest tree *Garcinia pseudoguttifera* (Clusiaceae).

The Vanuatu–Fiji disjunction is often explained by dispersal. Yet the two archipelagos formed a single archipelago before this was rifted apart and in many cases no dispersal is necessary (Heads, 2006a). In the centipede *Cryptops niuensis*, Murienne *et al.* (2011) found that the Vanuatu population was nested among the Fijian populations and so they inferred colonization of Vanuatu from Fiji. They suggested that this is concordant with the younger age of Efate (Vanuatu) compared to Fiji. In fact, the phylogeny is also compatible with vicariance of a widespread Vanuatu–Fiji ancestor following differentiation in and around the complex Fijian archipelago (cf. Fig. 1.6). As indicated above, it is likely that the age of the current islands is not related to the age of their clades. A similar pattern occurs in Trichoptera (Polycentropodidae), with the single Vanuatu species nested in a diverse, otherwise Fijian clade (Johanson *et al.*, 2012).

Disjunctions in Vanuatu

A strong biogeographic boundary exists between the islands of southern Vanuatu (Erromango, Tanna and Aneityum) and the northern islands (Hamilton *et al.*, 2010). In many groups, clades on the southern islands are closer to groups in New Caledonia than to others in northern Vanuatu (Heads, 2008b). Hamilton *et al.* (2010) suggested that the break between north and south must have been the result of dispersal rather than vicariance, as there is no evidence for a continuous landbridge connecting the southern islands with each other or with New Caledonia. Nevertheless, a continuous landbridge is not necessary for survival. Regional biotas in southern and northern Vanuatu have survived as dynamic metapopulations on ephemeral islands. In addition, there is evidence for submergence of entire islands in the archipelago in geological and even historical time (Nunn *et al.*, 2006).

Other patterns in Vanuatu remain almost unexplored. They include interesting disjunct affinities between Efate (central Vanuatu) and Banks Islands (northern Vanuatu) that miss intermediate islands. The pattern has been described in the passerine genera *Pachycephala* (Mayr, 1932b) and *Petroica* (Mayr, 1934). These particular cases have not yet been tested, but the connection itself was confirmed in sequencing studies of another passerine, *Zosterops flavifrons* (Phillimore *et al.*, 2008). (*Zosterops* is supposed to be a great colonizer of islands, but does not occur on any of the islands east of Samoa, including the Cook, Society, Marquesas and Hawaiian Islands. Small passerines that do occur in this region include a widespread central Pacific clade of Monarchidae.)

Fiji–New Guinea connections

The affinities of the Fijian biota include a direct link with New Guinea. This recalls the Fiji–Solomon Islands connection seen in *Cettia*, missing Vanuatu (Fig. 8.4). These connections follow the old plate boundary marked by the Vitiaz trench (subduction zone 1 in Fig. 7.1). The spreading centres (mid-ocean ridges) that have opened the North Fiji Basin are shown in Figure 7.2. The Fiji–New Guinea disjunction (~3000 km) does not follow the island chain distribution predicted in 'stepping-stone' dispersal theory. Nevertheless, it is documented in morphological studies; for example, an undescribed mayfly genus in Fiji (Baetidae, new genus 'A') is related to forms of New Britain and eastern New Guinea (Flowers, 1990). Flowers dismissed chance dispersal as an explanation and instead suggested a tectonic cause for the pattern.

Molecular studies also show the Fiji–New Guinea connection. In passerines (Corvida), *Lamprolia* of Fiji is sister to *Chaetorhynchus* of New Guinea (Irestedt *et al.*, 2008; Jönsson *et al.*, 2011a). The pair is sister to *Rhipidura* of Pakistan to Australasia and Samoa. Irestedt *et al.* (2008) cited three possibilities for the New Guinea–Fiji distribution: long-distance dispersal from New Guinea to Fiji, long-distance dispersal in the opposite direction or long-term survival on the ephemeral islands around the subduction zone. They accepted the third process, and this is supported here. (They also suggested a centre of origin and dispersal history, although this is not necessary if the group has survived in the long term in the region as a metapopulation.) Irestedt *et al.* (2008: 1222) concluded: 'We are unable to determine the area of origin of the *Lamprolia/Chaetorhynchus* lineage, whether it is Australo-Papuan or Pacific, although we consider an Australo-Papuan origin more likely'. A centre of origin for the lineage in either region is not needed; all that is required is a split of *Lamprolia* + *Chaetorhynchus* from the sister-group, the widespread *Rhipidura*, along an old plate margin (when this was closer to Gondwana), followed by local overlap of the two groups in the precursor terranes of New Guinea and Fiji.

The Fiji–New Guinea connection is also documented in elapid snakes, with the underground *Ogmodon* of Fiji sister to *Toxicocalamus* of New Guinea (Keogh *et al.*, 1998). The connection is well known in traditional taxonomy. For example, *Dacrydium nidulum* (Podocarpaceae) is in Fiji, New Guinea, Sulawesi, the Maluku Islands and Sumba Island (de Laubenfels, 1988).

In the invertebrate fauna, molecular studies have indicated Fiji–New Guinea affinities in ants. The genus *Lordomyrma* s.str. forms a great arc around the margin of the western Pacific: one clade is in Japan, the Philippines and Borneo, while the other is in New

Guinea, east coast Australia, New Caledonia and Fiji (Lucky and Sarnat, 2010). In the second clade, dispersal theory predicts migration from New Guinea to Australia, New Caledonia and then Fiji. Instead, there are two allopatric subclades, with one in eastern Australia and New Caledonia, and the other in New Guinea and Fiji. Again, the distribution of the last clade is unexpected in a dispersal model, but is a standard pattern shared with insects, birds, underground snakes and others, and coincides with the old plate margin.

The 11 species of *Lordomyrma* in Fiji are all endemic there and, despite the fact that all members of the genus possess winged queens and males, the restricted distributions of most species suggest limited dispersal capacity (Lucky and Sarnat, 2010). This makes a dispersal account for the Fiji–New Guinea disjunction even less likely. As the authors proposed, the breakup of the Vitiaz arc has probably led to disjunction and speciation in the group. Nevertheless, the breakup does not explain the origin of the Fiji–New Guinea clade itself. Its sister-group is the Australia/New Caledonia clade and the Paleogene Coral Sea break between the two clades pre-dates the breakup of the Vitiaz arc.

West of New Guinea: the Maluku Islands (the Moluccas) and Sulawesi

Maluku Islands

The Maluku Islands represent an important boundary zone for many groups. For example, in marsupials, *Strigocuscus pelengensis* (Sulawesi and Maluku Islands) and *Phalanger* (Maluku Islands to Solomon Islands) meet at a junction in the Maluku Islands (see p. 314).

In butterflies, *Papilio* subgen. *Achillides* comprises two groups (Condamine *et al.*, 2013b):

1. India and China to Maluku Islands (north and south) and Bird's Head Peninsula.
2. Maluku Islands (north and south), New Guinea, Solomon Islands, NE Queensland and New Caledonia.

This indicates vicariance of an India–New Caledonia group at the Maluku Islands/Bird's Head Peninsula. (A DEC analysis instead found a centre of origin in Sundaland and/or Wallacea; Condamine *et al.*, 2013b.)

As many as four tectonic plates – the Eurasian, Indo–Australian, Philippine and Pacific – meet at the the Maluku Islands/Bird's Head Peninsula and it is understandable that the local tectonics and biogeographic patterns are complex (Heads, 2001a; 2003). In the Banda Sea area (the southern Maluku Islands), Villeneuve *et al.* (2010) distinguished seven crustal blocks. Six are thought to be derived from the eastern Gondwanan margin (northern Australia and southern New Guinea). In contrast, the Halmahera block (the northern Maluku Islands) is thought to have originated on the Pacific plate and moved westward along the northern New Guinea margin to its present position. These separate histories account for the great difference between the biotas of northern and the southern Maluku Islands. The translation of the northern Maluku Islands block would also explain disjunct biogeographic affinities between the Maluku Islands and the Solomon Islands, referred to already (Fig. 9.10). Similar disjunctions occur between the Maluku Islands and the Milne Bay area of south-eastern New Guinea, the Bismarck Archipelago and parts of northern New Guinea.

Maluku Islands–Bismarck Archipelago disjunctions have been recorded in morphological studies of butterflies (*Eurema*), hawks (two separate links in *Accipiter*), ferns (*Christella*) and flowering plants (Araceae) (Heads, 2001a). A well-sampled molecular study of the butterfly *Cethosia* (Nymphalidae) has indicated a disjunct clade present in the Maluku Islands (*C. cydippe obiana*) and the Bismarck Archipelago (*C. vasilia* and *C. obscura*), with the gap on the New Guinea mainland filled by the sister-group (the rest of *C. cydippe*) (Müller and Beheregaray, 2010). The authors stressed the importance of understanding the relationship between mainland New Guinea and the Bismarck Archipelago, as the latter has such high numbers of endemic Lepidoptera. The Maluku Islands also form an important centre of endemism and its relationship with the Bismarck Archipelago/Solomon Islands coincides with the plate margin.

Australasian connections with Sulawesi

An intercontinental node is recorded at Sulawesi in many groups. For example, *Ficus* sect. *Oreosycea* (Moraceae) includes a clade: New Caledonia–New Guinea–Sulawesi (*F. racimigera* etc.). The group is sister to *F. vasculosa* of India, China, Sundaland, Sulawesi and the Philippines, with simple allopatry between the two at a break in Sulawesi (Cruaud *et al.*, 2012). In addition, the Sulawesi biota is well known for its many endemics and also its peculiar absences. It is often regarded as anomalous because it is not intermediate in composition between biotas to the west (in Borneo) and to the east (in New Guinea and Australia). The terranes of Sulawesi have diverse origins; some are derived from Sundaland, others from Australia and, possibly, the central Pacific (Heads, 2012a). The eastern parts of Sulawesi show immediate affinities with areas to the east, while Hall (2011) mapped south-western Sulawesi with East Java, and central-western Sulawesi with part of north-eastern Borneo. These western areas are underlain by blocks of continental crust that had rifted earlier from Australia and drifted north, colliding with Asia. The rifted blocks brought with them much of the deep geological structure now observed.

There is a standard connection between New Guinea and Sulawesi, and many groups of the Tasman–Coral Sea region include localities in Sulawesi. One example is a sister pair of monogeneric families in Magnoliales (Stevens, 2012):

Sulawesi, montane New Guinea and north-eastern Queensland (Himantandraceae: *Galbulimima*).

Fiji (Degeneriaceae: *Degeneria*).

The pair is sister to the Eupomatiaceae: eastern Australia and Papua New Guinea, plus the pantropical Annonaceae (Sauquet *et al.*, 2003). The phylogeny indicates an early phase of breaks in a pantropical complex around the Tasman–Coral Sea region.

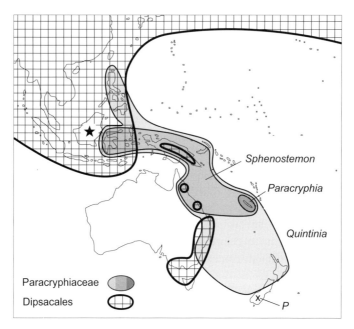

Fig. 9.13 Distribution of two sister-groups, Paracryphiales (grey) and Dipsacales (grid pattern). Star = the basal species in the widespread *Viburnum* (Dipsacales; northern hemisphere, South America, Asia). *P* = fossil record of *Paracryphia* in New Zealand. (Data from Winkworth and Donoghue, 2005; Pole, 2010; Stevens, 2012; GBIF, 2012). (From Heads, 2010b, reprinted by permission of John Wiley & Sons.)

Other Tasman groups with western extensions include the tree order Paracryphiales. This is made up of three genera, *Quintinia*, *Paracryphia* and *Sphenostemon*, distributed from New Zealand to Sulawesi and the Philippines (Fig. 9.13). The order is sister to the much larger Dipsacales (Dipsacaceae, Caprifoliaceae, Adoxaceae and Valerianaceae) (Tank and Donoghue, 2010). Dipsacales are more or less cosmopolitan but show an interesting absence from New Zealand, New Caledonia (where Paracryphiales are most diverse) and the Bismarck Archipelago. As Winkworth *et al.* (2008) suggested, a simple geographic split between the Paracryphiales and the Dipsacales seems likely. This break in a cosmopolitan ancestor has been followed by some local overlap of the two orders between the Philippines and New South Wales, but Paracryphiales occur on their own in New Zealand, New Caledonia and the Bismarck Archipelago. The star in Figure 9.13 indicates the basal species in the widespread, diverse *Viburnum* (Dipsacales; northern hemisphere, South America, Asia) in Borneo. This suggests further early breaks in Malesia.

Other Sulawesi endemics with links to Australasia include a freshwater gastropod, *Sulawesidrobia*, that proved to belong in the Australia–South Pacific family Tateidae (Zielske *et al.*, 2011). Members of Tateidae (formerly treated under Hydrobiidae) are now recorded from Sulawesi east into the central Pacific (New Guinea, Australia, Tasmania, Lord Howe and Norfolk Islands, New Zealand, New Caledonia, Vanuatu, Fiji and the Austral Islands). Similar distributions occur in groups such as the plant family Trimeniaceae: Sulawesi, Australasia and the central Pacific, east to the Marquesas. The sister-group of Tateidae is not known, but the Trimeniaceae, in the South Pacific, have an allopatric, North Pacific sister-group, Schisandraceae in eastern Asia and North America, suggesting vicariance of a pan-Pacific ancestor.

Anatomical studies suggested that the Tateidae of New Zealand, New Caledonia, Fiji and the Austral Islands form a monophyletic, central Pacific clade (Haase *et al.*, 2010). Haase *et al.* (2010) described the Vanuatu forms and wrote that 'In accordance with geography, [morphological] characters placed the species . . . between those from New Caledonia and Fiji, suggesting a stepping stone-like dispersal across the Pacific with an origin in New Zealand and the far end on the Austral Islands'. Instead, the standard distribution pattern suggests that the phylogeny has occurred by *in-situ* differentiation of a widespread Zealandia–Fiji–south-eastern Polynesia group. In the same way, the nesting of the Sulawesi clade among Australasian ones does not mean that it dispersed from Australasia. As mentioned already, at least part of modern Sulawesi is thought to have had an origin in the central Pacific.

Stelbrink *et al.* (2012) discussed Sulawesi biogeography and 'tectonic dispersal' of flora and fauna on the accreting terranes. The terranes have converged on the Sulawesi region from the west (Sundaland) and the east (New Guinea/Australia) and built up the modern island. Stelbrink *et al.* (2012) calculated fossil-calibrated clade ages for 20 clades found in Sulawesi and elsewhere. Even based on the fossil record, they found clade ages that are too old to refute vicariance ('tectonic dispersal') in four of the analysed taxa, one with affinities in the west, and three with affinities in the east.

In their discussion of tectonic dispersal in some Sulawesi clades, Stelbrink *et al.* (2012) considered how terrestrial groups survive in the long term on oceanic terranes. For Sulawesi, they suggested a convincing model of 'dynamic land connections'. In this model, at least some land was always present, but it was fragmented and had rapidly shifting coastlines. There may be no one spot that has stayed above sea level at all times; terrestrial and near-shore marine groups would have needed to colonize new ground, at least at a local scale, if they were to survive. As Stelbrink *et al.* (2012) stressed, estimating the location of smaller islands through geological time, using geological data, is 'notoriously difficult'. Characteristic signatures of larger land areas such as fluvial sediments will not be present and the existence of land is mainly inferred from negative evidence such as gaps in sediments of marine origin. Müller *et al.* (2010) concluded in a similar way: 'tectonic events in the Indo-Pacific during the Tertiary are still poorly known and conjectural, and inferred sea levels are highly speculative beyond approximately 100 000 years ago (R. Hall, pers. comm.)'. Given this context, it is possible that biogeographic patterns can shed some light on aspects of tectonics that are still not well understood.

Sulawesi–Maluku Islands–New Guinea as a centre of absence

Meiogyne (Annonaceae: Miliuseae) has a common distribution pattern: India to Fiji, and has two main clades (Fig. 9.14; Thomas *et al.*, 2012b):

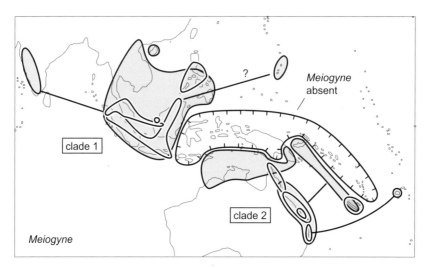

Fig. 9.14 Distribution of *Meiogyne* (Annonaceae) (Thomas *et al.*, 2012b). (In the Australian and New Caledonian centres of diversity not all species are mapped separately.)

Western: India to Philippines, Borneo, and Java. (The Micronesian form has not been sequenced; see query on Fig. 9.14.)

Eastern: northern Australia and eastern New Guinea to Fiji.

The two clades are separated by a belt where *Meiogyne* as such is absent, and this includes Sulawesi, the Maluku Islands, most of New Guinea and the Solomon Islands. The gap is not absolute, though, and is occupied by close relatives of *Meiogyne* in Miliuseae (*Miliusa, Phaeanthus, Mitrephora, Polyalthia* p.p. and *Popowia*). *Meiogyne* and its relatives form a trans-Indian Ocean group that is recorded in East Africa, Madagascar and from India to Fiji (Chatrou *et al.*, 2012). This trans-Indian group is sister to a trans-tropical Pacific group, present in Malesia and from central Mexico to western Colombia (Schatz and Maas, 2010).

In the eastern clade of *Meiogyne*, the single New Britain species (*M. glabra*) is well supported as sister to a mainland New Caledonian species (*M. lecardii*), not to the single species on mainland New Guinea (*M. cylindrocarpa*). This indicates a break between New Caledonia + New Britain, and mainland Gondwana, coinciding with the Coral Sea rift (Fig. 4.12) or a precursor. A similar rift is evident in *Meiogyne stenopetala*, which comprises one subspecies at the McPherson–Macleay Overlap and one in Fiji (only the former has been sequenced). This link, if confirmed, would be consistent with a Cretaceous break caused by the eastward retreat of the Pacific arc away from the Gondwana margin (Schellart *et al.*, 2006).

New ideas on Cenozoic deformation in eastern Indonesia (Sulawesi–Maluku Islands–Bird's Head Peninsula)

In eastern Indonesia, Hall (2011, 2012) concluded that young and rapid vertical movements of the crust have caused greater changes in paleogeography than have been recognized. GPS measurements show that the upper crust is deforming in a complex way that is largely independent of plate boundaries and plate movements. The region is responding to movements of the large plates, not by deforming as plates or even microplates, but by rapid, local uplift and subsidence.

Structures in many Southeast Asian mountain belts that were interpreted as thrusts are reinterpreted in a new model as major extensional detachments (Hall, 2011, 2012). Hall suggested that microcontinental fragments west of New Guinea were not sliced from New Guinea and translated during convergence and strike-slip, but formed during extension of a larger, coherent continental block, 'Sula Spur', which included the Bird's Head Peninsula, the southern Maluku Islands and south-eastern Sulawesi. It is possible that the biogeographic links among these do not reflect tectonic translation with strike-slip, as suggested above, but differentiation between the areas and others during extension.

As Hall (2012) suggested, the rapid vertical movements in eastern Indonesia through the Cenozoic imply that there was more land in the region than was thought. In dispersal theory, these former land areas and islands would be interpreted as short-lived stepping stones for 'dispersal' between Asia and Australasia. Instead, the dynamic vertical tectonics indicate that metapopulations persisted *in situ*, in the Sulawesi–Maluku Islands–Bird's Head area. This history requires local movement only, by normal means of dispersal and without range expansion. The dynamic landscape that is proposed will lead to biogeographic convergence and also divergence with vicariance. Hall (2012) concluded that evidence from the biota, particularly freshwater organisms, and from molecular techniques, 'could contribute in some areas to a better mapping of the past distribution of land where the geological record is silent'.

10   Biogeography of the Philippines

Wallace (1876) treated the Philippines as part of Asia, while Huxley (1868) instead saw the Philippines (except Palawan) as part of the Australia–New Guinea region. Molecular work confirms both connections, but also indicates that some Philippines clades do not have close ties with either Asia or Australia. Instead, these interesting clades are sister to widespread Old World groups, suggesting early breaks.

The Philippines archipelago includes more than 7000 islands, the main ones being Luzon in the north, Mindanao in the south, the Visayas between these two, and Palawan and Mindoro in the south-west (Fig. 10.1). The highest islands in the Philippines are Mindanao (2954 m), Luzon (2922 m), Mindoro (2582 m), Negros (2435 m), Panay (2117 m), Palawan (2085 m), and Sibuyan in the Romblon group (2057 m). Much of the sea in the central Philippines is shallow, as shown by the 120-m isobaths (Fig. 10.1).

The Philippines archipelago is a composite system of accreted island arcs, formed in much the same way as New Zealand, New Caledonia and New Guinea. All have developed by accretion at past or present plate boundaries. The Philippines occupy a narrow plate lying between the Eurasian (Sundaland) and Philippine Sea plates (Fig. 10.2). The last two plates both converge obliquely on the Philippines and subduct beneath the islands at trenches to the west and east. The area between the subduction zones includes most of the Philippines (except Palawan) and makes up the deformed 'Philippine mobile belt', well known for its seismicity and active volcanism. A major strike-slip feature, the Philippine fault zone, traverses the archipelago and takes up stress that is not accommodated at the surrounding trenches (Yumul *et al.*, 2008, 2009a).

The continental Palawan block (including Mindoro and the Sulu blocks) forms a stable, aseismic region in the south-west of the Philippines, while the other islands comprise the 'mobile belt' or Philippines orogen (Bird, 2003). In regional terms, the Philippines composite arc has become incorporated into a collision zone, an accretionary orogen, that surrounds the continental core of Southeast Asia (Sundaland). This collision zone stretches from Sumatra into eastern Indonesia and the Philippines (Hall *et al.*, 2009).

As with New Guinea, New Caledonia and New Zealand, the Philippines archipelago has been built up from older, continental crust in the south and south-west, and younger oceanic terranes that have accreted from the north or east. The various crustal blocks in the Philippines have continental, mid-oceanic, island arc or ophiolitic affinities. Igneous and sedimentary rocks are widespread thoughout, metamorphics occur in the south-western Palawan block, and ophiolite bodies, slices of ocean floor mainly formed in the Cretaceous, occur as dismembered massifs still arranged, more or less, in arcs. One

Fig. 10.1 The Philippines and the 120-m isobath.

possible configuration of the ophiolites is shown in Figure 10.3 (cf. Yumul *et al.*, 2008). The Sibuyan Sea forms a distinctive indentation in the geological structure and is also important in biogeography. The small islands in the sea (the Romblon group) have high levels of terrestrial endemism for their area (Siler *et al.*, 2012b). In marine groups, the Sibuyan Sea is the site of maximum species diversity in global clades such as the coneshells, *Conus* (Vallejo, 2005).

Growth of the modern Philippines accelerated with magmatism in the Late Eocene–Oligocene. The composite arc forms part of a belt (the volcanic arc of the Philippine Sea Plate) that extends from Taiwan to New Guinea (Pubellier *et al.*, 2003). The main geological features, including numerous ophiolites, are identical along the length of this belt.

The Philippine mobile belt or orogen is a collage of blocks that amalgamated before being carried westward on the Philippine Sea Plate and colliding with the thinned, rifted

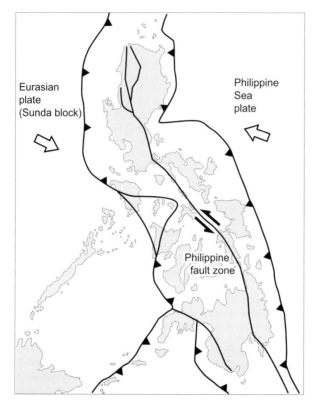

Fig. 10.2 The Philippines, with subduction zones (barbs on the over-riding plate) and the Philippine fault zone (Yumul *et al.*, 2008; Yu *et al.*, 2013).

margin of Eurasia (Sundaland) in the Miocene. The collision led to volcanism, uplift and the development of the Philippine fault zone. One of the best examples of a modern arc-continent collision is between the Luzon arc and the South China Sea margin in Taiwan (Whattam, 2009), and Philippines–Taiwan connections have been mentioned already (e.g. in the moss *Elmeriobryum*; p. 107).

The traditional model of Philippines biogeography

In traditional dispersal theory, plant and animal groups have dispersed into the Philippines from centres of origin outside the region – in Borneo, Sulawesi or Taiwan – via present-day islands such as Palawan and the Sulu Islands. The model is related to the idea that young islands on oceanic crust are colonized from areas of continental crust, either by trans-oceanic long-distance dispersal or by island hopping. Pacific islands in general are seen as the recipients of colonizers from the Pacific margins. In the Philippines, the model predicts that groups in the archipelago will be nested in groups from source areas such as northern Borneo, Sulawesi or Taiwan. Nevertheless, as discussed below, this prediction is contradicted in many groups.

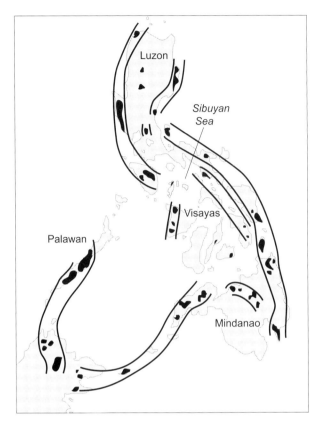

Fig. 10.3 Ophiolites and inferred ophiolite belts in the Philippines (Andal *et al.*, 2005).

Biogeographers have often accepted that groups on continental crust have had a different history from those on oceanic island arcs. Lohman *et al.* (2011) wrote: 'Today, many of the Philippine and Wallacean islands are generally regarded as oceanic islands because they have had no terrestrial connection to any surrounding land since their emergence ... Consequently, their biota arose predominantly via dispersal and not vicariance ... '. Yet terrestrial and reef clades on oceanic crust persist as metapopulations in dynamic archipelagos and so there is no real difference between evolution on continents and on oceanic islands. Metapopulations can undergo vicariance, whether or not the land masses they inhabit were originally continuous. A metapopulation on an island arc can be sundered by rifting in the same way that continental groups can (Fig. 2.1). Whether or not an island has been joined to other land would only be important for biogeography if plants and animals never crossed small sea gaps by normal ecological dispersal (without differentiating), for example, colonizing new islands or ecological successions nearby. There are many observations showing that this process does occur, though; for example, on Krakatoa in Indonesia.

Dispersal theory proposes that the Philippines have received all their groups through dispersal from other, extant land masses outside the Philippines, but in a metapopulation model the extant islands have received at least some of their biota from prior, non-extant islands in the region. For island biogeography, Heaney (2007) rejected both the equilibrium model of MacArthur and Wilson (1967) and also the vicariance model. He wrote: 'vicariance biogeography . . . implicitly requires great stability of island biotas and emphasizes the process of diversification, with persistence of species over millions of years with little or no colonization/dispersal and static biotic composition'. This is not quite accurate, though; vicariance in an archipelago setting can involve dynamic metapopulations, with individual populations constantly going extinct in areas of submergence, and constantly colonizing new islands that erupt in the vicinity. With seafloor spreading, strike-slip displacement, uplift and subsidence, archipelagos can be rifted and metapopulations can undergo vicariance. Heaney (2007) proposed that 'Geologically younger islands often have younger species than nearby, geologically older islands . . . '. Nevertheless, many molecular studies have shown that young islands host endemic clades that are much older than the island, and the relationship between the age of an island and the age of its taxa is probably more complex than has been thought.

The traditional model proposes that terrestrial groups have arrived in the Philippines and then differentiated in the Pleistocene. During the phases of cooler climate in the Pleistocene, sea levels were lowered and islands coalesced to form 'Pleistocene aggregate island complexes' (PAICs) delimited by the present 120-m isobath. There were three large PAICs – Greater Luzon, Greater Negros–Panay and Greater Mindanao (Fig. 10.1). The 'late Pleistocene model of speciation' dominated studies of Philippines birds and mammals for many decades, but recent studies have rejected it and instead favoured earlier divergence times (Heaney, 2007). In Philippines birds, for example, degrees of molecular divergence contradict the idea of a major role for Pleistocene evolution (Jones and Kennedy, 2008a). The authors concluded that 'new hypotheses' are needed, in particular models that incorporate ideas on pre-Pleistocene geography.

In *Apomys* (Muridae), endemic to the Philippines, the clades were dated as older than the Pleistocene and 'a Pleistocene diversification model in which isolation is driven by sea-level changes is inconsistent with the data' (Steppan *et al.*, 2003). Steppan *et al.* discussed alternative models and wrote that 'The archipelago fails to meet the criteria necessary for application of tectonic vicariance models . . . because its tectonic pattern is not one of separation or subdivision of a landmass but one of coalescence'. Yet tectonic coalescence by accretion at plate margins is a complex process and often involves uplift, volcanism and strike-slip movement. Elsewhere these processes are common modes of subdivision and vicariance. In the Philippines, arc formation and accretion have been widespread throughout Cenozoic and late Mesozoic time, as indicated by the ophiolites.

Geology of the Philippines

The Philippine archipelago is a composite, formed by the fusion of continental blocks in the south and west with the accreted arc and ophiolite terranes of the Philippine mobile belt. The Palawan block in the south-western Philippines is formed from continental crust and includes several terranes. These make up parts of northern Palawan, Mindoro, north-western Panay, the Romblon group and western Mindanao (Yumul *et al.*, 2008; 2009b; Zamoras *et al.*, 2008). Arc magmatism in Mindoro dates back at least to the Permian (Knittel *et al.*, 2010).

Ophiolites and intrusives in the mobile belt date back to the Late Jurassic. Some of the present land in the Philippines first appeared above sea level during the Eocene or Oligocene, although most of it resulted from tectonic and volcanic activity during the Miocene and Pliocene. Nevertheless, if groups survived in earlier times on other, former islands that existed in the region, these Cenozoic dates for current land will have no special significance for the origin of the present biota.

The composite basement of the Philippines contains numerous ophiolites, slices of seafloor that formed in ocean basins as these opened above subduction zones (Pubellier *et al.*, 2004; Dimalanta *et al.*, 2006). These supra-subduction ophiolites were associated with arcs that would have maintained living metapopulations, even if no trace of the arcs themselves, let alone individual islands, now remains. The ophiolites date from Jurassic (northern Luzon, Panay) to Oligocene, but most are Cretaceous, as is the Papuan ultramafic belt (Bowutu terrane). The many Cretaceous ophiolites in the region suggest that this period was dominated by ocean basin formation (and vicariance), and was followed through the Cenozoic by basin closure (with juxtaposition of populations) (Dimalanta and Yumul, 2006).

Van Welzen *et al.* (2011: 533) discussed the Philippines–Sulawesi region and wrote:

For geologists, it is difficult to find evidence for the exact time of sub/emergence of areas (R. Hall, pers. comm.) and opposed views, allowing for more or earlier emerged areas, certainly exist (Michaux, 2010) . . . it is very likely that more areas were already above water than described by Hall (2009), although these areas were still isolated (surrounded by water) but could have acted as stepping stones and perhaps as rafts (Michaux, 2010; R. Hall, pers. comm.).

Tectonic reconstructions for the Philippines suggest prior positions of plate margins and blocks of crust rather than individual islands. Yet plate margins and supra-subduction ophiolites are associated with island arcs and so tectonic reconstructions are of great interest for biogeography. Plate margins in the Philippines–New Guinea region, where four plates meet, have changed their position, along with their arcs, and this is relevant for modern island biogeography. Hall (2002) and Hall *et al.* (2009) suggested a model in which northern Luzon, Negros and the Zamboanga Peninsula of Mindanao developed in the mid-Eocene along one subduction zone. The eastern Philippines (Bicol Peninsula of Luzon, Samar and eastern Mindanao), northern Sulawesi and the Maluku Islands developed along a separate subduction zone. In the mid-Eocene the two zones were located north of New Guinea, and lay more or less parallel, facing each other. By mid-Miocene the two arcs had each rotated 90° away from each other and lay along a single subduction zone, running NW–SE. By the end of the Miocene, at 5 Ma, the two arcs had converged and collided with each other and also with the continental Palawan and Mindoro blocks. Collision and strike-slip movement has continued until the present.

The tectonic (and biogeographic) history of the Philippines before the Eocene is more enigmatic, with areas such as the main parts of Luzon and Mindanao representing composites. Hall's (2002) model began at the Early Eocene and Hall *et al.* (2009) wrote that 'At the moment it is impossible to reconstruct the Cretaceous–Paleocene West Pacific in detail but it is clear that there were many intra-oceanic arcs within the Pacific basin'.

Moores (2009) emphasized the challenging nature of the regional tectonics and wrote that: 'In the western Pacific, ophiolite/arc sequences may result from complex intra-arc rifting and closure (e.g. Philippines . . .) or multiple ophiolites, sutures, collisions, rapid Neogene plate margin changes, and extensive transform faulting (Indonesia and Philippines) . . . Some active collisional sutures, e.g., arc–arc, may be difficult or even impossible to detect when completed, especially in older deformed belts'. Biogeography has much more data on spatial differentiation than geology, and so, if the assumption of chance dispersal is suspended, biogeography can help resolve aspects of the history that are otherwise inaccessible to geologists.

In Taiwan, the Philippines, northern New Guinea, New Caledonia and New Zealand, ophiolites and associated arc terranes have docked with continental crust in the west and south-west. The many ophiolites in the Philippines suggest that many basin fragments have 'piled up' there, and the tectonics of the archipelago are even more complex than in the other countries mentioned. Much of the crust in the Philippines is 31–65 km thick (Yumul *et al.*, 2009a), although it is formed from oceanic crust. The great thickness, equivalent to that of continental crust, has developed with repeated phases of magmatism, ophiolite obduction and the integration of different arcs. Pubellier *et al.* (2004) emphasized the discrepancies in age and nature of the Philippines ophiolites (Fig. 10.3). For example, in north-western Luzon, the Zambales ophiolite cannot be correlated with any of the neighbouring basins, and has probably been carried over a large distance. It has been thrust over relicts of continental basement present along the coast. The Zambales ophiolite is Eocene, while in central and north-eastern Luzon the ophiolites are Cretaceous, and in Ilocos Norte, far northern Luzon, they are Jurassic.

The Philippine Sea plate, north of the Australian plate, began its growth in the Eocene, but it also includes reworked Mesozoic rocks. For example, reworked pebbles in a rock from the Mariana forearc include fossils of a Late Jurassic–Early Cretaceous ciliate, *Calpionella*, and Late Cretaceous radiolarians and foraminiferans have been found in the same area (Pubellier *et al.*, 2003). Johnson *et al.* (1991) proposed that these rocks were allochthonous, accreted fragments of the old Pacific Plate. They represent pieces of a late Mesozoic arc that was dismantled and reworked into the Eocene–Oligocene Mariana arc.

On the opposite side of the Philippine Sea Plate, in its north-western corner, the Huatung Basin is a small basin in oceanic crust east of Taiwan, and gabbros dredged there are Early Cretaceous (Deschamps *et al.*, 2000). This is consistent with Early Cretaceous ages for radiolarians in cherts from Lanyu Island, off Taiwan. These Cretaceous cherts and gabbros are part of the former oceanic plate on which the Luzon Arc (including Lanyu and the Babuyan Islands) has been developing since Late Oligocene time. Deschamps *et al.* (2000) suggested that the Huatung Basin opened during the Early Cretaceous, from 131 to 119 Ma. They proposed that the basin is a trapped fragment of the 'proto-south China Sea' or possibly the 'New Guinea Basin'.

Many aspects of Philippines biogeography are older than the present islands, and intercontinental distributions suggest they are also older than the Cenozoic arcs and plate margins. Intra-arc rifting and arc–arc fusion have added to the tectonic and biotic complexity. Late Mesozoic arcs have been destroyed and the material reworked into the Eocene–Oligocene arc, along with the subduction zone weeds of the earlier flora and fauna. Late Mesozoic events are probably necessary to explain the great intercontinental breaks in and around the Philippines; for example, between Philippines endemics and widespread Indian Ocean sister-groups.

The accreted arc/metapopulation model of Philippines biogeography

Supra-subduction zone ophiolites are significant for biogeography as they originated in association with island arcs (Polhemus, 1996; Michaux, 2010). The ophiolites have been uplifted during emplacement on land, and this means they can preserve the arc biota, even if the arc itself has since been destroyed. The most mature phase of arc collision is seen in old arc fragments and ophiolites now deeply embedded in modern mainlands, as in the mountains of New Guinea, New Zealand and the Philippines. These mountains are all well known for their high levels of endemism. Studies on a diverse Philippines group of *Hydropsyche* (Trichoptera) concluded that a vicariance model, based on tectonic accretion in the Late Miocene, is the most likely (Mey, 2003). Mey wrote: 'This vicariance explanation is fascinating and impressive because it does not need the additional assumptions of the dispersal scenario, i.e. dispersal to Borneo, dispersal over the sea to the Proto-Philippines, and finally, extinction in Borneo'.

Apart from its great diversity of terrestrial groups, the Indo–Australian archipelago is well known as the site of maximum marine biodiversity on Earth. The two phenomena are probably related, as many of the marine groups are reef organisms that require shallow water. Within the Indo–Australian archipelago, Carpenter and Springer (2005) analysed the distributions of 2983 marine species (including 1775 bony fishes) and found a peak of diversity in the central Philippine Islands, between southern Luzon and northern Mindanao. The authors concluded that the great diversity in the Philippines, which they termed the 'center of the center', can be explained by the 'numerous vicariant and island integration events . . . ' in the tectonic history of the region. They wrote:

The Philippines are integrated from at least three major island systems that were widely separated during much of the Cenozoic . . . Each of these three major elements were displaced over 1000 km to reach their current locations, . . . In addition to potentially generating vicariant events [with uplift, strike-slip etc.], the accretion of the archipelago would also have concentrated diversity, assuming that the different elements of the Philippines developed their own endemic biotas (Carpenter and Springer, 2005: 475).

This 'bioconcentration' caused by island arc integration is consistent with the conclusions of Santini and Winterbottom (2002). These authors considered that the modern fauna of the Indo–Australian Archipelago is a consequence of different faunas rafting into the region in association with diverse terranes.

In a tectonic/vicariance model, the paucity of groups such as cursorial mammals in the Philippines is not because they were unable to disperse there from the mainland, but because they were poorly pre-adapted for survival on ephemeral islands as metapopulations. In contrast, terrestrial vertebrates such as rats and bats have been able to persist in this way. Endemic Philippines clades of these and other groups have not colonized the region from outside by one-off, long-distance dispersal events; instead, the groups were able to persist and evolve in the region because of their normal means of survival.

Some of the main biogeographic patterns of Philippines groups are discussed next. Records from tetrapods, especially birds, are stressed here as these groups' distributions and phylogenies are the best known. Many of the groups are mapped by IUCN (2013) and mammals are mapped by Heaney *et al.* (2010).

Philippines groups with widespread sister-groups

Several endemic Philippines clades have diverse sister-groups with widespread, intercontinental distributions, and examples occur in freshwater gastropods, owls, passerines and mammals (Heads, 2012a; Chapter 1). Although these groups have very different ecology and means of dispersal, the phylogeny suggests that the clades have not simply colonized the Philippines from Borneo, Sulawesi or Taiwan, otherwise they would be nested in the groups endemic there. Dispersal theory suggests that the Philippines clades with widespread sister-groups evolved somewhere else, then migrated (often alone) to the Philippines, then went extinct where they came from, while keeping their sister-groups out of the Philippines by competition. However, this is a convoluted process and many groups show the pattern, so a single allopatric break that affected all groups is much simpler. In any case, the ad hoc dispersal–extinction scenario leaves the original question unanswered: Why or how did the Philippines group diverge from the rest? Again, the depth of the geographic structure – the allopatry of the distributions and the repetition of the pattern in unrelated groups – suggests a simple explanation.

No Philippines endemics with global sisters are known. The Philippines appear to lie outside the Tasman–Coral Sea region of endemics with global sisters – the most northwesterly of these endemics is the fish *Protanguilla* from Palau (see below). Nevertheless, there are many Philippines endemics with widespread Old World sister-groups and in one example, discussed next, the sister-group ranges from Madagascar to Central America.

Philippines sister to Madagascar–Asia–Central America

The Pachychilidae are freshwater gastropods that are widespread in Madagascar, Asia and Central America (Fig. 10.4). The following phylogeny has been presented (Köhler and Glaubrecht, 2007):

1. Philippines (*Jagora*).
   2. Madagascar.
      3. Nepal to southern India, southern China, Indochina and Java.
      4. Sulawesi, Torres Strait, Central America.

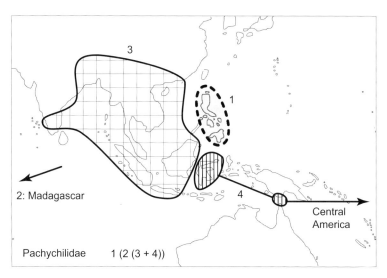

Fig. 10.4 Distribution of Pachychilidae. The phylogeny is indicated (Köhler and Glaubrecht, 2007).

A study focusing on the Madagascar species (Köhler and Glaubrecht, 2010) found the alternative phylogeny: 3 (1 (2 + 4)), but the Philippines remained basal in a widespread tropical group. In either case, the sequence of differentiation appears to 'jump' across the Indian Ocean, from around the Philippines to around Madagascar, and then to Torres Strait and a trans-tropical Pacific affinity (cf. pp. 271–275). Neverthless, all the clades are allopatric, with no overlap at all, suggesting vicariance of an ancestor that already had a wide distribution through the tropical Indo–Pacific. Whatever the exact phylogeny, there is no need to infer any chance migration or even range expansion of the current groups, only normal dispersal within their respective ranges.

The first break in this vast Indo–Pacific complex was between *Jagora*, endemic to the Philippine mobile belt, and the rest. Köhler and Glaubrecht (2002) noted that the eastern Philippine belt has been built up by island arc integration from the Late Jurassic–Cretaceous onwards. They concluded that the distribution of Pachychilidae reflects 'relatively ancient vicariance events like the migration, separation or amalgamation of terranes rather than younger events related to the Pleistocene sea-level changes, while active dispersal is believed to play a minor role if at all . . .'.

In the phylogeny, the first break, between *Jagora* of the Philippines and the rest, is followed by a break between Madagascar and southern India, where Cretaceous rifting is well documented. *Jagora* comprises *J. asperata* on Luzon, Tablas (Romblon group), Samar and Leyte (with unconfirmed records on Mindanao), and *J. dactylus* on Bohol, Cebu and Guimaras (between Panay and Negros) (Köhler and Glaubrecht, 2007). The two species are allopatric, but neither distribution corresponds with Pleistocene islands. Instead, the distribution conforms to a common pattern in which a group in the north, east and south forms an eastern arc around a group in the western Visayas (see Appendix to this chapter).

Philippines sister to Africa–Asia

Groups with this pattern include *Pithecophaga* (Accipitridae), the monkey-eating eagle of the Philippines. It is sister to *Terathopius*: Africa, and *Circaetus* (including *Dryotriorchis*): all Africa, Europe and Asia to India and China (Lerner and Mindell, 2005).

Coracina mcgregori (Campephagidae) of the Philippines (Mindanao) is sister to a clade ranging from West Africa to India, China, Australia and Samoa (Jønsson *et al.*, 2010c).

Hypocryptadius (Passeridae) of Mindanao is sister to *Passer* (true sparrows), *Montifringilla* and *Petronia* (the central Asian *Carpospiza* was not sampled), a widespread group known from South Africa to Socotra, Europe, central Asia and India, Japan and Java (introduced in the Philippines and many parts of the world) (Fjeldså *et al.*, 2010). The authors discussed *Hypocryptadius* and its restricted distribution in the montane cloud-forest of the island of Mindanao (not in Zamboanga). They wrote that, considering the phylogenetic length of the *Hypocryptadius* branch, 'it seems plausible to refer to it as a relictual form and representative of a clade that originated somewhere else'. Nevertheless, this does not account for the allopatry between *Hypocryptadius*, in the Philippines, and its sister-group, widespread in the Old World to Sumatra and Java, but in the Philippines present only as human introductions.

Fjeldså *et al.* (2010) concluded that *Hypocryptadius* 'is difficult to explain without assuming an over-water dispersal event … [Its] isolated occurrence … on Mindanao is even more remarkable when we take into account that the southern Philippine islands originated as a group of volcanic islands far out in the Pacific Ocean, with no near contact with Gondwana or other subaerial land areas (Hall, 2002)'. Yet the paleogeographic details of the early arcs north of New Guinea are poorly known and were probably complex (Hall, 2002, 2011). Their bird populations probably included the ancestors of *Hypocryptadius*.

Another passerine, *Robsonius* of Luzon (Locustellidae), is sister to a clade (*Bradypterus*, *Megalurus*, *Malia* and *Locustella*) distributed from sub-Saharan Africa and Europe throughout central and tropical Asia to Japan, New Guinea, Australia and Tasmania (Oliveros *et al.*, 2012). Oliveros *et al.* noted that *Robsonius* joins *Hypocryptadius* and also *Rhabdornis* (see below) among Philippine endemic genera that represent early offshoots of widespread Old World families.

In the rodent subfamily Murinae, the basal clade (*Phloeomys* etc.) is endemic to the Philippines (not Palawan, Zamboanga Peninsula or Samar). Its sister-group, including true rats and mice, is diverse and widespread through Africa and Eurasia to the Solomon Islands (there are also two species introduced worldwide) (Rowe *et al.*, 2008).

Philippines sister to Madagascar/Mascarenes–Asia

In the scops owls, *Otus* (Strigidae), a Philippines montane clade (*O. longicornis* and *O. mirus*) is sister to a group ranging from Madagascar and the Seychelles to India, Sakhalin, Japan and Borneo (Fuchs *et al.*, 2008; Miranda *et al.*, 2011).

The bird *Lalage melanoleuca* (Campephagidae) of the Philippines (not the western Visayas or Palawan) is sister to a clade ranging from the Mascarenes to Sri Lanka, India, north-eastern China, Sulawesi and the Philippines (Jønsson *et al.*, 2010c).

In the magpie-robins, *Copsychus* s.str. (Muscicapidae), the Philippines *C. mindanensis* is sister to a clade in Madagascar (*C. albospecularis*), the Seychelles (*C. sechellarum*), Sri Lanka, Pakistan, India, to China, Borneo and Java (*C. saularis*) (Sheldon *et al.*, 2009; Lohman *et al.*, 2010; Lim *et al.*, 2010). Lohman *et al.* (2010) pointed out that in *Copsychus* poor taxonomy had 'hidden' the important Philippines endemic in *C. saularis*, and that this is probably also the case in other groups. The earlier taxonomy reflected a tendency to see Philippines taxa as having secondary importance, as they were interpreted as the products of recent, local dispersal from Borneo or Taiwan. Thus they have been 'shoe horned' into Asian groups.

In *Copsychus* s.str., Lim *et al.* (2010) found a phylogeny: (1. Philippines (2. Madagascar (3. Seychelles (4. India to China and Borneo)))). The second break, between 2 and 3+4, coincides with the rifting between Madagascar and the Seychelles + India, which took place in the Late Cretaceous. This gives a minimum date for the first break, between the Philippines and the rest. It is possible that the first and second breaks occurred about the same time. Sheldon *et al.* (2009) called the distribution of *Copsychus* s.str. in Madagascar, Asia and the Philippines, 'unusual' and proposed 'fortuitous' trans-oceanic dispersal to Madagascar, but the same overall pattern is repeated in *Otus* and others.

Philippines sister to southern Asia and northern hemisphere

In the duck genus *Anas* (Anatidae), a Philippines species is sister to a group that is widespread throughout southern Asia and the northern hemisphere. The clade is structured as follows (Johnson and Sorenson, 1999):

Philippines (all main islands) (*Anas luzonica*).

Eastern Russia, Japan, south in mainland Asia to Indochina (*A. zonorhyncha*).

India, Sri Lanka (*A. poecilorhyncha*).

Widespread throughout the northern hemisphere, including India (but not the Philippines) (*A. platyrhynchos*).

Philippines sister to India–Southeast Asia clades

These patterns are not quite as a stark as the ones described above, but the sister-groups are still widespread and there is no indication of a local centre of origin for the Philippines clades. In the first case, a group in *Dicrurus* (*D. balicassius*, etc. Dicruridae) has the following phylogeny (Fig. 10.5; Pasquet *et al.*, 2007):

Northern Philippines + (south and south-east Asia, **southern** Philippines).

New Ireland + (Australasia, **southern** Philippines).

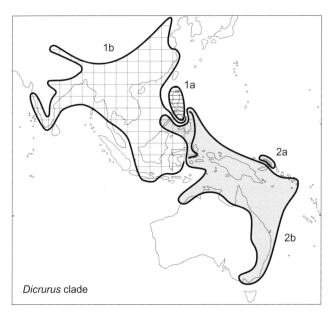

Fig. 10.5 Distribution of a clade in *Dicrurus* (Dicruridae). The phylogeny is: (1a + 1b) (2a + 2b). (Pasquet *et al.*, 2007.)

The Philippines members do not make up a monophyletic entity. Three of the four groups have a distribution margin between the northern Philippines (Luzon to Panay, Negros and Cebu) and southern Philippines (Greater Mindanao and Palawan). This indicates repeated basal differentiation around an early break. This would have occurred when the arcs were located further towards the north of New Guinea and the other main break in the *Dicrurus* group, around New Ireland. Another New Ireland clade sister to a widespread Australasian group is documented in *Citrus* (Heads, 2012a, Fig. 9.4).

Dicrurus balicassius of the northern Philippines might be a relic (cf. Fjeldså *et al.*, 2010, cited above), but this would not explain the precise vicariance with its sister-group (and with their joint sister-group) at a break in the central Philippines between Cebu and Leyte. Instead, the break is probably a direct result of the phylogeny.

The phylogeny of the shamas, a clade in *Copsychus* s.lat. (Muscicapidae), is as follows (Lim *et al.*, 2010):

Philippines (*C. luzoniensis*).

Philippines (*C. niger* and *C. cebuensis*).

India and Sri Lanka to Borneo and Java (*C. malabaricus*).

The sequence is the opposite of that predicted in the traditional model, in which dispersal moves from Asia into the Philippines via Palawan or the Sulu Archipelago. Instead, the result can be attributed to vicariance around the Philippines, before any other differentiation occurred through Asia.

A similar pattern to that of the shamas occurs in the bee *Apis cerana* (Smith *et al.*, 2000). This has a phylogeny:

Philippines.

Sulawesi and Sangihe (between the Philippines and Sulawesi).

Philippines.

India, Nepal, Korea, Japan, Indochina, Borneo, Java, Lesser Sunda Islands.

The main difference between this group and the shamas is the interpolation of a Sulawesi–Sangihe clade.

In the sunbird *Arachnothera longirostra* (Nectariniidae), Moyle *et al.* (2011) found the phylogeny:

Greater Mindanao (*A. l. flammifera*).

Palawan (*A. l. dilutior*).

Southern India, Nepal and Bangladesh to Borneo and Java (*A. l. longirostra*).

Deep divergences among the clades led the authors to propose species status for the two Philippines members. This repeats the pattern of Philippines basal endemics being hidden by the earlier taxonomy, as in *Copsychus mindanensis* (above).

Another passerine, *Irena* (the only genus in Irenidae), has a phylogeny and distribution that is almost identical to that of *Arachnothera longirostra* (Moltesen *et al.*, 2012):

Greater Mindanao and Luzon (*I. cyanogastra*).

Palawan (*I. puella tweeddalei*).

Southern India, Nepal and Bangladesh to Borneo and Java (*I. puella* subspp.).

Other Philippines groups with Indian–Southeast Asian sisters include the following:

The snake *Oligodon modestus* (Colubridae), endemic to the Philippines (Panay, Negros and the Romblon group), is sister to a diverse clade from Pakistan, India and Sri Lanka east to southern China, Taiwan, Indochina and Borneo (Green, 2010).

The snake *Lycodon* (Colubridae) ranges from eastern Iran to Timor. A sample of 16 of the 37 species showed a Philippines clade sister to one distributed from India to Japan, Sulawesi and the Philippines (Siler *et al.*, 2013).

In birds, the flower-pecker clade *Dicaeum bicolor* + *D. anthonyi* (Dicaeidae) of the Philippines (except Palawan) is sister to the rest of *Dicaeum*, recorded from the Himalayas, India and Sri Lanka to the Philippines, Australia and the Solomon Islands (Nyári *et al.*, 2009b).

Dicaeum nigrilore + *D. hypoleucum* of the Philippines (eastern track) is sister to a clade ranging from southern India to Australia and the Solomon Islands (Nyári *et al.*, 2009b).

Dicaeum pygmaeum of the Philippines (except Panay) is sister to *D. concolor* of southern India to Java and Borneo (Nyári *et al.*, 2009b).

Spizaetus philippensis + *S. pinskeri* (Accipitridae) of the Philippines is sister to
   S. circaetus, of India and Sri Lanka to Sulawesi and the Lesser Sunda Islands (not
   the Philippines) (Gamauf *et al.*, 2005).
Pernis steerei (Accipitridae) of the Philippines is sister to a clade of India and central
   Asia (Russian Altai) to Sulawesi and the Philippines (Gamauf and Haring, 2004).
Rhabdornis (Sturnidae) of the Philippines (not Palawan or Mindoro) is sister to a
   clade of nine genera (*Gracula*, *Aplonis*, etc.) distributed from India to Southeast
   Asia and the central Pacific islands (Lovette and Rubenstein, 2007).
The fruit bat *Ptenochirus* (Pteropodidae) is endemic to the Philippines (except
   Palawan) and is sister to *Cynopterus*: India and Sri Lanka to Sulawesi, Borneo and
   the Philippines (Almeida *et al.*, 2011).

Philippines sister to Bangladesh–Southeast Asia

The passerine *Pachycephala* (Pachycephalidae) (Jønsson *et al.*, 2010a) includes the
following clade:

Eastern Philippines (Luzon, Samar, Leyte, Bohol and Mindanao) (*P. philippinensis*).
   Sulawesi (*P. sulfuriventer*).
      Borneo (*P. hypoxantha*).
         Luzon and Mindoro (*P. albiventris*).
            Widespread from Bangladesh to Java, Borneo and Palawan (*P. cinerea* (=
            P. grisola), the mangrove whistler).

The distribution suggests an ancestral complex that was widespread from Bangladesh
to the Philippines, in mangrove and other lowland rainforest. Subsequent uplift of
populations has occurred in the Philippines, Sulawesi and Borneo.
   Other Philippines groups with Bangladesh–Southeast Asian sisters include the
following:

Polyplectron emphanum (now *P. napoleonis*) (Phasianidae) of Palawan is sister to
   the remaining species of *Polyplectron*, distributed from Bangladesh to Hainan,
   Sumatra and Borneo (Kimball *et al.*, 2001).
Dicaeum australe (Dicaeidae) of the Philippines (eastern track) is sister to *D. trigono-*
   stigma of Bangladesh to the Philippines (widespread) (Nyári *et al.*, 2009b).
Eurylaimus steerii (Eurylaimidae) of Mindanao is sister to a clade ranging from
   Bangladesh to Sumatra and Borneo (Irestedt *et al.*, 2006).
Pycnonotus urostictus (Pycnonotidae) of the Philippines (not Palawan, Mindoro or
   Panay) is sister to *P. atriceps*: Bangladesh to Borneo, Java and Palawan (Oliveros
   and Moyle, 2010).

Philippines sister to Indochina/China

This pattern is less common than a connection with Borneo–Peninsular Malaysia–Sumatra, and involves a direct connection between the Philippines and the Asian mainland. In pines, for example, *Pinus kesiya* is endemic in Indochina and northern Luzon (de Laubenfels, 1988). *Pinus merkusii* is in Indochina and Hainan (but not Malaysia), disjunct in Sumatra, and also in north-western Luzon (Zambales) and Mindoro. An example of the Philippines–China connection in animals is the tie between the frog *Barbourula* (Bombinatoridae) of Palawan and south-western Borneo and its sister *Bombina* of Europe to south-eastern China. Blackburn *et al.* (2010) found that the two genera diverged in the Eocene (47 Ma) and described the date as 'unexpected' and 'surprisingly ancient'. They suggested that some portion of the North Palawan terrane remained above water after it rifted from southern China and that components of the terrestrial fauna have survived in this way. The clock in the study was fossil-calibrated and so the Eocene date is only a minimum age, although a useful one. The North Palawan terrane rifted from Indochina/South China in the Cretaceous or Paleogene, and it is likely that the two frog genera diverged as a result of this event.

Other Philippines–Indochina/China groups include the following:

In the butterfly *Cethosia* (Nymphalidae), *C. luzonica* of the Philippines is sister to *C. cyane* of Indochina (Müller and Beheregaray, 2010), with the divergence assigned a date of 19.5 Ma (a fossil-calibrated minimum).

The butterfly *Charaxes bajula* (Nymphalidae) of the Philippines is sister to *C. marmax* + *C. aristogeiton* of Indochina (Müller *et al.*, 2010).

In *Eutropis* (Scincidae), a clade of the Philippines and Palau is sister to *E. quadricarinata* of India and Burma, plus *E. macularia* of Pakistan to Indochina (Mausfeld and Schmitz, 2003).

A clade in the passerine genus *Parus* (Paridae) comprises *P. elegans* of the Philippines, *P. amabilis* of Palawan and *P. venustulus* of eastern mainland China (Martens *et al.*, 2006).

There also appears to be a direct connection between the Philippines and Hainan. Examples include *Dacrydium pectinatum*: Borneo, Philippines, Hainan, and *Podocarpus pilgeri*: New Guinea, Maluku Islands, Sulawesi, Philippines, Hainan and adjacent parts of mainland China (both in Podocarpaceae; de Laubenfels, 1988).

Philippines sister to Sundaland (Peninsular Malaysia, Sumatra, Borneo, Java)

The term Sundaland originally referred to the former land in what is now the area of sea separating Peninsular Malaysia, Sumatra, Borneo and Java. Biogeographers extended the meaning of the term to include these land areas themselves. The geological concept of the Sundaland plate (a largely submerged continent, like Zealandia) also includes Indochina. Philippines–Sundaland groups include the following:

Cethosia mindanensis (Nymphalidae) of the Philippines is sister to *C. hypsea* p.p. of Sundaland (Müller and Beheregaray, 2010).

In the pit viper *Trimeresurus* (Viperidae), the Philippines clade *T. flavomaculatus* + *T. mcgregori* (Luzon, Mindanao, Panay, Bohol) is sister to a clade in Palawan, Peninsular Malaysia, Sumatra and Borneo (Malhotra and Thorpe, 2004).

In the mangrove snake *Cerberus rhynchops* (Colubridae), a Philippines clade is sister to a group ranging from Thailand to Borneo and Sulawesi (Alfaro *et al.*, 2004).

Oriolus steerei + *O. isabellae* + *O. albiloris* (Oriolidae) of the Philippines is sister to *O. xanthonotus* of Peninsular Malaysia, Sumatra, Borneo and Java (Jønsson *et al.*, 2010b).

Arachnothera clarae (Nectariniidae), of Luzon, Samar, Leyte and Mindanao, is sister to *A. chrysogenys* of southern Thailand, Peninsular Malaysia, Sumatra, western Java and Borneo (Moyle *et al.*, 2011).

The leafbird *Chloropsis palawanensis* (Chloropseidae) of Palawan is sister to *C. cyanopogon cyanopogon* of Peninsular Malaysia (just into Burma and Thailand) to Sumatra and Borneo (Moltesen *et al.*, 2012).

Bullimus (Muridae) of the Philippines (Luzon and Greater Mindanao) is sister to *Sundamys* of Peninsular Malaysia, Sumatra, Java, Borneo and Palawan (Jansa *et al.*, 2006; Heaney *et al.*, 2009).

Alionycteris + *Otopterus* + *Haplonycteris* (Pteropodidae) of the Philippines (except Palawan) is sister to *Penthetor* + *Chironax*, of Peninsular Malaysia, Sumatra and Borneo (Almeida *et al.*, 2011).

The gliding mammal *Cynocephalus* (Dermoptera) of the Philippines is sister to the only other dermopteran, *Galeopterus* of Thailand, Vietnam, Sumatra, Java and Borneo.

Philippines sister to Asia–Queensland/Melanesia

In Rubiaceae, Alejandro *et al.* (2011) reported a clade with three members (the relationships among the three are unresolved):

Villaria: Philippines.
Hypobathrum: Asia (Taiwan and Thailand) to NE Queensland.
Pouchetia: Africa.

The Philippines clade is sister to a sister-group in Asia + Australasia, Africa, or Asia + Australasia + Africa. Other groups with sister-groups in Asia + Australasia include the following:

In the spider *Nephila pilipes* (Tetragnathidae) a Philippines clade (PH43 and PH39) is sister to the rest of the species, which is distributed from India and Sri Lanka to eastern Australia and Vanuatu (Su *et al.*, 2007).

The butterfly clade *Charaxes sangana* + *antonius* + *amycus* (Nymphalidae) of the Philippines is sister to a clade (*C. ocellatus* etc.) known from Peninsular Malaysia to the Bismarck Archipelago and northern Queensland (not in the Philippines) (Müller and Beheregaray, 2010).

In *Rhipidura* (Rhipiduridae), the Philippines endemic clade (*R. cyaniceps* etc.) is sister to a clade distributed from Burma to the Solomon Islands, including the Philippines (Sánchez-González and Moyle, 2011).

Philippines–Borneo

The Philippines biota has close connections with Borneo. For example, the Opiliones suborder Cyphophthalmi is a group known for its locally endemic species and low vagility. The only Philippines representatives are from central Mindanao and were recovered as members of a clade found almost exclusively in Borneo. Clouse *et al.* (2011) wrote that 'Their deep placement within this clade suggests a very old origin and colonization that perhaps involved the mysterious landmass now underlying Mindanao's Zamboanga Peninsula'. The authors proposed a tectonic explanation, related to the opening of the South China Sea and the southward movement of Palawan, Mindoro and parts of Zamboanga from the Chinese coastline.

Philippines groups often have relatives in the Mount Kinabalu/Crocker Range region of northern Borneo. For example, the bird *Harpactes ardens* (Trogonidae) of the Philippines (Luzon, Samar, Leyte, Bohol, Mindanao) is sister to *H. whiteheadi* of northern Borneo (Mount Kinabalu etc.; weak support) (Hosner *et al.*, 2010). Here the connection appears to be via Mindanao–Kinabalu.

A clade in the frog *Staurois* (Ranidae) has the following phylogeny (Arifin *et al.*, 2011):

Mindanao, Samar and Leyte (*S. natator*).

Northern and western Borneo: Mount Kinabalu etc. (*S. parvus*, *S. tuberilinguis*).

Palawan (*S. nubilus*).

Central Borneo (*S. guttatus*).

In this group neither Borneo, the Philippines, nor both together, form a monophyletic unit. This is consistent with the idea that the distributions developed before the assembly of modern 'Borneo' and 'Philippines' from their component terranes (for notes on Borneo geology, see Heads, 2012a, p. 253).

A clade in another frog, *Leptobrachium* (Megophryidae), has a similar pattern (Brown *et al.*, 2009b; IUCN, 2013):

Mindanao.

Sumatra, Borneo.

Kinabalu (*L. gunungense*).

Palawan (*L. tagbanorum*) + Mindoro (*L. mangyanorum*).

This pattern shows that Palawan is not just an extension of 'Borneo', while 'Borneo' again appears as a biogeographic composite (Heads, 2003; 2012a). There are direct affinities between Palawan and central Borneo (see *Staurois*), between Palawan and Kinabalu, and, as the next group shows, between Mindanao and Kinabalu.

In *Ansonia* (Bufonidae) the Philippines taxa are all on Mindanao and form a clade whose sister-group is composed of two species endemic to Mount Kinabalu. Matsui *et al.* (2010) concluded 'that the Philippine taxa diverged from the Bornean lineages at 39.0 Ma (95% HPD: 25.3–53.5 Ma)'. Sanguila *et al.* (2011) considered these estimates to be too old. They wrote: 'the geological components of Mindanao were widely separated and largely still submerged at 20 Ma, casting doubt on the possibility that vertebrates of low relative dispersal abilities could achieve long-distance dispersal necessary to invade the separate paleoisland precursors of today's Mindanao'. Mindanao may have been largely submerged, but smaller islands would have been sufficient to preserve populations. The *Ansonia* clades in Mindanao are all strictly allopatric, consistent with a vicariance history within the island, as well as between it and the Kinabalu region.

Philippines–Taiwan–Ryukyu Islands

The volcanic arcs that run north from Luzon to Taiwan and the Ryukyu Islands mark earlier boundaries of the Eurasian and Philippine Sea plates (the Longitudinal Valley of eastern Taiwan forms part of the current convergent margin). Island arcs are often seen as dispersal routes from area A at one end of the arc (for example, Taiwan) to area B at the other (for example, the Philippines). Instead, A and B already have their own biotas, and the arcs also have their own biota of metapopulations, often including many endemics. An arc biota is more likely to have moved with the arc and its trench as this migrated, than from one end to the other. Examples of endemism along the arcs north of the Philippines include the following:

> *Treron formosae* (Columbidae): Batanes and Babuyan Islands north of Luzon, Taiwan, Ryukyu Islands.
> *Pteropus dasymallus* (Pteropodidae): Batanes and Babuyan Islands, Taiwan, Ryukyu Islands.

Other groups in the region but absent on mainland China include *Terpsiphone atrocaudata* (Monarchidae): Korean Peninsula, Japan, Ryukyu Islands, Taiwan, Batanes and Babuyan Islands, Luzon, Mindoro, Palawan, Peninsular Malaysia and Sumatra.

A clade of leaf warblers in the genus *Phylloscopus* (Phylloscopidae) has the phylogeny (Olsson *et al.*, 2005):

> Luzon, Negros, Cebu (*P. cebuensis*).
>> Luzon, Taiwan, Ryukyu Islands, Japan (not on mainland Asia) (*P. ijimae*).
>> Java and mainland Asia to Sakhalin and Japan (*P. coronatus*).

The northernmost islands in the Philippines, the Batanes, lie about halfway between Luzon and Taiwan and their biota includes many endemics; for example, ferns (Barcelona, 2003). The *Crocidura* shrews from the Batanes are more closely related to forms from Taiwan and the Asian mainland than to other Philippines groups (Esselstyn and Oliveros, 2010). The authors inferred that shrews colonized the northern Philippines from Taiwan or its immediate vicinity. Nevertheless, this interpretation depends on centre of origin theory and does not consider the possibility of metapopulation vicariance along the subduction zone.

Philippines–Lanyu Island

Lanyu (= Orchid or Botel Tobago) Island lies 70 km south-east of Taiwan and the biota is known for its diversity, as well as its close affinities with the Philippines. One hundred and ten plant species on Lanyu occur in the Philippines but are not in Taiwan proper, and the genus *Vanoverberghia* (Zingiberaceae) is known only from Lanyu and northern Luzon (Funakoshi and Ohashi, 2000). Likewise, the endemic *Gekko kikuchii* of Lanyu Island is more closely related to Philippine taxa than to other Taiwanese and Ryukyu Archipelago species (Brown *et al.*, 2009a). *Podocarpus costalis* (Podocarpaceae) is endemic to islets from Lanyu to Luzon (de Laubenfels, 1988). *Acalypha grandibracteata* (Euphorbiaceae) is on the Batanes and Babuyan islands north of Luzon, also Lanyu and other small islands off Taiwan (Sagun *et al.*, 2010). In terms of its ecology, this *Acalypha* is a small island specialist, but the biogeographic question is: Why does it occur only on *these* small islands? Many of the groups on Lanyu and other parts of this arc are related to groups in the Pacific. For example, in *Cyrtandra* (Gesneriaceae), a clade distributed along a narrow arc: southern Ryukyu Islands, Lanyu, Philippines, Sulawesi and Java, is sister to a widespread central Pacific clade (Fig. 8.5).

A tropical Pacific group of stick insects, Stephanacridini, has the total range: Lanyu Island, Peninsular Malaysia, all Indonesia, the Philippines, Micronesia, Melanesia, northeastern Queensland, Fiji, Tonga, Samoa and east to the Tuamotu Islands (Buckley *et al.*, 2010, distributions from Hennemann and Conle, 2006b, 2008, 2009). In its morphology the tribe is closely related to Achriopterini of Madagascar and the Comoros.

Philippines–Sulawesi

Since the nineteenth century, biogeographers have treated the Indo–Malay archipelago (Malesia) in two parts, with the western half belonging to 'Oriental region' and the eastern half to the 'Australian region' (Wallace, 1876). The central region, including the Philippines, Sulawesi and the Lesser Sunda Islands, has been interpreted as a transitional zone. In contrast, Dickerson (1928) treated it as a distinct region and named it Wallacea. This broad concept of Wallacea, including the Philippines, has often been overlooked, with most writers treating the Philippines separately from Sulawesi. Yet many conspicuous groups are endemic here, including the large trees *Agathis celebica* and *A. philippensis* (Araucariaceae), both endemic to the Philippines, Sulawesi and the Maluku Islands (de Laubenfels, 1988). In a study of the region's flora, Van Welzen *et al.* (2011) suggested a division of Malesia into: (1) the Malay Peninsula, Sumatra and Borneo in the west; (2) a central region, Wallacea (Philippines, Sulawesi, Lesser Sunda Islands, Maluku Islands and Java); and (3) New Guinea in the east. This scheme adopts a classificatory approach with non-overlapping regions (and so does not accommodate Philippines–Borneo, or Philippines–New Guinea groups, for example) but the recognition of the central region is important.

New molecular evidence for a general zoogeographic link between the Philippines and Sulawesi is accumulating (Brown and Diesmos, 2009). One example is the parrot *Prioniturus* (Psittacidae), endemic to the Philippines and Sulawesi (plus southern Maluku Islands). Its sister-group (*Psittacula*, etc.) ranges from Africa to the Solomon Islands, and includes one genus, *Tanygnathus*, endemic in the Philippines and Sulawesi (plus the Lesser Sundas and Maluku Islands) (Schweizer *et al.*, 2012). The pattern indicates two splits in or around the Philippines–Sulawesi region, and one phase of overlap there.

In the pig genus *Sus* (Suidae), four species endemic to the Philippines plus a Sulawesi species form a clade that is sister to the rest of the genus (*S. barbatus* and *S. scrofa*), known from Borneo and the Lesser Sunda Islands to Asia and Europe (Lucchini *et al.*, 2005).

Other Philippines–Sulawesi groups include the following:

Charaxes solon lampedo + *C. s. hannibal* (Nymphalidae) of the Philippines and Sulawesi is sister to *C. s. sulphureus* + *C. s. cunctator*: India to Sundaland (Müller *et al.*, 2010).

Tropidophorus grayi (Scincidae) of the Philippines (except Palawan and Mindanao) is sister to *T. baconi* of Sulawesi (Honda *et al.*, 2006).

Lophozosterops goodfellowi (Zosteropidae) of Mindanao and *L. squamiceps* of Sulawesi form a clade (Moyle *et al.*, 2009).

Penelopides (Bucerotidae), of the Philippines (except Palawan and the Sulu Islands) and Sulawesi, is sister to *Rhyticeros corrugatus* of Borneo, Peninsular Malaysia and Sumatra (Viseshakul *et al.*, 2011). (The only bucerotid on Palawan and the Sulu Islands is an unsequenced member of *Anthracoceros*, a genus also known from Borneo to Asia.)

Basilornis + *Streptocitta* + *Sarcops* (Sturnidae) of the Philippines (except Palawan), Sulawesi and Maluku Islands is sister to *Gracula* + *Ampeliceps* + *Mino* of India and Sri Lanka to the Solomon Islands (in the Philippines only on Palawan) (Lovette and Rubenstein, 2007).

Tadarida (*Mops*) *sarasinorum* (Molossidae) is in the Philippines and Sulawesi.

Crunomys (Muridae) of the Philippines and central Sulawesi (probably with *Sommeromys*: central Sulawesi) is sister to *Maxomys*: Laos to Borneo, Sulawesi and Java (Jansa *et al.*, 2006).

As Van Welzen *et al.* (2011) indicated, some groups define areas that include the Philippines–Sulawesi region and also Java, the Lesser Sunda Islands and Maluku Islands. Examples include:

Eumyias panayensis (Muscicapidae): Philippines, Sulawesi, Maluku Islands.

Ptilinopus melanospizus (Columbidae): Philippines (Palawan, Mindanao), Sulawesi, Java, Lesser Sunda Islands, Maluku Islands.

Aplonis minor (Sturnidae): Mindanao, Sulawesi, Java and Lesser Sunda Islands.

In fanged frogs (Dicroglossidae), the main clade of *Limnonectes* is an interesting example of Philippines–Sulawesi distribution. The phylogeny (from Evans *et al.*, 2003) is:

1. Northern and eastern Borneo, NW Sulawesi, **Mindanao** (*L. parvus, L. finchi, L. palavanensis*).
2. Indochina and Sundaland to Sulawesi and Maluku Islands (**not Philippines**).
3. **Philippines (9 species)–Sulawesi (13 species)**.

Clades 2 and 3 overlap in Sulawesi, but even there the distributions differ, as clade 3 occurs throughout the main island, while clade 2 occurs only in the east-central, west-central and south-east regions.

Clade 3 is endemic to the Philippines and Sulawesi and has the phylogeny:

South-eastern Philippines: Mindanao and all Visayas (*L. leytensis* complex).

South-western Philippines: Palawan and Mindoro (*L. acanthi*).

Northern and south-eastern Philippines, also Sulawesi (*L. magnus* etc.).

Evans *et al.* (2003) suggested an origin of the whole group 1–3 in Borneo, followed by an invasion of the Philippines, then Sulawesi, and back-invasion of the Philippines. Nevertheless, clades 2 and 3 are largely allopatric, as noted, and the three members of clade 3 are allopatric, apart from the overlap of two clades in the south-eastern Philippines. This is consistent with simple vicariance and with dispersal restricted to subsequent, local overlap in the south-east. The three regions occupied by the three clades do not correspond to the Pleistocene islands.

Philippines connections with areas further east

Wallace (1876) included the Philippines with Asia in his biogeographic system. Other authors have stressed eastern biogeographic connections, and have linked the Philippines with Australasia and the Pacific (Huxley, 1868). Examples of the eastern patterns are discussed next.

Philippines–Sulawesi and eastward

Many of the Philippines–Sulawesi groups discussed in the last section have sisters in Asia. Other Philippines–Sulawesi clades have sister-groups in Australasia. For example, fruit bats (Pteropodidae) include the following clade (Almeida *et al.*, 2011):

Harpionycteris: southern Philippines (not in Palawan or most of Luzon), Sulawesi, and Timor, + *Boneia* of Sulawesi.

Dobsonia: Philippines (Negros and Cebu) and Sulawesi to New Guinea, Bismarck Archipelago and the Solomon Islands, + *Aproteles*: New Guinea.

Most frog genera present in the Philippines are related to Asian groups. The exceptions include *Oreophryne* (Microhylidae subfam. Asterophryinae), with Australasian affinities. Its subfamily is present in the Philippines (Mindanao, Samar, Leyte), Sulawesi, Lesser Sunda Islands, New Guinea, New Britain, the Top End of Australia and northeastern Queensland. (All 18 genera occur in New Guinea; of these, *Oreophryne* extends to the west, and three genera extend to Australia.)

A similar distribution is seen in the cockatoos, Cacatuidae: Philippines, Sulawesi and the Lesser Sunda Islands, east to the Solomon Islands.

Seven genera in Myrtaceae make up the eucalypt clade. This ranges from the Philippines (Mindanao) and Sulawesi east through New Guinea, New Britain, Australia and New Caledonia, with fossils in New Zealand and South America (Gandolfo *et al.*, 2011).

The parrot subfamily Loriinae is in the Philippines, Sulawesi, New Guinea (most genera) and Australia east to the Marquesas Islands (Wright *et al.*, 2008).

Coracina mindanensis (Campephagidae) of the Philippines (not the western Visayas or Palawan) is sister to a clade of Sulawesi, Talaud Islands, Palau, New Guinea, eastern Australia and the Solomon Islands (Jønsson *et al.*, 2010c).

Gallirallus calayanensis (Rallidae) is endemic to the Babuyan Islands, north of Luzon. In different analyses its sister is a clade ranging from India to New Zealand (including the Philippines), or to a widespread Pacific clade (from Okinawa, the Philippines, Sulawesi and Australia east through the Pacific to the Cook Islands) (Kirchman, 2012). In either case, the endemism at the Babuyan Islands is local, but reflects the basal break in a widespread group.

Green tree skinks (*Lamprolepis*): does deep phylogenetic and geographic structure indicate the great powers of chance dispersal?

The green tree skink *Lamprolepis* has a single, variable 'species', *L. smaragdina*, which is widespread from the Philippines and Sulawesi to the Solomon Islands. It occurs in mangrove and coastal forest, and is also associated with human settlements, coconut plantations and other disturbed habitats. It is rarely seen in dense, primary rainforest.

In a detailed, well-sampled molecular study of *Lamprolepis*, Linkem *et al.* (2013) wrote that: 'Widespread species found in disturbed habitats are often expected to be human commensals. In island systems, this association predicts that dispersal will be mediated by humans'. Nevertheless, Linkem *et al.* (2013) concluded that this does not apply to *Lamprolepis* as it originated at 39 Ma and is composed of lineages with high levels of genetic differentiation. In addition, the groups are highly structured in their geography, with a jigsaw pattern of interlocking allopatry (Fig. 10.6). These results repeat the unexpected discoveries of deep geographic structure in the lizard *Gehyra* (Fig. 3.18) and other 'weeds' of anthropogenic habitats. Groups capable of surviving in disturbed environments such as mangrove and secondary forest often thrive in new habitats such as gardens and plantations, but this does not mean they are young groups. The authors wrote that the deep phylogenetic structure in *Lamprolepis* is 'contrary to expectations derived from its ecological preference' and it does contradict the traditional idea that weeds are recent groups with random, unstructured distributions.

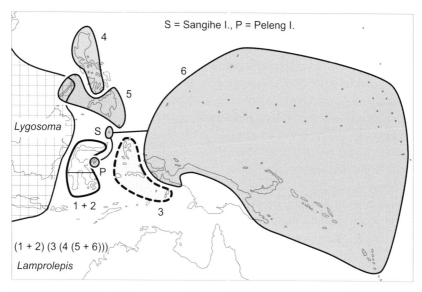

Fig. 10.6 Distribution of *Lygosoma* and *Lamprolepis* (Scincidae) (unsampled Taiwan and Timor populations not shown). (Linkem *et al.*, 2013; New Guinea distribution from www.nsdb.bishopmuseum.org.)

With respect to the dating of *Lamprolepis*, Linkem *et al.* (2013) wrote that the ages of various islands could provide ages for populations. They regarded this as problematic, though, and instead relied on a rough molecular rate for divergence dating. However, if the source of this rate is traced back it leads to Gübitz *et al.* (2000), who used the island ages in the Canary Islands to date an endemic gecko there. Thus the calculated date of 39 Ma for the *Lamprolepis/Lygosoma* split is a minimum, island-calibrated age.

The distribution of *Lamprolepis* and its boundary with its sister-group, *Lygosoma*, is shown in Figure 10.6. (*Lamprolepis* is also in Taiwan and Timor but these populations were not sampled by Linkem *et al.*, 2013, and are not mapped here.) *Lygosoma* extends to Africa and is based around the Indian Ocean while *Lamprolepis* is in the Pacific. The two genera are allopatric throughout their widespread Indo–Pacific range, except for minor overlap on Palawan (C. Linkem, personal communication, 17 November 2012). Linkem *et al.* (2012) wrote that the discovery of *Lygosoma* as the sister of *Lamprolepis smaragdina* is 'consistent with a western origin' of the latter, but it is also consistent with simple vicariance. The authors also suggested that 'The broad geographical range of these lizards provides clear evidence of the species' excellent dispersal capabilities'. Yet if the two genera originated by vicariance, the only range expansion in the genera themselves (not their ancestors) has been in a single island, Palawan.

The many interesting patterns shown by the clades include the standard Wallace's line break between the two genera, and the outliers of the Pacific clade of *Lamprolepis* on the Sangihe (Sangir) Islands and Peleng Island, off Sulawesi. These islands are not well known, but their small size belies their biogeographic importance. For example, Sangihe and Peleng (in the Banggai group) mark the eastern limit of non-human primates and they are also the only localities in the world where all the primates are tarsiers. In addition, the islands are the western limit for many Australasian and Pacific groups, such as the parrot genus *Eos* which extends from New Guinea west to the Sangihe Islands (not on mainland Sulawesi). The Sangihe Islands biota includes generic-level endemism, such as the monarchid bird *Eutrichomyias*, while Peleng Island endemics include the tarsier *Tarsius pelengensis*. Peleng and Sangihe together mark part of the western boundary in many diverse Australasian groups that are not on mainland Sulawesi, such as the frog *Platymantis* (discussed below).

Peleng and the Sangihe Islands lie along a system of thrust faults associated with the Molucca Sea plate boundary (Heads, 2012a; Fig. 5.16) and the Sangihe group forms part of a volcanic arc running between Sulawesi and the Philippines. The arc is caused by an active subduction zone and the endemics and others are traces of subducted ocean basins and their islands. Tarsiers are often recorded in mangrove and disturbed forest, and the biota of the Sangihe–Peleng strip can be interpreted as one of subduction zone weeds, including many local endemics. The populations have 'floated' on the surface of the geological maelstrom as a 'froth', with populations constantly colonizing new islands along the arc as they appeared.

Linkem *et al.* (2013) described the affinity of the Peleng and Sangihe *Lamprolepis* with the Pacific group (rather than with adjacent Sulawesi, Maluku Islands or Philippine populations) as 'unexpected' and 'most surprising from a geographical perspective'. This would be true if dispersal were a chance, stochastic process, but the repeated patterns at Sangihe and Peleng in frogs, birds and primates indicate that it is not. Despite the sharp genetic differentiation and strong geographic structure of the clades, Linkem *et al.* (2013) concluded that *Lamprolepis smaragdina* is a 'highly dispersive species . . . capable of long-distance dispersal, and has a clear history of over-water dispersal'. They did not cite any actual evidence for this, though. The authors inferred 'sweepstakes dispersal' and wrote that 'random waif dispersal has been a pervasive ongoing phenomenon throughout the evolutionary history of this species . . . stochastic dispersal will produce pronounced genetic differences among islands, and substantial geographical structure with clear historical signal'. But it is difficult to see how random dispersal would produce the highly structured clades seen in the group, or boundaries such as Peleng–Sangihe that are repeated in other, unrelated groups.

In the current molecular literature, more and more clades are being shown to have unexpected, high levels of geographic structure. In the traditional biogeographic model this is interpreted as evidence of the great powers of chance dispersal. However, in biology in general, unexpected, high levels of structure are not usually regarded as evidence for chance and the traditional approach is unproductive. If distribution really is due to chance, rather than normal biological and geological processes, there would be little point in mapping it in any detail or comparing it with other groups. Instead, the high levels of structure already found suggest that further investigation will reveal even more. A detailed field study of the Sangihe–Peleng groups and their boundaries with adjacent clades would be especially interesting.

Linkem *et al.* (2013) argued that 'Very few, if any, of the oceanic islands in Southeast Asia and the Pacific have a geological history of fragmentation that could serve as the basis of vicariant scenarios to explain current species distributions. Dispersal across marine barriers must therefore have occurred multiple times in *L. smaragdina*'. Nevertheless, populations on individual islands have been fragmented by processes such as volcanism, while at larger scales, vicariance of metapopulations on archipelagos has developed at plate margins and other large faults in the region. For example, the break in *Lamprolepis* between the Philippines (clade 5) and Palau Island (clade 6) correlates with the seafloor spreading in the Philippine Sea that formed the Philippine plate. At convergent margins in the region, archipelago populations have been fragmented during accretion by processes such as uplift of mountain chains and strike-slip movement. Outliers of Pacific groups at the Peleng–Sangihe islands, for example, can be explained by westward translation of Sulawesi and Maluku terranes, along with breaks at the thrust faults mentioned above.

Linkem *et al.* (2013) described the distribution of *Lamprolepis* and its clades as 'unique' and 'anomalous'. In its details, every group is unique, but the main breaks in *Lamprolepis* – at Wallace's line, at Sangihe–Peleng, between Mindanao–Palawan and the rest of the Philippines, and on both sides of the Maluku Islands – are all seen in many groups. Linkem *et al.* (2013) wrote: 'the convoluted patterns of relatedness that we have elucidated demonstrate that *L. smaragdina* has relied primarily on stochastic waif dispersal'. Yet while the distributions show interdigitation and 'folding', especially at Sangihe and Peleng, they are highly structured and no more complex than the regional tectonics.

Philippines sister to Australasia

Members of the Philippines orogen biota have close relatives in continental Asia, Taiwan and Sulawesi, as well as eastern connections with New Guinea and the rest of Australasia. These eastern connections are correlated with tectonic trends, as a 'deformed belt' (Pubellier *et al.*, 2003, Fig. 1) characterized by 'highly displaced ophiolites' extends from the Philippines to the Talaud Islands, northern Maluku Islands, Waigeo and northern New Guinea (Pubellier *et al.*, 2004, Fig. 7).

In the diverse subfamily of rats, Murinae, a sample (63 of 130 genera) had the following phylogeny (Rowe *et al.*, 2008):

Philippines (not Palawan, Zamboanga Peninsula or Samar) (*Phloeomys, Batomys, Carpomys, Musseromys, Crateromys*).

  Iran to Pacific islands (plus two species introduced worldwide) (*Rattus* and eight other sampled genera).

    Southeast Asia to Eurasia and Africa (18 sampled genera).

      Southeast Asia: Bangladesh to Java, Borneo and Palawan (*Chiropodomys*).

      Philippines (*Chrotomys, Rhynchomys, Archboldomys, Apomys*).

      New Guinea, Australia, Solomon Islands (26 sampled genera).

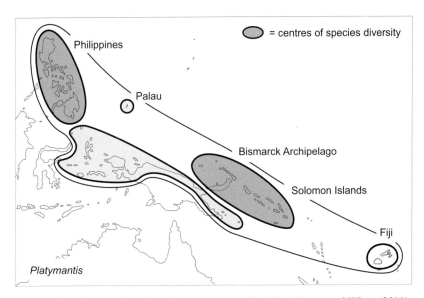

Fig. 10.7 Distribution of the frog *Platymantis* s.lat. (Ranidae) (Pyron and Wiens, 2011).

In this phylogeny, the Philippines has two diverse clades, both with interesting sister-groups. The first Philippines clade is sister to a widespread Old World group, mentioned already as it highlights the intercontinental significance of the Philippines. The second Philippines clade is sister to a diverse Australasian clade, the biogeographic connection that T.H. Huxley stressed.

Stick insects include similar distributions, with *Phasmotaenia* recorded from Lanyu Island, Luzon, Micronesia (Ponape and Truk), New Guinea (Mount Doorman), Solomon Islands and Fiji. Each locality has a single endemic species except the Solomon Islands (with seven species) and Luzon (with two) (Hennemann and Conle, 2009).

Philippines–Maluku Islands, and connections east of there

The frog genus *Platymantis* (Ceratobatrachidae) is well known to zoogeographers because, unlike most amphibians, it is diverse on oceanic islands (Fig. 10.7). It has two, well-marked centres of diversity and endemism, one in the Philippines with 27 species, and one in the Solomon–Bismarck–Admiralty group with 40 species (including *Batra-chylodes*, *Ceratobatrachus* and *Discodeles*; Pyron and Wiens, 2011). Outside these two centres of diversity, there are seven or eight *Platymantis* species in northern New Guinea, two to four in eastern Indonesia, two in Fiji and one in Palau (Siler *et al.*, 2009). The genus is not on mainland Sulawesi, but reaches a limit at Peleng and Sangihe Islands, as in the Pacific clade of the green tree skink *Lamprolepis* (IUCN, 2013). Many groups, such as birds of paradise, are diverse on mainland New Guinea but are absent on the Bismarck Archipelago; *Platymantis* is one of the groups that shows a contrasting pattern, with a great *increase* in species diversity in the archipelago.

Many endemic reptiles and frogs in the Philippines, including some *Platymantis* species, are restricted to forest on limestone karst (Siler *et al.*, 2009; 2010a). Local endemism in karst rainforest is also well known in places such as New Guinea. The Philippines frogs in this habitat live on the ground, not in trees, and in drier periods retreat to caves and crevices. As in New Caledonia, New Guinea and Fiji, limestone uplifted out of the sea has been immediately colonized by pioneer, coastal biota, but in many areas the land has then been uplifted further, 'trapping' endemics. Groups that can survive on limestone, ultramafics, or both, will be pre-adapted to survive on and around island arcs, both before and after they are accreted and integrated with other arcs, and even if the islands are reduced to atolls or uplifted to form mountains.

In other Philippines–Maluku Islands–Australasia groups, a clade of monitor lizards (*Varanus*: Varanidae) is present in north-eastern and southern Luzon (*V. bitatawa* and *V. olivaceus*) and sister to a clade (the *prasinus* group) of the southern Maluku Islands, New Guinea, Solomon Islands and Australia (Welton *et al.*, 2010).

The pigeon *Columba vitiensis* (Columbidae) occurs in the Philippines, Maluku Islands, Lesser Sunda Islands, New Guinea, east to New Caledonia, Lord Howe Island and Samoa (not Australia).

Philippines–Melanesia

Many groups show a direct disjunction between the Philippines and New Guinea, although the pattern is often neglected in biogeographic studies. Nevling (1961) described the disjunction in *Enkleia* (Thymelaceae) and wrote that it was 'puzzling indeed'. Vallejo (2011) stressed the importance of the Philippines–New Guinea connection for biogeography in general. In the palms of Malesia, peaks of species diversity occur in New Guinea and the Philippines (Baker and Couvreur, 2012). The palm genus *Heterospathe* exemplifies this connection; it is in the Philippines, Maluku Islands, New Guinea, New Britain, the Solomon Islands and Fiji.

In a similar case, the tree *Sararanga* (Pandanaceae) has two species, one in Luzon, and one in Japen Island (by the Bird's Head Peninsula), the north coast of New Guinea (Jayapura, Vanimo), Manus, New Ireland and throughout the Solomon Islands (Baker *et al.*, 1998). *Sararanga* is sister to the rest of the Pandanaceae, widespread through the warmer regions of the Old World (Buerki *et al.*, 2012).

In *Leea* (Vitaceae), 'clade 3' comprises *L. amabilis* of Borneo and its sister, a clade of 11 species endemic to the Philippines and New Guinea (Molina *et al.*, 2013).

The *Gallicolumba* group (Columbidae) is present throughout Southeast Asia and Australasia, and has its basal clades in the Philippines, Sulawesi and New Guinea (Fig. 10.8). These basal clades illustrate the deconstruction of the Philippines, with a Luzon clade related to one in Sulawesi, species in Mindoro, Panay and Negros related to a New Guinea group, and a separate clade on Greater Mindanao (Jønsson *et al.*, 2011b). As the authors concluded: 'it is possible that the diversification of *Gallicolumba* may in part have been shaped by the tectonic movements and corresponding extensive re-arrangements of land masses within the Philippine and Sulawesi region'.

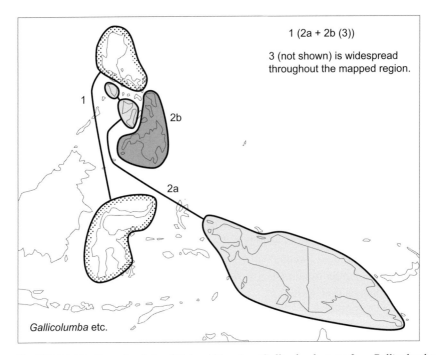

Fig. 10.8 Distribution of a clade of Columbidae. 1 = *Gallicolumba* p.p.; 2 = *Gallicolumba* p.p.; 3 = *Alopecoenas* etc. The phylogeny is indicated (Jønsson *et al.*, 2011b).

In other Philippines–New Guinea clades, Meredith *et al.* (2011) found a group of freshwater crocodiles structured as follows:

Northern Australia (*Crocodylus johnstoni*).

New Guinea (*C. novaeguineae*).

Philippines (*C. mindorensis*).

In the plant *Begonia*, a Philippines–New Guinea clade is sister to one from the Lesser Sunda Islands (Thomas *et al.*, 2012a).

Other examples of Philippines–New Guinea groups include the following (see also Heads, 2003):

The bird *Mearnsia* (Apodidae): Philippines and New Guinea.

The plant *Mearnsia* (Myrtaceae): Philippines, New Guinea, New Caledonia, and New Zealand (Wilson, 1996).

The *Rhagovelia novacaledonica* group (Hemiptera): Philippines (Mindanao), New Guinea and New Caledonia (Polhemus, 1996).

In morphological revisions, a keyed group in the plant genus *Vaccinium* (Ericaceae) shows an interesting parallel arc structure that requires further study (Fig. 10.9). The Philippines archipelago again appears to be a composite, with two separate affinities overlapping there. One connects Negros with the Papuan Peninsula; the other connects

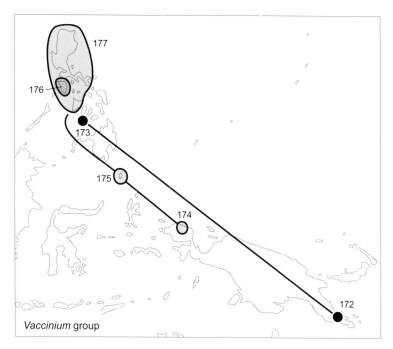

Fig. 10.9 Distribution of a keyed group in *Vaccinium* (Ericaceae). Key 119a: 172 *V. amphoterum*; 173 *V. banksii*. Key 119b: 174 *V. crenatifolium*; 175 *V. apophysatum*; 176 *V. woodianum*; 177 *V. cumingianum* (Sleumer, 1967; Heads, 2003, Fig. 63).

Luzon, Mindoro and Panay with the Talaud Islands and the Bird's Head Peninsula. The Talaud Islands include Eocene ophiolites and are part of the deformed belt that runs from the Philippines to the Bird's Head Peninsula.

Many groups connect the Philippines with New Guinea and the Mariana Islands. An example is the tree *Schleinitzia* (Fabaceae), in the Philippines, the Marianas and New Guinea to the Society Islands (not Australia). The genus is in a well-supported clade with *Kanaloa* of Hawaii and *Desmanthus*: warm America and the Caribbean (Lewis *et al.*, 2005). This is another example of a trans-tropical Pacific group (cf. Apocynaceae and others, cited on pp. 106–115).

Pteropodidae (fruit bats)

As with the birds and plants in the last two figures, fruit bats illustrate the deconstruction of the Philippines. The three allies of *Pteropus* in the Philippines – *Styloctenium*, *Acerodon* and *Desmalopex* – show Philippines–Sulawesi and Philippines–Melanesia connections (Fig. 10.10; Almeida *et al.*, 2011). The three genera have quite different distributions in the Philippines and their affinities outside the Philippines also have different ranges. This is compatible with the fusion of arc faunas owing to tectonic arc integration. In a parallel with the Philippines–New Guinea pattern (Fig. 10.9), *Desmalopex* and allies show a connection between the Philippines and the Bismarck Archipelago/Solomon Islands. Similar disjunctions are documented in plants such as

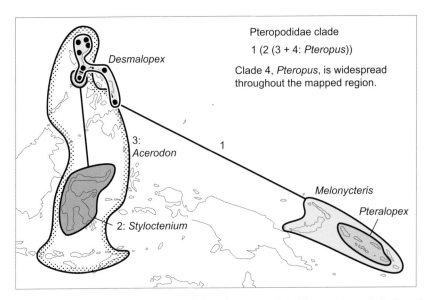

Fig. 10.10 Distribution of a clade of fruit bats (Pteropodidae). The phylogeny is indicated (Almeida *et al.*, 2011).

the *Schefflera* clade 'Elmeria' (Araliaceae), endemic to the Philippines, Sulawesi and New Britain (Frodin *et al.*, 2010).

In other Australasian fruit bats, *Dobsonia* is in the Philippines (Negros and Cebu) and Sulawesi, and extends east to the Solomon Islands. *Nyctimene* has a similar range in the Philippines (Negros, Cebu and Sibuyan), Sulawesi and east to the Solomon Islands; its sister-group is *Notopteris* of New Caledonia, Vanuatu and Fiji.

Blind snakes (Typhlopidae)

In blind snakes, Wallach *et al.* (2007) reported a 'strikingly distinct' new species of *Acutotyphlops* from Luzon. The genus is otherwise known only from 4000 km away, at the eastern extremity of Papua New Guinea (Alotau), the eastern Papua New Guinea islands and the Solomon Islands (Fig. 10.11). The authors described the range as 'peculiar' and suggested that it constitutes 'a curious new systematic and biogeographical problem'. The disjunct range is dramatic, but similar connections skirting northern New Guinea are recorded in Pteropodidae (Fig. 10.10) and in several plant groups (Heads, 2003, Figs. 18, 51, 63, 64, 125). Wallach *et al.* (2007) also noted that the *Acutotyphlops* pattern can be compared with that of *Platymantis* (Fig. 10.7).

Diversity centres in cowries

The cowries (Cypraeidae) are a diverse, well-known group of marine gastropods. Their highest species diversity is in the Indo–Australian archipelago, from Borneo to Fiji. Within this broad region, the group's maximum diversity peaks in two disjunct localities: the Philippines (plus Taiwan) and outer Melanesia (Bismarck Archipelago, Solomon Islands, Vanuatu and New Caledonia) (Bellwood *et al.*, 2012). Both regions lie along

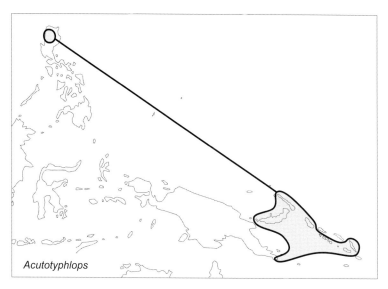

Fig. 10.11 Distribution of the snake *Acutotyphlops* (Typhlopidae) (Wallach *et al.*, 2007). The sister-group, *Ramphotyphlops*, is widspread throughout the mapped region.

plate margins that have been active as zones of accretion and ophiolite obduction. The concentration of cowries and many other groups along the margin suggests that the great diversity was built up by biological accretion during subduction (cf. Marshall, 2001). Later the Pacific margin, and with it the Philippines–Melanesia region of high diversity, has been disrupted by outward growth of the Pacific plate.

Philippines–Solomon Islands

In the damselfly family Platycnemididae, morphological studies indicated a disjunct Philippines–Solomon Islands clade, with the gap filled by the sister-group in central New Guinea (Fig. 10.12; Gassmann, 2005; Orr and Kalkman, 2010).

A similar disjunction occurs in the beetle *Allorthorhinus* (Curculionidae), present in the Philippines, the Solomon Islands (the main group and also the Santa Cruz Islands) and Fiji (morphological studies of Kuschel, 2008).

Philippines–Micronesia

Platystictidae are damselflies in which most species have localized distributions in rainforest. Van Tol *et al.* (2009) wrote that 'The occurrence of a species of *Drepanosticta* on Palau, presently ~800 km east of Mindanao, presumably the nearest founder population, is [an] enigma . . .'. (For the location of Palau, north of the Bird's Head Peninsula, see Figure 10.7.)

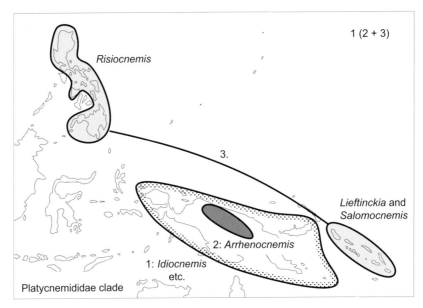

Fig. 10.12 Distribution of a clade in Platycnemididae. The phylogeny is indicated (Gassmann, 2005; Orr and Kalkman, 2010).

In other links with Micronesia, the passerine *Sterrhopteris* (= *Stachyris* p.p.) (Zosteropidae) is in the Philippines (not Palawan or Mindoro) and is sister to *Cleptornis* of the Mariana Islands, north-east of Palau (Moyle *et al.*, 2009).

The Philippines form one margin of the Philippine Sea plate, and the opposite margin is formed by the Mariana–Yap trenches (Palau lies near the latter). The plate has developed since the Eocene to form the largest back-arc basin on the eastern margin of Asia (Yin, 2010). The distributions of *Drepanosticta*, *Sterrhopteris* + *Cleptornis* and many others suggest that the seafloor spreading that opened the basin also ruptured biological clades.

Palau has already been mentioned several times in this book as a distribution limit, and in addition it includes many endemics; for example, *Protanguilla*. This fish is known only from a reef on Palau and is sister to all other Anguilliformes (eels) (G. D. Johnson *et al.*, 2012). The authors' molecular clock study dated the origin of *Protanguilla* to ~200 Ma, much older than the Palau–Kyushu Ridge itself, which formed as an island arc at ~60–70 Ma. The clade is older than the geomorphic structure to which it is endemic, a pattern shown in many groups.

Other groups disjunct across the Philippine Sea plate include a clade of *Eutropis* (Scincidae) endemic to the Philippines and Palau (Mausfeld and Schmitz, 2003). In morphological studies, the plant *Badusa* (Rubiaceae) occurs in the Philippines, Palau, Biak Island (off north-western New Guinea), the Solomon Islands, Vanuatu, Fiji and Tonga (Jansen, 1984). (Apart from the Palau record this repeats the pattern of the palm *Adonidia*; Fig. 9.11.) *Badusa* belongs to the Catesbaeae–Chiococceae complex, distributed in the Philippines, Micronesia, Melanesia (including mainland New Guinea and New Caledonia) and throughout tropical America (Motley *et al.*, 2005). The distribution conforms to a standard trans-tropical Pacific pattern (cf. Pachychilidae; Fig. 10.4, also Figs 8.6–8.9).

The *Sphenomorphus* group (Scincidae): a clade illustrating the multiple connections of the Philippines

The *Sphenomorphus* group of forest skinks is widespread from Burma and China, via Australasia and the Pacific islands, to southern Mexico and Central America. Philippines members belong to six separate clades that are named, or will be named, as separate genera (Linkem *et al.*, 2011). Each of the Philippines clades is also found in different areas outside the country, as follows (reading from west to east):

> *Tytthoscincus atrigularis* is 'nested within a clade of species [otherwise known] from Borneo, Sulawesi, and peninsular Malaysia' (Linkem *et al.* (2011).
> *Sphenomorphus variegatus* is nested 'within a clade of [otherwise] Bornean species'.
> *Lipinia noctua* belongs to a genus that is widespread in Southeast Asia and the Pacific islands.
> *Insulasaurus* is sister to *Lipinia*: Southeast Asia–Pacific, + *Papuascincus* of Papua New Guinea.
> *Sphenomorphus fasciatus* (greater Mindanao) is nested within a clade of species also known from Palau, Papua New Guinea and the Solomon Islands.
> *Sphenomorphus diwata* + *S. acutus*, *Pinoyscincus*, *Otosaurus* and *Parvoscincus*, all from the Philippines, belong to a diverse clade that is otherwise endemic to New Guinea and Australia.

Linkem *et al.* (2011) concluded: 'our results unequivocally demonstrate that the complex southern and western Philippine communities of forest skinks are assembled from multiple regions of South-East Asia and the Papuan realm'. The pattern is also consistent with the clades representing old groups that the modern Philippines archipelago has inherited with its component terranes from the west and the east.

Distribution within the Philippines

The Philippine mobile belt is a collage of disparate, fault-bound blocks that amalgamated in the Paleogene before colliding with Eurasia (Sundaland) in the Miocene. Rangin and Pubellier (1990) accepted 16 blocks. These are six in Luzon (Sierra Madre Oriental, Angat, Zambales, Central Cordillera, Bicol Peninsula and Catanduanes Island), four in the Central Philippines (Mindoro, Panay, Cebu and Bohol) and six in Mindanao (Pacific Cordillera, Surigao, Pujada Peninsula, Central Cordillera, Daguma Range and Zamboanga Peninsula). Luzon, Mindanao and the Visayas are geological composites, and considerable strike-slip movement has occurred among the component blocks throughout the Cenozoic. The main Philippine fault is shown in Figure 10.2, but there are many other strike-slip faults. Many of these features related to the local distributions.

The Pleistocene island model: distributions and divergence dates that are incongruent with it

Heaney (1986) suggested that 'The limits of late Pleistocene islands, defined by the 120 m bathymetric line, are highly concordant with the limits of faunal regions'. The Pleistocene islands (Fig. 10.1) do delimit areas of endemism, but there are many other areas of endemism in the Philippines and these can be larger or smaller than the Pleistocene islands. For example, *Ansonia* (Bufonidae) in the Philippines is restricted to Mindanao, and Sanguila *et al.* (2011) cited its 'curious absence' from the northern islands of the Greater Mindanao Pleistocene island. Yet this distribution pattern is a common one and was described as 'curious' only because it does not conform to the Pleistocene model.

The larger Philippines islands all 'decompose' into smaller centres of endemism; this is one of the reasons for the great diversity in the country. For example, in Luzon the Bicol Peninsula is a separate terrane and centre of endemism, as is the Zamboanga Peninsula in Mindanao. In this way, the islands themselves are biogeographic as well as tectonic composites. Other areas of endemism in the Philippines include parts of two or more different Pleistocene islands and, again, these are incongruent with the traditional model.

Some Philippines groups are endemic to Pleistocene islands but have their closest relatives outside the Philippines, not on other Pleistocene islands. This is true for groups in Palawan and Greater Mindanao (*Dicrurus*, Fig. 10.5), Luzon (*Gallicolumba*, Figs. 10.8; *Acutotyphlops*, Fig. 10.11), and Mindoro, Panay and Negros (*Gallicolumba*, Fig. 10.8; *Vaccinium*, Fig. 10.9; *Styloctenium*, Fig. 10.10). In other examples, the fruit bats *Dobsonia* and *Nyctimene* (Pteropodidae) both extend from the Philippines to Sulawesi and east to the Solomon Islands, but in the Philippines *Dobsonia* is restricted to Negros and Cebu, while *Nyctimene* is only on Negros, Cebu and Sibuyan. The snake *Oligodon modestus* (Colubridae) is endemic to the western Visayas (Panay, Negros) and the Romblon group, and is sister to an Asian clade (Pakistan to Borneo) (Green, 2010). These examples are consistent with an independent geological/biological history of the different crustal blocks that make up the Philippines.

Pre-Pleistocene evolution

As shown, distributions of clades raise questions about the Pleistocene island model of speciation, and this model has also been contradicted in work on chronology. Many clock studies have indicated that clades endemic to one of the Pleistocene islands diverged before the Pleistocene. Most of these studies have used fossil-calibrated clocks, and so the dates are valid as estimates of minimum ages, assuming a rough evolutionary clock. These minimum ages can be used to rule out earlier events as relevant; problems only arise when they are used to rule out earlier dates. For the Indo–Australian archipelago as a whole, Lohman *et al.* (2011) concluded that the 'Present distribution patterns of species have been shaped largely by pre-Pleistocene dispersal and vicariance events . . . '. For example, species of spiderhunters (*Arachnothera*: Nectariniidae) showed substantial differentiation at mitochondrial loci (Moyle *et al.*, 2011). The authors concluded that 'the forces behind the diversification of major groups of spiderhunters . . . are ancient and not the result of recent geological and environmental events such as the eustatic sea-level changes caused by Pleistocene glaciation'.

Esselstyn and Brown (2009) noted that changing Pleistocene connections and disconnections among modern islands were proposed in early work as an important factor for evolution in the Philippines (Dickerson, 1928; Kloss, 1929; Delacour and Mayr, 1946). Yet in *Crocidura* they found that some divergence dates almost certainly pre-date the last glacial maximum. They commented that:

Because the role of Pleistocene geography has long been recognized, there is a risk that taxonomic decisions could have been based in part on PAIC [Pleistocene aggregate island complexes], and PAIC importance then inferred from taxonomy, thereby resulting in an over-emphasis of the importance of Pleistocene sea-level fluctuations . . . Overall, our results suggest Pleistocene sea-level fluctuations have been an important, but not dominant factor shaping shrew diversity . . . (Esselstyn and Brown, 2009: 171).

Molecular dating in the lowland Philippines birds *Ixos philippinus* and *Periparus elegans* placed major diversification events before the Pleistocene (Jones and Kennedy, 2008a). The four other species that the authors studied have inter-island genetic distances similar to *P. elegans*, 'suggesting similar time scales for these species' colonization events'. Jones and Kennedy (2008a) rejected Pleistocene geography as a major cause of diversification in the Philippines.

Case-studies of groups pre-dating the Pleistocene

The fruit bat *Haplonycteris* (Pteropodidae) is endemic to the Philippines and previous treatments have recognized a single species. Nevertheless, Roberts (2006) found that 'Each Pleistocene island complex had a single resident monophyletic lineage; these five groups were separated by approximately 6–8% sequence divergence and apparently have been diverging for 4–6 Myr'. (This is a fossil-calibrated, minimum age.) Thus, 'most divergence among lineages greatly predates the Pleistocene . . . in some cases, multiple allopatric clades were present on single islands . . . '. Roberts (2006) recovered the following phylogeny for the genus:

Mindoro (Greater Luzon (Panay + Negros)).
Sibuyan in the Romblon group (Greater Mindanao).

The distribution of the two main clades is not explained by Pleistocene geography. With respect to divergence dates, the Sibuyan clade was dated at 3.9–6.5 Ma, older than the island itself (2 Ma; Steppan *et al.*, 2003) and much older than Pleistocene sea level changes.

In passerines, the populations of *Turdus poliocephalus* (Turdidae) from the Philippines do not form a monophyletic group. Instead, clades from the southern Philippines are closer to forms from Borneo, Sulawesi and Melanesia than to those from northern Philippines (cf. *Dicrurus*, Fig. 10.5; Jones and Kennedy, 2008b). The phylogeny is:

Northern Philippines: Luzon and Sibuyan.

Central Philippines: Mindoro.

Southern Philippines: Panay, Negros and Mindanao. Also Sulawesi, Borneo, Solomon Islands and Vanuatu.

The pattern is not consistent with Pleistocene island vicariance, and Jones and Kennedy (2008b) proposed a double invasion of the Philippines. As an alternative, the pattern could reflect *in-situ*, mid-Cenozoic differentiation, first around Luzon and then in the Visayas.

In their study of the *Sphenomorphus jagori* complex (now *Pinoyscincus*; Scincidae), Linkem *et al.* (2010b) wrote: 'the two large clades revealed in our phylogenetic analysis strongly contradict predictions derived from the PAIC model . . .'. Of the two clades, the *S. abdictus aquilonius* + *S. coxi divergens* clade is on Luzon, Mindoro and Panay, while the *S. jagori* group is widespread in the Philippines, and has a basal clade on Polillo Island and Bicol Peninsula. Linkem *et al.* (2010b) concluded: 'our findings contribute to a growing body of literature that demonstrates the tendency of empirical data to deviate markedly from a widely accepted quarter-century-old theoretical paradigm on the dominating processes of evolutionary diversification [in the Philippines]'.

The shamas (a clade of *Copsychus*: Muscicapidae) have already been mentioned with reference to their overall biogeography. The distribution and the phylogeny of the Philippines members (from Lim *et al.*, 2010) are as follows:

Luzon, Panay, Negros (*C. luzoniensis*).

Palawan (*C. niger*) and Cebu (*C. cebuensis*).

India and Sri Lanka to Borneo and Java (*C. malabaricus*).

The first break has occurred in the central Philippines, around the Visayas and Tañon Strait, between Negros and Cebu. The second has developed between Palawan and Borneo. The bee *Apis cerana*, also cited above (Smith *et al.*, 2000), has a similar pattern:

Luzon, Negros, Leyte, Mindanao.

Sulawesi, Sangihe ('*A. nigrocincta*').

Palawan, Cebu, Mindanao.

India, Nepal, Korea, Japan, Indochina, Borneo, Java, Lesser Sunda Is.

The main difference between this group and the shamas is the interpolation of the Sulawesi clade, but the first node, at Tañon Strait etc., and the third node, between Palawan + Cebu, and Borneo, occur in similar localities in both groups. The break at Tañon Strait – a locality well known for its diverse marine mammals – is not predicted in the Pleistocene island model.

In the fantail genus *Rhipidura* (Rhipiduridae), a clade of Luzon (*R. cyaniceps*), Tablas in the Romblon group (*R. sauli*), and Panay, Negros and Masbate (*R. albiventris*), is sister to a clade in Samar, Leyte, Bohol (*R. samarensis*) and Mindanao (*R. superciliaris*). This whole Philippines group (not in Palawan or Mindoro) is sister to a montane clade in

Mindanao east of the Zamboanga Peninsula (Sánchez-González and Moyle, 2011). The authors found that, while some of the distributions are consistent with the Pleistocene islands model, substantial structure within the PAIC clades is not. The distributions in some clades are equivalent to Pleistocene islands, whereas others span multiple Pleistocene islands or only a portion of an island. In addition, the deep divergence among species 'probably predates the late Pleistocene sea level changes'. Overall, Sánchez-González and Moyle (2011) concluded that the differentiation is 'not consistent with the PAIC paradigm'; the biogeography is complex and 'the PAIC paradigm may not be enough in many patterns'. Instead, they proposed 'multiple processes at several time scales . . . The complexity of the patterns indicates that additional processes influencing diversity should be investigated'.

Studies of *Cyrtodactylus* geckos also found results that contradict the Pleistocene island model (Siler *et al.*, 2010). For example, the main break in the Philippines is between Palawan, Mindanao, Bohol, Cebu and the islands north of here. Again, the break is at Tañon Strait, between Cebu and Negros, not at the Pleistocene break between Cebu and Leyte. Likewise, the bird *Coracina coerulescens* (Campephagidae) of Luzon and Cebu is sister to *C. ostenta* of Panay and Negros (Jønsson *et al.*, 2010c), with the break at Tañon Strait. Siler *et al.* (2010) concluded: 'Contrary to many classic studies of Philippine vertebrates, we find complex patterns that are only partially explained by past island connectivity'. For example, greater genetic diversity is found within islands than between them. 'Among the topological patterns inconsistent with the Pleistocene model, we note some similarities with other lineages, but no obviously shared causal mechanisms are apparent'. Their results contribute to 'a nascent body of literature suggesting that the current paradigm for Philippine biogeography is an oversimplification requiring revision' (cf. Vallejo, 2011).

Brown and Diesmos (2009) discussed recent work that 'threatens to topple the prevailing view of diversification in the Philippines'. They cited an 'emerging consensus' that suggests Philippine biodiversity is 'far more complex than previously thought'. Brown and Diesmos proposed that ancestors of today's endemic species became isolated on palaeo-islands 'several to many tens of millions of years before present . . . These events appear to have given rise to deep phylogenetic diversity and the presence of numerous "old endemics" . . .'. Brown and Diesmos (2009) also interpreted microendemism on small islands and speciation within larger islands as undermining the 'PAIC-centric tradition'. Other studies have blurred the traditional dichotomy between 'landbridge' islands formed on continental crust (Palawan) and the other Philippines islands formed on oceanic crust.

The molecular clock studies and the broader biogeographic distributions both suggest that the phylogenetic breaks that do coincide with Pleistocene channels are because of earlier tectonic breaks. For example, there was a Pleistocene gap between Greater Luzon and Greater Mindanao, but the two regions have had a different tectonic history and both developed far apart from each other. Thus the 120-m contour and the distributions both have a common cause – the tectonic history – rather than the first causing the second. The 120-m contour is an aspect of present-day geography, while the distributions have developed instead on a very different, prior geography. The groups are probably older than their current islands and could have evolved on earlier islands along earlier plate margins in the vicinity. The Pleistocene islands and the map of the present are useful for descriptive purposes, but do not provide an adequate framework for biogeographic analysis. Instead, this can focus on the clade distributions.

The Palawan terrane

Unlike most of the Philippines, northern Palawan has developed on a block of continental crust, the Palawan terrane. This rifted from Indochina/South China in the mid-Cretaceous (Yumul *et al.*, 2009a; Metcalfe, 2010) or in the Oligocene (Hall, 2002; Hall *et al.*, 2009). The Palawan terrane underlies Palawan, southern Mindoro, western Panay, the Romblon group, Zamboanga Peninsula and southern Mindanao (Yumul *et al.*, 2008).

Many Philippines groups are widespread through the modile belt but absent from Palawan, and many of these have sisters on Palawan + Sundaland (see above). Thus biologists have often considered Palawan to be a biogeographic extension of Borneo. In the traditional model, invasions of the oceanic Philippine islands from the Sunda Shelf have occurred along a western arc: Palawan–Mindoro, and an eastern arc: Sulu Islands–Mindanao–Samar–Leyte–Luzon (Brown and Guttman, 2002). Nevertheless, Linkem *et al.* (2010a) regarded Palawan as 'biogeographically enigmatic', and Brown and Guttman (2002) concluded that 'Palawan is not at all the simple extension of north Borneo (or the Sunda Shelf) as the prevailing zoogeographical perspective would suggest....'. Esselstyn *et al.* (2010) showed that, while some groups link Palawan with Borneo, others link it with the Philippines. In addition, these authors cited Palawan taxa (*Crocidura batakorum*: Soricidae, and *Kaloula baleata*: Microhylidae) that have their sister-groups in Sulawesi, and one (*Bronchocela cristatella*: Agamidae) with its sister-group in the Philippines plus Sulawesi. All these suggest that Palawan is more than just a landbridge between northern Borneo and the central Philippines.

The squirrel *Sundasciurus* (Sciuridae) shows a distinctive pattern; the Palawan clade (five allopatric species) is sister to a clade from Greater Mindanao, Borneo, Sumatra and Peninsular Malaysia (den Tex *et al.*, 2010). This indicates that the Palawan clade is not directly related to either the Philippines or to Sundaland, as proposed in dispersal models, but to both. The diversity of the Palawan clade suggests that its endemism is not just the result of extinction elsewhere. Den Tex *et al.* (2010) noted that its estimated age (6.9 Ma, a minimum) is older than Palawan itself (5 Ma).

Other interesting Palawan endemics include the peacock-pheasant *Polyplectron napoleonis* (Phasianidae). This is sister to the remaining species of the genus, which range from Bangladesh to Borneo (see p. 370).

As mentioned already, the frog *Barbourula* (Bombinatoridae) of Palawan and south-western Borneo is sister to *Bombina* of Europe to south-eastern China. The North Palawan terrane rifted from Indochina/South China either in the Cretaceous or Paleogene, and this would account for the divergence of the two frog genera. Likewise, Siler *et al.* (2012a) suggested the Philippines endemic clade of *Gekko* species originated when Palawan separated from Asia, that the group survived on the Palawan 'raft', and that it later diversified in the oceanic Philippine islands. The authors concluded: 'Our results reveal the need to consider deeper time geological processes and their potential role in the evolution of some Philippine terrestrial organisms'.

The Palawan terrane shows geological affinities with other continental fragments further east in the Philippines (Fig. 10.2) and these affinities help to account for the following patterns.

Palawan–Mindoro

Geologists have often regarded the North Palawan and Mindoro terranes as closely related (Yumul *et al.*, 2009b) and there are also biological connections; for example, the frog *Leptobrachium tagbanorum* of Palawan is sister to *L. mangyanorum* of Mindoro (Brown *et al.*, 2009b).

Palawan–western Visayas

Examples of this pattern include:

> *Insulasaurus* (Scincidae): Palawan–western Visayas (Panay, Negros, Masbate) (Linkem *et al.*, 2011). Its sister-group (*Lipinia* and *Papuascincus*) is widespread from Burma to the Pacific Islands, which is incompatible with an origin of *Insulasaurus* simply by dispersal from Borneo.
> *Prionailurus* (Felidae): Siberia to the Lesser Sunda Islands, in the Philippines only on Palawan and Negros.
> *Copsychus* (Muscicapidae) clade: Palawan (*C. niger*) and Cebu (*C. cebuensis*) (Lim *et al.*, 2010).

Palawan–Mindanao etc.

Examples include:

> *Mops sarasinorum* (Molossidae): Palawan, south-central Mindanao (west of the western fault on the island shown in Figure 10.2), Sulawesi (Heaney *et al.*, 2010).
> A clade in the bee *Apis*: Palawan, Cebu, Mindanao (Smith *et al.*, 2000).

The Visayas

The western Visayan islands of Panay, Negros, Cebu and Masbate formed a single island in the Pleistocene, but this does not explain the important break at Tañon Strait between Cebu and Negros that has already been mentioned. In other cases of this break Diesmos *et al.* (2002) cited squamates such as *Brachymeles cebuensis* and *Typhlops hypogius* that are present on Cebu but not Negros or Panay.

The ophiolites of the Philippines are shown in Figure 10.3. Yumul *et al.* (2008) recognized a distinct central belt that includes Cebu and Bohol (but not Panay or Negros), along with parts of central Mindanao and north-western Luzon. This might be relevant to the biogeographic break between Cebu and Negros at Tañon Strait.

Siler and Brown (2010) described another pattern deconstructing the Pleistocene island of Greater Panay/Negros in the skink *Brachymeles* (Scincidae): 'Although we expected Masbate populations to be more closely related to Visayan (Negros + Panay)

populations [Pleistocene island model], all analyses strongly supported the sister relationship between *B. tungaoi* (Masbate) and *B. kadwa* (Luzon)...Comparison of the
systematic affinities of other Masbate species may provide interesting exceptions to the
prevailing PAIC-oriented perspective of Masbate as a faunistic extension of the central
Visayas'. (In Siler *et al.*, 2011, *B. tungaoi* is 'sp. A' and *B. kadwa* is 'sp. B'.)

Mindanao

Mindanao, Samar, Leyte and Bohol were linked in the Pleistocene as the island Greater
Mindanao. This is an area of endemism for groups such as the mammals *Cynocephalus* (Dermoptera), *Exilisciurus* (Sciuridae) and *Carlito* (Tarsiidae). Nevertheless,
these groups are probably much older than the Pleistocene. For example, *Exilisciurus*
is sister to the clade *Callosciurus + Sundasciurus + Tamiops + Dremomys*, distributed
from Nepal to Borneo (Steppan *et al.*, 2004). *Sundasciurus* itself has a phylogeny:
Palawan (Greater Mindanao (Sundaland)), indicating two different phases of differentiation in Sciuridae around Greater Mindanao.

   In other Mindanao groups, *Podogymnura* (Erinaceidae) is endemic to eastern Mindanao and is sister to *Echinosorex* of Peninsular Malaysia, Sumatra and Borneo (Grenyer
and Purvis, 2003). *Urogale* (Tupaiidae) of Mindanao (central Mindanao and Zamboanga
Peninsula) is sister to *Tupaia*, of India to Java, Borneo and Palawan (Olson *et al.*, 2005).

   Within Mindanao, the frog genus *Ansonia* (Bufonidae) forms a group with three basal
clades in the west (two on Zamboanga Peninsula), and a clade with five subclades in
the east (Sanguila *et al.*, 2011). One of the eastern groups occurs throughout the strip
of Mindanao east of the Philippine fault (Fig. 10.2), that is, the Pacific Cordillera block
(Rangin and Pubellier, 1990).

Strike-slip displacement in the Philippines

Many distributions in the Philippines cannot be explained by Pleistocene events and
authors have suggested that earlier tectonic history has been important. Hall (2002: 380)
wrote: 'the Philippines are composed of a collage of terranes, of uncertain origin, and
it is difficult to specify their relationship with one another. These terranes have been
fragmented and moved along strike-slip faults...'. Karig (1983) wrote that strike-slip
displacement has 'played a very important, if not dominant, role' in the assembly of the
Philippines, disrupting, transporting and amalgamating Philippines terranes throughout
much of Cenozoic time (Karig *et al.*, 1986). The possible effects of this strike-slip
movement on biological distributions have not been examined, but they would explain
several 'leapfrog' disjunctions that are otherwise puzzling.

   Although strike-slip movement in the Philippines has had profound effects, details are
still uncertain. Hall (2002) wrote: 'in almost all cases it is very difficult to find evidence
for the amounts of movement on these faults even over very short and recent periods of
time'. Nevertheless, for the main Philippine fault, long-term, left-lateral fault slip rates
of 24–40 mm/year have been inferred (Yu *et al.*, 2013).

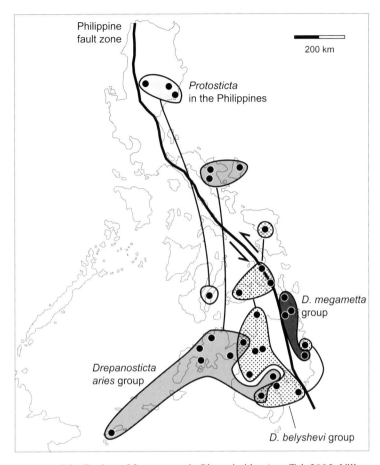

Fig. 10.13 Distribution of four groups in Platystictidae (van Tol, 2005; Villanueva and Gil, 2011).

Small-scale biogeographic patterns are best studied using groups in which the clades are abundant, conspicuous and diverse, and have localized distributions. They also need to be well collected and documented with detailed dot maps. The few Philippines groups that fulfil all these requirements include the damselflies in the family Platystictidae. These live mainly around water courses in rainforest, and have a typical trans-tropical Pacific distribution, in Southeast Asia (from India to the Solomon Islands) and tropical America (Central America and north-western South America) (van Tol *et al.*, 2009). As yet, no molecular studies have been published for the family, but the informal groups used by van Tol (2005) for the Philippines species have intriguing distributions in disjunct, parallel arcs (Fig. 10.13). The gaps in the Visayas seen in these groups are not absolute, as they are filled by related groups (Fig. 10.14). The gaps are spatially related to the Philippines fault zone, which developed during the Miocene accretion of the Philippine composite arc. The gaps are consistent with the large-scale, sinistral (left-lateral) strike-slip of the fault zone that has occurred since the Miocene. Fault-slip of ~24–40 mm/year since the origin of the fault in the mid-Miocene (15 Ma) would give a total displacement

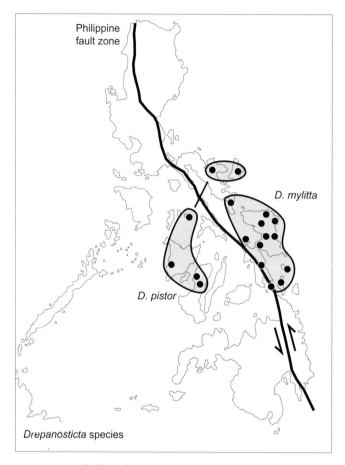

Fig. 10.14 Distribution of two species in *Drepanosticta* (van Tol, 2005).

of ~360–600 km, equivalent to the observed disjunctions. (The smaller disjunctions in the *D. belyshevi* group and in *D. pistor* could reflect the different prior distributions of these clades.)

Other disjunctions associated with the fault include *Villaria* (Rubiaceae), a genus of shrubs or small trees found mainly in coastal forest. The genus shows a 500-km disjunction between north-eastern and south-central Mindanao, and the break corresponds with the fault (Alejandro *et al.*, 2011). In Luzon, an equivalent 600-km sinistral disjunction separates a clade of eight *Gekko* species present on the Babuyan Islands – central eastern Luzon, and also on the Romblon Islands (Siler *et al.*, 2012a).

While this faulting can explain the deformation of the groups' distributions, along with local speciation, it does not account for the origin of the groups or their mutual allopatry. This has developed before strike-slip movement occurred on the fault during Miocene accretion events. Van Tol *et al.* (2009) acknowledged that most islands in Southeast Asia consist of amalgamations of island arc and continental fragment terranes that have coalesced since the Late Cretaceous. Considered as a whole, these terranes have provided a subaerial environment for millions of years. Van Tol *et al.* (2009: 37) concluded:

... floral and faunal elements that dispersed to these fragments have been able to survive, and evolve in isolation. The composition of the fauna of such islands as Mindanao, Sulawesi, and especially New Guinea, may thus be a mixture of clades that became separated up to 40 to 50 million years ago, and re-assembled on one island only ten to twenty million years ago.

A note on marine diversification in the Philippines

Most of the examples cited in this chapter are terrestrial, but molecular work is also showing interesting patterns in marine groups. For example, the marine red alga *Portieria* was thought to be represented on coral reefs in the Philippines by one to three species, but a molecular survey found 21, with distributions that are highly structured (Payo *et al.*, 2013). Most species are restricted to particular island groups. The narrow species ranges and high levels of endemism within the archipelago contrast with the widely held belief that most marine organisms have large geographical ranges. The work on *Portieria* shows that speciation in the marine environment may occur at spatial scales smaller than 100 km, comparable with some terrestrial groups.

Appendix: eastern and western Philippines tracks

The following examples document the eastern and western distribution patterns that underlie much of Philippines biogeography.

Eastern Philippines track

Many groups occur along the eastern part of the Philippine mobile belt and these often have a vicariant in the western Visayas. For example, the owl *Otus megalotis* has two clades, one in Luzon and Greater Mindanao, the other in Panay and Negros (Miranda *et al.*, 2011).

The pycnonotid bird *Ixos philippinus philippinus* (including *I. p. saturatior*) occurs on the eastern arc: Luzon, Polillo, Samar, Leyte. Its sister-group comprises the pair *I. siquijorensis*, distributed along a distinctive arc: Tablas (Romblon group)–Cebu–Siquijor, plus *I. philippinus mindorensis* from Mindoro (Oliveros and Moyle, 2010).

Dicaeum australe (Dicaeidae) occurs along the eastern track plus Cebu. *D. haematostictum* of Panay and Negros is a possible western vicariant (it was not sequenced in the study by Nyári *et al.* (2009b), but was traditionally treated as a sub-species of *D. australe*). The break between the two is at Tañon Strait. The clade *Dicaeum nigrilore* + *D. hypoleucum* also occurs along the eastern track: Luzon and Greater Mindanao plus the Sulu Islands (Nyári *et al.*, 2009b).

Pachycephala philippensis (Pachycephalidae) is distributed along the eastern arc: Luzon, Samar, Leyte, Bohol and Mindanao. *P. homeyeri* has a western distribution:

Mindanao, Sulu Islands and western Visayas (Sibuyan Islands, Masbate, Panay, Negros, Cebu).

Corvus enca (Corvidae) in the Philippines is in Luzon, Samar and Mindanao, also Palawan and Mindoro (but not in Panay, Negros, Cebu, Masbate and Leyte). The eastern Asian *C. macrorhynchos* occurs throughout the Philippines.

The pig *Sus philippensis* (Suidae) is in Luzon and Greater Mindanao, and is replaced on Panay, Negros and Masbate by *S. cebifrons*.

The deer genus *Cervus* (Cervidae) has two Philippines species, both endemic: *C. mariannus* in Luzon, Mindoro, Leyte and Mindanao, and *C. alfredi* in Panay, Negros, Cebu and Masbate.

In *Crocidura* (Soricidae) (Esselstyn *et al.*, 2011) the main Philippines clade has a phylogeny: (Palawan (Sibuyan, Panay, Negros (eastern track: Luzon, Mindoro, Leyte, Mindanao))).

Some groups on the eastern track have no obvious vicariant in the western Visayas. For example, the cobra genus *Naja* (Elapidae) is present throughout the Philippines except the western Visayas (Romblon group, Panay, Negros, Cebu). *Otosaurus* (Scincidae) is in Luzon and greater Mindanao, but is absent from the western Visayas (Panay, Negros, Cebu and Masbate, also Palawan and Mindoro).

Western Philippines track

In the Philippines part of its range *Eumyias panayensis* (Muscicapidae) illustrates an important western track: Luzon–western Visayas–Mindanao, and the species is absent from the eastern Visayas (Masbate, Samar, Leyte, Bohol, Cebu and Camiguin Sur). *Ficedula hyperythra* (Muscicapidae), *Zosterops montanus* (Zosteropidae) and *Phylloscopus trivirgatus* (Phylloscopidae) have similar distributions through the western Philippines (also in Palawan, Borneo, etc.), and are absent from the eastern Visayas (IUCN, 2013). In some montane groups the western track probably reflects the higher elevations of the mountains there, compared with the east. This does not seem to apply to the birds just cited, as their localities in the west include sites at lower elevations than the uplands in the east.

Jones and Kennedy (2008a) sequenced each of the four cited bird species. Their results contrasted with expectations from Pleistocene geography, as Luzon and Mindoro populations were indistinguishable. In addition, in three of the four species, Negros clades and Panay clades did not appear as sister-groups.

In *Eumyias panayensis* Jones and Kennedy (2008a) found the phylogeny: (Panay + Negros (Mindanao (Luzon + Mindoro))). Again, the pattern is incompatible with Pleistocene geography, as the central Panay and Negros clade is basal in the phylogeny and there is a wide, unexplained disjunction between the Mindanao and Luzon clades.

The beetle *Neptosternus* (Dytiscidae) has three species in the Philippines: *N. montalbanensis* in Luzon, *N. hydaticoides* in Luzon, Mindoro, Palawan, Negros and Mindanao, and *N. cebuensis* in Cebu, with the break between the last two at Tañon Strait (Hendrich and Balke, 2000).

Other examples of western groups include the fruit bat *Acerodon jubatus* (Pteropodidae). It occurs in a broad belt through the central Philippines from Luzon to Mindanao, in mangrove as well as primary and secondary dry land forest, but is absent from the eastern Bicol Peninsula and Samar on the outer, eastern belt (IUCN, 2013). *Pteropus pumilus* has a similar distribution; it is also absent in Luzon and eastern Mindanao.

Aethopyga flagrans (Nectariniidae) is in Luzon, Panay, Negros. In contrast, *A. pulcherrima* has an eastern range: Luzon, Samar, Leyte, Bohol, Mindanao (IUCN, 2013).

The frog *Hylarana everetti* (Ranidae) has a western distribution (Mindanao, Negros, Panay and Masbate), while *H. albotuberculata* in the same species group has an eastern distribution (north-eastern Mindanao, Samar and Leyte) (IUCN, 2013).

The bat *Cheiromeles parvidens* (Molossidae) is recorded in central Mindanao, Negros and Mindoro, also in the Sulu Islands and Sulawesi (IUCN, 2013).

11 Conclusions

Molecular work has brought about revolutions in all the life sciences. In biogeography it has already contributed a huge amount of valuable new information, as summarized in phylogenies. Nevertheless, most interpretations of the phylogenies and explanations of the evolution in space and time still rely on old concepts inherited from the Modern Synthesis. Many papers written in this framework use a program such as DIVA or DEC in LAGRANGE to find a centre of origin for a group, and a Bayesian program such as BEAST to transmogrify estimates of minimum clade age into estimates of maximum clade age. Both these procedures and the concepts upon which they are based are flawed and can be dropped; outside the Modern Synthesis there is no need to find a centre of origin or to transmogrify fossil dates.

Although it still relies on older ideas about evolutionary space and time, modern molecular study is moving away from the Modern Synthesis model in its approach to structural evolution. The Modern Synthesis assumed monomorphic, undifferentiated ancestors, with evolution proceeding by the selection of random mutation, and groups defined by uniquely derived characters. Instead, new ideas propose polymorphic ancestral complexes, descent with incomplete lineage sorting, and evolution determined by prior genomic architecture and mutational bias, rather than selection (Heads, 2012a, Chapter 10).

Evolution and space

The idea that structure develops by growing out from a centre of origin, rather than by a series of differentiation events in a widespread entity, is basic to the neo-darwinian concepts of evolution in space and time. Methods and programs developed in this research tradition, such as DIVA and DEC/LAGRANGE, assume a centre of origin model for most distribution patterns. In both programs an area phylogeny A(A(A,B)) will always be interpreted as dispersal from A to B. Yet the pattern can also be explained by a widespread ancestor already present in A and B, a centre of differentiation (not a centre of origin) in A, and subsequent, local overlap there. The phylogeny does not require any dispersal between A and B in the group (although this could have taken place in an ancestor).

If groups do not have a centre of origin, evolution does not proceed by radiation from a centre, but by differentiation at phylogenetic and geographic nodes. Subsequent range expansion and overlap may or may not occur. Molecular data have turned out to have unexpected, high levels of biogeographic structure at all scales of space and time. In addition, the repetition of breaks at the same localities in many different groups indicates that these localities are not distributed at random. Nevertheless, most of the breaks do not correspond with the obvious breaks in modern topography or ecology. Instead, they show a general relationship with features that have been zones of tectonic instability in the past, such as spreading ridges, subduction zones, basins, transform faults, mobile belts and orogens.

Evolution and time

The oldest fossils of many modern groups, such as the orders of mammals and of birds, are Cenozoic, and if the groups themselves originated at this time they would be too young to have been affected by Gondwana breakup. There are two main possibilities. Either biogeography is due to chance dispersal and the fossil record is a more or less accurate representation of evolution; or, biogeographic patterns have not developed at random, and a literal reading of the fossil record, treating fossil ages as clade ages, is misleading for most terrestrial groups. There is plenty of direct evidence for large gaps in the fossil record; for example, in extant groups that have no Cenozoic record at all but are represented by Mesozoic fossils (S.W. Heads, 2008). Fossils indicate minimum ages and so they are useful, not as positive estimates of actual clade age, but because they can eliminate origins related to younger geological events. There is no real reason to assume that groups with their oldest fossils in the Cenozoic could not have existed in the Mesozoic. Many chronograms of evolution produced using Bayesian methods assign 95% credibility intervals (highest posterior densities) to clade ages. However, the analysis is based on the prior transmogrification of fossil-based estimates of minimum clade ages into estimates of maximum clade ages (with narrow confidence intervals), and the use of these for calibration.

Another problem with fossil-based calibration is fossil identification. In many cases molecular studies have contradicted morphological analyses of phylogeny in living groups, but morphological analysis of fossil groups is even more difficult, and even more likely to be flawed. In many groups the placement of pre-Neogene fossils in a phylogeny is dubious and controversial. Doubts increase with fossils that are older or fragmentary, or show great differences from modern forms. Given the dramatic phylogenetic rearrangements of the molecular revolution, especially in groups acknowledged as 'difficult', it is likely that many of the more unusual Mesozoic forms are misplaced and belong to extant orders.

The method adopted here is not 'against' the fossil record and does not neglect it. Fossil records must be integrated with the living ones and many fossils are cited in this book. As already mentioned, biogeographic evidence suggests that some brachiopods and crustaceans have fossil records that do indicate the approximate ages of their clades. Nevertheless, the general conclusions reached here conflict with a literal reading of the fossil record, in which the oldest fossil date gives the clade age. As

A.B. Smith (2007) wrote: 'The fossil record is direct evidence but it cannot be taken at face value'.

Panbiogeography, tectonics and evolution

Modern analyses rely on clade ages, calibrated using the fossil record (which is misleading in most groups) and fossil identifications (although morphology-based extant groups were often misconstrued). Instead, a panbiogeographic approach synthesizes the best of the molecular data – the clades and their distribution – with revolutionary developments in hard-rock geology – the radiometric dates for tectonic events. The synthesis involves calibrating a range of nodes with tectonic events. The process can be complicated by the fact that tectonic events are often episodic or of long duration. For example, some of the main events in Australasia include the rifting of continental crust to form the Tasman and Coral Sea basins. The rifting zone propagated northwards, from 84 to 55 Ma, and communities were slowly torn apart. The breakup itself was just the final step in a long process of extension, and many distributions suggest that pre-breakup deformation and magmatism caused much of the rift-associated differentiation. Failed rifts are important for biogeography; for example, at the Moonlight tectonic zone in New Zealand, the Lebombo monocline in southern Africa and the Cretaceous rift system in central Africa.

Other events in the Tasman–Coral Sea region have resulted from shortening and thickening in the crust rather than extension, and, again, some of these processes have had a long duration. For example, the Paleocene (58 Ma) collision along the northern margin of the Australian plate in New Guinea, the Eocene (44–34 Ma) collision preserved in New Caledonia, and the Oligocene (26–25 Ma) collision in the North Island of New Zealand may all be parts of a single collision event that migrated southwards along the plate boundary (Glen and Meffre, 2009). Here, collision between forearc crust and continental fragments has led to rollback of the subduction zone towards the Pacific; this has created new basins and prevented arc–continent collision.

Life, as the uppermost geological stratum, has proved to be stickier than was thought, and populations have survived episodes of marine flooding, volcanism, glaciation, desertification, uplift, subsidence and other events. Remnants of weedy taxa have persisted in gullies, in rock cracks, on nunataks, underground and in many other marginal habitats. At the other extreme, even local tectonic changes, such as faulting, can affect living populations of groups in a non-weedy phase of their evolution and lead to differentiation. Strike-slip movement, developing under transpressional regimes, is important in New Zealand, New Caledonia, New Guinea and the Philippines, and has also led to biogeographic differentiation and disjunction.

Darwinian and neo-darwinian models of evolution and ecology

Darwin (1859) reasoned that as new, highly adapted species evolve, they will force their less adapted, less competitive relatives to migrate away. When recent, advanced forms evolved in the great evolutionary workshops of the north, this caused ancient, primitive forms, such as marsupials and ratites, to migrate to the 'ends of the Earth': southern Africa, Patagonia and the Tasman region. This is a logical account that integrates adaptation, community-level ecology, speciation and global biogeography. Nevertheless, the 'neo-darwinian' Modern Synthesis (Matthew, 1915; Mayr, 1942) supported a model of evolution that was the complete opposite of Darwin's view, and is now almost universal. (Frey, 1993, and Briggs, 2005, were perhaps the last authors to support Darwin's, 1859, early analysis of centres of origin and dispersal.) In the 'neo-darwinian' model, new clades – not old clades – have moved into new habitat by chance dispersal across a barrier and have thus been separated from their relatives. The migrants evolve in response to extrinsic *needs* imposed by their new conditions (either habits or habitats) and adapt by natural selection acting on random mutation.

Darwin adopted the pan-selectionist viewpoint, as supported by the neo-darwinians, in his earlier publications, including the first edition of the *Origin of Species* (1859). But in his later, more profound work, the emphasis shifted as he began to investigate 'laws of growth'. Examples of these trends are the reduction and fusion of organs in phylogenetic series, and the concomitant development of fivefold and bilateral symmetry (Heads, 2009c). Twenty-three years after the *Origin*, Darwin (1882: 61) wrote:

I now admit that . . . in the earlier editions of my 'Origin of species' I perhaps attributed too much to the action of natural selection . . . I was not . . . able to annul the influence of my former belief, then almost universal, that each species had been purposely created; and this led to my tacit assumption that every detail of structure, excepting rudiments, was of some special, though unrecognised service. Any one with this assumption on his mind would naturally extend too far the action of natural selection . . .

Thus in Darwin's mature view, an organ's structure is determined not by its current function, but by prior evolutionary trends. Contemporary work in genomic evolution has reached a similar conclusion, rejecting the selectionist model of evolution and instead stressing intrinsic aspects of genomic architecture and mutational effects (Stoltzfus, 2006; Lynch, 2007; Heads, 2012a, Chapter 10). These are equivalent to Darwin's laws of growth.

To summarize, the 'neo-darwinian' Modern Synthesis reversed the key points of Darwin's own approach:

- Darwin suggested that advanced groups occur at the centre of origin. The neo-darwinian synthesis instead accepted that primitive, basal groups occur there.
- Darwin adopted a relativistic species concept that assumed no qualitative difference between species and clades with other ranks – species are not special. He rejected the reproductive isolation criterion as essentialistic (see pp. 183–184). The neo-darwinian synthesis instead proposed that only species are real and that they have an absolute value; they are defined by reproductive isolation.
- Darwin argued that prior, non-random trends in variation (laws of growth) provide material upon which selection can work. The neo-darwinian synthesis instead assumed that adaptation is caused by selection of random mutations and so the required function of an organ explains its structure.

Darwin and the neo-darwinian synthesis may both have been wrong about the first point, but modern molecular work suggests it was Darwin, not the neo-darwinians, who was correct about the second and third points.

Beyond the CODA model of evolution and ecology

The Modern Synthesis view of evolution and ecology is linked with the idea that structural evolution takes place by adaptation to extrinsic needs. The approach has been termed the CODA paradigm, as it is based on centres of origin, dispersal and adaptation (Lomolino and Brown, 2009). In the CODA model, evolution proceeds by adaptation and the 'penetration' or 'wedging' of new species into an 'empty niche'. Since the 1950s, ecology 'has been dominated by the paradigm of diversity being limited by species packing and the filling of available niche space' (Ricklefs and Jenkins, 2011).

In this neo-darwinian model, the environmental conditions determine the morphology and physiology of a group. A new group invades a new habitat and survives there before it becomes adapted to the new conditions. In an alternative model, a group evolves at the same time as its distribution, while its ecology and habitat are determined by its biogeography, morphology and physiology. Non-random mutation and long-term trends in genome evolution, such as biased gene conversion, lead to morphological trends, such as reduction–fusion–suppression series. These structural changes will lead to shifts in habitat and habits. For example, if an ungulate's neck grows longer, it will tend to feed on taller trees. Animals whose sight is compromised by reduction of the optic system (or whose eyes become over-sensitive) may be forced to adopt a nocturnal or subterranean lifestyle. If animals' teeth are reduced or lost they are forced to live on softer food.

Ecology can often be determined by regional biogeography and tectonics. For example, the elevational range of a community determines many aspects of its ecology and is itself the result of uplift, subsidence and erosion. The date palms and their fan-palm relatives (Chapter 8) have found themselves stranded around desert oases, in montane rainforest and on Pacific atolls. They have inherited their distribution and, within this, an ecology. In the CODA model the palms have invaded the extreme habitats of the Arabian desert or the Pacific atolls and have then developed suitable adaptive features to survive there.

Notoryctes, the 'marsupial mole', is the only extant member of the order Notoryctemorphia and is restricted to sandy desert in central Australia. In a CODA model, the animal invaded the underground habitat in sandy desert from a centre of origin. It then developed its bizarre, adaptive features in response to the needs imposed by its new conditions. Instead, Queensland fossils indicate that rainforest members of the order, dated at ∼20 Ma, possessed defining features of the order long before sandy desert developed (Archer et al., 2011). Instead of the order dispersing into the desert and then adapting, earlier forms probably occupied the region that became desert before the deserts or the modern clades existed.

In this model, polymorphic ancestral complexes already occupied ancestral landscapes, and aspects of modern biogeography and biodiversity were already established, long before the extant groups or their landscapes existed. Parallelisms concentrated in a particular area are often thought to be responses to particular needs imposed by the environment, but are just as likely to reflect ancestral polymorphism. For example, many unrelated groups that are herbs elsewhere produce trees in the Pacific, indicating a prior concentration of 'tree genes' in the Pacific basin, before the modern clades existed. Another example concerns the centre of diversity of divaricate plants around the Indian and Atlantic Ocean, together with their rarity on and east of New Guinea and the Chatham Islands.

The ideas discussed here can be tested by the reader in groups with which he or she is familiar. Any pattern can be explained by either chance dispersal or vicariance and the predictions made by the two models can be explored. A vicariance model predicts that the distribution of clades in a group will show high levels of structure and will reflect the phylogeny in a simple way. It also predicts that the phylogenetic/geographic breaks found in one group will be shared with other groups that have different ecology and means of dispersal. 'Old' groups (such as bryophytes or arthropods) and 'young' groups (such as angiosperms, birds or mammals) have all been exposed to the same Mesozoic revolutions. If means of dispersal are not directly relevant to biogeography, distribution in groups of invertebrates or lower plants could shed light on evolution in other groups, including mammals, birds and angiosperms.

There are two possibilities for biogeography. All the patterns discussed in this book could be the result of chance dispersal and chance extinction. If this were true there would be little left to say about biogeography; all coincident patterns in distribution would be illusory pseudopatterns, as the different components would have developed at different times by different chance events. Means of dispersal would be seen as puzzling and enigmatic, and any biogeographic distribution not related to ecology would be more or less meaningless. This model would be compatible with a literal reading of the clade ages in the fossil record.

As an alternative to the chance dispersal model, evolution at intercontinental, regional and local scales could be caused by simple vicariance and range expansion of communities by normal means of dispersal (geodispersal), with both processes triggered by geological change. This view implies that the fossil record does not provide accurate estimates of evolutionary chronology for most terrestrial groups. In this model, distribution and dispersal are neither enigmatic nor due to chance, but can be treated as evidence in their own right, and can be analysed using normal observations and comparisons.

The molecular phylogenies themselves indicate that distribution and dispersal are not random; despite the kaleidoscopic variation in the details, coherent patterns are emerging and these are shared between groups with different ecology. Sequencing studies are revealing beautiful, interlocking, geographic structure at all scales. Many of the patterns are based on simple recombinations of just a few main nodes. In the light of the new patterns, the underlying paradigm of chance dispersal can be rejected as a conceptual roadblock impeding further progress. Everyone now agrees that geographic distribution has turned out to have a special phylogenetic importance. It is, in fact, what Darwin (1887) thought it was – a 'keystone of the laws of creation'.

Abbreviations and terms

Some key geological terms are explained in the Glossary. The abbreviations used for geological time are: B.P. = before present, Ma = a million years ago, and m.y. = a million years. The following are used to describe biological groups: p.p. = in part, incl. = including, s.str. = in a narrow sense, s.lat. = in a broad sense. The term 'basal group' sometimes causes confusion. As used here it refers to a small (less diverse) sister-group and there is no implication that it is primitive or ancestral.

Glossary

Allochthonous Not formed in place, formed elsewhere.

Arc A chain of volcanoes at a convergent plate margin.

Autochthonous Formed in place.

Back-arc basin An extensional basin formed by spreading behind an *arc*. The formation of the basin may separate an arc from a mainland (e.g. Japan from mainland Asia) or may split an island arc lengthwise along its axis (e.g. the Lau arc separating from the Tonga arc).

Dextral (= right-lateral) fault A *strike-slip fault* with dextral (right-lateral) slip. (Contrast with *sinistral fault*.)

Dextral movement (right-lateral movement) Relative movement in which the block on the other side of the fault from the observer moves to the right. (Contrast with *sinistral movement*.)

Dip The dip of a tilted bed, fault surface or other planar feature is the steepest angle of descent of the feature relative to a horizontal plane. The dip direction is at 90° to the *strike*.

Dip-slip fault A fault where the relative movement (slip) on the fault plane parallels the *dip* of the fault. (Contrast with *strike-slip fault*.)

Foot wall If a *dip-slip fault* plane is not exactly vertical, the lower block is the *foot wall*. (Contrast with *hanging wall*.)

Forearc In front of an *arc* (between the arc and its subduction zone).

Graben A depression between two adjacent *normal faults* that dip towards each other. (Contrast with *horst*.)

Hanging wall If a *dip-slip fault* plane is not exactly vertical, the upper block is the *hanging wall*. (Contrast with *foot wall*.)

Horst A high *foot wall* block formed between two *normal faults* that dip away from each other – the land on either side of the *horst* has dropped. (Contrast with *graben*.)

Island arc (or arc) A chain of volcanic islands formed in oceanic crust along the convergent plate margin.

Metamorphic core complexes Domes of metamorphic rock found in the core of an orogen. They form as the result of major continental extension, when the middle and lower crust is dragged out from beneath the fracturing, extending upper crust. The rocks are formed at depth and move up with the arching of the shear zone, and the effect on the topography is localized uplift in a region of extension. Examples occur in the Paparoa Range in New Zealand and the D'Entrecasteaux Archipelago in Papua New Guinea.

Metamorphic facies Groups of mineral assemblages found in rocks metamorphosed under particular pressure–temperature regimes. The facies sequence: prehnite-pumpellyite, lawsonite, blueschist, eclogite is seen in north-eastern New Caledonia. The facies are named after the rock that is produced by 'cooking' basalt under the same conditions. The facies sequence in New Caledonia represents metamorphism under increasing pressure but comparatively low temperatures.

Nappe A sheet-like body of rock that is *allochthonous* and has been moved by more than 5 km from its original position.

Normal fault A *dip-slip fault* in which the *hanging wall* moves downward, relative to the *foot wall*. Normal faults develop when the crust is extended. (Contrast with *reverse fault*.)

Obduction The overthrusting of continental crust by rocks of the oceanic crust or mantle.

Ophiolite A suite of rocks from oceanic crust and the mantle that has been uplifted and often emplaced onto continental crust.

Peridotite An *ultramafic* rock formed in the mantle and found in *ophiolite* suites.

Reverse fault A *dip-slip fault* in which the *hanging wall* moves up, relative to the *foot wall*. Reverse faults develop where the crust is shortened or contracted. (Contrast with *normal fault*.)

Serpentinite A rock formed by the weathering of *ultramafic* rock.

Slab roll-back (= trench roll-back) The retrograde migration of a *subduction* zone towards the subducting plate. The descending slab and its trench fall or roll back away from the *arc*, as well as subducting, like a retreating wave on a beach. This can cause extension in the crust above the subduction zone.

Strike The strike line of a bed, fault or other planar feature is a line representing the intersection of the feature with a horizontal plane.

Strike-slip fault (= 'transcurrent fault', 'wrench fault') A fault in which the slip is parallel to the strike of the fault. The fault surface is usually more or less vertical. (Contrast with *dip-slip fault*.)

Subduction The sinking of one tectonic plate under another at a convergent plate margin.

Transcurrent fault A *strike-slip fault* that is confined to the crust. (Compare with *transform fault*.)

Transform fault A *strike-slip fault* that cuts the lithosphere, forming a plate boundary.

Transpression A combination of strike-slip (transcurrent) movement and compression.

Ultramafic rock A rock including <45% silica.

References

Aagesen, L., D. Medan, J. Kellermann, and H.H. Hilger. (2005). Phylogeny of the tribe Colletieae (Rhamnaceae) – a sensitivity analysis of the plastid region *trn*L-*trn*F combined with morphology. *Plant Systematics and Evolution*, **250**, 197–214.

Abbott, R.J., L.H. Smith, R.I. Milne, *et al.* (2000). Molecular analysis of plant migration refugia in the Arctic. *Science*, **289**, 1343–1346.

Adamowicz, S.J., S. Menu-Marque, S.A. Halse, *et al.* (2010). The evolutionary diversification of the Centropagidae (Crustacea, Calanoida): a history of habitat shifts. *Molecular Phylogenetics and Evolution*, **55**, 418–430.

Adamowicz, S.J., A. Petrusek, J.K. Colbourne, P.D.N. Hebert, and J.D.S. Witt. (2009). The scale of divergence: a phylogenetic appraisal of intercontinental allopatric speciation in a passively dispersed freshwater zooplankton genus. *Molecular Phylogenetics and Evolution*, **50**, 423–436.

Adams, C.J. (2010). Lost terranes of Zealandia: possible development of late Paleozoic and early Mesozoic sedimentary basins at the southwest Pacific margin of Gondwanaland, and their destination as terranes in southern South America. *Andean Geology*, **37**, 442–454.

Adams, C.J., H.J. Campbell, and W.L. Griffin. (2007). Provenance comparisons of Permian to Jurassic tectonostratigraphic terranes in New Zealand: perspectives from detrital zircon age patterns. *Geological Magazine*, **144**, 701–729.

Ai, H.-A., J.M. Stock, R. Clayton, and B. Luyendyk. (2008). Vertical tectonics of the High Plateau region, Manihiki Plateau, Western Pacific, from seismic stratigraphy. *Marine Geophysical Research*, **29**, 13–26.

Airy Shaw, H.K. (1953). Thymelaeaceae-Gonystyloideae. *Flora Malesiana* I, **4**, 349–365.

Aitchison, J.C., G.L. Clarke, S. Meffre, and D. Cluzel. (1995). Eocene arc-continent collision in New Caledonia and implications for regional southwest Pacific tectonic evolution. *Geology*, **23**, 161–164.

Aitchison, J.C., T.R. Ireland, G.L. Clarke, *et al.* (1998). Regional implications of U/Pb SHRIMP age constraints on the tectonic evolution of New Caledonia. *Tectonophysics*, **299**, 333–343.

Albach, D.C., and M.W. Chase. (2001). Paraphyly of *Veronica* (Veroniceae; Scrophulariaceae): evidence from the internal transcribed spacer (ITS) sequences of nuclear ribosomal DNA. *Journal of Plant Research*, **114**, 9–18.

Albach, D.C., and H.A. Meudt. (2010). Phylogeny of *Veronica* in the southern and northern hemispheres based on plastid, nuclear ribosomal and nuclear low-copy DNA. *Molecular Phylogenetics and Evolution*, **54**, 457–471.

Albrecht, D.E., C.T. Owens, C.M. Weiller, and C.J. Quinn. (2010). Generic concepts in Ericaceae: Styphelioideae – the *Monotoca* group. *Australian Systematic Botany*, **23**, 320–332.

Alejandro, G.J.D., U. Meve, A. Mouly, M. Thiv, and S. Liede-Schumann. (2011). Molecular phylogeny and taxonomic revision of the Philippine endemic *Villaria* Rolfe (Rubiaceae). *Plant Systematics and Evolution*, **296**, 1–20.

Alfaro, M.E., D.R. Karns, H.K. Voris, E. Abernathy, and S.L. Sellins. (2004). Phylogeny of *Cerberus* (Serpentes: Homalopsinae) and phylogeography of *Cerberus rynchops*: diversification of a coastal marine snake in Southeast Asia. *Journal of Biogeography*, **31**, 1277–1292.

Alfaro, M.E., D.R. Karns, H.K. Voris, C.D. Brock, and B.L. Stuart. (2008). Phylogeny, evolutionary history, and biogeography of Oriental–Australian rear-fanged water snakes (Colubroidea: Homalopsidae) inferred from mitochondrial and nuclear DNA sequences. *Molecular Phylogenetics and Evolution*, **46**, 576–593.

Allan, H.H. (1961). *Flora of New Zealand. Vol. 1*. Wellington: Government Printer.

Allegrucci, G., S.A. Trewick, A. Fortunato, G. Carchini, and V. Sbordoni. (2010). Cave crickets and cave weta (Orthoptera, Rhaphidophoridae) from the southern end of the world: a molecular phylogeny test of biogeographical hypotheses. *Journal of Orthoptera Research*, **19**, 121–130.

Allibone, A.H., and A.J. Tulloch. (2004). Geology of the plutonic basement rocks of Stewart Island, New Zealand. *New Zealand Journal of Geology and Geophysics*, **47**, 233–256.

Allwood, J., D. Gleeson, G. Mayer, *et al.* (2010). Support for vicariant origins of the New Zealand Onychophora. *Journal of Biogeography*, **37**, 669–681.

Almeida, E.A.B., M.R. Pie, S.G. Brady, and B.N. Danforth. (2012). Biogeography and diversification of colletid bees (Hymenoptera: Colletidae): emerging patterns from the southern end of the world. *Journal of Biogeography*, **39**, 526–544.

Almeida, F.C., N.P. Giannini, R. DeSalle, and N.B. Simmons. (2011). Evolutionary relationships of the old world fruit bats (Chiroptera, Pteropodidae): another star phylogeny? *BMC Evolutionary Biology*, **11** (281), 1–17.

Alonso, M.A., and M.B. Crespo. (2008). Taxonomic and nomenclatural notes on South American taxa of *Sarcocornia* (Chenopodiaceae). *Annales Botanici Fennici*, **45**, 241–254.

Ambrose, L., C. Riginos, R.D. Cooper, *et al.* (2012). Population structure, mitochondrial polyphyly and the repeated loss of human biting ability in anopheline mosquitoes from the southwest Pacific. *Molecular Ecology*, **21**, 4327–43.

Andal, E.S., S. Arai, and G.P. Yumul Jr. (2005). Complete mantle section of a slow-spreading ridge-derived ophiolite: an example from the Isabela ophiolite in the Philippines. *Island Arc*, **14**, 272–294.

Anderson, C.L., K. Bremer, and E.M. Friis. (2005). Dating phylogenetically basal eudicots using *rbcL* sequences and multiple fossil reference points. *American Journal of Botany*, **92**, 1737–48.

Aoki, S., and M. Ito. (2000). Molecular phylogeny of *Nicotiana* (Solanaceae) based on the nucleotide sequence of the *matK* gene. *Plant Biology* [Stuttgart], **2**, 316–324.

Arancibia, G. (2004). Mid-Cretaceous crustal shortening: evidence from a regional-scale ductile shear zone in the Coastal Range of central Chile (32° S). *Journal of South American Earth Sciences*, **17**, 209–226.

Archer, M., R. Beck, M. Gott, S. Hand, H. Godthelp, and K. Black. (2011). Australia's first fossil marsupial mole (Notoryctemorphia) resolves controversies about their evolution and palaeoenvironmental origins. *Proceedings of the Royal Society of London, B* **278**, 1498–1506.

Arias, J.S., C.A. Szumik, and P.A. Goloboff. (2011). Spatial analysis of vicariance: a method for using direct geographical information in historical biogeography. *Cladistics*, **27**, 1–12.

Arifin, U., D.T. Iskandar, D.P. Bickford, *et al.* (2011). Phylogenetic relationship within the genus *Staurois* (Anura, Ranidae) based on 16S rRNA sequences. *Zootaxa*, **2744**, 29–52.

Arnold, E.N., and G. Poinar. (2008). A 100 million year old gecko with sophisticated adhesive toe pads, preserved in amber from Myanmar. *Zootaxa*, **1847**, 62–68.

Atlas of Living Australia. (2012). www.ala.org.au

Aubréville, A., J.-F. Leroy, P. Morat, and H.S. MacKee (eds). (1967–present). *Flore de la Nouvelle-Calédonie et dépendances.* Paris: Muséum National d'Histoire Naturelle.

Austin, A.D., D.K. Yeates, G. Cassis, *et al.* (2004). Insects 'Down Under' – diversity, endemism and evolution of the Australian insect fauna: examples from select orders. *Australian Journal of Entomology*, **43**, 216–234.

Austin, J.J., V. Bretagnolle, and E. Pasquet. (2004). A global molecular phylogeny of the small *Puffinus* shearwaters and implications for systematics of the Little-Audubon's Shearwater complex. *Auk*, **121**, 847–864.

Australia's Virtual Herbarium. (2011). http://avh.ala.org.au/

Avise, J.C. (1992). Molecular population structure and the biogeographic history of a regional fauna: a case history with lessons for conservation biology. *Oikos*, **63**, 62–76.

Avise, J.C. (2000). *Phylogeography: The History and Formation of Species.* Cambridge, MA: Harvard University Press.

Avise, J.C. (2007). Twenty-five key evolutionary insights from the phylogeographic revolution in population genetics. In *Phylogeography of Southern European Refugia*, eds. S. Weiss and N. Ferrand. Dordrecht: Springer, pp. 7–21.

Azuma, Y., Y. Kumazawa, M. Miya, K. Mabuchi, and M. Nishida. (2008). Mitogenomic evaluation of the historical biogeography of cichlids: toward reliable dating of teleostean divergences. *BMC Evolutionary Biology*, **8** (215), 1–13.

Bacon, C.D, W.J. Baker, and M.P. Simmons. (2012). Miocene dispersal drives island radiations in the palm tribe Trachycarpeae (Arecaceae). *Systematic Biology*, **61**, 426–442.

Báez, A.M., and N.G. Basso. (1996). The earliest known frogs of the Jurassic of South America: review and cladistic appraisal of their relationships. *Münchner Geowissenschaftliche Abhandlungen. Reihe A. Geologie und Paläontologie*, **30**, 131–158.

Baker, A.J., C.H. Daugherty, R. Colbourne, and J.L. McLennan. (1995). Flightless brown kiwis of New Zealand possess extremely subdivided population structure and cryptic species like small mammals. *Proceedings of the National Academy of Sciences USA*, **92**, 8254–8258.

Baker, C.H., G.C. Graham, K.D. Scott, *et al.* (2008). Distribution and phylogenetic relationships of Australian glow-worms *Arachnocampa* (Diptera, Keroplatidae). *Molecular Phylogenetics and Evolution*, **48**, 506–514.

Baker, W.J., and T.L.P. Couvreur. (2012). Biogeography and distribution patterns of Southeast Asian palms. In *Biotic Evolution and Environmental Change in Southeast Asia*, eds. D. Gower, K. Johnson, J. Richardson, *et al.* Cambridge: Cambridge University Press.

Baker, W. J., and T.L.P. Couvreur. (2013a). Global biogeography and diversification of palms sheds light on the evolution of tropical lineages. I. Historical biogeography. *Journal of Biogeography*, **40**, 274–285.

Baker, W. J., and T.L.P. Couvreur, (2013b). Global biogeography and diversification of palms sheds light on the evolution of tropical lineages. II. Diversification history and origin of regional assemblages. *Journal of Biogeography*, **40**, 286–298.

Baker, W.J., M.J.E. Coode, J. Dransfield, *et al.* (1998). Patterns of distribution of Malesian vascular plants. In *Biogeography and Geological Evolution of SE Asia*, ed. R. Hall and J.D. Holloway. Leiden: Backhuys, pp. 243–58.

Baker, W.J., V. Savolainen, C.B. Asmussen-Lange, *et al.* (2009). Complete generic-level phylogenetic analyses of palms (Arecaceae) with comparisons of supertree and supermatrix approaches. *Systematic Biology*, **58**, 240–256.

Baldwin, B.G., and W.L. Wagner. (2010). Hawaiian angiosperm radiations of North American origin. *Annals of Botany*, **105**, 849–879.

Baldwin, B.G., D.J. Crawford, J. Francisco-Ortega, *et al.* (1998). Molecular phylogenetic insights on the origin and evolution of oceanic island plants. In *Molecular Systematics of Plants II. DNA Sequencing*, eds. D.E. Soltis, P.S. Soltis, and J.J. Doyle. Boston: Kluwer, pp. 410–441.

Baldwin, S.L., P.G. Fitzgerald, and L.E. Webb. (2012). Tectonics of the New Guinea region. *Annual Review of Earth and Planetary Sciences*, **40**, 495–520.

Baldwin, S.L., T. Rawling, and P.G. Fitzgerald. (2007). Thermochronology of the New Caledonian high-pressure terrane: implications for middle Tertiary plate boundary processes in the southwest Pacific. *Geological Society of America Special Papers*, **419**, 1–18.

Baldwin, S.L., L.E. Webb, and B.D. Monteleone. (2008). Late Miocene coesite-eclogite exhumed in the Woodlark Rift. *Geology*, **36**, 735–738.

Balke, M., I. Ribera, L. Hendrich, *et al.* (2009). New Guinea highland origin of a widespread arthropod supertramp. *Proceedings of the Royal Society B*, **276**, 2359–2367.

Banks, H., S. Feist-Burkhart, and B. Klitgaard. (2006). The unique pollen morphology of *Duparquetia* (Leguminosae: Caesalpinioideae): developmental evidence of aperture orientation using confocal microscopy. *Annals of Botany*, **98**, 107–115.

Barcelona, J.F. (2003). The taxonomy and ecology of the pteridophytes of Mt. Iraya and vicinity, Batan Island, Batanes province, northern Philippines. In *Pteridology in the New Millennium*, eds. S. Chandra and M. Srivastava. Dordrecht: Kluwer, pp. 299–325.

Barker, F.K., G.F. Barrowclough, and J.G. Grouth. (2002). A phylogenetic hypothesis for passerine birds: taxonomic and biogeographic implications of analysis of nuclear DNA sequence data. *Proceedings of the Royal Society of London B*, **269**, 295–308.

Barker, F.K., A. Cibois, P. Schikler, J. Feinstein, and J. Cracraft. (2004). Phylogeny and diversification of the largest avian radiation. *Proceedings of the National Academy of Sciences USA*, **101**, 11040–11045.

Barker, M.S., and R.J. Hickey. (2006). A taxonomic revision of the Caribbean *Adiantopsis* (Pteridaceae). *Annals of the Missouri Botanical Garden*, **93**, 371–401.

Barker, N.P., P.H. Weston, F. Rutschmann, and H. Sauquet. (2007). Molecular dating of the 'Gondwanan' plant family Proteaceae is only partially congruent with the break-up of Gondwana. *Journal of Biogeography*, **34**, 2010–2027.

Barker, W.R. (1982). Taxonomic studies in *Euphrasia* L. (Scrophulariaceae). A revised infrageneric classification and a revision of the genus in Australia. *Journal of the Adelaide Botanical Garden*, **5**, 1–304.

Barker, W.R., and P.J.M. Greenslade. (1982). Preface. In *Evolution of the Flora and Fauna of Arid Australia*, eds. W.R. Barker and P.J.M. Greenslade. Frewville, SA: Peacock, pp. iii–iv.

Barrabé, L., S. Buerki, A. Mouly, *et al.* (2012). Delimitation of the genus *Margaritopsis* (Rubiaceae) in the Asian, Australasian and Pacific region, based on molecular phylogenetic inference and morphology. *Taxon*, **61**, 1251–1268.

Bartish, I.V., A. Antonelli, J.E. Richardson, and U. Swenson. (2011). Vicariance or long-distance dispersal: historical biogeography of the pantropical subfamily Chrysophylloideae (Sapotaceae). *Journal of Biogeography*, **38**, 177–190.

Batianoff, G.N., G.C. Naylor, J. Olds, and V.J. Neldner. (2009). Distribution patterns, weed incursions and origins of terrestrial flora at the Capricorn-Bunker Islands, Great Barrier Reef, Australia. *Cunninghamia*, **11**, 107–121.

Bauer, A.M., T. Jackman, R.A. Sadlier, and A.H. Whitaker. (2006). A revision of the *Bavayia validiclavis* group (Squamata: Gekkota: Diplodactylidae), a clade of New Caledonian geckos exhibiting microendemism. *Proceedings of the California Academy of Sciences*, 4th Series, **57**, 503–547.

Bauer, A.M., T.R. Jackman, R.A. Sadlier, and A.H. Whitaker. (2012). Revision of the giant geckos of New Caledonia (Reptilia: Diplodactylidae: *Rhacodactylus*). *Zootaxa*, **3404**, 1–52.

Baum, D.A., R. L. Small, and J.F. Wen. (1998). Biogeography and floral evolution of baobabs (*Adansonia*, Bombacaceae) as inferred from multiple data sets. *Systematic Biology*, **47**, 181–207.

Bayer, R.J., I. Breitwieser, J.M. Ward, and C. Puttock. (2006). Tribe Gnaphalieae (Cass.) Lecoq and Juillet. In *The Families and Genera of Vascular Plants, Vol. 8. Flowering Plants, Eudicots: Asterales*, eds. J.W. Kadereit and C. Jeffrey. Berlin: Springer, pp. 246–284.

Bayer, R.J., D.G. Greber, and N.H. Bagnall. (2002). Phylogeny of Australian Gnaphalieae (Asteraceae) based on chloroplast and nuclear sequences, the *trnL* intron, *trnL/trnF* intergenic spacer, *matK*, and ETS. *Systematic Botany*, **27**, 801–814.

Bayly, M., and A. Kellow. (2006). *An Illustrated Guide to New Zealand Hebes*. Wellington: Te Papa Press.

Bayly, M.J., P.J. Garnock-Jones, K.A. Mitchell, K.R. Markham, and P.J. Brownsey. (2000). A taxonomic revision of the *Hebe parviflora* complex (Scrophulariaceae), based on morphology and flavonoid chemistry. *New Zealand Journal of Botany*, **38**, 165–190.

Beauchamp, A.J., and G.R. Parrish. (1999). Bird use of the sediment settlement ponds and roost areas at Port Whangarei. *Notornis*, **46**, 470–483.

Beaumont, A.J., T.J. Edwards, J. Manning, *et al.* (2009). *Gnidia* (Thymelaeaceae) is not monophyletic: taxonomic implications for Thymelaeoideae and a partial new generic taxonomy for *Gnidia*. *Botanical Journal of the Linnean Society*, **160**, 402–417.

Beehler, B.M., D.M. Prawiradilaga, Y. de Fretes, and N. Kemp. (2007). A new species of Smoky honeyeater (Meliphagidae: *Melipotes*) from western New Guinea. *Auk*, **124**, 1000–1009.

Bell, T.P., and G.B. Patterson. (2008). A rare alpine skink *Oligosoma pikitanga* n. sp. (Reptilia: Scincidae) from Llawrenny Peaks, Fiordland, New Zealand. *Zootaxa*, **1882**, 57–68.

Bellwood, D.R., W. Renema, and B.R. Rosen. (2012). Biodiversity hotspots, evolution and coral reef biogeography: a review. In *Biotic Evolution and Environmental Change in Southeast Asia*, eds. D. Gower, K. Johnson, J. Richardson, *et al.* Cambridge: Cambridge University Press.

Benton, M.J., and P.C.J. Donoghue. (2007). Paleontological evidence to date the tree of life. *Molecular Biology and Evolution*, **24**, 26–53.

Bergh, N., and H.P. Linder. (2009). Cape diversification and repeated out-of-southern-Africa dispersal in paper daisies (Asteraceae-Gnaphalieae). *Molecular Phylogenetics and Evolution*, **51**, 5–18.

Berry, P.E., W.J. Hahn, K.J. Sytsma, J.C. Hall, and A. Mast. (2004). Phylogenetic relationships and biogeography of *Fuchsia* (Onagraceae) based on noncoding nuclear and chloroplast DNA data. *American Journal of Botany*, **91**, 601–614.

Bialas, R.W., W.R. Buck, and M. Studinger. (2007). Plateau collapse model for the Transantarctic Mountains–West Antarctic Rift System: insights from numerical experiments. *Geology*, **35**, 687–690.

Bickel, D. (2009). Diversity, relationships and biogeography of Australian flies. In *Diptera Diversity: Status, Challenges and Tools*, eds. T. Pape, D. Bickel, and R. Meier. Brill: Koninklijke, pp. 238–286.

Biffin, E., J.G. Conran, and A.J. Lowe. (2011). Podocarp evolution: a molecular phylogenetic perspective. *Smithsonian Contributions to Botany*, **95**, 1–20.

Biffin, E., R.S. Hill, and A.J. Lowe. (2010). Did Kauri (*Agathis*: Araucariaceae) really survive the Oligocene drowning of New Zealand? *Systematic Biology*, **59**, 594–602.

Birch, J.L., S.C. Keeley, and C.W. Morden. (2012). Molecular phylogeny and dating of Asteliaceae (Asparagales): *Astelia* s.l. evolution provides insight into the Oligocene history of New Zealand. *Molecular Phylogenetics and Evolution*, **65**, 102–115.

Bird, P. (2003). An updated digital model of plate boundaries. *Geochemistry Geophysics Geosystems*, **4**(3), 1027, doi:10.1029/2001GC000252.

Bitner, M.A. (2009). Recent Brachiopoda from the Norfolk Ridge, New Caledonia, with description of four new species. *Zootaxa*, **2235**, 1–39.

Bitner, M.A. (2010). Biodiversity of shallow-water brachiopods from New Caledonia, SW Pacific, with description of a new species. *Scientia Marina*, **74**, 643–657.

Blackburn, D.C., D.P. Bickford, A.C. Diesmos, D.T. Iskandar, and R.M. Brown. (2010). An ancient origin for the enigmatic flat-headed frogs (Bombinatoridae: *Barbourula*) from the islands of Southeast Asia. *PLoS One*, **5**, e12090, 1–10.

Bond, W.J., and J.A. Silander. (2007). Springs and wire plants: anachronistic defences against Madagascar's extinct elephant birds. *Proceedings of the Royal Society, B*, **274**, 1985–1992.

Bond, W.J., W.G. Lee, and J.M. Craine. (2004). Plant structural defences against avian browsers: the legacy of New Zealand's extinct moas. *Oikos*, **104**, 500–508.

Bone, T.S., S.R. Downie, J.M. Affolter, and K. Spalik. (2011). A phylogenetic and biogeographic study of the genus *Lilaeopsis* (Apiaceae tribe Oenantheae). *Systematic Botany*, **36**, 789–805.

Bonifacino, J.M., and V.A. Funk. (2012). Phylogenetics of the *Chiliotrichum* group (Compositae: Astereae): the story of the fascinating radiation in the paleate Astereae genera from southern South America. *Taxon*, **61**, 180–196.

Bostock, P.D., and A.E. Holland (eds). (2010). *Census of the Queensland Flora 2010*. Brisbane: Queensland Herbarium, Department of Environment and Resource Management.

Botello, A., T.M. Iliffe, F. Alvarez, *et al.* (2012). Historical biogeography and phylogeny of *Typhlatya* cave shrimps (Decapoda: Atyidae) based on mitochondrial and nuclear data. *Journal of Biogeography*, **40**, 594–607.

Bouchet, P., and Y.I. Kantor. (2003). New Caledonia: the major center of biodiversity for volutomitrid molluscs (Mollusca: Neogastropoda: Volutomitridae). *Systematics and Biodiversity*, **1**, 467–502.

Bouchet, P., and J.-P. Rocroi. (2005). Classification and nomenclat of gastropod families. *Malacologia* **47**, 1–397.

Bouchet, P., P. Lozouet, P. Maestrati, and V. Heros. (2002). Assessing the magnitude of species richness in tropical marine environments: exceptionally high numbers of mollusks at a New Caledonia site. *Biological Journal of the Linnean Society*, **75**, 421–436.

Bouetard, A., P. Lefeuvre, R. Gigant, *et al.* (2010). Evidence of transoceanic dispersion of the genus *Vanilla* based on plastid DNA phylogenetic analysis. *Molecular Phylogenetics and Evolution*, **55**, 621–630.

Boyer, S.L., and G. Giribet. (2007). A new model Gondwanan taxon: systematics and biogeography of the harvestman family Pettalidae (Arachnida, Opiliones, Cyphophthalmi), with a taxonomic revision of genera from Australia and New Zealand. *Cladistics*, **23**, 337–361.

Boyer, S.L., J.M. Baker, and G. Giribet. (2007a). Deep genetic divergences in *Aoraki denticulata* (Arachnida, Opiliones, Cyphophthalmi): a widespread 'mite harvestman' defies DNA taxonomy. *Molecular Ecology*, **16**, 4999–5016.

Boyer, S.L., R.M. Clouse, L.R. Benavides, *et al.* (2007b). Biogeography of the world: a case study from cyphophthalmid Opiliones, a globally distributed group of arachnids. *Journal of Biogeography*, **34**, 2070–2085.

Braby, M.F., J.W.H. Trueman, and R. Eastwood. (2005). When and where did troidine butterflies (Lepidoptera : Papilionidae) evolve? Phylogenetic and biogeographic evidence suggests an origin in remnant Gondwana in the Late Cretaceous. *Invertebrate Systematics*, **19**, 113–143.

Bradford, J.C. (2002). Molecular phylogenetics and morphological evolution in Cunonieae (Cunoniaceae). *Annals of the Missouri Botanic Garden*, **89**, 491–503.

Breitwieser I., D.S. Glenny, A. Thorne, and S.J. Wagstaff. (1999). Phylogenetic relationships in Australasian Gnaphalieae (Compositae) inferred from ITS sequences. *New Zealand Journal of Botany*, **37**, 399–412.

Brelsford, A., B. Milá, and D.E. Irwin. (2011). Hybrid origin of Audubon's warbler. *Molecular Ecology*, **20**, 2380–2389.

Bremer, K. (1992). Ancestral areas: a cladistic reinterpretation of the center of origin concept. *Systematic Biology*, **41**, 436–445.

Bremer, K. (1994). *Asteraceae: Cladistics and Classification*. Portland, OR: Timber Press.

Breure, A.S.H. (1979). Systematics, phylogeny and zoogeography of Bulimulinae (Mollusca). *Zoologische Verhandelingen, Leiden* **168**, 1–215.

Breure, A.S.H., and P. Romero. (2012). Support and surprises: molecular phylogeny of the land snail superfamily Orthalicoidea using a three-locus gene analysis with a divergence time analysis and ancestral area reconstruction (Gastropoda: Stylommatophora). *Archiv für Molluskenkunde*, **141**, 1–20.

Briggs, J.C. (1974). Operation of zoogeographic barriers. *Systematic Zoology*, **23**, 248–256.

Briggs, J.C. (2005). The biogeography of otophysan fishes (Ostariophysi: Otophysi): a new appraisal. *Journal of Biogeography*, **32**, 287–294.

Brochu, C.A. (2004). Calibration age and quartet divergence date estimation. *Evolution*, **58**, 1375–1382.

Brothers, R.N., and M.C. Blake Jr. (1973). Tertiary plate tectonics and high-pressure metamorphism in New Caledonia. *Tectonophysics*, **17**, 337–358.

Brower, A.V.Z. (1994). Rapid morphological radiation and convergence among races of the butterfly *Heliconius erato* inferred from patterns of mitochondrial DNA evolution. *Proceedings of the National Academy of Sciences USA*, **91**, 6491–95.

Brown, G.K., G. Nelson, and P.Y. Ladiges. (2006). Historical biogeography of *Rhododendron* section *Vireya* and the Malesian Archipelago. *Journal of Biogeography*, **33**, 1929–1944.

Brown, R.M., and A. Diesmos. (2009). Philippines, biology. In *Encyclopedia of Islands*, eds. R. Gillespie and D. Clague. Berkeley: University of California Press, pp. 723–732.

Brown, R.M., and S.I. Guttman. (2002). Phylogenetic systematic of the *Rana signata* complex of Philippine and Bornean stream frogs: reconsideration of Huxley's modification of Wallace's line at the Oriental–Australian faunal zone interface. *Biological Journal of the Linnean Society*, **76**, 393–461.

Brown, R.M., C. Oliveros, C.D. Siler, and A.C. Diesmos. (2009a). Phylogeny of *Gekko* from the northern Philippines, and description of a new species from Calayan Island. *Journal of Herpetology*, **43**, 620–635.

Brown, R.M., C.D. Siler, A.C. Diesmos, and A.C. Alcala. (2009b). Philippine frogs of the genus *Leptobrachium* (Anura; Megophryidae): phylogeny-based species delimitation, taxonomic review, and descriptions of three new species. *Herpetological Monographs*, **23**, 1–44.

Brownsey, P.J. (1977). A taxonomic revision of the New Zealand species of Asplenium. *New Zealand Journal of Botany*, **15**, 39–86.

Brownsey, P.J., and J.C. Smith-Dodsworth. (1989). *New Zealand Ferns and Allied Plants*. Auckland: David Bateman.

Brundin, L. (1965). On the real nature of transantarctic relationships. *Evolution*, **19**, 496–505.

Brundin, L. (1966). Transantarctic relationships and their significance as evidenced by chironomid midges: with a monograph of the sub-families Podonominae and Aphroteniinae and the Austral Heptagyiae. *Kunglica Svenska Vetenskapsakademiens Handlingar*, **11**, 1–472 + plates.

Brunken, U., and A.N. Muellner. (2012). A new tribal classification of Grewioideae (Malvaceae) based on morphological and molecular phylogenetic evidence. *Systematic Botany*, **37**, 699–711.

Brüstle, L., D. Alaruikka, J. Muona, and M. Teräväinen. (2010). The phylogeny of the Pantropical genus *Arrhipis* Bonvouloir (Coleoptera, Eucnemidae). *Cladistics*, **26**, 14–22.

Bryan, S., and R. Ernst. (2007). Revised definition of large igneous province (LIP). *Earth Science Reviews*, **86**, 175–202.

Bryan, S.E., A.G. Cook, C.M. Allen, *et al.* (2012). Early-mid Cretaceous tectonic evolution of eastern Gondwana: from silicic LIP magmatism to continental rupture. *Episodes*, **35**, 142–152.

Bryan, S.E., T.R. Riley, D.A. Jerram, C.J. Stephens, and P.T. Leat. (2002). Silicic volcanism: an undervalued component of large igneous provinces and volcanic rifted margins. *Geological Society of America Special Papers*, **362**, 99–120.

Buck, W.R., and B.C. Tan. (2008). A review of *Elmeriobryum* (Hypnaceae). *Telopea*, **12**, 251–256.

Buckeridge, J.S., and W.A. Newman. (2006). A revision of the Iblidae and the stalked barnacles (Crustacea: Cirripedia: Thoraica), including new ordinal, familial and generic taxa, and two new species from New Zealand and Tasmanian waters. *Zootaxa*, **113**, 1–38.

Buckley, T.R., D. Attanayake, and S. Bradler. (2009). Extreme convergence in stick insect evolution: phylogenetic placement of the Lord Howe Island tree lobster. *Proceedings of the Royal Society, B*, **276**, 1055–1062.

Buckley, T.R., D. Attanayake, J.A.A. Nylander, and S. Bradler. (2010). The phylogenetic placement and biogeographical origins of the New Zealand stick insects (Phasmatodea). *Systematic Entomology*, **35**, 207–225.

Buckley, T.R., S. James, J. Allwood, *et al.* (2011). Phylogenetic analysis of New Zealand earthworms (Oligochaeta: Megascolecidae) reveals ancient clades and cryptic taxonomic diversity. *Molecular Phylogenetics and Evolution*, **58**, 85–96.

Buckley, T.R., C. Simon, and G.K. Chambers. (2001). Phylogeography of the New Zealand cicada *Maoricicada campbelli* based on mitochondrial DNA sequences: ancient clades associated with Cenozoic environmental change. *Evolution*, **55**, 1395–1407.

Buerki, S., M.W. Callmander, D.S. Devey, *et al.* (2012). Straightening out the screw-pines: a first step in understanding phylogenetic relationships within Pandanaceae. *Taxon*, **61**, 1010–1020.

Buerki, S., F. Forest, N. Alvarez, *et al.* (2011). An evaluation of new parsimony-based versus parametric inference methods in biogeography: a case study using the globally distributed plant family Sapindaceae. *Journal of Biogeography*, **38**, 531–550.

Burbridge, N.T. (1960). The phytogeography of the Australian region. *Australian Journal of Botany*, **8**, 75–212.

Burbrink, F.T., and R. Lawson. (2007). How and when did Old World rat snakes disperse into the New World? *Molecular Phylogenetics and Evolution*, **43**, 173–189.

Burckhardt, D. (2009). Taxonomy and phylogeny of the Gondwanan moss bugs or Peloridiidae (Hemiptera, Coleorrhyncha). *Deutsche Entomologische Zeitschrift*, **56**, 173–235.

Burckhardt, D., E. Bochud, J. Damgaard, *et al.* (2011). A review of the moss bug genus *Xenophyes* (Hemiptera: Coleorrhyncha: Peloridiidae) from New Zealand: systematics and biogeography. *Zootaxa*, **2923**, 1–26.

Burleigh, J.G., M.S. Bansal, O. Eulenstein, *et al.* (2011). Genome-scale phylogenetics: inferring the plant tree of life from 18,896 gene trees. *Systematic Biology*, **60**, 117–125.

Burridge, C.P. (1999). Molecular phylogeny of *Nemadactylus* and *Acantholatris* (Perciformes Cirrhitoidea: Cheilodactylidae), with implications for taxonomy and biogeography. *Molecular Phylogenetics and Evolution*, **13**, 93–109.

Burridge, C.P., R.M. McDowall, D. Craw, M.V.H. Wilson, and J.M. Waters. (2012). Marine dispersal as a pre-requisite for Gondwanan vicariance among elements of the galaxiid fish fauna. *Journal of Biogeography*, **39**, 306–321.

Burridge, C.P., R. Meléndez, and B.S. Dyer. (2006). Multiple origins of the Juan Fernández kelpfish fauna, and evidence for frequent and unidirectional dispersal of cirrhitoid fishes across the South Pacific. *Systematic Biology*, **55**, 566–578.

Burton, D.W. (1963). A revision of the New Zealand and subantarctic Athoracophoridae. *Transactions of the Royal Society of New Zealand*, **3**, 47–75.

Burton, D.W. (1980). Anatomical studies on Australian, New Zealand, and subantarctic Athoracophoridae (Gastropoda: Pulmonata). *New Zealand Journal of Zoology*, **7**, 173–98.

Bush, C.M., S.J. Wagstaff, P.W. Fritsch, and K.A. Kron. (2009). The phylogeny, biogeography and morphological evolution of *Gaultheria* (Ericaceae) from Australia and New Zealand. *Australian Systematic Botany*, **22**, 229–242.

Butler, A.D., G.D. Edgecombe, A.D. Ball, and G. Giribet. (2010). Resolving the phylogenetic position of enigmatic New Guinea and Seychelles Scutigeromorpha (Chilopoda): a molecular and morphological assessment of Ballonemini. *Invertebrate Systematics*, **24**, 539–559.

Cabezas, P., I. Sanmartín, G. Paulay, E. Macpherson, and A. Machordom. (2012). Deep under the sea: unraveling the evolutionary history of the deep-sea squat lobster *Paramunida* (Decapoda, Munididae). *Evolution*, **66**, 1878–1896.

Cabrera, J., S.W.L. Jacobs, and G. Kadereit. (2011). Biogeography of Camphorosmeae (Chenopodiaceae): tracking the Tertiary history of Australian aridification. *Telopea*, **13**, 313–326.

Cadotte, M.W., and T.J. Davies. (2010). Rarest of the rare: advances in combining evolutionary distinctiveness and scarcity to inform conservation at biogeographical scales. *Diversity and Distributions*, **16**, 376–385.

Cameron, E.K., P.J. de Lange, L.R. Perrie, *et al.* (2006). A new location for the Poor Knights spleenwort (*Asplenium pauperequitum*, Aspleniaceae) on the Forty Fours, Chatham Islands, New Zealand. *New Zealand Journal of Botany*, **44**, 199–209.

Cameron, K.M. (2000). Gondwanan biogeography of Vanilloideae (Orchidaceae). *Southern Connections Congress, Programme and Abstracts*. Lincoln, New Zealand, pp. 25–26.

Cameron, K.M. (2003). Tribe Vanilleae: distribution, *Clematepistephium, Epistephium, Eriaxis*. In *Genera Orchidacearum. Vol. 3. Orchidoideae (Part 2). Vanilloideae*, eds. A.M. Pridgeon, P.J. Cribb, M.W. Chase, and F.N. Rasmussen. New York: Oxford University Press, pp. 299–302, 306–311.

Campbell, H.J. (2008). Geology. In *Chatham Islands: Heritage and Conservation*, ed. C. Miskelly. Christchurch: Canterbury University Press, pp. 35–53.

Campbell, H.J. and G. Hutching. (2007). *In Search of Ancient New Zealand*. Auckland Penguin and Lower Hutt Institute of Geological and Nuclear Sciences.

Campbell, H.J., J. Begg, A. Beu, *et al.* (2009). Geological considerations relating to the Chatham Islands, mainland New Zealand and the history of New Zealand terrestrial life. *Geological Society of New Zealand Miscellaneous Publication*, **126**, 5–7.

Campo, D., G. Machado-Schiaffino, J. Perez, and E. Garcia-Vazquez. (2007). Phylogeny of the genus *Merluccius* based on mitochondrial and nuclear genes. *Gene*, **406**, 171–179.

Cantino, P.D., S.J. Wagstaff, and R.G. Olmstead. (1999). *Caryopteris* (Lamiaceae) and the conflict between phylogenetic and pragmatic considerations in botanical nomenclature. *Systematic Botany*, **23**, 369–386.

Carlquist, S. (1974). *Island Biology*. New York: Columbia University Press.

Carpenter, K.E., and V.G. Springer. (2005). The center of the center of marine shore fish biodiversity: the Philippine Islands. *Environmental Biology of Fishes*, **72**, 467–480.

Carpenter, R.J., G.J. Jordan, D.E. Lee, and R.S. Hill. (2010). Leaf fossils of *Banksia* (Proteaceae) from New Zealand: an Australian abroad. *American Journal of Botany*, **97**, 288–297.

Castro, L.R., and M. Dowton. (2006). Molecular analyses of the Apocrita (Insecta: Hymenoptera) suggest that the Chalcidoidea are sister to the diaprioid complex. *Invertebrate Systematics*, **20**, 603–614.

Catullo, R.A., P. Doughty, J.D. Roberts, and J.S. Keogh. (2011). Multi-locus phylogeny and taxonomic revision of *Uperoleia* toadlets (Anura: Myobatrachidae) from the western arid zone of Australia, with a description of a new species. *Zootaxa*, **2902**, 1–43.

Cavalcanti, M.J., and V. Gallo. (2007). Panbiogeographic analysis of distribution patterns in hagfishes (Craniata: Myxinidae). *Journal of Biogeography*, **35**, 1258–1268.

Chakrabarty, P. (2004). Cichlid biogeography: comment and review. *Fish and Fisheries*, **5**, 97–119.

Chakrabarty, P. (2010). Status and phylogeny of Milyeringidae (Teleostei: Gobiiformes), with the description of a new blind cave-fish from Australia, *Milyeringa brooksi*, n. sp. *Zootaxa*, **2557**, 19–28.

Chakrabarty, P., M.P. Davis, and J.S. Sparks. (2012). The first record of a trans-oceanic sister-group relationship between obligate vertebrate troglobites. *PLoS One*, **7**(8), e44083, 1–8.

Chambers, K.L. (1955). A biosystematic study of the annual species of *Microseris*. *Contributions from the Dudley Herbarium*, **4**, 207–312.

Chandler, G.T., G.M. Plunkett, S.M. Pinney, L.W. Cayzer, and C.E.C. Gemmill. (2003). Molecular and morphological agreement in Pittosporaceae: phylogenetic analysis with nuclear ITS and plastid *trnL*–*trnF* sequence data. *Australian Systematic Botany*, **20**, 390–401.

Chapple, D.G., and G.B. Patterson. (2007). A new skink species (*Oligosoma taumakae* sp. nov.; Reptilia: Scincidae) from the Open Bay Islands, New Zealand. *New Zealand Journal of Zoology*, **34**, 347–357.

Chapple, D.G., J.S. Keogh, and M.N. Hutchinson. (2004). Molecular phylogeography and systematics of the arid-zone members of the *Egernia whitii* (Lacertilia: Scincidae) species group. *Molecular Phylogenetics and Evolution*, **33**, 549–561.

Chapple, D.G., P.A. Ritchie, and C.H. Daugherty. (2009). Origin, diversification, and systematics of the New Zealand skink fauna. *Molecular Phylogenetics and Evolution*, **52**, 470–487.

Chase, M.W., A.Y. De Bruijn, A.V. Cox, *et al.* (2000). Phylogenetics of Asphodelaceae (Asparagales): an analysis of plastid *rbcL* and *trnL-F* DNA sequences. *Annals of Botany*, **86**, 935–951.

Chase, M.W., S. Knapp, A.V. Cox, *et al.* (2003). Molecular systematics, GISH and the origin of hybrid taxa in *Nicotiana* (Solanaceae). *Annals of Botany*, **92**, 107–127.

Chatrou, L.W., M.D. Pirie, R.H.J. Erkens, *et al.* (2012). A new subfamilial and tribal classification of the pantropical flowering plant family Annonaceae informed by molecular phylogenetics. *Botanical Journal of the Linnean Society*, **169**, 5–40.

Chatterjee, H., S. Ho, I. Barnes, and C. Groves. (2009). Estimating the phylogeny and divergence times of primates using a supermatrix approach. *BMC Evolutionary Biology*, **9** (259), 1–19.

Chen, L.-Y., J.-M. Chen, R.W. Gituru, T.D. Temam, and Q.-F. Wang. (2012). Generic phylogeny and historical biogeography of Alismataceae, inferred from multiple DNA sequences. *Molecular Phylogenetics and Evolution*, **63**, 407–416.

Chenoweth, L.B. and M.P. Schwarz. (2011). Biogeographical origins and diversification of the exoneurine allodapine bees of Australia (Hymenoptera, Apidae). *Journal of Biogeography*, **38**, 1471–1483.

Chicangana, G. (2005). The Romeral fault system: a shear and deformed extinct subduction zone between oceanic and continental lithospheres in northwestern South America. *Earth Science Research Journal*, **9**, 51–66.

Chin, N., M. Brown, and M. Heads. (1991). The biogeography of *Macrocystis* (Lessoniaceae). *Hydrobiologia*, **215**, 1–11.

Chinnock, R.J. (2007). Eremophila *and Allied Genera: A Monograph of the Myoporaceae*. Dural, NSW: Rosenberg Publishing.

Chintauan-Marquier, I.C., S. Jordan, P. Berthier, C. Amédégnato, and F. Pompanon. (2011). Evolutionary history and taxonomy of a short-horned grasshopper subfamily: The Melanoplinae (Orthoptera: Acrididae). *Molecular Phylogenetics and Evolution*, **58**, 22–32

Cho, G.Y., F. Rousseau, B. de Reviers, and S.M. Boo. (2006). Phylogenetic relationships within the Fucales (Phaeophyceae) assessed by the photosystem I coding *psa*A sequences. *Phycologia*, **45**, 512–519.

Choi, H.-G., G.T. Kraft, I.K. Lee, and Saunders, G.W. (2002). Phylogenetic analyses of anatomical and nuclear SSU rDNA sequence data indicate that the Dasyaceae and Delesseriaceae (Ceramiales, Rhodophyta) are polyphyletic. *European Journal of Phycology*, **37**, 551–569.

Christenhusz, M.J.M., and M.W. Chase. (2013). Biogeographical patterns of plants in the Neotropics – dispersal rather than plate tectonics is most explanatory. *Botanical Journal of the Linnean Society*, **171**, 277–286.

Christidis, L., and J.A. Norman. (2010). Evolution of the Australasian songbird fauna. *Emu*, **110**, 21–31.

Christidis, L., M. Irestedt, D. Rowe, W.E. Boles, and J.A. Norman. (2011). Mitochondrial and nuclear DNA phylogenies reveal a complex evolutionary history in the Australasian robins (Passeriformes: Petroicidae). *Molecular Phylogenetics and Evolution*, **61**, 726–738.

Christin, P.-A., G. Besnard, E. Samaritani, *et al.* (2008). Oligocene CO_2 decline promoted C_4 photosynthesis in grasses. *Current Biology*, **18**, 37–43.

Chung, K.-F. (2007). Inclusion of the South Pacific alpine genus *Oreomyrrhis* (Apiaceae) in *Chaerophyllum* based on nuclear and chloroplast DNA sequences. *Systematic Botany*, **32**, 671–681.

Chung, K.-F., C.-I. Peng, S. Downie, K. Spalik, and B. Schaal. (2005). Molecular systematics of the trans-Pacific alpine genus *Oreomyrrhis* (Apiaceae): phylogenetic affinities and biogeographic implications. *American Journal of Botany*, **92**, 2054–2071.

Cifelli, R.L., and B.M. Davis. (2003). Marsupial origins. *Science*, **302**, 1899–1900.

Clark, J.R., R.H. Ree, M.G. King, W.L. Wagner, and E.H. Roalson. (2008). A comparative study in ancestral range reconstruction methods: retracing the uncertain histories of insular lineages. *Systematic Biology*, **57**, 693–707.

Clark, J.R., W.L. Wagner, and E.H. Roalson. (2009). Patterns of diversification and ancestral range construction in the southeast Asian angiosperm lineage *Cyrtandra* (Gesneriaceae). *Molecular Phylogenetics and Evolution*, **53**, 982–994.

Clouse, R.M., D.M. General, A.C. Diesmos, and G. Giribet. (2011). An old lineage of Cyphoph-thalmi (Opiliones) discovered on Mindanao highlights the need for biogeographical research in the Philippines. *Journal of Arachnology*, **39**, 147–153.

Cluzel, D. and S. Meffre. (2002). L'unité de la Boghen (Nouvelle-Calédonie, Pacifique sudouest): un complexe d'accrétion jurassique. Données radiochronologiques préliminaires U-Pb sur les zircons détritiques. *Comptes Rendus Geoscience*, **334**, 867–874.

Cluzel, D., C.J. Adams, S. Meffre, H. Campbell, and P. Maurizot. (2010a). Discovery of Early Cretaceous rocks in New Caledonia; new geochemical and U-Pb zircon age constraints on the transition from subduction to marginal breakup in the Southwest Pacific. *Journal of Geology*, **118**, 381–397.

Cluzel, D., J. Aitchison, G. Clarke, S. Meffre, and C. Picard. (1994). Point de vue sur l'évolution tectonique et géodynamique de la Nouvelle-Calédonie (Pacifique, France). *Comptes Rendus de l'Académie des Sciences, Paris*, sér. II, **319**, 683–690.

Cluzel, D., J. Aitchison, and C. Picard. (2001). Tectonic accretion and underplating of mafic terranes in the Late Eocene intraoceanic fore-arc of New Caledonia (Southwest Pacific): geo-dynamic implications. *Tectonophysics*, **340**, 23–59.

Cluzel, D., P.M. Black, C. Picard, and K.N. Nicholson. (2010b). Geochemistry and tectonic setting of Matakaoa Volcanics (East Cast Allochthon, New Zealand): supra-subduction zone affinity, regional correlations and origin. *Tectonics*, **29**, TC2013, 1–21. doi: 10.1029/2009TCC002454.

Cluzel, D., D. Bosch, J.-L. Paquette, *et al.* (2005). Late Oligocene post-obduction granitoids of New Caledonia: a case for reactivated subduction and slab break-off. *Island Arc*, **14**, 254–271.

Cluzel, D., P. Maurizot, J. Collot, and B. Sevin. (2012). An outline of the geology of New Caledonia; from Permian–Mesozoic southeast Gondwanaland active margin to Cenozoic obduction and supergene evolution. *Episodes*, **35**, 72–86.

Cluzel, D., S. Meffre, P. Maurizot, and A.J. Crawford. (2006). Earliest Eocene (53 Ma) convergence in the Southwest Pacific: evidence from pre-obduction dikes in the ophiolite of New Caledonia. *Terra Nova*, **18**, 395–402.

Cohen, B.L., M.A. Bitner, E.M. Harper, *et al.* (2011). Vicariance and convergence in Magel-lanic and New Zealand long-looped brachiopod clades (Pan-Brachiopoda: Terebratelloidea). *Zoological Journal of the Linnean Society*, **162**, 631–645.

Colenso, W. (1885). Notes on the bones of a species of *Sphenodon* (*S. diversum*, Col.) appar-ently distinct from the species already known. *Transactions of the New Zealand Institute*, **18**, 118–128.

Collar, D.C., J.A. Schulte II, B.C. O'Meara, and J.B. Losos. (2010). Habitat use affects morpho-logical diversification in dragon lizards. *Journal of Evolutionary Biology*, **23**, 1033–1049.

Colley, H., and W.H. Hindle. (1984). Volcano-tectonic evolution of Fiji and adjoining marginal basins. *Geological Society of London Special Publication*, **16**, 151–162.

Collot, J.Y., S. Lallemand, B. Pelletier, *et al.* (1992). Geology of the d'Entrecasteaux-New Hebrides Arc collision zone: results from a deep submersible survey. *Tectonophysics*, **212**, 213–241.

Condamine, F.L., F.A.H. Sperling, and G.J. Kergoat. (2013a). Global biogeographical pattern of swallowtail diversification demonstrates alternative colonization routes in the Northern and Southern Hemispheres. *Journal of Biogeography*, **40**, 9–23.

Condamine, F.L., E.F.A. Toussaint, A.M. Cotton, *et al.* (2013b). Fine-scale biogeographical and temporal diversification processes of peacock swallowtails (*Papilio* subgenus *Achillides*) in the Indo-Australian Archipelago. *Cladistics*, **29**, 88–111.

Conran, J.G., and H.T. Clifford. (1985). The taxonomic affinities of the genus *Ripogonum*. *Nordic Journal of Botany*, **5**, 215–219.

Conran, J.G., R.J. Carpenter, and G.J. Jordan. (2009). Early Eocene *Ripogonum* (Liliales: Ripogonaceae) leaf macrofossils from southern Australia. *Australian Systematic Botany*, **22**, 219–228.

Corner, E.J.H. (1963). *Ficus* in the Pacific region. In *Pacific Basin Biogeography: A Symposium*, ed. J.L. Gressitt. Honolulu: Bishop Museum, pp. 233–245.

Corner, E.J.H. (1967). *Ficus* in the Solomon Islands and its bearing on the post-Jurassic history of Melanesia. *Philosophical Transactions of the Royal Society of London B*, **253**, 23–159.

Corner, E.J.H. (1969). Introduction to 'A discussion on the results of the Royal Society Expedition to the British Solomon Islands Protectorate, 1965'. *Philosophical Transactions of the Royal Society of London B*, **255**, 187–189.

Cornet, B. (1989). Late Triassic angiosperm-like pollen from the Richmond rift basin of Virginia, USA. *Palaeontographica. Abteilung. B Paläophytologie*, **213**, 37–87.

Cosacov, A., A.N. Sérsic, V. Sosa, *et al.* (2009). New insights into the phylogenetic relationships, character evolution, and phytogeographic patterns of *Calceolaria* (Calceolariaceae). *American Journal of Botany*, **96**, 2240–2255.

Couper, P.J., R.A. Sadlier, G.M. Shea, and J. Worthington Wilmer. (2008). A reassessment of *Saltuarius swaini* (Lacertilia: Diplodactylidae) in southeastern Queensland and New South Wales; two new taxa, phylogeny, biogeography and conservation. *Records of the Australian Museum*, **60**, 87–118.

Couri, M.S. and C.J.B. Carvalho. (2003). Systematic relations among *Philornis* Meinert, *Passeromyia* Rodhain & Villeneuve and allied genera (Diptera, Muscidae). *Brazilian Journal of Biology*, **63**, 223–232.

Couri, M.S., A.C. Pont, and C. Daugeron. (2010). The Muscidae (Diptera) of New Caledonia. *Zootaxa*, **2503**, 1–61.

Couvreur, T.L.P., M.D. Pirie, L.W. Chatrou, *et al.* (2011). Early evolutionary history of the flowering plant family Annonaceae: steady diversification and boreotropical geodispersal. *Journal of Biogeography*, **38**, 664–680.

Coyer, J.A., G.J. Smith, and R.A. Andersen. (2001). Evolution of *Macrocystis* spp. (Phaeophyceae) as determined by ITS1 and ITS2 sequences. *Journal of Phycology*, **37**, 574–585.

Coyne, J.A. (2009). (Re)reading The Origin. *Current Biology*, **19**, page R96.

Coyne, J.A., and H.A. Orr. (2004). *Speciation*. Sunderland, MA: Sinauer Associates.

Cracraft, J., and F.K. Barker. (2009). Passerine birds (Passeriformes). In *Timetree of Life*, eds. S.B. Hedges and S. Kumar. New York: Oxford University Press, pp. 423–431.

Cracraft, J., F.K. Barker, M. Braun, *et al.* (2004). Phylogenetic relationships among modern birds (Neornithes): toward an avian tree of life. In *Assembling the Tree of Life*, eds. J. Cracraft and M.J. Donoghue. New York: Oxford University Press, pp. 468–489.

Craft, K.J., S.U. Pauls, K. Darrow, *et al.* (2010). Population genetics of ecological communities with DNA barcodes: an example from New Guinea Lepidoptera. *Proceedings of the National Academy of Sciences USA*, **107**, 5041–5046.

Craig, D.A., D.C. Currie, and D.A. Joy. (2001). Geographical history of the central-western Pacific black fly subgenus *Inseliellum* (Diptera: Simuliidae: *Simulium*) based on a reconstructed phylogeny of the species, hot-spot archipelagoes and hydrological considerations. *Journal of Biogeography*, **28**, 1101–1127.

Craig, D.A., R.A. Englund, and H. Takaoka. (2006). Simuliidae (Diptera) of the Solomon Islands: new records and species, ecology, and biogeography. *Zootaxa*, **1328**, 1–26.

Crame, J.A. (1999). An evolutionary perspective on marine faunal connections between southern-most South America and Antarctica. *Scientia Marina*, **63**, (Suppl. 1), 1–14.

Cranston, P.S., D.H.D. Edward, and D.H. Colless. (1987). *Archaeochlus* Brundin: a midge out of time (Diptera: Chironomidae). *Systematic Entomology*, **12**, 313–334.

Cranston, P.S., D.H. Edward, and L.G. Cook. (2002). New status, species, distribution records and phylogeny for Australian mandibulate Chironomidae (Diptera). *Australian Journal of Entomology*, **41**, 357–366.

Cranston, P.S., N.B. Hardy, and G.E. Morse. (2012). A dated molecular phylogeny for the Chironomidae (Diptera). *Systematic Entomology*, **37**, 172–188.

Cranston, P.S., N.B. Hardy, G.E. Morse, L. Puslednik, and S.R. McCluen. (2010). When molecules and morphology concur: the 'Gondwanan' midges (Diptera: Chironomidae). *Systematic Entomology*, **35**, 636–648.

Craw, R.C., J.R. Grehan, and M.J. Heads. (1999). *Panbiogeography: Tracking the History of Life.* New York: Oxford University Press.

Crawford, A.J., S. Meffre, and P.A. Symonds. (2003). 120 to 0 Ma tectonic evolution of the southwest Pacific and analogous geological evolution of the 60 to 220 Ma Tasman Fold Belt System. *Geological Society of America Special Paper*, **372**, 383–404.

Crayn, D.M., M. Rossetto, and D.J. Maynard. (2006). Molecular phylogeny and dating reveals an Oligo-Miocene radiation of dry-adapted shrubs (former Tremandraceae) from rainforest tree progenitors (Elaeocarpaceae) in Australia. *American Journal of Botany*, **93**, 1328–1342.

Cribb, P., A.M. Pridgeon, and M.W. Chase. (2003). *Pachyplectron.* In *Genera Orchidacearum. Vol. 3. Orchidoideae (Part 2). Vanilloideae*, eds. A.M. Pridgeon, P.J. Cribb, M.W. Chase, and F.N. Rasmussen. New York: Oxford University Press, pp. 131–133.

Crisp, M.D., and L.G. Cook. (2007). A congruent molecular signature of vicariance across multiple plant lineages. *Molecular Phylogenetics and Evolution*, **43**, 1106–1117.

Crisp, M.D., L.G. Cook, and D. Steane. (2004). Radiation of the Australian flora: what can comparisons of molecular phylogenies across multiple taxa tell us about the evolution of diversity in present-day communities? *Philosophical Transactions of the Royal Society, London, B*, **359**, 1551–1571.

Crisp, M.D., Y. Isagi, Y. Kato, L.G. Cook, and D.M.J.S. Bowman. (2010). *Livistona* palms in Australia: ancient relics or opportunistic immigrants? *Molecular Phylogenetics and Evolution*, **54**, 512–523.

Crisp, M.D., S. Laffan, H.P. Linder, and A. Monro. (2001). Endemism in the Australian flora. *Journal of Biogeography*, **28**, 183–198.

Crisp, M.D., S.A. Trewick, and L.G. Cook. (2011). Hypothesis testing in biogeography. *Trends in Ecology and Evolution*, **26**, 66–72.

Croizat, L. (1958). *Panbiogeography.* Caracas: Published by the author.

Croizat, L. (1964). *Space, Time, Form: The Biological Synthesis.* Caracas: Published by the author.

Cronin, S.J., M.A. Ferland, and T.P. Terry. (2003). Nabukelevu volcano (Mt. Washington), Kadavu – a source of hitherto unknown volcanic hazard in Fiji. *Journal of Volcanology and Geothermal Research*, **131**, 371–396.

Cronk, Q.C.B., M. Kiehn, W.L. Wagner, and J.F. Smith. (2005). Evolution of *Cyrtandra* (Gesneriaceae) of the Pacific Ocean: the origin of a supertramp clade. *American Journal of Botany*, **92**, 1017–1024.

Cross, E.W., C.J. Quinn, and S.J. Wagstaff. (2002). Molecular evidence for the polyphyly of *Olearia* (Astereae: Asteraceae). *Plant Systematics and Evolution*, **235**, 99–120.

Crow, J. F. (2008). Mid-century controversies in population genetics. *Annual Reviews in Genetics*, **42**, 1–16.

Crowley, L.E.L.M. (1990). Biogeography of the endemic freshwater fish *Craterocephalus* (family Atherinidae). *Memoirs of the Queensland Museum*, **28**, 89–98.

Cruaud, Λ., R. Jabbour-Zahab, G. Genson, S. Ungricht, and J.-Y. Rasplus. (2012). Testing the emergence of New Caledonia: fig wasp mutualism as a case study and a review of evidence. *PLoS One*, **7**(2), e30941, 1–9.

Cryan, J.R., and J.M. Urban. (2011). Higher-level phylogeny of the insect order Hemiptera: is Auchenorrhyncha really paraphyletic? *Systematic Entomology*, **37**, 7–21.

Cuenca, A., C.B. Asmussen-Lange, and F. Borchsenius. (2008). A dated phylogeny of the palm tribe Chamaedoreae supports Eocene dispersal between Africa, North and South America. *Molecular Phylogenetics and Evolution*, **46**, 760–775.

Culver, D.C., and T. Pipan. (2009). *The Biology of Caves and Other Subterranean Habitats*. Oxford: Oxford University Press.

Cutter, A.D., A. Dey, and R.L. Murray. (2009). Evolution of the *Caenorhabditis elegans* genome. *Molecular Biology and Evolution*, **26**, 1199–1234.

Daczko, N.R., P. Caffi, and P. Mann. (2011). Structural evolution of the Dayman dome metamorphic core complex, eastern Papua New Guinea. *Bulletin of the Geological Society of America*, **123**, 2335–2351.

Damborenea, S.E., and M.O. Manceñido. (1992). A comparison of Jurassic marine benthonic faunas from South America and New Zealand. *Journal of the Royal Society of New Zealand*, **22**, 131–152.

Darlington, P.D. [1957] (1966). *Zoogeography: The Geographical Distribution of Animals*. New York: Wiley.

Darwin, C. (1859). *On the Origin of Species by Means of Natural Selection or the Preservation of Favoured Races in the Struggle for Life*. 1st edn. London: Murray.

Darwin, C. (1882). *The Descent of Man*. 2nd edn. London: Murray.

Darwin, C. (1887). Letter to J.D. Hooker, February 10th, 1845. In *The Life and Letters of Charles Darwin. Vol. 1*, ed. F. Darwin. London: Murray, p. 336.

Daugeron, C., C.A. d'Haese, and A.R. Plant. (2009). Phylogenetic systematic of the Gondwanan *Empis macrorrhyncha* group (Diptera, Empididae, Empidinae). *Systematic Entomology*, **34**, 635–648.

Davies, H.L. (2012). The geology of New Guinea: the cordilleran margin of the Australian continent. *Episodes*, **35**, 87–102.

Davis, C.C., C.D. Bell, P.W. Fritsch, and S. Mathews. (2002). Phylogeny of *Acridocarpus-Brachylophon* (Malpighiaceae): implications for Tertiary tropical floras and Afroasian biogeography. *Evolution*, **56**, 2395–2405.

Davis, E.B., M.S. Koo, C. Conroy, J.L. Patton, and C. Moritz. (2008). The California Hotspots Project: identifying regions of rapid diversification of mammals. *Molecular Ecology*, **17**, 120–138.

Dawson, J.W. (1981). The species rich, highly endemic serpentine flora of New Caledonia. *Tuatara*, **25**, 1–6.

Dawson, J.W. (1993). *Forest Vines to Snow Tussocks: The Story of New Zealand Plants*. Wellington: Victoria University Press.

Dawson, M.I., P.J. Brownsey, and J.D. Lovis. (2000). Index of chromosome numbers of indigenous New Zealand pteridophytes. *New Zealand Journal of Botany*, **38**, 25–46.

de Boer, A.J. (1995). Islands and cicadas adrift in the West-Pacific. Biogeographic patterns related to plate tectonics. *Tijdschrift voor Entomologie*, **138**, 169–244.

de Boer, A.J., and J.P. Duffels. (1996a). Historical biogeography of the cicadas of Wallacea, New Guinea and the West Pacific: a geotectonic explanation. *Palaeogeography, Palaeoclimatology, Palaeoecology*, **124**, 153–177.

de Boer, A.J., and J.P. Duffels. (1996b). Biogeography of Indo-Pacific cicadas east of Wallace's Line. In *The Origin and Evolution of Pacific Island Biotas, New Guinea to Eastern Polynesia: Patterns and Processes*, eds. A. Keast and S.E. Miller. Amsterdam: SPB Academic Publishing, pp. 297–330.

Deiner, K., A.R. Lemmon, A.L. Mack, R.C. Fleischer, and J.P. Dumbacher. (2011). A passerine bird's evolution corroborates the geologic history of the island of New Guinea. *PLoS One*, **6**(5), e19479, 1–15.

de Jong, R. (2004). Phylogeny and biogeography of the genus *Taractrocera* Butler, 1870 (Lepidoptera: Hesperiidae), an example of Southeast Asian–Australasian interchange. *Zoologische Mededelingen*, **78**, 383–415.

de Jong, R., and K. van Dorp. (2006). The African copper connection. *Entomologische Berichten*, **66**, 124–129.

De Kok, R.P.J. (2002). Are plant adaptations to growing on serpentine soil rare or common? A few case studies from New Caledonia. *Adansonia*, **24**, 229–238.

Delacour, J., and E. Mayr. (1946). *Birds of the Philippines*. New York: MacMillan.

de Lange, P.J., R.D. Smissen, S.J. Wagstaff, *et al*. (2010). A molecular phylogeny and infrageneric classification for *Kunzea* (Myrtaceae) inferred from rDNA ITS and ETS sequences. *Australian Systematic Botany*, **23**, 309–319.

de Laubenfels, D.J. (1988). Coniferales. *Flora Malesiana* I, **10**, 367–442.

de Laubenfels, D.J. (1996). Gondwanan conifers on the Pacific Rim. In *The Origin and Evolution of Pacific Island Biotas, New Guinea to Eastern Polynesia: Patterns and Processes*, eds. A. Keast and S.E. Miller. Amsterdam: SPB Academic, pp. 261–265.

Dempewolf, H., and L.H. Reiseberg. (2007). Adaptive evolution: the legacy of past giants. *Current Biology*, **17**, 773–774.

den Tex, R.-J., R. Thorington, J.E. Maldonado, and J.A. Leonard. (2010). Speciation dynamics in the SE Asian tropics: putting a time perspective on the phylogeny and biogeography of Sundaland tree squirrels, *Sundasciurus*. *Molecular Phylogenetics and Evolution*, **55**, 711–720.

Der, J.P., and D.L. Nickrent. (2008). A molecular phylogeny of Santalaceae. *Systematic Botany*, **33**, 107–116.

Der, J.P., J.A. Thompson, J.K. Stratford, and P.G. Wolf. (2009). Global chloroplast phylogeny and biogeography of bracken (*Pteridium*: Dennstaedtiaceae). *American Journal of Botany*, **96**, 1041–1049.

Désamoré, A., A. Vanderpoorten, B. Laenen, P. Kok, and S.R. Gradstein. (2010). Biogeography of the Lost World (Pantepui area, South America): insights from bryophytes. *Phytotaxa*, **9**, 254–265.

Deschamps, A., P. Monié, S. Lallemand, S.K. Hsu, and J. Yeh. (2000). Evidence for Early Cretaceous oceanic crust trapped in the Philippine Sea Plate. *Earth and Planetary Science Letters*, **179**, 503–516.

Dettmann, M.E., and H.T. Clifford. (2005). Biogeography of Araucariaceae. In *Australian and New Zealand Forest Histories: Araucarian Forests*. Australian Forest History Society Inc. Occasional Publication 2, Kingston, ACT, Australia: Australian Forest History Society Inc., pp. 1–9.

Devey, D.S., I. Leitch, P.J. Rudall, J.C. Pires, Y. Pillon, and M.W. Chase. (2006). Systematics of Xanthorrhoeaceae sensu lato, with an emphasis on *Bulbine*. *Aliso*, **22**, 345–351.

d'Horta, F.M., A.M. Cuervo, C.C. Ribas, R.T. Brumfield, and C.Y. Miyaki. (2013). Phylogeny and comparative phylogeography of *Sclerurus* (Aves: Furnariidae) reveal constant and cryptic diversification in an old radiation of rain forest understorey specialists. *Journal of Biogeography*, **40**, 37–49.

Diamond, J.M. (1972). *Avifauna of the Eastern Highlands of New Guinea*. Cambridge, MA: Nuttall Ornithological Club.

Diamond, J.M. (1985). New distributional records and new taxa from the outlying mountain ranges of New Guinea. *Emu*, **85**, 65–91.

DiCaprio, L., M. Gurnis, R.D. Müller, and E. Tan. (2011). Mantle dynamics of continentwide Cenozoic subsidence and tilting of Australia. *Lithosphere*, **3**, 3–11.

Dickerson, R.E. (1928). Distribution of life in the Philippines. *Philippine Bureau of Sciences Monograph*, **21**, 1–322.

Dickinson, E.C. (ed.) (2003). *The Howard and Moore Complete Checklist of the Birds of the World*, 3rd edn. Princeton, NJ: Princeton University Press.

Diesmos, A.C., R.M. Brown, A.C. Alcala, *et al.* (2002). Philippine amphibians and reptiles. In *Philippine Biodiversity Conservation Priorities: A Second Iteration of the National Biodiversity Strategy and Action Plan*, eds. P. Ong, L. Afuang, and R. Rosell-Ambal. Quezon City: Philippines Department of the Environment and Natural Resources – Protected Areas and Wildlife Bureau, pp. 26–44.

Dimalanta, C.B., and G.P. Yumul, Jr. (2006). Magmatic and amagmatic contributions to crustal growth in the Philippine island arc system: comparison of the Cretaceous and post-Cretaceous periods. *Geosciences Journal*, **10**, 321–329.

Dimalanta, C.B., L.O. Suerte, G.P. Yumul Jr., R.A. Tamayo Jr., and E.G.L. Ramos. (2006). A Cretaceous supra-subduction oceanic basin source for Central Philippine ophiolitic basement complexes: geological and geophysical constraints. *Geosciences Journal*, **10**, 305–320.

Dingle, H. (2008). Bird migration in the southern hemisphere: a review comparing continents. *Emu*, **108**, 341–259.

Dixon, D., M.J. Benton, A. Kingsley, and J. Baker. (2001). *Atlas of Life on Earth*. New York: Barnes and Noble.

Doan, T.M. (2003). A south-to-north biogeographic hypothesis for Andean speciation: evidence from the lizard genus *Proctoporus* (Reptilia, Gymnophthalmidae). *Journal of Biogeography*, **30**, 361–374.

Dobzhansky, T. (1935). A critique of the species concept in biology. *Philosophy of Science*, **2**, 344–355.

Donoghue, P.C.J., and M.J. Benton. (2007). Rocks and clocks: calibrating the Tree of Life using fossils and molecules. *Trends in Ecology and Evolution*, **22**, 424–431.

Donovan, S.K., and C.R.C. Paul (eds.) (1998a). *The Adequacy of the Fossil Record*. Chichester, UK: Wiley.

Donovan, S.K., and C.R.C. Paul. (1998b). Introduction: adequacy vs. incompleteness. In *The Adequacy of the Fossil Record*, eds. S.K. Donovan and C.R.C. Paul. Chichester: Wiley, pp. ix–x.

Doughty, P., and P.M. Oliver. (2011). A new species of *Underwoodisaurus* (Squamata: Gekkota: Carphodactylidae) from the Pilbara region of Western Australia. *Zootaxa*, **3010**, 20–30.

Dowe, J.L. (2009). A taxonomic account of *Livistona* R.Br. (Arecaceae). *Gardens' Bulletin Singapore*, **60**, 185–344.

Doyle, J.A. (2005). Early evolution of angiosperm pollen as inferred from molecular and morphological phylogenetic analyses. *Grana*, **44**, 227–251.

Dransfield, J. (2012). *Paschalococos disperta*. Available at: www.pacsoa.org.au/palms/Paschalococos/disperta.html

Dress, C., A. Matern, G. von Oheimb, T. Reimann, and T. Assmann. (2010). Multiple glacial refuges of unwinged ground beetles in Europe: molecular data support classical phylogeographic models. In *Relict Species: Phylogeography and Conservation Biology*, ed. J.C. Habel and T. Assmann. Berlin: Springer, pp.199–215.

Driskell, A., L. Christidis, B.J. Gill, *et al.* (2007). A new endemic family of New Zealand passerine birds: adding heat to a biodiversity hotspot. *Australian Journal of Zoology*, **55**, 73–78.

Driskell, A.C., J.A. Norman, S. Pruett-Jones, *et al.* (2011). A multigene phylogeny examining evolutionary and ecological relationships in the Australo-Papuan wrens of the subfamily Malurinae (Aves). *Molecular Phylogenetics and Evolution*, **60**, 480–485.

Drummond, A.J., and A. Rambaut. (2007). BEAST: Bayesian evolutionary analysis by sampling trees. *BMC Evolutionary Biology* **7**(214), 1–8.

Duangjai, S., R. Samuel, J. Munzinger, *et al.* (2009). A multi-locus plastid phylogenetic analysis of the pantropical genus *Diospyros* (Ebenaceae), with an emphasis on the radiation and biogeographic origins of the New Caledonian endemic species. *Molecular Phylogenetics and Evolution*, **52**, 602–620.

Duchene, S., A. Frey, A. Alfaro-Núñez, *et al.* (2012). Marine turtle mitogenome phylogenetics and evolution. *Molecular Phylogenetics and Evolution*, **65**, 241–250.

Dugdale, J.S. (1996). *Chrysorthenches* new genus, conifer-associated plutellid moths (Yponomeutoidea, Lepidoptera) in New Zealand and Australia. *New Zealand Journal of Zoology*, **23**, 33–59.

Dumbacher, J.P., T.K. Pratt, and R.C. Fleischer. (2003). Phylogeny of the owlet-nightjars (Aves: Aegothelidae) based on mitochondrial DNA sequence. *Molecular Phylogenetics and Evolution*, **29**, 540–549.

Duret, L., and N. Galtier. (2009). Biased gene conversion and the evolution of mammalian genomic landscapes. *Annual Review of Genomics and Human Genetics*, **10**, 285–311.

Early, J.W., L. Masner, I.D. Naumann, and A.D. Austin. (2001). Maamingidae, a new family of proctotrupoid wasp (Insecta: Hymenoptera) from New Zealand. *Invertebrate Taxonomy*, **15**, 341–352.

Ebach, M., and C.J. Humphries. (2002). Cladistic biogeography and the art of discovery. *Journal of Biogeography*, **29**, 427–444.

Eberhard, S.M., S.A. Halse, and W.F. Humphreys. (2005). Stygofauna in the Pilbara region, north-west Western Australia: a review. *Journal of the Royal Society of Western Australia*, **88**, 167–176.

Edgar, E., and H.E. Connor. (2000). *Flora of New Zealand. Vol. 5. Gramineae*. Wellington: Government Printer.

Edgecombe, G.D. (2001). Revision of *Paralamyctes* (Chilopoda: Lithobiomorpha: Henicopidae), with six new species from eastern Australia. *Records of the Australian Museum*, **53**, 201–241.

Edgecombe, G.D., and L. Barrow. (2007). A new genus of scutigerid centipedes (Chilopoda) from Western Australia, with new characters for morphological phylogenetics of Scutigeromorpha. *Zootaxa*, **1409**, 23–50.

Edgecombe, G.D., and G. Giribet. (2008). A New Zealand species of the trans-Tasman centipede order Craterostigmomorpha (Arthropoda: Chilopoda) corroborated by molecular evidence. *Invertebrate Systematics*, **22**, 1–15.

Edwards, D.L., and J. Melville. (2010). Phylogeographic analysis detects congruent biogeographic patterns between a woodland agamid and Australian wet tropics taxa despite disparate evolutionary trajectories. *Journal of Biogeography*, **37**, 1543–1556.

Edwards, S.V. (2009). Is a new and general theory of molecular systematics emerging? *Evolution*, **63**, 1–19.

Edwards, S.V., and W.E. Boles. (2002). Out of Gondwana: the origin of passerine birds. *Trends in Ecology and Evolution*, **17**, 347–349.

Ehrendorfer, F., and R. Samuel. (2000). Comments on S. B. Hoot's interpretation of Southern Hemisphere relationships in Anemone (Ranunculaceae) based on molecular data [Am. J. Bot. 2000; 87(6, Suppl.), 154–155]. *Taxon*, **49**, 781–784.

Eiting, T.P., and G.F. Gunnell. (2009). Global completeness of the fossil record. *Journal of Mammalian Evolution*, **16**, 151–173.

Eken, G., L. Bennun, T.M. Brooks, *et al.* (2004). Key Biodiversity Areas as site conservation targets. *BioScience*, **54**, 1110–1118.

Eldredge, N., J.N. Thompson, P.M. Brakefield, *et al.* (2005). The dynamics of evolutionary stasis. *Paleobiology*, **31**, 133–145.

Ellis, D.H., C.B. Kepler, A.K. Kepler, and K. Teebaki. (1990). Occurrence of the Longtailed Cuckoo *Eudynamis taitensis* on Caroline Atoll, Kiribati. *Emu*, **90**, 202–203.

Elórtegui Francioli, S., and A. Moreira Muñoz. (2002). *Parque Nacional La Campana: Orígen de Una Reserva de la Bíosfera en Chile Central*. Santiago: Taller la Era.

Engel, J.J., and J. Heinrichs. (2008). Studies of New Zealand Hepaticae. 39. *Dinckleria* Trevis, an older name for *Proskauera*. J. Heinrichs and J.J. Engel. *Cryptogamie, Bryologie*, **29**, 193–194.

Ericson, P.G.P., L. Christidis, A. Cooper, *et al.* (2002). A Gondwanan origin of passerine birds supported by DNA sequences of the endemic New Zealand wrens. *Proceedings of the Royal Society, London, B*, **269**, 235–241.

Eriksen, B. (1993). Phylogeny of the Polygalaceae and its taxonomic implications. *Plant Systematics and Evolution*, **186**, 33–55.

Espeland, M., and K.A. Johanson. (2010). The effect of environmental diversification on species diversification in New Caledonian caddisflies (Insecta: Trichoptera: Hydropsychidae). *Journal of Biogeography*, **37**, 879–890.

Espeland, M., and J. Murienne. (2011). Diversity dynamics in New Caledonia: towards the end of the museum model? *BMC Evolutionary Biology*, **11**(254), 1–13.

Esselstyn, J.A., and R.M. Brown. (2009). The role of repeated sea-level fluctuations in the generation of shrew (Soricidae: *Crocidura*) diversity in the Philippine Archipelago. *Molecular Phylogenetics and Evolution*, **53**, 171–181.

Esselstyn, J.A., and C.H. Oliveros. (2010). Colonization of the Philippines from Taiwan: a multilocus test of the biogeographic and phylogenetic relationships of isolated populations of shrews. *Journal of Biogeography*, **37**, 1504–1514.

Esselstyn, J.A., S.P. Maher, and R.M. Brown. (2011). Species interactions during diversification and community assembly in an island radiation of shrews. *PLoS One*, **6**(7), e21885, 1–13.

Esselstyn, J.A., C.H. Oliveros, R.G. Moyle, *et al.* (2010). Integrating phylogenetic and taxonomic evidence illuminates complex biogeographic patterns along Huxley's modification of Wallace's Line. *Journal of Biogeography*, **37**, 2054–2066.

Evans, B.J., R.M. Brown, J.A. McGuire, *et al.* (2003). Phylogenetics of fanged frogs: testing biogeographical hypotheses at the interface of the Asian and Australian faunal zones. *Systematic Biology*, **52**, 794–819.

Evans, K.M., V.A. Chepurnov, H.J. Sluiman, *et al.* (2009). Highly differentiated populations of the freshwater diatom *Sellaphora capitata* suggest limited dispersal and opportunities for allopatric speciation. *Protist*, **160**, 386–396.

Evenhuis, N.L. (2006). Catalog of the Keroplatidae of the world. *Bishop Museum Bulletin in Entomology*, **13**, 1–177.

Ewald, J. (2003). The calcareous riddle: why are there so many calciphilous species in the central European flora? *Folia Geobotanica*, **38**, 357–366.

Ewen, J.G., I. Flux, and P.G.P. Ericson. (2006). Systematic affinities of two enigmatic New Zealand passerines of high conservation priority, the hihi or stitchbird *Notiomystis cincta* and the kokako *Callaeas cinerea*. *Molecular Phylogenetics and Evolution*, **40**, 281–284.

Fain, M.G., and P. Houde. (2007). Multilocus perspectives on the monophyly and phylogeny of the order Charadriiformes (Aves). *BMC Evolutionary Biology*, **7** (35), 1–15.

Faith, D.P. (1992). Conservation evaluation and phylogenetic diversity. *Biological Conservation*, **61**, 1–10.

Faith, D.P., C.A.M. Reid, and J. Hunter. (2004). Integrating phylogenetic diversity, complementarity, and endemism for conservation assessment. *Conservation Biology*, **18**, 255–261.

Farrell, B.D. (1998). 'Inordinate fondness' explained: why are there so many species of beetles? *Science*, **281**, 555–559.

Faulks, L.K., D.M. Gilligan, and L.B. Beheregaray. (2010). Clarifying an ambiguous evolutionary history: range-wide phylogeography of an Australian freshwater fish, the golden perch (*Macquaria ambigua*). *Journal of Biogeography*, **37**, 1329–1340.

Fay, M.F., M.W. Chase, N. Rønsted, *et al.* (2006). Phylogenetics of Liliales: summarized evidence from combined analyses of five plastid and one mitochondrial loci. *Aliso*, **22**, 559–565.

Fay, M.F., P.J. Rudall, S. Sullivan, *et al.* (2000). Phylogenetic studies of Asparagales based on four plastid DNA regions. In *Monocots: Systematics and Evolution*, eds. K.L. Wilson and D.A. Morrison. Collingwood: CSIRO, pp. 360–371.

Feng, M. (2005). *Floral Morphogenesis and Molecular Systematics of the Family Violaceae*. PhD Dissertation, Ohio University, Athens, OH.

Ferris, A., G.A. Abers, B. Zelt, B. Taylor, and S. Roecker. (2006). Crustal structure across the transition from rifting to spreading: the Woodlark rift system of Papua New Guinea. *Geophysical Journal International*, **166**, 622–634.

Fiedler, P.L. (1985). Heavy metal accumulation and the nature of edaphic endemism in the genus *Calochortus* (Liliaceae). *American Journal of Botany*, **72**, 1712–1718.

Filardi, C.E., and R.G. Moyle. (2005). Single origin of a pan-Pacific bird group and upstream colonization of Australasia. *Nature*, **438**, 216–219.

Filardi, C.E., and C.E. Smith. (2005). Molecular phylogenetics of monarch flycatchers (genus *Monarcha*) with emphasis on Solomon Island endemics. *Molecular Phylogenetics and Evolution*, **37**, 776–788.

Fitton, J.G., J.J. Mahoney, P.J. Wallace, and A.D. Saunders. (2004). Origin and evolution of the Ontong Java Plateau: introduction. In *Origin and Evolution of the Ontong Java Plateau*, eds. J.G. Fitton, J.J. Mahoney, P.J. Wallace, and A.D. Saunders. *Geological Society of London Special Publications*, **229**, 1–8.

Fjeldså, J., M. Irestedt, P.G.P. Ericson, and D. Zuccon. (2010). The Cinnamon Ibon *Hypocryptadius cinnamomeus* is a forest canopy sparrow. *Ibis*, **152**, 747–760.

Flecks, M., A. Schmitz, W. Böhme, F.W. Henkel, and I. Ineich. (2012). A new species of *Gehyra* Gray, 1834 (Squamata, Gekkonidae) from the Loyalty Islands and Vanuatu, and phylogenetic relationships in the genus *Gehyra* in Melanesia. *Zoosystema*, **34**, 203–221.

Fleming, C.A. (1979). *The Geological History of New Zealand and its Life*. Auckland: Auckland University Press and Oxford University Press.

Florin, A-B. (2001). Bottlenecks and blowflies: speciation, reproduction and morphological variation in Lucilia. *Acta Universitatis Upsaliensis. Comprehensive Summaries of Uppsala Dissertations from the Faculty of Science and Technology*, **660**, 1–40.

Flowers, R.W. (1990). Ephemeroptera of the Fiji Islands. In *Mayflies and Stoneflies: Life Histories and Biology*, ed. I.C. Campbell. Dordrecht: Kluwer, pp. 125–133.

Ford, K.A., J.M. Ward, R.D. Simssen, S.J. Wagstaff, and I. Breitwieser. (2007). Phylogeny and biogeography of *Craspedia* (Asteraceae: Gnaphalieae) based on ITS, ETS and *psbA-trnH* sequence data. *Taxon*, **56**, 783–794.

Forest, F. (2009). Calibrating the Tree of Life: fossils, molecules and evolutionary timescales. *Annals of Botany*, **104**, 789–794.

Forest, F., M.W. Chase, C. Persson, P.R. Crane, and J.A. Hawkins. (2007a). The role of biotic and abiotic factors in evolution of ant dispersal in the milkwort family (Polygalaceae). *Evolution*, **61**, 1675–1694.

Forest, F., R. Grenyer, M. Rouget, *et al.* (2007b). Preserving the evolutionary potential of floras in biodiversity hotspots. *Nature*, **445**, 757–760.

Forman, L. L., R.W.J.M. van der Ham, M.M. Harley, and T.J. Lawrence. (1994). *Rosselia*, a new genus of Burseraceae from the Louisiade Archipelago, Papua New Guinea. *Kew Bulletin*, **49**, 601–621.

Fortey, R.A., and R.H. Thomas. (1993). The case of the velvet worm. *Nature*, **361**, 205–206.

Foulger, G.R., and D.M. Jurdy, eds. (2007). Plates, plumes, and planetary processes. *Geological Society of America Special Papers*, **430**, 1–998.

Foulger, G. R., and J. H. Natland. (2003). Is 'hotspot' volcanism a consequence of plate tectonics? *Science*, **300**, 921–922.

Foulger, G.R., J.H. Natland, D.C. Presnall, and D.L. Anderson, eds. (2005). Plates, plumes, and paradigms. *Geological Society of America Special Paper*, **388**, 1–881.

Frey, J.K. (1993). Modes of peripheral isolate formation and speciation. *Systematic Zoology*, **42**, 373–381.

Frey, W., T. Pfeiffer, and M. Stech. (2010). Geomolecular divergence patterns of Gondwanan and Palaeoaustral bryophytes – an overview. Studies in austral temperate rain forest bryophytes 34. *Nova Hedwigia*, **91**, 317–348.

Fries, M., W. Chapco, and D. Contreras. (2007). A molecular phylogenetic analysis of the Oedipodinae and their intercontinental relationships. *Journal of Orthoptera Research*, **16**, 115–125.

Friis, E.M., K.R. Pedersen, and P.R. Crane. (2006). Cretaceous angiosperm flowers: innovation and evolution in plant reproduction. *Palaeogeography, Palaeoclimatology, Palaeoecology*, **232**, 251–293.

Fritsch, P.W., L. Lu, C.M. Bush, *et al.* (2011). Phylogenetic analysis of the wintergreen group (Ericaceae) based on six genic regions. *Systematic Botany*, **36**, 990–1003.

Fritz, S.A., K.A. Jønsson, J. Fjeldså, and C. Rahbek. (2012). Diversification and biogeographic patterns in four island radiations of passerine birds. *Evolution*, **66**, 179–190.

Frodin, D.G., P.P. Lowry II, and G.M. Plunkett. (2010). *Schefflera* (Araliaceae): taxonomic history, overview and progress. *Plant Diversity and Evolution*, **128**, 561–595.

Fuchs, J., J.-M. Pons, S.M. Goodman, *et al.* (2008). Tracing the colonization history of the Indian Ocean scops-owls (Strigiformes: *Otus*) with further insight into the spatio-temporal origin of the Malagasy avifauna. *BMC Evolutionary Biology*, **8**(197), 1–15.

Funakoshi, H., and H. Ohashi. (2000). *Vanoverberghia sasakiana* H. Funak. & H. Ohashi (Zingiberaceae), a new species and a new generic record for the flora of Taiwan. *Taiwania*, **45**, 270–275.

Funk, V.A., R. Chan, and A. Holland. (2007). *Cymbonotus* (Compositae: Arctotideae, Arctotidinae): an endemic Australian genus embedded in a southern African clade. *Botanical Journal of the Linnean Society*, **153**, 1–8.

Gadek, P.A., D.L. Alpers, M.M. Heslewood, and C.J. Quinn. (2000). Relationships within Cupressaceae sensu lato: a combined morphological and molecular approach. *American Journal of Botany*, **87**, 1044–1057.

Gaina, C., R.D. Müller, C. Heine, and G. Borel. (2002). Mesozoic–Cenozoic plate boundaries between Neo-Tethys/Indian and Pacific Oceans. *EOS Transactions of the American Geophysical Union* **83**(22). Western Pacific Geophysics Meeting Supplement Abstract #SE41D-01.

Gaina, C., R.D. Müller, J.Y. Royer, and P. Symonds. (1999). Evolution of the Louisiade triple junction. *Journal of Geophysical Research*, **104**, 12927–12940.

Galley, C., and H.P. Linder. (2006). Geographical affinities of the Cape flora, South Africa. *Journal of Biogeography*, **33**, 236–250.

Gallo, V., M.J. Cavalcanti, and H.M.A. da Silva. (2007). Track analysis of the marine palaeofauna from the Turonian (Late Cretaceous). *Journal of Biogeography*, **34**, 1167–1172.

Galloway, D.J. (1994). Studies in *Pseudocyphellaria* (Lichens) IV. Palaeotropical species (excluding Australia). *Bulletin of the Natural History Museum, London, Botany Series*, **24**, 115–159.

Galloway, D.J. (2007). *Flora of New Zealand: Lichens*. 2nd edn. Lincoln: Manaaki Whenua.

Gamauf, A., and E. Haring. (2004). Molecular phylogeny and biogeography of Honey-buzzards (genera *Pernis* and *Henicopernis*). *Journal of Zoological Systematics and Evolutionary Research*, **42**, 145–153.

Gamauf, A., J.-O. Gjershaug, N. Røv, K. Kvaløy, and E. Haring. (2005). Species or subspecies? The dilemma of taxonomic ranking of some South-East Asian hawk-eagles (genus *Spizaetus*). *Bird Conservation International*, **15**, 99–117.

Gamble, T., A.M. Bauer, G.R. Colli, *et al.* (2010). Coming to America: multiple origins of New World geckos. *Journal of Evolutionary Biology*, **24**, 231–244.

Gamerro, J.C., and V. Barreda. (2008). New fossil record of Lactoridaceae in South America: a palaeobiogeographical approach. *Botanical Journal of the Linnean Society*, **158**, 41–50.

Gandolfo, M.A., M.C. Dibbern, and E.J. Romero. (1988). *Akania patagonica* n. sp. and additional material on *Akania americana* Romero & Hickey (Akaniaceae) from Paleocene sediments of Patagonia. *Bulletin of the Torrey Botanical Club*, **115**, 83–88.

Gandolfo, M.A., E.J. Hermsen, M.C. Zamaloa, *et al.* (2011). Oldest known Eucalyptus macrofossils are from South America. *PLoS One*, **6**(6), e21084, 1–9.

Gao, K., and N.H. Shubin. (2003). Earliest known crown group salamanders. *Nature*, **422**, 424–426.

Gardner, J.L., J.W.H. Trueman, D. Ebert, L. Joseph, and R.D. Magrath. (2010). Phylogeny and evolution of the Meliphagoidea, the largest radiation of Australasian songbirds. *Molecular Phylogenetics and Evolution*, **55**, 1087–1102.

Gardner, M.G., A.F. Hugall, S.C. Donnellan, M.N. Hutchinson, and R. Foster. (2008). Molecular systematics of social skinks: phylogeny and taxonomy of the *Egernia* group (Reptilia: Scincidae). *Zoological Journal of the Linnean Society*, **154**, 781–794.

Gardner, R.O., and P.J. de Lange. (2002). Revision of *Pennantia* (Icacinaceae), a small isolated genus of southern hemisphere trees. *Journal of the Royal Society of New Zealand*, **32**, 669–695.

Gassmann, D. (2005). The phylogeny of Southeast Asian and Indo-Pacific Calicnemiinae (Odonata, Platycnemididae). *Bonner Zoologische Beiträge*, **53**, 37–80.

Gaston, K.J., P.H. Williams, P. Eggleton, C.J. Humphries. (1995). Large scale patterns of biodiversity: spatial variation in family richness. *Proceedings of the Royal Society, London, B*, **260**, 149–154.

Gauld, I.D., and D.B. Wahl. (2000). The Labeninae (Hymenoptera: Ichneumonidae): a study in phylogenetic reconstruction and evolutionary biology. *Zoological Journal of the Linnean Society*, **129**, 271–347.

Gauthier, J.A., R. Estes, and K. de Queiroz. (1988). A phylogenetic analysis of Lepidosauromorpha. In *Phylogenetic Relationships of the Lizard Families*, eds. R. Estes and G. Pregill. Palo Alto, CA: Stanford University Press, pp. 15–98.

GBIF. (2012). Global Biodiversity Information Facility website. http://data.gbif.org/

Genner, M.J., O. Seehausen, D.H. Lunt, *et al.* (2007). Age of cichlids: new dates for ancient fish radiation. *Molecular Biology and Evolution*, **24**, 1269–1282.

George, A. (2009). Introduction. In *Wildflowers of Southern Western Australia*, 3rd edn., eds. M.G. Corrick and B.A. Fuhrer. Kenthurst NSW: Rosenberg Publishing, pp. 9–15.

Gibbs, G.W. (1990). Local or global? Biogeography of some primitive Lepidoptera in New Zealand. *New Zealand Journal of Zoology*, **16**, 689–698.

Gibbs, G.W. (2001). Habitats and biogeography of New Zealand's deinacridine and tusked weta species. In *The Biology of Wetas, King Crickets and their Allies*, ed. L.H. Field. Oxford: CABI Publishing, pp. 35–56.

Gibbs, G.W. (2006). *Ghosts of Gondwana: The History of Life in New Zealand*. Nelson: Craig Potton.

Gibbs, G.W. (2010). Micropterigidae (Lepidoptera) of the Southwestern Pacific: a revision with the establishment of five new genera from Australia, New Caledonia and New Zealand. *Zootaxa*, **2520**, 1–48.

Gibson, R., and A. Baker. (2012). Multiple gene sequences resolve phylogenetic relationships in the shorebird suborder Scolopaci (Aves: Charadriiformes). *Molecular Phylogenetics and Evolution*, **64**, 66–72.

Gill, B.J. (1998). Behavior and ecology of the shining cuckoo, *Chrysococcyx lucidus*. In *Parasitic Birds and their Hosts*, eds. S.I. Rothstein and S.K. Robinson. Oxford: Oxford University Press, pp. 143–151

Gillespie, E.L., and K.A. Kron. (2012). A new tribe, Bryantheae (Ericoideae: Ericaceae), composed of disjunct genera from South America and Japan. *Brittonia*, **64**, 73–75.

Gillespie, R.G., B.G. Baldwin, J.M. Waters, *et al.* (2012). Long-distance dispersal: a framework for hypothesis testing. *Trends in Ecology and Evolution*, **27**, 47–56.

Gilmore, S., and K.D. Hill. (1997). Relationships of the Wollemi pine (*Wollemia nobilis*) and a molecular phylogeny of the Araucariaceae. *Telopea*, **7**, 275–291.

Gingerich, P.D., and M.D. Uhen. (1994). Time of origin of primates. *Journal of Human Evolution*, **27**, 443–445.

Giribet, G., and G.D. Edgecombe. (2006a). Conflict between datasets and phylogeny of centipedes: an analysis based on seven genes and morphology. *Proceedings of the Royal Society, B*, **269**, 235–241.

Giribet, G., and G.D. Edgecombe. (2006b). The importance of looking at small-scale patterns when inferring Gondwanan biogeography: a case study of the centipede *Paralamyctes*

(Chilopoda, Lithobiomorpha, Henicopidae). *Biological Journal of the Linnean Society*, **89**, 65–78.

Givnish, T.J., and S.S. Renner. (2004a). Tropical intercontinental disjunctions: Gondwana breakup, immigration from the boreal tropics, and transoceanic dispersal. *International Journal of Plant Science*, **165** (Suppl. 4), S1–S6.

Givnish, T.J., and S.S. Renner, eds. (2004b). Tropical intercontinental disjunctions. *International Journal of Plant Science*, **165** (Suppl. 4).

Glen, R.A., and S. Meffre. (2009). Styles of Cenozoic collisions in the western and southwestern Pacific and their applications to Palaeozoic collisions in the Tasmanides of eastern Australia. *Tectonophysics*, **479**, 130–149.

Glenny, D. (1997). Evolution and biogeography of New Zealand *Anaphalis* (Asteraceae: Inuleae: Gnaphaliinae) inferred from rDNA sequences. *New Zealand Journal of Botany*, **35**, 441–449.

Glenny, D. (2004). A revision of the genus *Gentianella* in New Zealand. *New Zealand Journal of Botany*, **42**, 361–530.

Goetsch, L.A., L.A. Craven, and B.D. Hall. (2011). Major speciation accompanied the dispersal of Vireya Rhododendrons (Ericaceae, *Rhododendron* sect. *Schistanthe*) through the Malayan archipelago: evidence from nuclear gene sequences. *Taxon*, **60**, 1015–1028.

Goldblatt, P., A. Rodriguez, M.P. Powell, *et al.* (2008). Iridaceae 'out of Australasia'? Phylogeny, biogeography and divergence time based on plastid DNA sequences. *Systematic Botany*, **33**, 495–508.

González, F., and P.J. Rudall. (2007). Floral morphology of the neotropical family Metteniusaceae, an isolated member of the lamiids. *Plant Biology and Botany 2007. Program and Abstract Book*. Chicago, IL.

González, F., J. Betancur, O. Maurin, J.V. Freudenstein, and M.W. Chase. (2007). Metteniusaceae, an early-diverging family in the lamiid clade. *Taxon*, **56**, 795–800.

Gonzalez, L.A., R.O. Bustamante, R.M. Navarro, M.A. Herrera, and M.T. Ibañez. (2009). Ecology and management of the Chilean palm (*Jubaea chilensis*): history, current situation and perspectives. *Palms*, **53**, 68–74.

Good, R. (1957). Australasian floras. *Nature*, **179**, 926.

Good, R. (1974). *The Geography of Flowering Plants*, 4th edn. London: Longman.

Goswami, A., and P. Upchurch. (2010). The dating game: a reply to Heads. *Zoologica Scripta*, **39**, 406–409.

Gould, S. (1989). *The Burgess Shale and the Nature of History*. London: Penguin.

Gove, A.D., M.C. Fitzpatrick, J.D. Majer, and R.R. Dunn. (2009). Dispersal traits linked to range size through range location, not dispersal ability, in Western Australian angiosperms. *Global Ecology and Biogeography*, **18**, 596–606.

Graf, D.L., and D. Ó Foighil. (2000). Molecular phylogenetic analysis of 28S rDNA supports a Gondwanan origin for Australasian Hyriidae (Mollusca: Bivalvia: Unionoida). *Vie et Milieu*, **50**, 245–254.

Graham, S.A. (2002). Phylogenetic relationships and biogeography of the endemic Caribbean genera *Crenea*, *Ginoria*, and *Haitia* (Lythraceae). *Caribbean Journal of Science*, **38**, 195–204.

Graham, S.W., J.M. Zgurski, M.A. McPherson, *et al.* (2006). Robust inference of monocot deep phylogeny using an expanded multigene plastid data set. *Aliso*, **22**, 3–21.

Grandcolas, P., J. Murienne, T. Robillard, *et al.* (2008). New Caledonia: a very old Darwinian island? *Philosophical Transactions of the Royal Society B*, **363**, 3309–3317.

Grant-Mackie, J.A., J.S. Buckeridge, and P.M. Johns. (1996). Two new Upper Jurassic arthropods from New Zealand. *Alcheringa*, **20**, 31–39.

Green, M.D. (2010). *Molecular Phylogeny of the Snake Genus* Oligodon *(Serpentes: Colubridae), with an Annotated Checklist and Key*. MSc Thesis, University of Toronto.

Green, P.S. (1979). Observations on the phytogeography of the New Hebrides, Lord Howe Island and Norfolk Island. In *Plants and Islands*, ed. D. Bramwell. London: Academic Press, pp. 41–53.

Green, P.S. (1994). Oceanic islands I. *Flora of Australia*, Vol. 49. Canberra: Australian Government Publishing Service.

Grenyer, R., and A. Purvis. (2003). A composite species-level phylogeny of the 'Insectivora' (Mammalia: Order Lipotyphla Haeckel, 1866). *Journal of Zoology, London*, **260**, 245–257.

Grimaldi, D., and M.S. Engel. (2005). *Evolution of the Insects*. New York: Cambridge University Press.

Grobys, J.W.G., K. Gohl, G. Uenzelmann-Neben, B. Davy, and D. Barker. (2009). Extensional and magmatic nature of the Cambell Plateau and Great South Basin. *Tectonophysics*, **472**, 213–225.

Groppo, M., J.A. Kallunki, J.R. Pirani, and A. Antonelli. (2012). Chilean *Pitavia* more closely related to Oceania and Old World Rutaceae than to Neotropical groups: evidence from two cpDNA non-coding regions, with a new subfamilial classification of the family. *PhytoKeys*, **19**, 9–29.

Grose, S.O., and R.G. Olmstead. (2007). Evolution of a charismatic neotropical clade: molecular phylogeny of *Tabebuia* s.l., Crescentieae, and allied genera (Bignoniaceae). *Systematic Botany*, **32**, 650–659.

Gübitz, T., R.S. Thorpe, and A. Malhotra. (2000). Phylogeography and natural selection in the Tenerife gecko *Tarentola delalandii*: testing historical and adaptive hypotheses. *Molecular Ecology*, **9**, 1213–1221.

Gullan, P.J., and A.W. Sjaarda. (2001). Trans-Tasman *Platycoelostoma* Morrison (Hemiptera: Coccoidea: Margarodidae) on endemic Cupressaceae, and the phylogenetic history of margarodids. *Systematic Entomology*, **26**, 257–278.

Gunnell, G.F., and N.B. Simmons. (2005). Fossil evidence and the origin of bats. *Journal of Mammalian Evolution*, **12**, 209–246.

Guo, Y.-Y., Y.-B. Luo, Z.-J. Liu, and X.-Q. Wang. (2012). Evolution and biogeography of the slipper orchids: Eocene vicariance of the conduplicate genera in the Old and New World tropics. *PLoS One* **7**(6), e38788, 1–13.

Haase, M., B. Fontaine, and O. Gargominy. (2010). Rissooidean freshwater gastropods from the Vanuatu archipelago. *Hydrobiologia*, **637**, 53–71.

Hackett, S.J., R.T. Kimball, S. Reddy, *et al.* (2008). A phylogenetic study of birds reveals their evolutionary history. *Science*, **320**, 1763–1768.

Hagen, K.B. von, and J.W. Kadereit. (2001). The phylogeny of *Gentianella* (Gentianaceae) and its colonization of the southern hemisphere as revealed by nuclear and chloroplast DNA sequence variation. *Organisms Diversity and Evolution*, **1**, 61–79.

Hall, R. (2002). Cenozoic geological and plate tectonic evolution of SE Asia and the SW Pacific: computer-based reconstructions, model and animations. *Journal of Asian Earth Sciences*, **20**, 353–431.

Hall, R. (2009). SE Asia's changing palaeogeography. *Blumea*, **54**, 148–161.

Hall, R. (2011). Australia–SE Asia collision: plate tectonics and crustal flow. In *The SE Asian Gateway: History and Tectonics of Australia–Asia Collision*, eds. R. Hall, M.A. Cottam, and M.E.J. Wilson. *Geological Society of London Special Publications*, **355**, 75–109.

Hall, R. (2012). Sundaland and Wallacea: geology, plate tectonics and palaeogeography. In *Biotic Evolution and Environmental Change in Southeast Asia*, ed. D. Gower, K. Johnson, J. Richardson, *et al.* Cambridge: Cambridge University Press, pp. 32–78.

Hall, R., B. Clements, and H.R. Smyth. (2009). Sundaland: basement character, structure and plate tectonic development. *Proceedings, Indonesian Petroleum Association 33rd Annual Convention*, IPA09-G-134, 1–27.

Hamilton, A.M., E.R. Klein, and C.C. Austin. (2010). Biogeographic breaks in Vanuatu, a nascent oceanic archipelago. *Pacific Science*, **64**, 149–159.

Hamilton, W.B. (1988). Plate tectonics and island arcs. *Bulletin of the Geological Society of America*, **100**, 1503–1527.

Hampe, A., and R.J. Petit. (2007). Ever deeper phylogeographies: trees retain the genetic imprint of Tertiary plate tectonics. *Molecular Ecology*, **16**, 5113–5114.

Harbaugh, D.T. (2008). Polyploid and hybrid origins of Pacific island sandalwoods (*Santalum*, Santalaceae) inferred from low-copy nuclear and flow cytometry data. *International Journal of Plant Science*, **169**, 677–685.

Hardy, A. (1965). *The Living Stream*. London: Collins.

Hardy, N.B., P.J. Gullan, R.C. Henderson, and L.G. Cook. (2008). Relationships among felt scale insects (Hemiptera: Coccoidea: Eriococcidae) of southern beech, *Nothofagus* (Nothofagaceae), with the first descriptions of Australian species of the *Nothofagus*-feeding genus *Madarococcus* Hoy. *Invertebrate Systematics*, **22**, 1–41.

Hare, K.M., C.H. Daugherty, and D.G. Chapple. (2008). Comparative phylogeography of three skink species (*Oligosoma moco, O. smithi, O. suteri*; Reptilia: Scincidae) in northeastern New Zealand. *Molecular Phylogenetics and Evolution*, **46**, 303–315.

Harrington, H.J., and R.J. Korsch. (1985). Late Permian to Cainozoic tectonics of the New England Orogen. *Australian Journal of Earth Sciences*, **32**, 161–203.

Hartley, T.G. (2001). Morphology and biogeography in Australasian-Malesian Rutaceae. *Malayan Nature Journal*, **55**, 197–219.

Hawkins, B.A., M.Á. Rodríguez, and S.G. Weller. (2011). Global angiosperm family richness revisited: linking ecology and evolution to climate. *Journal of Biogeography*, **38**, 1253–1266.

Hay, J.M., C.H. Daugherty, A. Cree, and L.R. Maxson. (2003). Low genetic divergence obscures phylogeny among populations of *Sphenodon*, remnant of an ancient reptile lineage. *Molecular Phylogenetics and Evolution*, **29**, 1–19.

Heads, M. (1990a). A taxonomic revision of *Kelleria* and *Drapetes* (Thymelaeaceae). *Australian Systematic Botany*, **3**, 595–652.

Heads, M. (1990b). Integrating earth and life sciences in New Zealand natural history: the parallel arcs model. *New Zealand Journal of Zoology*, **16**, 549–585.

Heads, M. (1994a). Morphology, architecture and taxonomy in the *Hebe* complex (Scrophulariaceae). *Bulletin du Muséum national d'Histoire naturelle* Paris 4e sér., **16**, sect. B., *Adansonia*, 163–191.

Heads, M. (1994b). Biogeographic studies in New Zealand Scrophulariaceae: tribes Rhinantheae, Calceolarieae and Gratioleae. *Candollea*, **49**, 55–80.

Heads, M. (1996). Biogeography, taxonomy and evolution in the Pacific genus *Coprosma* (Rubiaceae). *Candollea*, **51**, 381–405.

Heads, M. (1998a). Biodiversity in the New Zealand divaricating tree daisies: *Olearia* sect. nov. (Compositae). *Botanical Journal of the Linnean Society*, **127**, 239–285.

Heads, M. (1998b). Biogeographic disjunction along the Alpine fault, New Zealand. *Biological Journal of the Linnean Society*, **63**, 161–176.

Heads, M. (1999). Vicariance biogeography and terrane tectonics in the South Pacific: an analysis of the genus *Abrotanella* (Compositae). *Biological Journal of the Linnean Society*, **67**, 391–432.

Heads, M. (2000). A new species of *Hoheria* (Malvaceae) from northern New Zealand. *New Zealand Journal of Botany*, **38**, 373–377.

Heads, M. (2001a). Regional patterns of biodiversity in New Guinea plants. *Botanical Journal of the Linnean Society*, **136**, 67–73.

Heads, M. (2001b). Birds of paradise, biogeography and ecology in New Guinea: a review. *Journal of Biogeography*, **28**, 1–33.

Heads, M. (2001c). Birds of paradise (Paradisaeidae) and bowerbirds (Ptilonorhynchidae): regional levels of biodiversity and terrane tectonics in New Guinea. *Journal of Zoology*, **255**, 331–339.

Heads, M. (2002a). Birds of paradise, vicariance biogeography and terrane tectonics in New Guinea. *Journal of Biogeography*, **29**, 261–284.

Heads, M. (2002b). Regional patterns of biodiversity in New Guinea animals. *Journal of Biogeography*, **29**, 285–294.

Heads, M. (2003). Ericaceae in Malesia: vicariance biogeography, terrane tectonics and ecology. *Telopea*, **10**, 311–449.

Heads, M. (2005a). Towards a panbiogeography of the seas. *Biological Journal of the Linnean Society*, **84**, 675–723.

Heads, M. (2005b). Dating nodes on molecular phylogenies: a critique of molecular biogeography. *Cladistics*, **21**, 62–78.

Heads, M. (2005c). The history and philosophy of panbiogeography. In *Regionalización biogeográfica en Iberoamérica y tópicos afines*, eds. J. Llorente and J.J. Morrone. Mexico City: Universidad Nacional Autónoma de México, pp. 67–123.

Heads, M. (2006a). Seed plants of Fiji: an ecological analysis. *Biological Journal of the Linnean Society*, **89**, 407–431.

Heads, M. (2006b). Panbiogeography of *Nothofagus* (Nothofagaceae): analysis of the main species massings. *Journal of Biogeography*, **33**, 1066–1075.

Heads, M. (2008a). Biological disjunction along the West Caledonian fault, New Caledonia: a synthesis of molecular phylogenetics and panbiogeography. *Botanical Journal of the Linnean Society*, **158**, 470–488.

Heads, M. (2008b). Panbiogeography of New Caledonia, southwest Pacific: basal angiosperms on basement terranes, ultramafic endemics inherited from volcanic island arcs, and old taxa endemic to young islands. *Journal of Biogeography*, **35**, 2153–2175.

Heads, M. (2009a). Globally basal centres of endemism: the Tasman–Coral Sea region (southwest Pacific), Latin America and Madagascar/South Africa. *Biological Journal of the Linnean Society*, **96**, 222–245.

Heads, M. (2009b). Inferring biogeographic history from molecular phylogenies. *Biological Journal of the Linnean Society*, **98**, 757–774.

Heads, M. (2009c). Darwin's changing ideas on evolution: from centres of origin and teleology to vicariance and incomplete lineage sorting. *Journal of Biogeography*, **36**, 1018–1026.

Heads, M. (2010a). The biogeographical affinities of the New Caledonian biota: a puzzle with 24 pieces. *Journal of Biogeography*, **37**, 1179–1201.

Heads, M. (2010b). The endemic plant families and the palms of New Caledonia: a biogeographical analysis. *Journal of Biogeography*, **37**, 1239–1250.

Heads, M. (2011). Using island age to estimate the age of island-endemic clades and calibrate molecular clocks. *Systematic Biology*, **60**, 204–218.

Heads, M. (2012a). *Molecular Panbiogeography of the Tropics*. Berkeley: University of California Press.

Heads, M. (2012b). Bayesian transmogrification of clade divergence dates: a critique. *Journal of Biogeography*, **39**, 1749–1756.

Heads, M. (2012c). South Pacific biogeography, tectonic calibration, and pre-drift tectonics: cladogenesis in *Abrotanella* (Asteraceae). *Biological Society of the Linnean Society*, **107**, 938–952.

Heads, M., and R.C. Craw. (2004). The Alpine fault biogeographic hypothesis revisited. *Cladistics*, **20**, 184–190.

Heads, S.W. (2008). The first fossil Proscopiidae (Insecta, Orthoptera, Eumastacoidea) with comments on the historical biogeography and evolution of the family. *Palaeontology*, **51**, 499–507.

Heaney, L.R. (1986). Biogeography of mammals in SE Asia: estimates of rates of colonization, extinction and speciation. *Biological Journal of the Linnean Society*, **28**, 127–165.

Heaney, L.R. (2007). Is a new paradigm emerging for oceanic island biogeography? *Journal of Biogeography*, **34**, 753–757.

Heaney, L.R., D.S. Balete, E.A. Rickart, M.J. Veluz, and S.A. Jansa. (2009). A new genus and species of small 'tree-mouse' (Rodentia, Muridae) related to the Philippine giant cloud rats. *Bulletin of the American Museum of Natural History*, **331**, 205–229.

Heaney, L.R., M.L. Dolar, D.S. Balete, *et al.* (2010). Synopsis of Philippine Mammals. http://archive.fieldmuseum.org/philippine_mammals/

Hedges, S.B., and N. Vidal. (2009). Lizards, snakes and amphisbaenians (Squamata). In *The Timetree of Life*, eds. S.B. Hedges and S. Kumar. New York: Oxford University Press, pp. 383–389.

Heenan, P.B., and B.P.J. Molloy. (2006). A new species of *Oreomyrrhis* (Apiaceae) from southern South Island, New Zealand, and comparison of its limestone and ultramafic habitats. *New Zealand Journal of Botany*, **44**, 99–106.

Heenan, P.B., M.I. Dawson, and S.J. Wagstaff. (2004). The relationship of *Sophora* sect. *Edwardsia* (Fabaceae) to *Sophora tomentosa*, the type species of the genus *Sophora*, observed from DNA sequence data and morphological characters. *Botanical Journal of the Linnean Society*, **146**, 439–446.

Heenan, P.B., A.D. Mitchell, P.J. de Lange, J. Keeling, and A.M. Paterson. (2010). Late-Cenozoic origin and diversification of Chatham Islands endemic plant species revealed by analyses of DNA sequence data. *New Zealand Journal of Botany*, **48**, 83–136.

Heinicke, M.P., E. Greenbaum, T.R. Jackman, and A.M. Bauer. (2011). Phylogeny of a trans-Wallacean radiation (Squamata, Gekkonidae, *Gehyra*) supports a single early colonization of Australia. *Zoologica Scripta*, **40**, 584–602.

Heinrichs, J., M. Lindner, H. Groth, *et al.* (2006). Goodbye or welcome Gondwana? – Insights into the phylogenetic biogeography of the leafy liverwort *Plagiochila* with a description of *Proskauera*, gen. nov. (Plagiochilaceae, Jungermanniales). *Plant Systematics and Evolution*, **258**, 227–250.

Hendrich, L., and M. Balke. (2000). The genus *Neptosternus* SHARP 1882 in the Philippines: taxonomy and biogeography (Coleoptera: Dytiscidae). *Linzer Biologische Beiträge*, **32**, 1291–1299.

Hennemann, F.H., and O.V. Conle. (2006a). *Papuacocelus papuanus* n. gen. n. sp. – a new Eurycanthinae from Papua New Guinea, with notes on the genus *Dryococelus* Gurney, 1947 and description of the egg (Phasmatodea: Phasmatidae: Eurycanthinae). *Zootaxa*, **1375**, 31–49.

Hennemann, F.H., and O.V. Conle. (2006b). Studies on New Guinean giant stick-insects of the tribe Stephanacridini Günther, 1953, with the descriptions of a new genus and three new species of *Stephanacris* Redtenbacher, 1908 (Phasmatodea: "Anareolatae"). *Zootaxa*, **1283**, 1–24.

Hennemann, F.H., and O.V. Conle. (2008). Revision of Oriental Phasmatodea: the tribe Pharnaciini Günther, 1953, including the description of the world's longest insect, and a survey of the family Phasmatidae Gray, 1835 with keys to the subfamilies and tribes (Phasmatodea: "Anareolatae": Phasmatidae). *Zootaxa*, **1906**, 1–316.

Hennemann, F.H., and O.V. Conle. (2009). Studies on the genus *Phasmotaenia* Navás, 1907, with the descriptions of five new species from the Solomon Islands, a revised key to the species and notes on its geographic distribution (Phasmatodea: "Anareolatae": Phasmatidae *s.l.*: Stephanacridini). *Zootaxa*, **2011**, 1–46.

Hennig, W. (1966). *Phylogenetic Systematics*. Urbana, IL: University of Illinois Press.

He-Nygrén, X., A. Juslén, I. Ahonen, D. Glenny, and S. Piippo. (2006). Illuminating the evolutionary history of liverworts (Marchantiophyta) – towards a natural classification. *Cladistics*, **22**, 1–31.

Herbert, D.G., and A. Mitchell. (2008). Phylogenetic relationships of the enigmatic land snail genus *Prestonella*: the missing African element in the Gondwanan superfamily Orthalicoidea (Mollusca: Stylommatophora). *Biological Journal of the Linnean Society*, **96**, 203–221.

Herbert, J., M.W. Chase, M. Möller, and R.J. Abbott. (2006). Nuclear and plastid DNA sequences confirm the placement of the enigmatic *Canacomyrica monticola* in Myricaceae. *Taxon*, **55**, 349–357.

Hermsen, E.J., M.A. Gandolfo, and M. del Carmen Zamaloa. (2012). The fossil record of *Eucalyptus* in Patagonia. *American Journal of Botany*, **99**, 1356–1374.

Herzer, R., P.G. Quilty, G.C.H. Chaproniere, and H.J. Campbell. (2009). Assisted passage – early Miocene transport of insular terrestrial biota towards New Zealand by a moving tectonic plate. *Geological Society of New Zealand Miscellaneous Publication*, **126**, 22–24.

Heuret, A., and S. Lallemand. (2005). Plate motions, slab dynamics and back-arc deformation. *Physics of the Earth and Planetary Interiors*, **149**, 31–51.

Hibbett, D.S., K. Hansen, and M.J. Donoghue. (1998). Phylogeny and biogeography of *Lentinula* inferred from an expanded rDNA dataset. *Mycological Research*, **102**, 1041–1049.

Hidas, E.Z., T.L. Costa, D.J. Ayre, and T.E. Minchinton. (2007). Is the species composition of rocky intertidal invertebrates across a biogeographic barrier in south-eastern Australia related to their potential for dispersal? *Marine and Freshwater Research*, **58**, 835–842.

Hill, R.S., and M.S. Pole. (1992). Leaf and shoot morphology of extant *Afrocarpus*, *Nageia* and *Retrophyllum* (Podocarpaceae) species, and species with similar leaf arrangement from Tertiary sediments in Australasia. *Australian Systematic Botany*, **5**, 337–358.

Ho, S.Y.W., and M.J. Phillips. (2009). Accounting for calibration uncertainty in phylogenetic estimation of evolutionary divergence times. *Systematic Biology*, **58**, 367–380.

Hoernle, K., F. Hauff, P. van den Bogaard, *et al.* (2010). Age and geochemistry of volcanic rocks from the Hikurangi and Manihiki oceanic plateaus. *Geochimica et Cosmochimica Acta*, **74**, 7196–7219.

Hoernle, K., R. Werner, F. Hauff, and P. van den Bogaard. (2004). The Hikurangi Oceanic Plateau: a fragment of the largest volcanic event on earth. *IFM-GEOMAR Report, 2002–2004*, 51–54.

Hoffmann, H. (2000). The rise of life on earth: Messel – window on an ancient world. *National Geographic Magazine*, 2000, **2**, 34–51.

Hoffmann, R.L. (1999). Checklist of the millipedes of North and Middle America. *Virginia Museum of Natural History Special Publications*, **8**, 1–584.

Holloway, B.A. (2007). Lucanidae (Insecta: Coleoptera). *Fauna of New Zealand*, **61**, 1–254.

Holttum, R.E. (1973). Posing the problems. In *The Phylogeny and Classification of the Ferns*, eds. A.C. Jermy, J.A. Crabbe, and B.A. Thomas. London: Linnean Society, pp. 1–10.

Holyoake, A., B. Waldmann, and N.J. Gemmell. (2001). Determining species status of one of the world's rarest frogs. *Animal Conservation*, **4**, 29–35.

Holzapfel, S., H.A. Robertson, J.A. McLennan, *et al.* (2008). *Kiwi* (Apteryx *spp.*) *Recovery Plan 2008–2018*. Wellington: Department of Conservation.

Honda, M., H. Ota, R.W. Murphy, and T. Hikida. (2006). Phylogeny and biogeography of *Tropidophorus* (Reptilia: Scincidae): a molecular approach. *Zoologica Scripta*, **35**, 85–95.

Honeycutt, R.L. (2009). Rodents (Rodentia). In *The Timetree of Life*, eds. S.B. Hedges and S. Kumar. New York: Oxford University Press, pp. 490–494.

Hong, L., J. Trusty, R. Oviedo, A. Anderberg, and J. Francisco-Ortega. (2004). Molecular phylogenetics of the Caribbean genera *Rhodogeron* and *Sachsia* (Asteraceae). *International Journal of Plant Science*, **165**, 209–217.

Hooker, J.D. (1860). *Botany of the Antarctic Voyage of H.M. Discovery-Ships* Erebus *and* Terror *in 1839–1843. III. Flora Tasmaniae*. London: Reeve.

Hooker, J.J. (2001). Tarsals of the extinct insectivoran family Nyctitheriidae (Mammalia): evidence for archontan relationships. *Zoological Journal of the Linnean Society*, **132**, 501–529.

Hopper, S.D. (2009). OCBIL theory: towards an integrated understanding of the evolution, ecology and conservation of biodiversity on old, climatically buffered, infertile landscapes. *Plant and Soil*, **322**, 49–86.

Hopper, S.D., and P. Gioia. (2004). The Southwest Australian Floristic Region: evolution and conservation of a global hot spot of biodiversity. *Annual Review of Ecology and Systematics*, **35**, 623–650.

Hörandl, E., and K. Emadzade. (2012). Evolutionary classification: a case study on the diverse plant genus *Ranunculus* L. (Ranunculaceae). *Perspectives in Plant Ecology, Evolution and Systematics*, **14**, 310–324.

Horsák, M., M. Chytrý, B.M. Pokryszko, *et al.* (2010). Habitats of relict terrestrial snails in southern Siberia: lessons for the reconstruction of palaeoenvironments of full-glacial Europe. *Journal of Biogeography*, **37**, 1450–1462.

Horton, T.W., R.N. Holdaway, A.N. Zerbini, *et al.* (2011). Straight as an arrow: humpback whales swim constant course tracks during long-distance migration. *Biology Letters*, **7**, 674–679.

Hosaka, K., M.A. Castellano, and J.W. Spatafora. (2008). Biogeography of Hysterangiales (Phallomycetidae, Basidiomycota). *Mycological Research*, **112**, 448–462.

Hosner, P.A., F.H. Sheldon, H.C. Lim, and R.G. Moyle. (2010). Phylogeny and biogeography of the Asian trogons (Aves: Trogoniformes) inferred from nuclear and mitochondrial DNA sequences. *Molecular Phylogenetics and Evolution*, **57**, 1219–1225.

Houde, P. (2009). Cranes, rails and allies (Gruiformes). In *The Timetree of Life*, eds. S.B. Hedges and S. Kumar. New York: Oxford University Press, pp. 440–444.

Hubbell, S.P. (2001). *A Unified Neutral Theory of Biodiversity and Biogeography*. Princeton, NJ: Princeton University Press.

Huelsenbeck J.P., B. Larget, R. E. Miller, and F. Ronquist. (2002). Potential applications and pitfalls of Bayesian inference of phylogeny. *Systematic Biology*, **51**, 673–688.

Hugall, A.F., and J. Stanisic. (2011). Beyond the prolegomenon: a molecular phylogeny of the Australian camaenid land snail radiation. *Zoological Journal of the Linnean Society*, **161**, 531–572.

Humphreys, A.M., M.D. Pirie, and H.P. Linder. (2009). A plastid tree can bring order to the chaotic generic taxonomy of *Rytidosperma* Steud. s.l. (Poaceae). *Molecular Phylogenetics and Evolution*, **55**, 911–928.

Humphreys, W.F. (2008). Rising from Down Under: developments in subterranean biodiversity in Australia from a groundwater fauna perspective. *Invertebrate Systematics*, **22**, 85–101.

Humphreys, W.F., L.S. Kornicker, and D.L. Danielopol. (2009). On the origin of *Danielopolina baltanasi* sp. n. (Ostracoda, Thaumatocypridoidea) from three anchialine caves on Christmas Island, a seamount in the Indian Ocean. *Crustaceana*, **82**, 1177–1203.

Humphries, E.M., and K. Winker. (2011). Discord reigns among nuclear, mitochondrial and phenotypic estimates of divergence in nine lineages of trans-Beringian birds. *Molecular Ecology*, **20**, 573–583.

Hurr, K.A., P.J. Lockhart, P.B. Heenan, and D. Penny. (1999). Evidence for the recent dispersal of *Sophora* (Leguminosae) around the Southern Oceans: molecular data. *Journal of Biogeography*, **26**, 565–577.

Hutchinson, M. (1993). Scincidae. *Fauna of Australia*, **31**, 1–45.

Hutton, F.W. (1872). On the geographic relations of the New Zealand fauna. *Transactions of the New Zealand Institute*, **5**, 227–256.

Huxley, T.H. (1868). On the classification and distribution of the Alectoromorphae and Heteromorphae. *Proceedings of the Zoological Society, London*, **1868**, 294–319.

Huxley, T.H. (1873). Palaeontology and the doctrine of evolution. (The Anniversary Address to the Geological Society, 1870.). In *Critiques and Addresses by Thomas Henry Huxley*. London: Macmillan, pp. 181–217.

IFM-GEOMAR. (2007). Research Project SO193 Manihiki. At: www.ifm-geomar.de. Accessed November 2007.

Ingle, S., J.J. Mahoney, H. Sato, *et al.* (2007). Depleted mantle wedge and sediment fingerprint in unusual basalts fom the Manihiki Plateau, central Pacific Ocean. *Geology*, **35**, 595–598.

Irestedt, M., and J.I. Ohlson. (2008). The division of the major songbird radiation into Passerida and 'core Corvoidea' (Aves: Passeriformes) – the species tree vs. the gene tree. *Zoologica Scripta*, **37**, 305–313.

Irestedt, M., J. Fuchs, K.A. Jønsson, *et al.* (2008). The systematic affinity of the enigmatic *Lamprolia victoriae* (Aves: Passeriformes) – an example of avian dispersal between New Guinea and Fiji over Miocene intermittent land bridges? *Molecular Phylogenetics and Evolution*, **48**, 1218–1222.

Irestedt, M., K.A. Jønsson, J. Fjeldså, L. Christidis, and P.G.P. Ericson. (2009). An unexpectedly long history of sexual selection in birds-of-paradise. *BMC Evolutionary Biology*, **9**(235), 1–11.

Irestedt, M., J.I. Ohlson, D. Zuccon, M. Källersjö, and P. Ericson. (2006). Nuclear DNA from old collections of avian study skins reveals the evolutionary history of the Old World suboscines (Aves, Passeriformes). *Zoologica Scripta*, **35**, 567–580.

Isaac, N.J.B., S.T. Turvey, B. Collen, C. Waterman, and J.E.M. Baillie. (2007). Mammals on the EDGE: conservation priorities based on threat and phylogeny. *PLoS One*, **3**, e296, 1–7.

Isozaki, Y., K. Aoki, T. Nakama, and S. Yanai. (2010). New insight into a subduction-related orogen: a reappraisal of the geotectonic framework and evolution of the Japanese Islands. *Gondwana Research*, **18**, 82–105.

IUCN. (2013). *The IUCN Redlist of Threatened Species*. www.iucnredlist.org.

Jabaily, R.S., and K.J. Sytsma. (2010). Phylogenetics of *Puya* (Bromeliaceae): placement, major lineages, and evolution of Chilean species. *American Journal of Botany*, **97**, 337–356.

Jaffré, T. (1980). *Étude Écologique du Peuplement Végétal des Sols Dérivés de Roches Ultra-basiques en Nouvelle-Calédonie*. Paris: ORSTOM.

Jaffré, T., P. Morat, J.-M. Veillon, and H.S. MacKee. (1987). Changements dans la végétation de la Nouvelle-Calédonie au cours du Tertiaire: la végétation et la flore des roches ultrabasiques. *Bulletin du Muséum National d'Histoire Naturelle*, Paris. 4e sér. 9, sect. B, *Adansonia*, **4**, 365–391.

Jaffré, T., P. Morat, J.-M. Veillon, F. Rigault, and G. Dagostini, G. (2001). *Composition et Caractérisation de la Flore Indigène de Nouvelle-Calédonie*. Nouméa: Institut de Recherche pour le Développement.

Jansa, S.A., F.K. Barker, and L.R. Heaney. (2006). The pattern and timing of diversification of Philippine endemic rodents: evidence from mitochondrial and nuclear gene sequences. *Systematic Biology*, **55**, 73–88.

Jansen, M.E. (1984). *Badusa. Pacific Plant Areas*, **4**, 224–225.

Janssen, T., and K. Bremer. (2004). The age of major monocot groups inferred from 800+ *rbc*L sequences. *Botanical Journal of the Linnean Society*, **146**, 385–398.

Jaramillo, M.A. (2006). Using *Piper* species diversity to identify conservation priorities in the Chocó region of Colombia. *Biodiversity and Conservation*, **15**, 1695–1712.

Jaume, D. (2008). Global diversity of spelaeogriphaceans & thermosbaenaceans (Crustacea; Spelaeogriphacea & Thermosbaenacea) in freshwater. *Hydrobiologia*, **595**, 219–224.

Jaume, D., G.A. Boxshall, and W.F. Humphreys. (2001). New stygobiont copepods (Calanoida; Misophrioida) from Bundera Sinkhole, an anchialine cenote in north-western Australia. *Zoological Journal of the Linnean Society*, **133**, 1–24.

Jeekel, C.A.W. (1986). Millipedes from Australia, 10: three interesting new species and a new genus (Diplopoda: Sphaerotheriida, Spirobolida, Polydesmida). *Beaufortia*, **36**, 35–50.

Jennings, W.B., and S.V. Edwards. (2005). Speciational history of Australian grass finches (*Poephila*) inferred from thirty gene trees. *Evolution*, **59**, 2033–2047.

Jennings, W.B., E.R. Pianka, and S. Donnellan. (2003). Systematics of the lizard family Pygopodidae with implications for the diversification of Australia temperate biotas. *Systematic Biology*, **52**, 757–780.

Jeon, M.-J., J.-H. Song, and K.-J. Ahn. (2012). Molecular phylogeny of the marine littoral genus *Cafius* (Coleoptera: Staphylinidae: Staphylininae) and implications for classification. *Zoologica Scripta*, **41**, 150–159.

Jestrow, B., J. Gutiérrez Amaro, and J. Francisco-Ortega. (2012). Islands within islands: a molecular phylogenetic study of the *Leucocroton* alliance (Euphorbiaceae) across the Caribbean Islands and within the serpentinite archipelago of Cuba. *Journal of Biogeography*, **39**, 452–464.

Johanson, K.A., and J. Oláh. (2012). Revision of the Fijian *Chimarra* (Trichoptera, Philopotamidae) with description of 24 new species. *Zootaxa*, **3354**, 1–58.

Johanson, K.A., and J.B. Ward. (2009). Twenty-one new *Polyplectropus* species from New Caledonia (Trichoptera: Polycentropidae). *Annales de la Société Entomologique de France* (n.s.), **45**, 11–47.

Johanson, K.A., T. Malm, M. Espeland, and E. Weingartner. (2012). Phylogeny of the Polycentropodidae (Insecta: Trichoptera) based on protein-coding genes reveal non-monophyletic genera. *Molecular Phylogenetics and Evolution*, **65**, 126–135.

Johansson, U.S., E. Pasquet, and M. Irestedt. (2011). The New Zealand thrush: an extinct oriole. *PLoS One*, **6**(9), e24317, 1–6.

Johnson, G.D., H. Ida, J. Sakaue, *et al.* (2012). A 'living fossil' eel (Anguilliformes: Protanguillidae, fam. nov.) from an undersea cave in Palau. *Proceedings of the Royal Society B*, **279**, 934–943.

Johnson, K.A., B.R. Holland, M.M. Heslewood, and D.M. Crayn. (2012). Supermatrices, supertrees and serendipitous scaffolding: inferring a well-resolved, genus-level phylogeny of Styphelioideae (Ericaceae) despite missing data. *Molecular Phylogenetics and Evolution*, **62**, 146–158.

Johnson, K.P., and M.D. Sorenson. (1999). Phylogeny and biogeography of dabbling ducks (genus: *Anas*): a comparison of molecular and morphological evidence. *Auk*, **116**, 792–805.

Johnson, L.E., P. Fryer, B. Taylor, *et al.* (1991). New evidence for crustal accretion in the outer Mariana forearc: Cretaceous radiolarian cherts and MORB-like lavas. *Geology*, **19**, 811–814.

Johnson, M.S., Z.R. Hamilton, and J. Fitzpatrick. (2006). Genetic diversity of *Rhagada* land snails on Barrow Island. *Journal of the Royal Society of Western Australia*, **89**, 45–50.

Johnson, M.S., Z.R. Hamilton, C.E. Murphy, *et al.* (2004). Evolutionary genetics of island and mainland species of *Rhagada* (Gastropoda: Pulmonata) in the Pilbara Region, Western Australia. *Australian Journal of Zoology*, **52**, 341–355.

Johnson, M.S., E.K. O'Brien, and J.J. Fitzpatrick. (2010). Deep, hierarchical divergence of mitochondrial DNA in *Amplirhagada* land snails (Gastropoda: Camaenidae) from the Bonaparte Archipelago, Western Australia. *Biological Journal of the Linnean Society*, **100**, 141–153.

Jolivet, P. (2008). La faune entomologique en Nouvelle-Calédonie. *Le Coléoptériste*, **11**, 35–47.

Jolivet, P., and K.K. Verma. (2008a). Eumolpinae – a widely distributed and much diversified subfamily of leaf beetles (Coleoptera, Chrysomelidae). *Terrestrial Arthropod Reviews*, **1**, 3–37.

Jolivet, P., and K.K. Verma. (2008b). On the origin of the chrysomelid fauna of New Caledonia. In *Research on Chrysomelidae, Vol. 1.* eds. P. Jolivet, J.A. Santiago-Blay and M. Schmitt, Leiden: Brill, pp. 309–319.

Jolivet, P., and K.K. Verma. (2010). Good morning Gondwana. *Annales de la Société Entomologique de France* (n.s.), **46**, 53–61.

Jones, A.W., and R.S. Kennedy. (2008a). Evolution in a tropical archipelago: comparative phylogeography of Philippine fauna and flora reveals complex patterns of colonization and diversification. *Biological Journal of the Linnean Society*, **95**, 620–639.

Jones, A.W., and R.S. Kennedy. (2008b). Plumage convergence and evolutionary history of the Island Thrush in the Philippines. *Condor*, **110**, 35–44.

Jones, M.E.H., A.J.D. Tennyson, J.P. Worthy, S.E. Evans, and T.H. Worthy. (2009). A sphenodontine (Rhynchocephalia) from the Miocene of New Zealand and palaeobiogeography of the tuatara (*Sphenodon*). *Proceedings of the Royal Society B*, **276**, 1385–1390.

Jønsson, K.A., R.C.K. Bowie, R.G. Moyle, *et al.* (2010a). Historical biogeography of an Indo-Pacific passerine bird family (Pachycephalidae): different colonization patterns in the Indonesian and Melanesian archipelagos. *Journal of Biogeography*, **37**, 245–257.

Jønsson, K.A., R.C.K. Bowie, R.G. Moyle, *et al.* (2010b). Phylogeny and biogeography of Oriolidae (Aves: Passeriformes). *Ecography*, **33**, 232–241.

Jønsson, K.A., R.C.K. Bowie, J.A.A. Nylander, *et al.* (2010c). Biogeographical history of cuckoo-shrikes (Aves: Passeriformes): transoceanic colonization of Africa from Australo-Papua. *Journal of Biogeography*, **37**, 1767–1781.

Jønsson, K.A., P.-H. Fabre, and M. Irestedt. (2012). Brains, tools, innovation and biogeography in crows and ravens. *BMC Evolutionary Biology*, **12** (72), 1–12.

Jønsson, K.A., P.-H. Fabre, R.E. Ricklefs, and J. Fjeldså. (2011a). Major global radiation of corvoid birds originated in the proto-Papuan archipelago. *Proceedings of the National Academy of Sciences USA*, **108**, 2328–2333.

Jønsson, K.A., M. Irestedt, R.C.K. Bowie, L. Christidis, and J. Fjeldså. (2011b). Systematics and biogeography of Indo-Pacific ground-doves. *Molecular Phylogenetics and Evolution*, **59**, 538–543.

Joseph, L. (2008). The changing faces of systematics and biogeography in Australasian ornithology: a young Turks's view. In *Contributions to the History of Australasian Ornithology*, eds. W.E. Davis Jr., H.F. Recher, W.E. Boles, and J.A. Jackson. Cambridge, MA: Nuttall Ornithology Club, pp. 253–303.

Joseph, L., A. Toon, E.E. Schirtzinger, and T.F. Wright. (2011). Molecular systematics of two enigmatic genera *Psittacella* and *Pezoporus* illuminate the ecological radiation of Australo-Papuan parrots (Aves: Psittaciformes). *Molecular Phylogenetics and Evolution*, **59**, 675–684.

Joseph, M.J., and V.W. Framenau. (2012). Systematic review of a new orb-weaving spider genus (Araneae: Araneidae), with special reference to the Australasian-Pacific and South-East Asian fauna. *Zoological Journal of the Linnean Society*, **166**, 279–341.

Judd, W.S., J.D. Skean Jr., and R.S. Beaman. (1988). *Miconia zanonii* (Melastomataceae: Miconieae), a new species from Hispaniola. *Brittonia*, **40**, 208–213.

Kadereit, G., and H. Freitag. (2011). Molecular phylogeny of Camphorosmeae (Camphorosmoideae, Chenopodiaceae): implications for biogeography, evolution of C4-photosynthesis and taxonomy. *Taxon*, **60**, 51–78.

Kadereit, G., D. Gotzek, S. Jacobs, and H. Freitag. (2005). Origin and evolution of Australian Chenopodiaceae. *Organisms Diversity and Evolution*, **5**, 59–80.

Kadereit, G., L. Mucina, and H. Freitag. (2006). Phylogeny of Salicornioideae (Chenopodiaceae): diversification, biogeography, and evolutionary trends in leaf and flower morphology. *Taxon*, **55**, 617–642.

Kamp, P.J.J., P.F. Green, and S.H. White. (1989). Fission track analysis reveals character of collisional tectonics in New Zealand. *Tectonics*, **8**, 169–195.

Kantvilas, G., and H.T. Lumbsch. (2012). Reappraisal of the genera of Megalosporaceae (Teloschistales, Ascomycota). *Australian Systematic Botany*, **25**, 210–216.

Karanovic, T., and S.M. Eberhard. (2009). Second representative of the order Misophrioida (Crustacea, Copepoda) from Australia challenges the hypothesis of the Tethyan origin of some anchialine faunas. *Zootaxa*, **2059**, 51–68.

Karig, D.E. (1983). Accreted terranes in the northern part of the Philippine Archipelago. *Tectonics*, **2**, 211–236.

Karig, D.E., D.R. Sarewitz, and G.D. Haeck. (1986). Role of strike-slip faulting in the evolution of allochthonous terranes in the Philippines. *Geology*, **14**, 852–855.

Kathriarachchi, H., R. Samuel, P. Hoffmann, *et al.* (2006). Phylogenetics of tribe Phyllantheae (Phyllanthaceae: Euphorbiaceae sensu lato) based on nrITS and *matK* DNA sequence data. *American Journal of Botany*, **93**, 637–655.

Kayaalp, P., M.P. Schwarz, and M.I. Stevens. (2013). Rapid diversification in Australia and two dispersals out of Australia in the globally distributed bee genus, *Hylaeus* (Colletidae: Hylaeinae). *Molecular Phylogenetics and Evolution*, **66**, 668–678.

Kearns, A.M., L. Joseph, and L.G. Cook. (2013). A multilocus coalescent analysis of the speciational history of the Australo-Papuan butcherbirds and their allies. *Molecular Phylogenetics and Evolution*, **66**, 941–952.

Keeley, S.C., Z.H. Forsman, and R. Chan. (2007). A phylogeny of the "evil tribe" (Vernonieae: Compositae) reveals Old/New World long distance dispersal: support from separate and combined congruent datasets (*trn*L-F, *ndh*F, ITS). *Molecular Phylogenetics and Evolution*, **44**, 89–103.

Keogh, J.S., I.A.W. Scott, M. Fitzgerald, and R. Shine. (2003). Molecular phylogeny of the Australian venomous snake genus *Hoplocephalus* (Serpentes, Elapidae) and conservation genetics of the threatened *H. stephensii*. *Conservation Genetics*, **4**, 57–63.

Keogh, J.S., R. Shine, and S. Donnellan S. (1998). Phylogenetic relationships of terrestrial Australo-Papuan elapid snakes (Subfamily Hydrophiinae) based on cytochrome *b* and 16S rRNA sequences. *Molecular Phylogenetics and Evolution*, **10**, 67–81.

Kerr, A.C., and J. Tarney. (2005). Tectonic evolution of the Caribbean and northwestern South America: the case for accretion of two Late Cretaceous oceanic plateaus. *Geology*, **33**, 269–272.

Key, K.H.L. (1976). A generic and suprageneric classification of the Morabinae (Orthoptera: Eumastacidae), with description of the type species and a bibliography of the subfamily. *Australian Journal of Zoology Supplementary Series*, **37**, 1–185.

Kim, H., S. Lee, and Y. Jang. (2011). Macroevolutionary patterns in the Aphidini aphids (Hemiptera: Aphididae): diversification, host association, and biogeographic origins. *PLoS One*, **6**(9), e24749, 1–17.

Kim, S., L. Chunghee, and J.A. Mejías. (2007). Analysis of chloroplast DNA *mat*K gene and ITS of nrDNA sequences reveals polyphyly of the genus *Sonchus* and new relationships among the subtribe Sonchinae (Asteraceae: Cichorieae). *Molecular Phylogenetics and Evolution*, **44**, 578–597.

Kimball, R.T., E.L. Braun, J.D. Ligon, V. Lucchini, and E. Randi. (2001). A molecular phylogeny of the peacock-pheasants (Galliformes: *Polyplectron* spp.) indicates loss and reduction of ornamental traits and display behaviours. *Biological Journal of the Linnean Society*, **73**, 187–198.

Kirchman, J.J. (2012). Speciation of flightless rails on islands: A DNA-based phylogeny of the typical rails of the Pacific. *Auk*, **129**, 56–69.

Kitching, R.L., and K.L. Dunn. (1999). Biogeography of Australian butterflies. In *Biology of Australian Butterflies*, eds. R.L. Kitching, E. Scheermeyer, R.E. Jones, and N.E. Pierce. Collingwood, Victoria: CSIRO, pp. 53–74.

Klaus, S., D.C.J. Yeo, and S.T. Ahyong. (2011). Freshwater crab origins – laying Gondwana to rest. *Zoologischer Anzeiger*, **250**, 449–456.

Kloss, B. (1929). The zoo-geographical boundaries between Asia and Australia and some Oriental sub-regions. *Bulletin of the Raffles Museum*, **2**, 1–10.

Knapp, M., R. Mudaliar, D. Havell, S.J. Wagstaff, and P.J. Lockhart. (2007). The drowning of New Zealand and the problem of *Agathis*. *Systematic Biology*, **56**, 862–870.

Knapp, S., M.W. Chase, and J.J. Clarkson. (2004). Nomenclatural changes and a new sectional classification in *Nicotiana* (Solanaceae). *Taxon*, **53**, 73–82.

Knesel, K.M., B.E. Cohen, P.M. Vasconcelos, and D.S. Thiede. (2008). Rapid change in drift of the Australian plate records collision with Ontong Java plateau. *Nature*, **454**, 754–757.

Knight, A.T., R.J. Smith, R.M. Cowling, *et al.* (2007). Improving the key biodiversity areas approach for effective conservation planning. *BioScience*, **57**, 256–261.

Knittel, U., C.-H. Hung, T.F. Yang, and Y. Iizuka. (2010). Permian arc magmatism in Mindoro, the Philippines: an early Indosinian event in the Palawan Continental Terrane? *Tectonophysics*, **493**, 113–117.

Knopf, P., C. Schulz, D.P. Little, T. Stützel, and D.W. Stevenson. (2012). Relationships within Podocarpaceae based on DNA sequence, anatomical, morphological, and biogeographical data. *Cladistics*, **28**, 271–299.

Kocyan, A., D.A. Snijman, F. Forest, *et al.* (2011). Molecular phylogenetics of Hypoxidaceae – evidence from plastid DNA data and inferences on morphology and biogeography. *Molecular Phylogenetics and Evolution*, **60**, 122–136.

Kodandaramaiah, U. (2010). Use of dispersal–vicariance analysis in biogeography – a critique. *Journal of Biogeography*, **37**, 3–11.

Kodandaramaiah, U., D.C. Lees, C.J. Müller, *et al.* (2010). Phylogenetics and biogeography of a spectacular Old World radiation of butterflies: the subtribe Mycalesina (Lepidoptera: Nymphalidae: Satyrini). *BMC Evolutionary Biology*, **10** (172), 1–13.

Köhler, F. (2010). Three new species and two new genera of land snails from the Bonaparte Archipelago in the Kimberley, Western Australia (Pulmonata, Camaenidae). *Molluscan Research*, **30**, 1–16.

Köhler, F. (2011). *Australocosmica*, a new genus of land snails from the Kimberley, Western Australia (Eupulmonata, Camaenidae). *Malacologia*, **53**, 199–216.

Köhler, F., and M. Glaubrecht. (2002). Morphology, reproductive biology and molecular genetics of ovoviviparous freshwater gastropods (Cerithioidea, Pachychilidae) from the Philippines, with description of a new genus *Jagora. Zoologica Scripta*, **32**, 35–59.

Köhler, F., and M. Glaubrecht. (2007). Out of Asia and into India: on the molecular phylogeny and biogeography of the endemic freshwater gastropod *Paracrostoma* Cossmann, 1990 (Caenogastropoda: Pachychilidae). *Biological Journal of the Linnean Society*, **91**, 627–651.

Köhler, F., and M. Glaubrecht. (2010). Uncovering an overlooked radiation: molecular phylogeny and biogeography of Madagascar's endemic river snails (Caenogastropoda: Pachychilidae: *Madagasikara* gen. nov.). *Biological Journal of the Linnaean Society*, **99**, 867–94.

Kohn, B.P., A.J.W. Gleadow, and S.J. Cox. (1999). Denudation history of the Snowy Mountains: constraints from apatite fission track thermochronology. *Australian Journal of Earth Sciences*, **46**, 181–198.

Kotlík, P., V. Deffontaine, S. Mascheretti, *et al.* (2006). A northern glacial refugium for bank voles (*Clethrionomys glareolus*). *Proceedings of the National Academy of Sciences USA*, **103**, 14860–14864.

Krause, D.W., S.E. Evans, and K.-Q. Gao. (2003). First definitive record of Mesozoic lizards from Madagascar. *Journal of Vertebrate Paleontology*, **23**, 842–856.

Kristensen, N.P., M. Scoble, and O. Karsholt. (2007). Lepidoptera phylogeny and systematics: the state of inventorying moth and butterfly diversity. *Zootaxa*, **1668**, 699–747.

Kroenke, L.W. (1996). Plate tectonic development of the western and Southwestern Pacific: Mesozoic to the present. In *The Origin and Evolution of Pacific Island Biotas*, eds. A. Keast and S.E. Miller. Amsterdam: SPB Academic, pp. 19–34.

Kron, K.A., and J.L. Luteyn. (2005). Origins and biogeographic patterns in Ericaceae: new insights from recent phylogenetic analyses. *Biologiske Skrifter*, **55**, 479–500.

Krosch, M.N., A.M. Baker, P.B. Mather, and P.S. Cranston. (2011). Systematics and biogeography of the Gondwanan Orthocladiinae (Diptera: Chironomidae). *Molecular Phylogenetics and Evolution*, **59**, 458–468.

Krosnick, S.E., A.J. Ford, and J.V. Freudenstein. (2009). Taxonomic revision of *Passiflora* subgenus *Tetrapathea* including the monotypic genera *Hollrungia* and *Tetrapathea* (Passifloraceae), and a new species of *Passiflora. Systematic Botany*, **34**, 375–385.

Krzeminska, E. (2001). *Nothotrichocera* Alexander (Diptera: Trichoceridae): new species, redescriptions and phylogenetic relationships within the genus. *Invertebrate Taxonomy*, **15**, 205–216.

Kuch, U., J.S. Keogh, J. Weigel, L.A. Smith, and D. Mebs. (2005). Phylogeography of Australia's king brown snake (*Pseudechis australis*) reveals Pliocene divergence and Pleistocene dispersal of a top predator. *Naturwissenschaften*, **92**, 121–127.

Kundu, S., C.G. Jones, R.P. Prys-Jones, and J.J. Groombridge. (2012). The evolution of the Indian Ocean parrots (Psittaciformes): Extinction, adaptive radiation and eustacy. *Molecular Phylogenetics and Evolution*, **62**, 296–305.

Kurata, K., T. Jaffré, and H. Setoguchi, H. (2008). Genetic diversity and geographical structure of the pitcher plant *Nepenthes vieillardii* in New Caledonia: a chloroplast DNA haplotype analysis. *American Journal of Botany*, **95**, 1632–1644.

Kuschel, G. (2008). Curculionoidea (weevils) of New Caledonia and Vanuatu: basal families and some Curculionidae. In *Zoologica Neocaledonica 6. Mémoires du Muséum national d'Histoire naturelle*, ed. P. Grandcolas, **197**, 99–249.

Kuschel, G., and R.A.B. Leschen. (2010). Phylogeny and taxonomy of the Rhinorhynchinae (Coleoptera: Nemonychidae). *Invertebrate Systematics*, **24**, 573–615.

Kutty, S.N., T. Pape, B.M. Wiegmann, and R. Meier. (2010). Molecular phylogeny of the Calyptratae (Diptera: Cyclorrhapha) with an emphasis on the superfamily Oestroidea and the position of Mystacinobiidae and McAlpine's fly. *Systematic Entomology*, **35**, 614–635.

Ladiges, P.Y. (1998). Biogeography after Burbidge. *Australian Systematic Botany*, **11**, 231–242.

Ladiges, P.Y. (2006). Interpreting biogeographic patterns. *Australian Systematic Botany Society Newsletter*, **128**, 7–9.

Ladiges, P.Y., and D. Cantrill. (2007). New Caledonia–Australian connections: biogeographic patterns and geology. *Australian Systematic Botany*, **20**, 383–389.

Ladiges, P.Y., M.J. Bayly, and G.J. Nelson. (2010). East-west continental vicariance in *Eucalyptus* subgenus *Eucalyptus*. In *Beyond Cladistics: The Branching of a Paradigm*, eds. D.M. Williams and S. Knapp. Berkeley, CA: University of California Press, pp. 267–302.

Ladiges, P.Y., M.J. Bayly, and G. Nelson. (2012). Searching for ancestral areas and artifactual centres of origin in biogeography: with comment on east-west patterns across southern Australia. *Systematic Biology*, **61**, 703–708.

Ladiges, P.Y., C.E. Marks, and G. Nelson. (2011). Biogeography of *Nicotiana* section *Suaveolentes* (Solanaceae) reveals geographical tracks in arid Australia. *Journal of Biogeography*, **38**, 2066–2077.

Ladiges, P.Y., F. Udovicic, and G. Nelson. (2003). Australian biogeographical connections and the phylogeny of large genera in the plant family Myrtaceae. *Journal of Biogeography*, **30**, 989–998.

Lane, C.E., C. Mayes, L.D. Druehl, and G.W. Saunders. (2006). A multi-gene molecular investigation of the kelp (Laminariales, Phaeophyceae) supports substantial taxonomic re-organization. *Journal of Phycology*, **42**, 493–512.

Larsen, C., M. Speed, N. Harvey, and H.A. Noyes. (2007). A molecular phylogeny of the nightjars (Aves: Caprimulgidae) suggests extensive conservation of primitive morphological traits across multiple lineages. *Molecular Phylogenetics and Evolution*, **42**, 789–796.

Larson, R.L., R.A. Pockalny, R.F. Viso, *et al.* (2002). Mid-Cretaceous tectonic evolution of the Tongareva triple junction in the southwestern Pacific Basin. *Geology*, **30**, 67–70.

Lecroy, M., and F.K. Barker. (2006). A new species of Bush-warbler from Bougainville Island and a monophyletic origin for Southwest Pacific *Cettia*. *American Museum Novitates*, **3511**, 1–20.

Lee, D.E., W.G. Lee, and N. Mortimer. (2001). Where and why have all the flowers gone? Depletion and turnover in the New Zealand Cenozoic angiosperm flora in relation to palaeogeography and climate. *Australian Journal of Botany*, **49**, 341–356.

Lee, J., B.G. Baldwin, and L.D. Gottlieb. (2003). Phylogenetic relationships among the primarily North American genera of Cichorieae (Compositae) based on analysis of 18S–26S nuclear rDNA ITS and ETS sequences. *Systematic Botany*, **28**, 616–626.

Lee, J.Y., L. Joseph, and S.V. Edwards. (2012). A species tree for the Australo-Papuan Fairy-wrens and allies (Aves: Maluridae). *Systematic Biology*, **61**, 253–271.

Lee, M.S.Y., and A. Skinner. (2011). Testing fossil calibrations for vertebrate molecular trees. *Zoologica Scripta*, **40**, 538–543.

Lee, M.S.Y., M.N. Hutchinson, T.H. Worthy, *et al.* (2009). Miocene skinks and geckos reveal long-term conservatism of New Zealand's lizard fauna. *Biology Letters*, **5**, 833–837.

Lee, W.G., J.R. Wood, and G.M. Rogers. (2010). Legacy of avian-dominated plant–herbivore systems in New Zealand. *New Zealand Journal of Ecology*, **34**, 28–47.

Lerner, H.R.L., and D.P. Mindell. (2005). Phylogeny of eagles, Old World vultures, and other Accipitridae based on nuclear and mitochondrial DNA. *Molecular Phylogenetics and Evolution*, **37**, 327–346.

Leschen R.A.B., and B. Michaux. (2005). Biogeography and evolution of New Zealand Priasilphidae (Coleoptera: Cucujoidea). *New Zealand Entomologist*, **28**, 55–64.

Leschen, R.A.B., J.F. Lawrence, G. Kuschel, S. Thorpe, and Q. Wang. (2003). Coleoptera genera of New Zealand. *New Zealand Entomologist*, **26**, 15–28.

Leschen R.A.B., J.F. Lawrence, and A. Ślipiński. (2005). Classification of basal Cucujoidea (Coleoptera: Polyphaga): cladistic analysis, keys and review of new families. *Invertebrate Systematics*, **19**, 17–73.

Lessios, H.A., and D.R. Robertson. (2006). Crossing the impassable: genetic connections in 20 reef fishes across the eastern Pacific barrier. *Proceedings of the Royal Society B*, **273**, 2201–2208.

Levin, D.A. (2000). *The Origin, Expansion, and Demise of Plant Species*. New York: Oxford University Press.

Lewis, G., B. Schrire, B. Mackinder, and M. Lock. (2005). *Legumes of the World*. Kew: Royal Botanic Gardens.

Li, J., R. Xia, R.M. McDowall, *et al.* (2010). Phylogenetic position of the enigmatic *Lepidogalaxias salamandroides* with comment on the orders of lower euteleostean fishes. *Molecular Phylogenetics and Evolution*, **57**, 932–936.

Li, L., J. Li, J.G. Rohwer, *et al.* (2011). Molecular phylogenetic analysis of the *Persea* group (Lauraceae) and its biogeographic implications on the evolution of tropical and subtropical amphi-Pacific disjunctions. *American Journal of Botany*, **98**, 1520–1536.

Lieberman, B.S. (2000). *Paleobiogeography: Using Fossils to Study Global Change, Plate Tectonics, and Evolution*. New York: Kluwer/Plenum.

Liebherr, J.K. (2008). *Mecyclothorax kavanaughi* sp.n. (Coleoptera: Carabidae) from the Finisterre Range, Papua New Guinea. *Tijdschrift voor Entomologie*, **151**, 147–154.

Liebherr, J.K., J.W.M. Marris, R.M. Emberson, P. Syrett, and S. Roig-Juñent. (2011). *Orthoglymma wangapeka* gen.n., sp.n. (Coleoptera: Carabidae: Broscini): a newly discovered relict from the Buller Terrane, north-western South Island, New Zealand, corroborates a general pattern of Gondwanan endemism. *Systematic Entomology*, **36**, 395–414.

Liggins, L., D.G. Chapple, C.H. Daugherty, and P.A. Ritchie. (2008a). Origin and post-colonization evolution of the Chatham Islands skink (*Oligosoma nigriplantare nigriplantare*). *Molecular Ecology*, **17**, 3290–3305.

Liggins, L., D.G. Chapple, C.H. Daugherty, and P.A. Ritchie. (2008b). A SINE of restricted gene flow across the Alpine Fault: phylogeography of the New Zealand common skink (*Oligosoma nigriplantare nigriplantare*). *Molecular Ecology*, **17**, 3668–3683.

Lim, H.C., F. Zou, S.S. Taylor, *et al.* (2010). Phylogeny of magpie-robins and shamas (Aves: Turdidae: *Copsychus* and *Trichixos*): implications for island biogeography in Southeast Asia. *Journal of Biogeography*, **37**, 1894–1906.

Linder, H.P., M. Baeza, N.P. Barker, *et al.* (2010). A generic classification of the Danthonioideae (Poaceae). *Annals of the Missouri Botanical Garden*, **97**, 306–364.

Linkem, C.W., R.M. Brown, C.D. Siler, *et al.* (2013). Stochastic faunal exchanges drive diversification in widespread Wallacean and Pacific island lizards (Squamata: Scincidae: *Lamprolepis smaragdina*). *Journal of Biogeography*, **40**, 507–520.

Linkem, C.W., A.C. Diesmos, and R.M. Brown. (2010a). A new species of scincid lizard (genus *Sphenomorphus*) from Palawan Island, Philippines. *Herpetologica*, **66**, 2010, 67–79.

Linkem, C.W., A.C. Diesmos, R.M. Brown. (2011). Molecular systematics of the Philippine forest skinks (Squamata: Scincidae: Sphenomorphus): testing morphological hypotheses of interspecific relationships. *Zoological Journal of the Linnean Society*, **163**, 1217–1243.

Linkem, C.W., K.M. Hesed, A.C. Diesmos, and R.M. Brown. (2010b). Species boundaries and cryptic lineage diversity in a Philippine forest skink complex (Reptilia; Squamata; Scincidae: Lygosominae). *Molecular Phylogenetics and Evolution*, **56**, 572–585.

Little, T.A., S.L. Baldwin, P.G. Fitzgerald, and B. Monteleone. (2007). Continental rifting and metamorphic core complex formation ahead of the Woodlark spreading ridge, D'Entrecasteaux Islands, Papua New Guinea. *Tectonics*, **26**, TC1002, 1–26. doi:10.1029/2005TC001911.

Liu, X., Y. Wang, C. Shih, D. Ren, and D. Yang. (2012). Early evolution and historical biogeography of fishflies (Megaloptera: Chauliodinae): implications from a phylogeny combining fossil and extant taxa. *PLoS One*, **7**(7), e40345, 1–12

Livshultz, T., J.V. Mead, D.J. Goyder, and M. Brannin. (2011). Climate niches of milkweeds with plesiomorphic traits (Secamonoideae; Apocynaceae) and the milkweed sister group link ancient African climates and floral evolution. *American Journal of Botany*, **98**, 1966–1977.

Livshultz, T., D.J. Middleton, M.E. Endress, and J.K. Williams. (2007). Phylogeny of Apocynoideae and the APSA clade (Apocynaceae s.l.). *Annals of the Missouri Botanical Garden*, **94**, 324–359.

Logares, R.E. (2006). Does the global microbiota consist of a few cosmopolitan species? *Ecología Austral*, **16**, 85–90.

Lohman, D.J., M. de Bruyn, T. Page, *et al.* (2011). Biogeography of the Indo-Australian Archipelago. *Annual Review of Ecology, Evolution, and Systematics*, **42**, 205–226.

Lohman, D.J., K.K. Ingram, D.M. Prawiradilaga, *et al.* (2010). Cryptic genetic diversity in "widespread" Southeast Asian bird species suggests that Philippine avian endemism is gravely underestimated. *Biological Conservation*, **143**, 1885–1890.

Lohse, K., J.A. Nicholls, and G.N. Stone. (2011). Inferring the colonization of a mountain range – refugia vs. nunatak survival in high alpine ground beetles. *Molecular Ecology*, **20**, 394–408.

Lohwasser, U., A. Granda, and F.R. Blattner. (2004). Phylogenetic analysis of *Microseris* (Asteraceae), including a newly discovered Andean population from Peru. *Systematic Botany*, **29**, 774–780.

Lomolino, M.V., and J.H. Brown. (2009). The reticulating phylogeny of island biogeography theory. *Quarterly Review of Biology*, **84**, 357–390.

Losos, J.B., and R.E. Ricklefs. (2009). Adaptation and diversification on islands. *Nature*, **457**, 830–836.

Lovette, I.J., and D.R. Rubenstein. (2007). A comprehensive molecular phylogeny of the starlings (Aves: Sturnidae) and mockingbirds (Aves: Mimidae): congruent mtDNA and nuclear trees for a cosmopolitan avian radiation. *Molecular Phylogenetics and Evolution*, **44**, 1031–1056.

Lowry, P.P. II. (1998). Diversity, endemism, and extinction in the flora of New Caledonia: a review. In *Rare, Threatened, and Endangered Floras of Asia and the Pacific Rim*, eds. C.I. Peng and P.P. Lowry II. (Academica Sinica Monograph 16). Taipei: Institute of Botany, pp. 181–206.

Lucchini, V., E. Meijaard, C.H. Diong, C.P. Groves, and E. Randi. (2005). New phylogenetic perspectives among species of South-east Asian wild pig (*Sus* sp.) based on mtDNA sequences and morphometric data. *Journal of Zoology, London*, **266**, 25–35.

Lucky, A. (2011). Molecular phylogeny and biogeography of the spider ants, genus *Leptomyrmex* Mayr (Hymenoptera: Formicidae). *Molecular Phylogenetics and Evolution*, **59**, 281–292.

Lucky, A., and E.M. Sarnat. (2010). Biogeography and diversification of the Pacific ant genus *Lordomyrma* Emery. *Journal of Biogeography*, **37**, 624–634.

Lundberg, J. (2001). The asteralean affinity of the Mauritian *Roussea* (Rousseaceae). *Botanical Journal of the Linnean Society*, **137**, 267–276.

Luo, Z.-X., Q. Ji, J.R. Wible, and C.-X. Yuan. (2003). An early Cretaceous tribosphenic mammal and metatherian evolution. *Science*, **302**, 1934–1940.

Lynch, M. (2007). *The Origins of Genome Architecture*. Sunderland, MA: Sinauer Associates.

Mabberley, D.J. (1998). On *Neorapinia* (*Vitex* sensu lato, Labiatae-Viticoideae). *Telopea*, **7**, 313–317.

MacArthur, R.H. (1965). Patterns of species diversity. *Biological Review*, **40**, 510–533.

MacArthur, R.H. (1972). *Geographical Ecology: Patterns in the Distribution of Species*. New York: Harper and Row.

MacArthur, R.H., and E.O. Wilson. (1967). *The Theory of Island Biogeography*. Princeton, NJ: Princeton University Press.

Macaya, E.C., and G.C. Zuccarello. (2010). Genetic structure of the giant kelp *Macrocystis pyrifera* along the southeastern Pacific. *Marine Ecology Progress Series*, **420**, 103–112.

Macpherson, E., B. Richer de Forges, K. Schnabel, *et al.* (2010). Biogeography of the deep-sea galatheid squat lobsters of the Pacific Ocean. *Deep-Sea Research. Part 1. Oceanographic Research Papers*, **57**, 228–238.

Maddison, D.R. (2012). Phylogeny of *Bembidion* and related ground beetles (Coleoptera: Carabidae: Trechinae: Bembidiini: Bembidiina). *Molecular Phylogenetics and Evolution*, **63**, 533–576.

Magallón, S. (2004). Dating lineages: molecular and paleontological approaches to the temporal framework of clades. *International Journal of Plant Science*, **165** (4, Suppl.), S7–S21.

Magallón, S. (2010). Using fossils to break long branches in molecular dating: a comparison of relaxed clocks applied to the origin of angiosperms. *Systematic Biology*, **59**, 384–399.

Magallón, S., and A. Castillo. (2009). Angiosperm diversity through time. *American Journal of Botany*, **96**, 349–365.

Magri, D., S. Fineschi, P. Bellarosa, *et al.* (2007). The distribution of *Quercus suber* chloroplast haplotypes matches the palaeogeographical history of the western Mediterranean. *Molecular Ecology*, **16**, 5259–5266.

Maiden, J.H. (1914). Australian vegetation. In *Federal Handbook on Australia*, ed. G.H. Knibbs. Melbourne: Government Printer, pp. 163–209.

Makinson, R.O. (ed.) (2000). *Flora of Australia, Vol. 17A. Proteaceae part 2, Grevillea*. Canberra: Australian Biological Resources Study and CSIRO.

Malhotra, A., and R.S. Thorpe. (2004). A phylogeny of four mitochondrial gene regions suggests a revised taxonomy for Asian pitvipers (*Trimeresurus* and *Ovophis*). *Molecular Phylogenetics and Evolution*, **32**, 83–100.

Mallet, J. (1995). A species definition for the Modern Synthesis. *Trends in Ecology and Evolution*, **10**, 294–299.

Mallet, J. (2007). Hybrid speciation. *Nature*, **446**, 279–283.

Mallet, J. (2008). Mayr's view of Darwin: was Darwin wrong about speciation? *Biological Journal of the Linnean Society*, **95**, 3–16.

Mallet, J. (2010a). Why was Darwin's view of species rejected by twentieth century biologists? *Biology and Philosophy*, **25**, 497–527.

Mallet, J. (2010b). Group selection and the development of the biological species concept. *Philosophical Transactions of the Royal Society B*, **365**, 1853–1863.

Manchester, S.R., and Z. Kvaček. (2009). Fruits of *Sloanea* (Elaeocarpaceae) in the Paleogene of North America and Greenland. *International Journal of Plant Sciences*, **170**, 941–950.

Manchester, S.R., Z.-D. Chen, A.-M. Lu, and K. Uemura. (2009). Eastern Asian endemic seed plant genera and their paleogeographic history throughout the northern hemisphere. *Journal of Systematics and Evolution*, **47**, 1–42.

Mann, P., and F.W. Taylor. (2002). Emergent Late Quaternary coral reefs of eastern Papua New Guinea constrain the regional pattern of oceanic ridge propagation. *American Geophysical Union*, Fall Meeting 2002, abstract #T52C-1206.

Markey, A. (2005). *The Evolution of Fruit Traits in* Coprosma *and the Subtribe Coprosminae.* PhD Thesis, Dunedin: University of Otago.

Marko, P.B., J.M. Hoffman, S.A. Emme, *et al.* (2010). The 'expansion-contraction' model of Pleistocene biogeography: rocky shores suffer a sea change? *Molecular Ecology*, **19**, 146–169.

Marks, C.E., E. Newbigin, and P.Y. Ladiges. (2011). Comparative morphology and phylogeny of *Nicotiana* section *Suaveolentes* (Solanaceae) in Australia and the South Pacific. *Australian Systematic Botany*, **24**, 61–86.

Marquínez, X., L.G. Lohmann, M.L. Faria Salatino, A. Salatino, and F. González. (2009). Generic relationships and dating of lineages in Winteraceae based on nuclear (ITS) and plastid (*rp*S16 and *psb*A-*trn*H) sequence data. *Molecular Phylogenetics and Evolution*, **53**, 435–449.

Marshall, B.A. (2001). Mollusca Gastropoda: Seguenziidae from New Caledonia and the Loyalty Islands. *Mémoires du Muséum national d'Histoire naturelle*, **150**, 41–109.

Martens, J., D.T. Tietze, and Y.-H. Sun. (2006). Molecular phylogeny of *Parus* (*Periparus*), a Eurasian radiation of tits (Aves: Passeriformes: Paridae). *Zoologische Abhandlungen*, **55**, 103–120.

Martens, K., and C. Rossetti. (2002). On the Darwinulidae (Crustacea: Ostracoda) from Oceania. *Invertebrate Systematics*, **16**, 195–208.

Martin, H.A. (2006). Cenozoic climatic change and the development of the arid vegetation in Australia. *Journal of Arid Environments*, **66**, 533–563.

Martin, J. E., and M. Fernández. (2007). The synonymy of the Late Cretaceous mosasaur (Squamata) genus *Lakumasaurus* from Antarctica with *Taniwhasaurus* from New Zealand and its bearing upon faunal similarity within the Weddellian Province. *Geological Journal*, **42**, 203–211.

Mast, A.R., and K. Thiele. (2007). The transfer of *Dryandra* R.Br. to *Banksia* L.f. (Proteaceae). *Australian Systematic Botany*, **20**, 63–71.

Mast, A.R., C.L. Willis, E.H. Jones, K.M. Downs, and P.H. Weston. (2008). A smaller *Macadamia* from a more vagile tribe: inference of phylogenetic relationships, divergence times, and diaspore evolution in *Macadamia* and relatives (tribe Macadamieae; Proteaceae). *American Journal of Botany*, **95**, 843–870.

Matheny, P.B., and N.L. Bougher. (2006). The new genus *Auritella* from Africa and Australia (Inocybaceae, Agaricales): molecular systematics, taxonomy and historical biogeography. *Mycological Progress*, **5**, 2–17.

Matheny, P.B., M.C. Aime, N.L. Bougher, *et al.* (2009). Out of the palaeotropics? Historical biogeography and diversification of the cosmopolitan mushroom family Inocybaceae. *Journal of Biogeography*, **36**, 577–592.

Mathews, S. (2009). Phylogenetic relationships among seed plants: persistent questions and the limits of molecular data. *American Journal of Botany*, **96**, 228–236.

Mathias, M.E., and L. Constance. (1955). The genus *Oreomyrrhis* (Umbelliferae), a problem in South Pacific distribution. *University of California Publications in Botany*, **27**, 347–416.

Mathiasen, P., and A.C. Premoli. (2010). Out in the cold: genetic variation of *Nothofagus pumilio* (Nothofagaceae) provides evidence for latitudinally distinct evolutionary histories in austral South America. *Molecular Ecology*, **19**, 371–385.

Matsui, M., A. Tominaga, W. Liu, *et al.* (2010). Phylogenetic relationships of *Ansonia* from Southeast Asia inferred from mitochondrial DNA sequences: systematic and biogeographic implications (Anura: Bufonidae). *Molecular Phylogenetics and Evolution*, **54**, 561–570.

Matthew, W.D. (1915). Climate and evolution. *Annals of the New York Academy of Science*, **24**, 171–318.

Matthews, E.G. (2000). Origins of Australian arid-zone tenebrionid beetles. *Invertebrate Systematics*, **14**, 941–951.

Matthews, K.J., A.J. Hale, M. Gurnis, R.D. Müller, and L. DiCaprio. (2011). Dynamic subsidence of Eastern Australia during the Cretaceous. *Gondwana Research*, **19**, 372–383.

Mausfeld, P., and A. Schmitz. (2003). Molecular phylogeography, intraspecific variation and speciation of the Asian scincid lizard genus *Eutropis* Fitzinger, 1843 (Squamata: Reptilia: Scincidae): taxonomic and biogeographic implications. *Organisms Diversity and Evolution*, **3**, 161–171.

Mayr, E. (1931). Birds collected during the Whitney South Sea Expedition. XIV. *American Museum Novitates*, **488**, 1–11.

Mayr, E. (1932a). Birds collected during the Whitney South Seas Expedition. XIX. Notes on the bronze cuckoo *Chalcites lucidus* and its subspecies. *American Museum Novitates*, **520**, 1–9.

Mayr, E. (1932b). Birds collected during the Whitney South Sea Expedition. XXI. Notes on thickheads (*Pachycephala*) from Polynesia. *American Museum Novitates*, **531**, 1–23.

Mayr E. (1934). Birds collected during the Whitney South Sea Expedition. XXIX. Notes on the genus *Petroica*. *American Museum Novitates*, **714**, 1–19.

Mayr, E. (1940). The origin and history of the bird fauna of Polynesia. *Proceedings of the Sixth Pacific Science Congress (California)*, **4**, 197–216.

Mayr, E. (1942). *Systematics and the Origin of Species*. New York: Columbia University Press.

Mayr, E. (1944). The birds of Timor and Sumba. *Bulletin of the American Museum of Natural History*, **83**, 127–194.

Mayr, E. (1954). Change of genetic environment and evolution. In *Evolution as a Process*, eds. J. Huxley, A.C. Hardy, and E.B. Ford. London: Allen and Unwin, pp. 157–180.

Mayr, E. (1963). *Animal Species and Evolution*. Cambridge, MA: Harvard University Press.

Mayr, E. (1965). Summary. In *The Genetics of Colonizing Species*, eds. H.G. Baker and G.L. Stebbins. New York: Academic Press, pp. 553–562.

Mayr, E. (1982). *The Growth of Biological Thought: Diversity, Evolution and Inheritance*. Cambridge, MA: Harvard University Press.

Mayr, E. (1999). Introduction. In *Systematics and the Origin of Species*, ed. E. Mayr. Cambridge MA: Harvard University Press, pp. xiii–xvi.

Mayr, E. (2004). 80 years of watching the evolutionary scenery. *Science*, **305**, 46–47.

Mayr, E., and J. Diamond. (2001). *The Birds of Northern Melanesia: Speciation, Ecology and Biogeography*. New York: Oxford University Press.

Mayr, E., and W.H. Phelps Jr. (1967). The origin of the bird fauna of the South Venezuelan highlands. *Bulletin of the American Museum of Natural History*, **136**, 274–327.

Mayr, G. (2004). Old World fossil record of modern-type hummingbirds. *Science*, **304**, 861–864.

McCann, C. (1964). A coincidental distributional pattern of some of the larger marine animals. *Tuatara*, **12**, 119–124.

McCarthy, P. (ed.) (1995). *Flora of Australia, Vol. 16, Elaeagnaceae, Proteaceae part 1*. Canberra: Australian Biological Resources Study and CSIRO.

McCoy, M. (2006). *Reptiles of the Solomon Islands*. Sofia: Pensoft.

McCulloch, G.A. (2010). *Evolutionary Genetics of Southern Stoneflies*. PhD thesis, Dunedin: University of Otago.

McDaniel, S.F., and A.J. Shaw. (2003). Phylogeographic structure and cryptic speciation in the trans-Antarctic moss *Pterobryum mnioides*. *Evolution*, **57**, 205–215.

McDowall, R.M. (2008). Process and pattern in the biogeography of New Zealand – a global microcosm? *Journal of Biogeography*, **35**, 197–212.

McDowall, R.M. (2010). Historical and ecological context, pattern and process, in the derivation of New Zealand's freshwater fish fauna. *New Zealand Journal of Ecology*, **34**, 185–194.

McGuigan, K., D. Zhu, G.R. Allen, and C. Moritz. (2000). Phylogenetic relationships and historical biogeography of melanotaeniid fishes in Australia and New Guinea. *Marine and Freshwater Research*, **51**, 713–23.

McKenzie, R.J., and N.P. Barker. (2008). Radiation of southern African daisies: biogeographic inferences for subtribe Arctotidinae (Asteraceae, Arctotideae). *Molecular Phylogenetics and Evolution*, **49**, 1–16.

McKinnon, J.S., and H.D. Rundle. (2002). Speciation in nature: the threespine stickleback model systems. *Trends in Ecology and Evolution*, **17**, 480–488.

McPherson, H. (2006). Molecular systematics of *Tetratheca* and allies. *Australian Systematic Botany Society Newsletter*, **126**, 20–21.

McQuillan, P.B. (2003). The giant Tasmanian 'pandani' moth *Proditrix nielseni*, sp. nov. (Lepidoptera: Yponomeutoidea: Plutellidae s.l.). *Invertebrate Systematics*, **17**, 59–66.

Meerow, A.W., and D.A. Snijman. (2006). The never-ending story: multigene approaches to the phylogeny of Amaryllidaceae. *Aliso*, **22**, 355–366.

Meerow, A.W., M.F. Fay, C.L. Guy, *et al.* (1999). Systematics of Amaryllidaceae based on cladistic analysis of plastid sequence data. *American Journal of Botany*, **86**, 1325–1345.

Meffre, S., J.C. Aitchison, and A.J. Crawford. (1996). Geochemical evolution and tectonic significance of boninites and tholeiites from the Koh Ophiolite, New Caledonia. *Tectonics*, **15**, 67–83.

Meffre, S., A.J. Crawford, and P.G. Quilty. (2007). Arc continent collision forming a large island between New Caledonia and New Zealand in the Oligocene. Extended Abstracts, *Australian Earth Sciences Convention 2006*, Melbourne. 3 pp.

Meléndez, R., and B.S. Dyer. (2010). Review of the southern hemisphere fish family Chironemidae (Perciformes: Cirrhitoidei). *Revista de Biología Marina y Oceanografía*, **45**, S1, 683–693.

Meredith, R.W., E.R. Hekkala, G. Amato, and J. Gatesy. (2011). A phylogenetic hypothesis for *Crocodylus* (Crocodylia) based on mitochondrial DNA: evidence for a trans-Atlantic voyage from Africa to the New World. *Molecular Phylogenetics and Evolution*, **60**, 183–191.

Meredith, R.W., M. Westerman, and M.S. Springer. (2009). A phylogeny of Diprotodontia (Marsupialia) based on sequences for five nuclear genes. *Molecular Phylogenetics and Evolution*, **51**, 554–571.

Mesibov, R. (2009). Revision of *Agathodesmus* Silvestri, 1910 (Diplopoda, Polydesmida, Haplodesmidae). *ZooKeys*, **12**, 87–110.

Metcalfe, I. (2010). Tectonic framework and Phanerozoic evolution of Sundaland. *Gondwana Research*, **19**, 3–21.

Metzger, G.A., F. Kraus, A. Allison, and C.L. Parkinson. (2010). Uncovering cryptic diversity in *Aspidomorphus* (Serpentes: Elapidae): evidence from mitochondrial and nuclear markers. *Molecular Phylogenetics and Evolution*, **54**, 405–416.

Meudt, H.M., and M.J. Bayly. (2008). Phylogeographic patterns in the Australasian genus *Chionohebe* (*Veronica* s.l., Plantaginaceae) based on AFLP and chloroplast DNA sequences. *Molecular Phylogenetics and Evolution*, **47**, 319–338.

Meudt, H.M., and B.B. Simpson. (2006). The biogeography of the austral, subalpine genus *Ourisia* (Plantaginaceae) based on molecular phylogenetic evidence: South American origin and dispersal to New Zealand and Tasmania. *Biological Journal of the Linnean Society*, **87**, 479–513.

Mey, W. (2003). Insular radiation of the genus *Hydropsyche* (Insecta, Trichoptera: Hydropsychidae) Pictet, 1834 in the Philippines and its implications for the biogeography of Southeast Asia. *Journal of Biogeography*, **30**, 227–236.

Michaux, B. (1994). Land movements and animal distributions in east Wallacea (eastern Indonesia, Papua New Guinea and Melanesia). *Palaeogeography, Palaeoclimatology, Palaeoecology*, **112**, 323–343.

Michaux, B. (2010). Biogeology of Wallacea: geotectonic models, areas of endemism, and natural biogeographical units. *Biological Journal of the Linnean Society*, **101**, 193–212.

Michaux, B., and R.A.B. Leschen. (2005). East meets west: biogeology of the Campbell Plateau. *Biological Journal of the Linnean Society*, **86**, 95–115.

Michener, C.D. (2007). *The Bees of the World*, 2nd edn. Baltimore, MD: Johns Hopkins University Press.

Michl, T., S. Huck, T. Schmitt, *et al.* (2010). The molecular population structure of the tall forb *Cicerbita alpina* (Asteraceae) supports the idea of cryptic glacial refugia in central Europe. *Botanical Journal of the Linnean Society*, **164**, 142–154.

Middleton, D.J. (2002) Revision of *Alyxia* (Apocynaceae). Part 2: Pacific Islands and Australia. *Blumea*, **47**, 1–93.

Millener, L.H. (1946). A Study of *Entelea arborescens* R. Br. ("Whau"). Part I. Ecology. *Transactions of the Royal Society of New Zealand*, **76**, 267–288.

Milsom, J. (2003). Forearc ophiolites: a view from the western Pacific. In *Ophiolites in Earth history*, eds. Y. Dilek and P.T. Robinson. *Geological Society, London, Special Publications* **218**, pp. 507–515.

Mincarone, M.M., and A.L. Stewart. (2006). A new species of giant seven-gilled hagfish (Myxinidae: *Eptatretus*) from New Zealand. *Copeia*, **2006**, 225–229.

Mirams, A.G.K., E. A. Treml, J. L. Shields, L. Liggins, and C. Riginos. (2011). Vicariance and dispersal across an intermittent barrier: population genetic structure of marine animals across the Torres Strait land bridge. *Coral Reefs*, **30**, 937–949.

Miranda, H.C. Jr., D.M. Brooks, and R.S. Kennedy. (2011). Phylogeny and taxonomic review of Philippine lowland Scops Owls (Strigiformes): parallel diversification of highland and lowland clades. *Wilson Journal of Ornithology*, **123**, 441–453.

Mishler, B.D. (2001). The biology of bryophytes: bryophytes aren't just small tracheophytes. *American Journal of Botany*, **88**, 2129–2131.

Mitchell, A.D., and P.B. Heenan. (2008). *Sophora* sect. *Edwardsia* (Fabaceae): further evidence from nrDNA sequence data of a recent and rapid radiation around the Southern Oceans. *Botanical Journal of the Linnean Society*, **140**, 435–441.

Mitchell, A.D., P.B. Heenan, B.G. Murray, B.P.J. Molloy, and P.J. de Lange. (2009). Evolution of the south-western Pacific genus *Melicytus* (Violaceae): evidence from DNA sequence data, cytology and sex expression. *Australian Systematic Botany*, **22**, 143–157.

Miura, T., Y. Roisin, and T. Matsumoto. (2000). Molecular phylogeny and biogeography of the nasute termite genus *Nasutitermes* (Isoptera: Termitidae) in the Pacific Tropics. *Molecular Phylogenetics and Evolution*, **17**, 1–10.

Molina, J.E., J. Wen, and L. Struwe. (2013). Systematics and biogeography of the non-viny grape relative *Leea* (Vitaceae). *Botanical Journal of the Linnean Society*, **171**, 354–376.

Moltesen, M., M. Irestedt, J. Fjeldså, P.G.P. Ericson, and K.A. Jønsson. (2012). Molecular phylogeny of Chloropseidae and Irenidae – cryptic species and biogeography. *Molecular Phylogenetics and Evolution*, **65**, 903–914.

Monaghan, M. T., M. Balke, J. Pons, and A.P. Vogler. (2006). Beyond barcodes: complex DNA taxonomy of a South Pacific Island radiation. *Proceedings of the Royal Society B*, **273**, 887–893.

Monaghan, M.T., J.L. Gattolliat, M. Sartori, *et al.* (2005). Transoceanic and endemic origins of the small minnow mayflies (Ephemeroptera, Baetidae) of Madagascar. *Proceedings of the Royal Society B*, **272**, 1829–1836.

Monaghan, M.T., D.J.G. Inward, T. Hunt, A.P. Vogler. (2007). A molecular phylogenetic analysis of the Scarabaeinae (dung beetles). *Molecular Phylogenetics and Evolution*, **45**, 674–692.

Moody, M.L., and D.H. Les. (2007). Phylogenetic systematic and character evolution in the angiosperm family Haloragaceae. *American Journal of Botany*, **94**, 2005–2025.

Mooers, A.Ø., and D.W. Redding. (2009). Where the rare species are. *Molecular Ecology*, **18**, 3955–3957.

Mooers, A.Ø., H.D. Rundle, and M.C. Whitlock. (1999). The effects of selection and bottlenecks on male mating success in peripheral isolates. *American Naturalist*, **153**, 437–444.

Moore, L.B. (1957). The species of *Xeronema* (Liliaceae). *Pacific Science*, **11**, 355–62.

Moores, E. (2009). Western Pacific tectonics and the suggested Sierra seven suture saga. *Proceedings of the 2009 Portland Geological Society of America Meeting*, Paper No. 162–13.

Moores, E.M., J. Wakabayashi, and J.R. Unruh. (2002). Crustal-scale cross-section of the U.S. Cordillera, California and beyond, its tectonic significance, and speculations on the Andean orogeny. *International Geology Reviews*, **44**, 479–500.

Morat, P., T. Deroin, and H. Couderc. (1994). Présence en Nouvelle-Calédonie d'une espèce endémique du genre *Oryza* L. (Gramineae). *Bulletin du Muséum national d'Histoire naturelle* 4e sér., sect. B *Adansonia*, **16**, 155–160.

Morat, P., T. Jaffré, and J.-M. Veillon. (2001). The flora of New Caledonia's calcareous substrates. *Adansonia* sér. 3, **23**, 109–127.

Morat, P., J.-M. Veillon, and H.S. MacKee. (1994). Floristic relationships of New Caledonian rain forest phanerogams. In *Biogeography of the Tropical Pacific*, eds. P. Raven, F. Radovsky, and S. Sohmer. Honolulu: Association of Systematics Collections and Bernice P. Bishop Museum, pp. 71–128.

Moreira-Muñoz, A. (2007). *Plant Geography of Chile: An Essay on Postmodern Biogeography.* Ph.D. thesis, Friedrich-Alexander University, Erlangen-Nuremberg.

Moreira-Muñoz, A. (2011). *Plant Geography of Chile.* Dordrecht: Springer.

Moreira-Muñoz, A., and M. Muñoz-Schick. (2007). Classification, diversity, and distribution of Chilean Asteraceae: implications for biogeography and conservation. *Diversity and Distributions*, **13**, 818–828.

Morgan M.J., J.D. Roberts, and J.S. Keogh. (2007). Molecular phylogenetic dating supports an ancient endemic speciation model in Australia's biodiversity hotspot. *Molecular Phylogenetics and Evolution*, **44**, 371–385.

Morley, R.J. (2011). Dispersal and paleoecology of tropical podocarps. In *Ecology of the Podocarpaceae in Tropical Forests*, eds. B.L. Turner and L.A. Cernusak. *Smithsonian Contributions to Botany*, **95**, pp. 21–41.

Mortimer, N. (2004). Basement gabbro from the Lord Howe Rise. *New Zealand Journal of Geology and Geophysics*, **47**, 501–507.

Mortimer, N. (2006). Zealandia. *Australian Earth Science Convention 2006.* Conference Handbook, p. 4, Melbourne.

Mortimer, N., F.J. Davey, A. Melhuish, J. Yu, and N.J. Godfrey. (2002). Geological interpretation of a deep seismic reflection profile across the Eastern Province and Median Batholith, New Zealand: crustal architecture of an extended Phanerozoic convergent orogen. *New Zealand Journal of Geology and Geophysics*, **45**, 349–363.

Mortimer, N., P. Gans, A. Calvert, and N. Walker. (1999). Geology and thermochronometry of the east edge of the Median Batholith (Median Tectonic Zone): a new perspective on Permian to Cretaceous crustal growth of New Zealand. *Island Arc*, **8**, 404–425.

Mortimer, N., I.J. Graham, C.J. Adams, A.J. Tulloch, and H.J. Campbell. (2005). Relationships between New Zealand, Australian and New Caledonian mineralised terranes: a regional geological framework. *New Zealand Minerals Conference*, pp. 151–159. New Zealand Petroleum and Minerals, Wellington. http://www.nzpam.govt.nz/cms/pdf-library/minerals/conferences-1/151_papers_42.pdf

Mortimer, N., F. Hauff, and A.T. Calvert. (2008). Continuation of the New England Orogen, Australia, beneath the Queensland Plateau and Lord Howe Rise. *Australian Journal of Earth Sciences*, **55**, 195–209.

Morton, C.M. (2011). Newly sequenced nuclear gene (*Xdh*) for inferring angiosperm phylogeny. *Annals of the Missouri Botanical Garden*, **98**, 63–89.

Motley, T.J., K.J. Wurdack, and P.G. Delprete. (2005). Molecular systematics of the Catesbaeeae–Chiococceae complex (Rubiaceae): flower and fruit evolution and biogeographic implications. *American Journal of Botany*, **92**, 316–329.

Moyle, R.G., M.J. Anderson, C.H. Oliveros, F. Steinheimer, and S. Reddy. (2012). Phylogeny and biogeography of the core Babblers (Aves: Timaliidae). *Systematic Biology*, **61**, 631–651.

Moyle, R.G., C.E. Filardi, C.E. Smith, and J. Diamond. (2009). Explosive Pleistocene diversification and hemispheric expansion of a "great speciator." *Proceedings of the National Academy of Sciences, USA*, **106**, 1863–1868.

Moyle, R.G., S.S. Taylor, C.H. Oliveros, *et al.* (2011). Diversification of an endemic Southeast Asian genus: phylogenetic relationships of the Spiderhunters (Nectariniidae: *Arachnothera*). *The Auk*, **128**, 777–788.

Muasya, A.M., D.A. Simpson, G.A. Verboom, *et al.* (2009). Phylogeny of Cyperaceae based on DNA sequence data: current progress and future prospects. *Botanical Review*, **75**, 2–21.

Müller, C.J., and L.B. Beheregaray. (2010). Palaeo island-affinities revisited – biogeography and systematics of the Indo-Pacific genus *Cethosia* Fabricius (Lepidoptera: Nymphalidae). *Molecular Phylogenetics and Evolution*, **57**, 314–326.

Müller, C.J., N. Wahlberg, and L.B. Beheregaray. (2010). 'After Africa': the evolutionary history and systematic of the genus *Charaxes* Ochsenheimer (Lepidoptera: Nymphalidae) in the Indo-Pacific region. *Biological Journal of the Linnean Society*, **100**, 457–481.

Müller, R.D., S. Dyksterhuis, and P. Rey. (2012). Australian paleo-stress fields and tectonic reactivation over the past 100 Ma. *Australian Journal of Earth Sciences*, **59**, 13–28.

Mummenhoff, K., H. Brüggemann, and J.L. Bowman. (2001). Chloroplast DNA phylogeny and biogeography of *Lepidium* (Brassicaceae). *American Journal of Botany*, **88**, 2051–2063.

Mummenhoff, K., P. Linder, N. Friesen, *et al.* (2004). Molecular evidence for bicontinental hybridogenous genomic constitution in *Lepidium* sensu stricto (Brassicaceae) species from Australia and New Zealand. *American Journal of Botany*, **91**, 254–261.

Muñoz, J., Á.M. Felicísimo, F. Cabezas, A.R. Burgaz, and I. Martínez. (2004). Wind as a long-distance dispersal vehicle in the southern hemisphere. *Science*, **304**, 1144–1147.

Muñoz Schick, M., and A. Moreira Muñoz. (2003). *Alstroemerias de Chile: Diversidad, Distribución y Conservación*. Santiago: Taller la Era.

Munro, J.B., J.M. Heraty, R.A. Burks, *et al.* (2011). A molecular phylogeny of the Chalcidoidea (Hymenoptera). *PLoS One*, **6**(11), e27023, 1–27.

Murienne, J. (2009a). New Caledonia, biology. In *Encyclopedia of Islands*, eds. R. Gillespie and D.A. Clague. Berkeley, CA: University of California Press, pp. 643–645.

Murienne, J. (2009b). Testing biodiversity hypotheses in New Caledonia using phylogenetics. *Journal of Biogeography*, **36**, 1433–1434.

Murienne, J., G.D. Edgecombe, and G. Giribet. (2011). Comparative phylogeography of the centipedes *Cryptops pictus* and *C. niuensis* (Chilopoda) in New Caledonia, Fiji and Vanuatu. *Organisms Diversity and Evolution*, **11**, 61–74.

Murienne, J., P. Grandcolas, M.D. Piulachs, *et al.* (2005). Evolution on a shaky piece of Gondwana: is local endemism recent in New Caledonia? *Cladistics*, **21**, 2–7.

Murienne, J., R. Pellens, R.B. Budinoff, E.C. Wheeler, and P. Grandcolas. (2008). Phylogenetic analysis of the endemic New Caledonian cockroach *Lauraesilpha*. Testing competing hypotheses of diversification. *Cladistics*, **24**, 1–11.

Murphy, D.J., G.K. Brown, J.T. Miller, and P.Y. Ladiges. (2010). Molecular phylogeny of *Acacia* Mill. (Mimosoideae: Leguminosae): evidence for major clades and informal classification. *Taxon*, **59**, 7–19.

Murphy, S.A., M.C. Double, and S.M. Legge. (2007). The phylogeography of palm cockatoos, *Probosciger aterrimus*, in the dynamic Australo-Papuan region. *Journal of Biogeography*, **34**, 1534–1545.

Musgrave, R.J. (2003). Early to Middle Miocene Pacific-Australia plate boundary in New Zealand: an alternative transcurrent-fault system. *Geological Society of Australia Special Publication*, **22**, and *Geological Society of America Special Paper*, **372**, 333–341.

Myers, A.A., and P.S. Giller, eds. (1988). *Analytical Biogeography: An Integrated Approach to the Study of Animal and Plant Distributions*. London: Chapman and Hall.

Myers, A.A., and J.K. Lowry. (2009). The biogeography of Indo-West Pacific tropical amphipods with particular reference to Australia. *Zootaxa*, **2260**, 109–127.

Myers, N., R. Mittermeier, G.C. Mittermeier, G.A.B. Dafonseca, and J. Kent. (2000). Biodiversity hotspots for conservation priorities. *Nature*, **403**, 853–858.

Nattier, R., T. Robillard, L. Desutter-Grandcolas, A. Couloux, and P. Grandcolas. (2011). Older than New Caledonia emergence? A molecular phylogenetic study of the eneopterine crickets (Orthoptera: Grylloidea). *Journal of Biogeography*, **38**, 2195–2209.

Nei, M. (2002). Review of 'Where do we come from? The molecular evidence for human descent' by J. Klein and N. Takahata. *Nature*, **417**, 899–900.

Neiber, M.T., T.R. Hartke, T. Stemme, *et al.* (2011). Global biodiversity and phylogenetic evaluation of Remipedia (Crustacea). *PLoS One*, **6**(5), e19627, 1–12.

Neiman, M., and C. Lively. (2004). Pleistocene glaciation is implicated in the phylogeographical structure of *Potamopyrgus antipodarum*, a New Zealand snail. *Molecular Ecology*, **13**, 3085–3098.

Nelson, G. (2004). Cladistics: its arrested development. In *Milestones in Systematics: The Development of Comparative Biology*, eds. D.M. Williams and P.L. Forey. London: Taylor and Francis, pp. 127–147.

Nelson, G., and P.Y. Ladiges. (2001). Gondwana, vicariance biogeography, and the New York school revisited. *Australian Journal of Botany*, **49**, 389–409.

Nevling, L.I., Jr. (1961). A revision of the Asiatic genus *Enkleia* (Thymelaeaceae). *Journal of the Arnold Arboretum*, **42**, 373–396.

Nickrent, D.L. (2012). The parasitic plant connection. At: www.science.siu.edu/parasitic-plants/

Nie, Z.-L., H. Sun, S.R Manchester, *et al.* (2012). Evolution of the intercontinental disjunctions in six continents in the *Ampelopsis* clade of the grape family (Vitaceae). *BMC Evolutionary Biology*, **12**(17), 1–13.

Nielsen, S.V., A.M. Bauer, T.R. Jackman, R.A. Hitchmough, and C.H. Daugherty. (2011). New Zealand geckos (Diplodactylidae): cryptic diversity in a post-Gondwanan lineage with trans-Tasman affinities. *Molecular Phylogenetics and Evolution*, **59**, 1–22.

Nilsson, M.A., G. Churakov, M. Sommer, *et al.* (2010). Tracking marsupial evolution using archaic genomic retroposon insertions. *PLoS Biology*, **8**(7), e1000436, 1–9.

Nonnotte, P., H. Guillou, B. Le Gall, *et al.* (2008). New K-Ar age determinations of Kilimanjaro volcano in the North Tanzanian diverging rift, East Africa. *Journal of Volcanology and Geothermal Research*, **173**, 99–112.

Noonan, B.P., and P.T. Chippindale. (2006). Dispersal and vicariance: the complex evolutionary history of boid snakes. *Molecular Phylogenetics and Evolution*, **40**, 347–358.

Noonan, B.P., and J.W. Sites Jr. (2010). Tracing the origins of iguanid lizards and boine snakes of the Pacific. *American Naturalist*, **175**, 61–72.

Norman, J.A., P.G.P. Ericson, K.A. Jønsson, J. Fjeldså, and L. Christidis. (2009). A multi-gene phylogeny reveals novel relationships for aberrant genera of Australo-Papuan core Corvoidea and polyphyly of the Pachycephalidae and Psophodidae (Aves: Passeriformes). *Molecular Phylogenetics and Evolution*, **52**, 488–497.

Norman, J.A., F.E. Rheindt, D.L. Rowe, and L. Christidis. (2007). Speciation dynamics in the Australo-Papuan *Meliphaga* honeyeaters. *Molecular Phylogenetics and Evolution*, **42**, 80–91.

Norup, M.V., J. Dransfield, M.W. Chase, *et al.* (2006). Homoplasious character combinations and generic delimitation: a case study from the Indo-Pacific arecoid palms (Arecaceae: Areceae). *American Journal of Botany*, **93**, 1065–1080.

Nowicke, J., V. Patel, and J.J. Skvarla. (1985). Pollen morphology and the relationships of *Aëtoxylon*, *Amyxa* and *Gonystylus* to the Thymelaeaceae. *American Journal of Botany*, **72**, 1106–1113.

Nunn, P.D., M. Baniala, M. Harrison, and P. Geraghty. (2006). Vanished islands in Vanuatu: new research and a preliminary geohazard assessment. *Journal of the Royal Society of New Zealand*, **36**, 37–50.

Nyári, Á.S. (2011). *Origin and Evolution of the Unique Australo-Papuan Mangrove-Restricted Avifauna: Novel Insights From Molecular Phylogenetic and Comparative Phylogeographic Analyses*. Ph.D. thesis, University of Kansas.

Nyári, Á.S., and L. Joseph. (2012). Evolution in Australasian mangrove forests: multilocus phylogenetic analysis of the *Gerygone* warblers (Aves: Acanthizidae). *PLoS One*, **7**(2), e31840, 1–9.

Nyári, Á.S., B.W. Benz, K.A. Jønsson, J. Fjeldså, and R.G. Moyle. (2009a). Phylogenetic relationships of fantails (Aves: Rhipiduridae). *Zoologica Scripta*, **38**, 553–561.

Nyári, Á.S., A.T. Peterson, N.H. Rice, and R.G. Moyle. (2009b). Phylogenetic relationships of flowerpeckers (Aves: Dicaeidae): novel insights into the evolution of a tropical passerine clade. *Molecular Phylogenetics and Evolution*, **53**, 613–19.

Nylinder, S., U. Swenson, C. Persson, S.B. Janssens, and B. Oxelman. (2012). A dated species-tree approach to the trans-Pacific disjunction of the genus *Jovellana* (Calceolariaceae, Lamiales). *Taxon*, **61**, 381–391.

O'Grady, P.M., G.M. Bennett, V.A. Funk, and T.K. Altheide. (2012). Retrograde biogeography. *Taxon*, **61**, 699–705.

Oliver, L.A., E.N. Rittmeyer, F. Kraus, S.J. Richards, and C.C. Austin. (2013). Phylogeny and phylogeography of *Mantophryne* (Anura: Microhylidae) reveals cryptic diversity in New Guinea. *Molecular Phylogenetics and Evolution*, **67**, 600–607.

Oliver, P.M., and K. Sanders. (2009). Molecular evidence for multiple lineages with ancient Gondwanan origins in a diverse Australian gecko radiation. *Journal of Biogeography*, **36**, 2044–2055.

Oliver, P.M., M. Adams, and P. Doughty. (2010). Molecular evidence for ten species and Oligo-Miocene vicariance within a nominal Australian gecko species (*Crenadactylus ocellatus*, Diplodactylidae). *BMC Evolutionary Biology*, **10** (386), 1–11.

Oliver, P.M., Richards, S.J., and Sistrom, M. (2012). Phylogeny and systematics of Melanesia's most diverse gecko lineage (Cyrtodactylus, Gekkonidae, Squamata). *Zoologica Scripta*, **41**, 437–454.

Oliver, W.R.B. (1928). A revision of the genus *Dracophyllum*. *Transactions of the New Zealand Institute*, **59**, 678–714.

Oliveros, C.H., and R.G. Moyle. (2010). Origin and diversification of Philippine bulbuls. *Molecular Phylogenetics and Evolution*, **54**, 822–832.

Oliveros, C.H., Reddy, S., Moyle, R.G. (2012). The phylogenetic position of some Philippine "babblers" spans the muscicapoid and sylvioid bird radiations. *Molecular Phylogenetics and Evolution*, **65**, 799–804.

Olmstead, R.G., L. Bohs, H.A. Migid, *et al.* (2008). A molecular phylogeny of the Solanaceae. *Taxon*, **57**, 1159–1181.

Olmstead, R.G., M.L. Zjhra, L.G., Lohmann, S.O. Grose, and A.J. Eckert. (2009). A molecular phylogeny and classification of Bignoniaceae. *American Journal of Botany*, **96**, 1731–1743.

Olson, L.E., E.J. Sargis, and R.D. Martin. (2005). Intraordinal phylogenetics of treeshrews (Mammalia: Scandentia) based on evidence from the mitochondrial *12S* rRNA gene. *Molecular Phylogenetics and Evolution*, **35**, 656–673.

Olsson, U., P. Alström, P.G.P. Ericson, and P. Sundberg. (2005). Non-monophyletic taxa and cryptic species—evidence from a molecular phylogeny of leaf-warblers (*Phylloscopus*, Aves). *Molecular Phylogenetics and Evolution*, **36**, 261–276.

Opgenoorth, L., G.G. Vendramin, K. Mao, *et al.* (2010). Tree endurance on the Tibetan Plateau marks the world's highest known tree line of the Last Glacial Maximum. *New Phytologist*, **185**, 332–342.

O'Quinn, R., and L. Hufford. (2005). Molecular systematics of Montieae (Portulacaceae): implications for taxonomy, biogeography and ecology. *Systematic Botany*, **30**, 314–331.

Orchard, A.E., and J.B. Davies. (1985). *Oreoporanthera*, a New Zealand 'endemic' plant genus discovered in Tasmania. *Papers and Proceedings of the Royal Society of Tasmania*, **119**, 62.

Orenstein, R.I., and H.D. Pratt. (1983). The relationships and evolution of the southwest Pacific warbler genera *Vitia* and *Psamathia* (Sylviinae). *Wilson Bulletin*, **95**, 184–198.

O'Riordan, R.M., A.M. Power, and A.A. Myers. (2010). Factors, at different scales, affecting the distribution of species of the genus *Chthamalus* Ranzani (Cirripedia, Balanomorpha, Chthamaloidea). *Journal of Experimental Marine Biology and Ecology*, **392**, 46–64.

Orr, A.G., and V.J. Kalkman. (2010). *Arrhenocnemis parvibullis* sp. nov. (Odonata: Platycnemididae), a new calicnemiine damselfly from Papua New Guinea, with a description of the female of *A. amphidactylis* Lieftinck, 1949. *Australian Entomologist*, **37**, 137–146.

Orr, H.A. (2005). The genetic basis of reproductive isolation: insights from *Drosophila*. *Proceedings of the National Academy of Sciences, USA*, **102**, 6522–6526.

O'Sullivan, P.B., M. Orr, A.J. O'Sullivan, and A.J.W. Gleadow. (1999). Episodic Late Palaeozoic to recent denudation of the Eastern Highlands of Australia: evidence from the Bogong High Plains, Victoria. *Australian Journal of Earth Sciences*, **46**, 199–216.

Oxelman, B., P. Kornhall, R.G. Olmstead, and B. Bremer. (2005). Further disintegration of Scrophulariaceae. *Taxon*, **54**, 411–425.

Ozawa, T., F. Köhler, D.G. Reid, and M. Glaubrecht. (2009). Tethyan relicts on continental coastlines of the northwestern Pacific Ocean and Australasia: molecular phylogeny and fossil record of batillariid gastropods (Caenogastropoda, Cerithioidea). *Zoologica Scripta*, **38**, 503–525.

Pacheco, M.A., F.U. Battistuzzi, M. Lentino, *et al.* (2011). Evolution of modern birds revealed by mitogenomics: timing the radiation and origin of major orders. *Molecular Biology and Evolution*, **28**, 1927–1942.

Padial, J.M., and I. de la Riva. (2006). Taxonomic inflation and the stability of species lists: the perils of ostrich's behavior. *Systematic Biology*, **55**, 859–867.

Page, T.J., W.F. Humphreys, and J.M. Hughes. (2008). Shrimps down under: evolutionary relationships of subterranean crustaceans from Western Australia (Decapoda: Atyidae: *Stygiocaris*). *PLoS One*, **3**(2), e1618, 1–12.

Panero, J.L., and V.A. Funk. (2008). The value of sampling anomalous taxa in phylogenetic studies: major clades of the Asteraceae revealed. *Molecular Phylogenetics and Evolution*, **47**, 757–782.

Papadopoulou, A., I. Anastasiou, and A.P. Vogler. (2010). Revisiting the insect mitochondrial molecular clock: the mid-Aegean trench calibration. *Molecular Biology and Evolution*, **27**, 1659–1672.

Papadopulos, A.S.T., W.J. Baker, D. Crayn, *et al.* (2011). Speciation with gene flow on Lord Howe Island. *Proceedings of the National Academy of Sciences, USA*, **108**, 13188–13193.

Papanicolaou, K., D. Babalonas, and S. Kokkini. (1983). Distribution patterns of some Greek mountain endemic plants in relation to geological substrate. *Flora*, **174**, 405–437.

Parham, J.F., P.C.J. Donoghue, C.J. Bell, *et al.* (2012). Best practices for justifying fossil calibrations. *Systematic Biology*, **61**, 346–359.

Pasquet, E., J.-M. Pons, J. Fuchs, C. Cruaud, and V. Bretagnolle. (2007). Evolutionary history and biogeography of the drongos (Dicruridae), a tropical Old World clade of corvoid passerines. *Molecular Phylogenetics and Evolution*, **45**, 158–167.

Patterson, G.B., and T.P. Bell. (2009). The Barrier skink *Oligosoma judgei* n. sp. (Reptilia: Scincidae) from the Darran and Takitimu Mountains, South Island, New Zealand. *Zootaxa*, **2271**, 43–56.

Payo, D.A., F. Leliaert, H. Verbruggen, *et al.* (2013). Extensive cryptic species diversity and fine-scale endemism in the marine red alga *Portieria* in the Philippines. *Proceedings of the Royal Society B*. doi: 10.1098/rspb.2012.2660.

Pellens, R. (2004). New species of *Angustonicus* Grandcolas, 1997 (Insecta, Dictyoptera, Blattaria, Tryonicinae) and the endemism of the genus in New Caledonia. *Zoosystema*, **26**, 307–314.

Pelletier, B. (2007). Geology of the New Caledonia region and its implications for the study of the New Caledonian biodiversity. In *Compendium of Marine Species from New Caledonia*, eds. C. Payri and B. Richer de Forges. Nouméa: IRD, pp. 19–32.

Pelser, P.B., A.H. Kennedy, E.J. Tepe, *et al.* (2010). Patterns and causes of incongruence between plastid and nuclear Senecioneae (Asteraceae) phylogenies. *American Journal of Botany*, **97**, 856–873.

Peña, R.C., L. Iturriaga, G. Montenegro, and B.K. Cassels. (2000). Phylogenetic and biogeographic aspects of *Sophora* sect. *Edwardsia* (Papilionaceae). *Pacific Science*, **54**, 159–167.

Pennington, R.T., and C.W. Dick. (2004). The role of immigrants in the assembly of the South American rainforest tree flora. *Philosophical Transactions of the Royal Society, London B*, **359**, 1611–1622.

Pennington, R.T., Q.C.B. Cronk, and J.A. Richardson, eds. (2004a). Plant phylogeny and the origin of major biomes. *Philosophical Transactions of the Royal Society, London B*, **359**, 1450.

Pennington, R.T., Q.C.B. Cronk, and J.A. Richardson. (2004b). Introduction and synthesis: plant phylogeny and the origin of major biomes. *Philosophical Transactions of the Royal Society, London B*, **359**, 1455–1464.

Pepper, M., P. Doughty, M.N. Hutchinson, and J.S. Keogh. (2011c). Ancient drainages divide cryptic species in Australia's arid zone: morphological and multi-gene evidence for four new species of beaked geckos (*Rhynchoedura*). *Molecular Phylogenetics and Evolution*, **61**, 810–822.

Pepper, M., P. Doughty, and J.S. Keogh. (2006). Molecular phylogeny and phylogeography of the Australian *Diplodactylus stenodactylus* (Gekkota; Reptilia) species-group based on mitochondrial and nuclear genes reveals an ancient split between Pilbara and non-Pilbara *D. stenodactylus*. *Molecular Phylogenetics and Evolution*, **41**, 539–555.

Pepper, M., M.K. Fujita, C.Moritz, and J.S. Keogh. (2011b). Palaeoclimate change drove diversification among isolated mountain refugia in the Australian arid zone. *Molecular Ecology*, **20**, 1529–1545.

Pepper, M.R., S.Y.W. Ho, M.K. Fujita, and J.S. Keogh. (2011a). The genetic legacy of aridification: climate cycling fostered lizard diversification in Australian montane refugia and left low-lying deserts genetically depauperate. *Molecular Phylogenetics and Evolution*, **61**, 810–822.

Pereira, S.L., K.P. Johnson, D.H. Clayton, and A.J. Baker. (2007). Mitochondrial and nuclear DNA sequences support a Cretaceous origin of Columbiformes and a dispersal-driven radiation in the Paleogene. *Systematic Biology*, **56**, 656–672.

Pérez-Losada, M., and K.A. Crandall, K.A. (2003). Can taxonomic richness be used as a surrogate for phylogenetic distinctness indices for ranking areas for conservation? *Animal Biodiversity and Conservation*, **26**, 77–84.

Perret, M., A. Chautems, A.O. De Araujo, and N. Salamin. (2013). Temporal and spatial origin of Gesneriaceae in the New World inferred from plastid DNA sequences. *Botanical Journal of the Linnean Society*, **171**, 61–79.

Perrie, L.R., and P.J. Brownsey. (2005). Insights into the biogeography and polyploid evolution of New Zealand *Asplenium* from chloroplast DNA sequence data. *American Fern Journal*, **95**, 1–21.

Perrie, L.R., and P.J. Brownsey. (2007). Molecular evidence for long-distance dispersal in the New Zealand pteridophyte flora. *Journal of Biogeography*, **34**, 2028–2038.

Perrie, L.R., D. J. Ohlsen, L.D. Shepherd, *et al.* (2010). Tasmanian and Victorian populations of the fern *Asplenium hookerianum* result from independent dispersals from New Zealand. *Australian Systematic Botany*, **23**, 387–392

Persano, C., P. Bishop, and F.M. Stuart. (2006). Apatite (U–Th)/He age constraints on the Mesozoic and Cenozoic evolution of the Bathurst region, New South Wales: evidence for antiquity of the continental drainage divide along a passive margin. *Australian Journal of Earth Sciences*, **53**, 1041–1050.

Peters, S.E. (2005). Geologic constraints on the macroevolutionary history of marine animals. *Proceedings of the National Academy of Sciences USA*, **102**, 12326–12331.

Peters, S.E., and M. Foote. (2001). Biodiversity in the Phanerozoic: a reinterpretation. *Paleobiology*, **27**, 583–601.

Peters, S.E., and M. Foote. (2002). Determinants of extinction in the fossil record. *Nature*, **416**, 420–424.

Petterson, M.G., T. Babbs, C.R. Neal, *et al.* (1999). Geological–tectonic framework of Solomon Islands, SW Pacific: crustal accretion and growth within an intra-oceanic setting. *Tectonophysics* **301**, 35–60.

Philippe, M., M. Bamford, S. McLoughlin, *et al.* (2004). Biogeographic analysis of Jurassic-Early Cretaceous wood assemblages from Gondwana. *Review of Palaeobotany and Palynology*, **129**, 141–173.

Phillimore, A.B., I.P.F. Owens, R.A. Black, *et al.* (2008). Complex patterns of genetic and phenotypic divergence in an island bird and the consequences for delimiting conservation units. *Molecular Ecology*, **17**, 2839–2853.

Phillips, L.E. (2001). Morphology and molecular analysis of the Australian monotypic genera *Lembergia* and *Sonderella* (Rhodomelaceae, Rhodophyta), with a description of the tribe Sonderelleae trib. nov. *Phycologia*, **40**, 487–499.

Phillips, M.J., G.C. Gibb, E.A. Crimp, and D. Penny. (2010). Tinamous and moa flock together: mitochondrial genome sequence analysis reveals two independent losses of flight among ratites. *Systematic Biology*, **59**, 90–107.

Pianka, E.R. (1972). Zoogeography and speciation of Australian desert lizards: an ecological perspective. *Copeia*, **1972**, 127–145.

Picard, C., D. Cluzel, and P. Black. (2002). Remnants of a Late Cretaceous arc-backarc system in New Caledonia and New Zealand, inference on the southwest Pacific geodynamic evolution. *2002 Western Pacific Geophysics Meeting*. Abstract SE41D-06. (www.agu.org).

Pignal, M., and J. Munzinger. (2011). Une nouvelle espèce de *Microtatorchis* (Orchidaceae, Vandeae, Aeridinae) en Nouvelle-Calédonie, et clé d'identification des Aeridinae aphylles du Territoire. *Adansonia*, sér. 3, **33**, 183–190.

Pigram, C.J., and P.J. Davies. (1987). Terranes and the accretion history of the New Guinea orogen. *BMR Journal of Australian Geology and Geophysics*, **10**, 193–212.

Pillon, Y. (2012). Time and tempo of diversification in the flora of New Caledonia. *Botanical Journal of the Linnean Society*, **170**, 288–298.

Pillon, Y., M.F. Fay, A.B. Shipunov, and M.W. Chase. (2006). Species diversity versus phylogenetic diversity: a practical study in the taxonomically difficult genus *Dactylorhiza* (Orchidaceae). *Biological Conservation*, **129**, 4–13.

Pillon, Y., H.C.F. Hopkins, J. Munzinger, and M.W. Chase. (2009). A molecular and morphological survey of generic limits of *Acsmithia* and *Spiraeanthemum* (Cunoniaceae). *Systematic Botany*, **34**, 141–148.

Pillon, Y., J. Munzinger, H. Amir, and M. Lebrun. (2010). Ultramafic soils and species sorting in the flora of New Caledonia. *Journal of Ecology*, **98**, 1108–1116.

Pintaud, J.-C., and W.J. Baker. (2008). A revision of the palm genera (*Arecaceae*) of New Caledonia. *Kew Bulletin*, **63**, 61–73.

Pintaud, J.-C., T. Jaffré, and H. Puig. (2001). Chorology of New Caledonian palms and possible evidence of Pleistocene rain forest refugia. *Comptes Rendus de l'Académie des Sciences. Sér. 3, Sciences de la Vie*, **324**, 453–463.

Pirie, M.D., A.M. Humphreys, N.P. Barker, and H.P. Linder. (2009a). Reticulation, data combination, and inferring evolutionary history: an example from Danthonioideae (Poaceae). *Systematic Biology*, **58**, 612–628.

Pirie, M.D., K.M. Lloyd, W.G. Lee, and H.P. Linder. (2009b). Diversification of *Chionochloa* (Poaceae) and biogeography of the New Zealand Southern Alps. *Journal of Biogeography*, **37**, 379–392.

Platnick, N.I. (2011). *The World Spider Catalog, version 11.5*. Available at: http://research.amnh.org/entomology/spiders/catalog/ (Accessed May 2011).

Poinar, G.O., Jr. (2000). Fossil onychophorans from Dominican and Baltic amber: *Tertiapatus dominicanus* n. g., n. sp. (Tertiapatidae n.fam.) and *Succinipatopsis balticus* n. g., n. sp. (Succinipatopsidae n. fam.) with a proposed classification of the subphylum Onychophora. *Invertebrate Biology*, **119**, 104–109.

Poinar, G.O., and B.N. Danforth. (2006). A fossil bee from Early Cretaceous Burmese amber. *Science*, **314**, 614.

Pole, M. (1994). The New Zealand flora – entirely long-distance dispersal? *Journal of Biogeography*, **21**, 625–635.

Pole, M. (2010). Was New Zealand a primary source for the New Caledonian flora? *Alcheringa*, **34**, 61–74.

Polhemus, D.A. (1996). Island arcs, and their influence on Indo-Pacific biogeography. In *The Origin and Evolution of Pacific Island Biotas, New Guinea to Eastern Polynesia: Patterns and Processes*, eds. A. Keast and S.E. Miller. Amsterdam: SPB Academic Publishing, pp. 51–66.

Polhemus, D.A., and J.T. Polhemus. (1998). Assembling New Guinea: 40 million years of island arc accretion as indicated by the distributions of aquatic Heteroptera (Insecta). In *Biogeography and Geological Evolution of SE Asia*, eds. R. Hall and J.D. Holloway. Leiden: Backhuys, pp. 327–340.

Pollock, D.A. (1995). Classification, reconstructed phylogeny and geographical history of genera of Pilipalpinae (Coleoptera: Tenebrionoidea: Pyrochroidae). *Invertebrate Taxonomy*, **9**, 563–708.

Posadas, P., D.R. Miranda Esquivel, and J.V. Crisci. (2001). Using phylogenetic diversity measures to set priorities in conservation: an example from southern South America. *Conservation Biology*, **15**, 1325–1334.

Potter, S., S.J.B. Cooper, C.J. Metcalfe, D.A. Taggart, and M.D.B. Eldridge. (2012). Phylogenetic relationships of rock-wallabies, *Petrogale* (Marsupialia: Macropodidae) and their biogeographic history within Australia. *Molecular Phylogenetics and Evolution*, **62**, 640–652.

Powell, A.W.B. (1979). *New Zealand Mollusca*. Auckland: Collins.

Powney, G.D., R. Grenyer, C.D.L. Orme, I.P.F. Owens, and S. Meiri. (2010). Hot, dry and different: Australian lizard richness is unlike that of mammals, amphibians and birds. *Global Ecology and Biogeography*, **19**, 386–396.

Pramuk, J.B., T. Robertson, J.W. Sites, and B.P. Noonan. (2007). Around the world in 10 million years: biogeography of the nearly cosmopolitan true toads (Anura: Bufonidae). *Global Ecology and Biogeography*, **17**, 72–83.

Prance, G. (1994). A comparison of the efficacy of higher taxa and species numbers in the assessment of biodiversity in the neotropics. *Philosophical Transactions of the Royal Society, London B*, **345**, 89–99.

Pratt, R.C., M. Morgan-Richards, and S.A. Trewick. (2008). Diversification of New Zealand weta (Orthoptera: Ensifera: Anostostomatidae) and their relationships in Australasia. *Philosophical Transactions of the Royal Society, B*, **363**, 3427–3437.

Pratt, S.J. (2013). *Evolution of the genera* Vitex *(Lamiaceae) and* Zygogynum *(Winteraceae) on New Caledonia*. M.Sc. thesis Hamilton: University of Waikato.

Preußing, M., S. Olsson, A. Schäfer-Verwimp, *et al.* (2010). New insights in the evolution of the liverwort family Aneuraceae (Metzgeriales, Marchantiophyta), with emphasis on the genus *Lobatiriccardia. Taxon*, **59**, 1424–1440.

Proctor, J. (2003). Vegetation and soil and plant chemistry on ultramafic rocks in the tropical Far East. *Perspectives in Plant Ecology, Evolution and Systematics*, **6**, 105–124.

Provan, J., and K.D. Bennett. (2008). Phylogeographic insights into cryptic glacial refugia. *Trends in Ecology and Evolution*, **23**, 564–571.

Pubellier, M., J. Ali, and C. Monnier. (2003). Cenozoic plate interaction of the Australia and Philippine Sea plates: 'hit-and-run' tectonics. *Tectonophysics*, **363**, 181–199.

Pubellier, M., C. Monnier, R. Maury, and R. Tamayo. (2004). Plate kinematics, origin and tectonic emplacement of supra-obduction ophiolites in SE Asia. *Tectonophysics*, **392**, 9–36.

Puente-Lelièvre, C., M. Harrington, E. Brown, M. Kuzmina, and D. Crayn. (2013). Cenozoic extinction and recolonization in the New Zealand flora: the case of the fleshy-fruited epacrids (Styphelieae, Styphelioideae, Ericaceae). *Molecular Phylogenetics and Evolution*, **66**, 203–214.

Puigbò, P., Y.I. Wolf, and E.V. Koonin. (2009). Search for a Tree of Life in the thicket of the phylogenetic forest. *Journal of Biology*, **8** (59), 1–17.

Pulvers, J.N., and D.J. Colgan. (2007). Molecular phylogeography of the fruit bat genus *Melonycteris* in northern Melanesia. *Journal of Biogeography*, **34**, 713–723.

Pyron, R.A., and F.T. Burbrink. (2010). Hard and soft allopatry: physically and ecologically mediated modes of geographic speciation. *Journal of Biogeography*, **37**, 2005–2015.

Pyron, R.A., and J.J. Wiens. (2011). A large-scale phylogeny of Amphibia including over 2800 species, and a revised classification of extant frogs, salamanders, and caecilians. *Molecular Phylogenetics and Evolution*, **61**, 543–583.

Pyron, R.A., F.T. Burbrink, G.R. Colli, *et al.* (2011). The phylogeny of advanced snakes (Colubroidea), with discovery of a new subfamily and comparison of support methods for likelihood trees. *Molecular Phylogenetics and Evolution*, **58**, 329–342.

Qiu, Y.L., L. Li, B. Wang, *et al.* (2007). A nonflowering land plant phylogeny inferred from nucleotide sequences of seven chloroplast, mitochondrial, and nuclear genes. *International Journal of Plant Science*, **168**, 691–708.

Quigley, M.C., D. Clark, and M. Sandiford. (2010). Tectonic geomorphology of Australia. In *Australian Landscapes*, eds. P. Bishop and B. Pillans. *Geological Society, London, Special Publications*, **346**, 243–265.

Quinn, C.J., E.A. Brown, M.M. Heslewood, and D.M. Crayn. (2005). Generic concepts in Styphelieae (Ericaceae): the *Cyathodes* group. *Australian Systematic Botany*, **18**, 439–454.

Quinn, C.J., D.M. Crayn, M.M. Heslewood, E.A. Brown, and P.A. Gadek. (2003). A molecular estimate of the phylogeny of Styphelieae (Ericaceae). *Australian Systematic Botany*, **16**, 581–594.

Rabosky, D.L., S.C. Donnellan, A.L. Talaba, and I.J. Lovette. (2007). Exceptional among-lineage variation in diversification rates during the radiation of Australia's most diverse vertebrate clade. *Proceedings of the Royal Society B*, **274**, 2915–2923.

Ragionieri, L., S. Fratini, M. Vannini, and C.D. Schubart. (2009). Phylogenetic and morphometric differentiation reveal geographic radiation and pseudo-cryptic speciation in a mangrove crab from the Indo-West Pacific. *Molecular Phylogenetics and Evolution*, **52**, 825–834.

Rand, A.L., and E.T. Gilliard. (1967). *Handbook of New Guinea Birds*. London: Weidenfeld and Nicolson.

Rangin, C., and M. Pubellier. (1990). Subduction and accretion of Philippine Sea Plate fragments along the Eurasian margin. In *Tectonics of Circum-Pacific Continental Margins*, eds. J. Aubouin and J. Bourgois. Zeist: VSP International, pp. 139–164.

Rapoport, E.H. (1982). *Areography: Geographical Strategies of Species*. Oxford: Pergamon.

Raven, P.H., and T.E. Raven. (1976). The genus *Epilobium* (Onagraceae) in Australasia: A systematic and evolutionary study. *New Zealand Department of Scientific and Industrial Research Bulletin*, **216**.

Rawling, T.J. (1998). *Oscillating Orogenesis and Exhumation of High-pressure Rocks in New Caledonia*. Unpublished PhD thesis. Melbourne: Department of Earth Sciences, Monash University.

Rawling, T.J., and G.S. Lister. (1997). The structural evolution of New Caledonia. Abstracts. *Geodynamics and Ore Deposits Conference*, Australian Geodynamics Cooperative Research Center, Ballarat, Victoria, 19–21 February 1997, pp. 62–64.

Rawling, T.J., and G.S. Lister. (1999). Oscillating modes of orogeny in the Southwest Pacific and the tectonic evolution of New Caledonia. In *Exhumation Processes: Normal Faulting, Ductile Flow and Erosion*, eds. G. Ring, M.T. Brandon, G.S. Lister, and S.D. Weillett. *Geological Society, London, Special Publication*, **154**, 109–127.

Rawling, T.J., and G.S. Lister. (2002). Large-scale structure of the eclogite-blueschist belt of New Caledonia. *Journal of Structural Geology*, **24**, 1239–1258.

Rawlings, L.H., and S.C. Donnellan. (2003). Phylogeographic analysis of the green python, *Morelia viridis*, reveals cryptic diversity. *Molecular Phylogenetics and Evolution*, **27**, 36–44.

Rawlings, L.H., D.L. Rabosky, S.C. Donnellan, and M.N. Hutchinson. (2008). Python phylogenetics: inference from morphology and mitochondrial DNA. *Biological Journal of the Linnean Society*, **93**, 603–619.

Raza, A., K.C. Hill, and R.J. Korsch. (2009). Mid-Cretaceous uplift and denudation of the Bowen and Surat Basins, eastern Australia: relationship to Tasman Sea rifting from apatite fission-track and vitrinite-reflectance data. *Australian Journal of Earth Sciences*, **56**, 501–531.

Read, K., J.S. Keogh, I.A.W. Scott, J.D. Roberts, and P. Doughty. (2001). Molecular phylogeny of the Australian frog genera *Crinia*, *Geocrinia*, and allied taxa (Anura: Myobatrachidae). *Molecular Phylogenetics and Evolution*, **21**, 294–308.

Redding, D.W., and Mooers, A. (2006). Incorporating evolutionary measures into conservation prioritization. *Conservation Biology*, **20**, 1670–1678.

Ree, R.H., and S.A. Smith. (2008). Maximum likelihood inference of geographic range evolution by dispersal, local extinction, and cladogenesis. *Systematic Biology*, **57**, 4–14.

Rehan, S.M., T.W. Chapman, A.I. Craigie, *et al.* (2010). Molecular phylogeny of the small carpenter bees (Hymenoptera: Apidae: Ceratinini) indicates early and rapid global dispersal. *Molecular Phylogenetics and Evolution*, **55**, 1042–1054.

Reinert, J.F., R.E. Harbach, and I.J. Kitching. (2006). Phylogeny and classification of *Finlaya* and allied taxa (Diptera: Culicidae: Aedini) based on morphological data from all life stages. *Zoological Journal of the Linnean Society*, **148**, 1–101.

Renner, S.S. (2004). Multiple Miocene Melastomataceae dispersal between Madagascar, Africa and India. *Philosophical Transactions of the Royal Society, London B*, **359**, 1485–1494.

Renner, S.S. (2005). Relaxed molecular clocks for dating historical plant dispersal events. *Trends in Plant Science*, **10**, 550–558.

Renner, S.S. (2009). Gymnosperms. In *The Timetree of Life*, eds. S.B. Hedges and S. Kumar. New York: Oxford University Press, pp. 157–160.

Renner, S.S., D.B. Foreman, and D. Murray. (2000). Timing transantarctic disjunctions in the Atherospermataceae (Laurales): evidence from coding and noncoding chloroplast sequences. *Systematic Botany*, **49**, 579–591.

Renner, S.S., J.S. Strijk, D. Strasberg, and C. Thébaud. (2010). Biogeography of the Monimiaceae (Laurales): a role for East Gondwana and long-distance dispersal, but not West Gondwana. *Journal of Biogeography*, **37**, 1227–1238.

Ribas, C.C., R.G. Moyle, C.Y. Miytaki, and J. Cracraft. (2007). The assembly of montane biotas: linking Andean tectonics and climatic oscillations to independent regimes of diversification in *Pionus* parrots. *Proceedings of the Royal Society B*, **274**, 2399–2408.

Ribeiro, G.C., and A. Eterovic. (2011). Neat and clear: 700 species of crane flies (Insecta: Diptera) link southern South America and Australasia. *Systematic Entomology*, **36**, 754–767.

Richardson, J.E., L.W. Chatrou, J.B. Mols, R.H.J. Erkens, and M.D. Pirie. (2004). Historical biogeography of two cosmopolitan families of flowering plants Annonaceae and Rhamnaceae. *Philosophical Transactions of the Royal Society, London, B*, **359**, 1495–1508.

Richardson, J.E., F.M. Weitz, M.F. Fay, *et al.* (2001). Rapid and recent origin of species richness in the Cape flora of South Africa. *Nature*, **412**, 181–183.

Ricklefs, R.E., and D.G. Jenkins. (2011). Biogeography and ecology: towards the integration of two disciplines. *Philosophical Transactions of the Royal Society, London, B*, **366**, 2438–2448.

Rivadavia, F., K. Kondo, M. Kato, and M. Hasebe. (2003). Phylogeny of the sundews, *Drosera* (Droseraceae), based on chloroplast *rbcL* and nuclear 18S ribosomal DNA sequences. *American Journal of Botany*, **90**, 123–130.

Rix, M.G., and M.S. Harvey. (2012). Phylogeny and historical biogeography of ancient assassin spiders (Araneae: Archaeidae) in the Australian mesic zone: evidence for Miocene speciation within Tertiary refugia. *Molecular Phylogenetics and Evolution*, **62**, 375–396.

Roalson, E.H., L.E. Skog, and E.A. Zimmer. (2008). Untangling Gloxinieae (Gesneriaceae). II. Reconstructing biogeographic patterns and estimating divergence times among New World continental and island lineages. *Systematic Botany*, **33**, 159–175.

Roberts, T.E. (2006). Multiple levels of allopatric divergence in the endemic Philippine fruit bat *Haplonycteris fischeri* (Pteropodidae). *Biological Journal of the Linnean Society*, **88**, 329–349.

Rocha, S., I. Ineich, and D.J. Harris. (2009). Cryptic variation and recent bipolar range expansion in stump-toed gecko *Gehyra mutilata* across Indian and Pacific Ocean islands. *Contributions to Zoology* [Leiden], **78** (1), 1–8.

Rodrigues, A.S.L., and K.J. Gaston. (2002). Maximising phylogenetic diversity in the selection of networks of conservation areas. *Biological Conservation*, **105**, 103–111.

Rodríguez-Trelles, F., R. Tarrio, and F. Ayala. (2002). A methodological bias toward overestimation of molecular evolutionary time scales. *Proceedings of the National Academy of Sciences, USA*, **99**, 8112–8115.

Roig-Juñent, S. (2000). The subtribes and genera of the tribe Broscini (Coleoptera: Carabidae): cladistic analysis, taxonomic treatment, and biogeographical considerations. *Bulletin of the American Museum of Natural History*, **255**, 1–90.

Ronquist, F. (1997). Dispersal–vicariance analysis: a new biogeographic approach to the quantification of historical biogeography. *Systematic Biology*, **46**, 195–203.

Rosauer, D., S.W. Laffan, M.D. Crisp, S.C. Donnellan, and L.G. Cook. (2009). Phylogenetic endemism: a new approach for identifying geographical concentrations of evolutionary history. *Molecular Ecology*, **18**, 4061–4072.

Roughgarden, J. (1995). *Anolis Lizards of the Caribbean: Ecology, Evolution and Plate Tectonics.* New York: Oxford University Press.

Rowe, K.C., E.J. Heske, P.W. Brown, and K.N. Paige. (2004). Surviving the ice: northern refugia and postglacial colonization. *Proceedings of the National Academy of Sciences, USA*, **101**, 10355–10359.

Rowe, K.C., M.L. Reno, D.M. Richmond, R.M. Adkins, and S.J. Steppan. (2008). Pliocene colonization and adaptive radiations in Australia and New Guinea (Sahul): multilocus systematic of the old endemic rodents (Muroidea: Murinae). *Molecular Phylogenetics and Evolution*, **47**, 84–101.

Rozefelds, A.C., R.W. Barnes, and B. Pellow. (2001). A new species and comparative morphology of *Vesselowskya* (Cunoniaceae). *Australian Systematic Botany*, **14**, 175–192.

Ruhfel, B.R., V. Bittrich, C.P. Bove, *et al.* (2011). Phylogeny of the clusioid clade (Malpighiales): evidence from the plastid and mitochondrial genomes. *American Journal of Botany*, **98**, 306–325.

Rundle, H.D. (2003). Divergent environments and population bottlenecks fail to generate premating isolation in *Drosophila pseudoobscura*. *Evolution*, **57**, 2557–2565.

Rundle, H.D., A.Ø. Mooers, and M.C. Whitlock. (1998). Single founder-flush events and the evolution of reproductive isolation. *Evolution*, **52**, 1850–1855.

Ryan, A.W., P.J. Smith, and J. Mork. (2002). Genetic differentiation between the New Zealand and Falkland Islands populations of southern blue whiting *Micromesistius australis*. *New Zealand Journal of Marine and Freshwater Research*, **36**, 637–643.

Sadlier, R.A., S.A. Smith, A.M. Bauer, and A.H. Whitaker. (2004). A new genus and species of live-bearing scincid lizard (Reptilia: Scincidae) from New Caledonia. *Journal of Herpetology*, **38**, 320–330.

Sagun, V.G., G.A. Levin, and P.C. van Welzen. (2010). Revision and phylogeny of *Acalypha* (Euphorbiaceae) in Malesia. *Blumea*, **55**, 21–66.

Samaniego, H., and Marquet, P.A. (2009). Mammal and butterfly species richness in Chile: taxonomic covariation and history. *Revista Chilena de Historia Natural*, **82**, 135–151.

Sánchez-González, L.A., and R.G. Moyle. (2011). Molecular systematics and species limits in the Philippine fantails (Aves: *Rhipidura*). *Molecular Phylogenetics and Evolution*, **61**, 290–299.

Sanders, K.L., M.S.Y. Lee, R. Leys, R. Foster, and J.S. Keogh. (2008). Molecular phylogeny and divergence dates for Australasian elapids and sea snakes (Hydrophiinae): evidence from seven genes for rapid evolutionary radiations. *Journal of Evolutionary Biology*, **21**, 682–695.

Sanderson, J.C. (1997). Subtidal macroalgal assemblages in temperate Australian coastal waters. *Australia: State of the Environment Technical Paper Series (Estuaries and the Sea)*. Canberra: Department of the Environment.

Sandiford, M., M. Quigley, P. de Broekert, and S. Jakica. (2009). Tectonic framework for the Cainozoic cratonic basins of Australia. *Australian Journal of Earth Sciences*, **56**, 5–18.

Sang, T., D.J. Crawford, S. Kim, and T.F. Stuessy. (1994). Radiation of the endemic genus *Dendroseris* (Asteraceae) on the Juan Fernandez Islands: evidence from sequences of the ITS regions of nuclear ribosomal DNA. *American Journal of Botany*, **81**, 1494–1501.

Sanguila, M.B., C.D. Siler, A.C. Diesmos, O. Nuñeza, and R.M. Brown. (2011). Phylogeography, geographic structure, genetic variation, and potential species boundaries in Philippine slender toads. *Molecular Phylogenetics and Evolution*, **61**, 333–350.

Sanmartín, I., and F. Ronquist. (2004). Southern hemisphere biogeography inferred by event-based models: plant versus animal patterns. *Systematic Biology*, **53**, 216–243.

San Mauro, D., M. Vences, M. Alcobendas, R. Zardoya, and A. Meyer. (2005). Initial diversification of living amphibians predated the breakup of Gondwana. *American Naturalist*, **165**, 590–599.

Santini, F., and R. Winterbottom. (2002). Historical biogeography of Indo-western Pacific coral reef biota: is the Indonesian region a center of origin? *Journal of Biogeography*, **29**, 189–205.

Saunders, R.M.K., and J. Munzinger. (2007). A new species of *Goniothalamus* (Annonaceae) from New Caledonia, representing a significant range extension for the genus. *Botanical Journal of the Linnean Society*, **155**, 497–503.

Sauquet, H., J.A. Doyle, T. Scharaschkin, *et al.* (2003). Phylogenetic analysis of Magnoliales and Myristicaceae based on multiple data sets: implications for character evolution. *Botanical Journal of the Linnean Society*, **142**, 125–186.

Sauquet, H., S.Y.W. Ho, M.A. Gandolfo, *et al.* (2012). Testing the impact of calibration of molecular divergence times using a fossil-rich group: the case of *Nothofagus* (Fagales). *Systematic Biology*, **61**, 289–313.

Sauquet, H., P.H. Weston, C.L. Anderson, *et al.* (2009a). Contrasted patterns of hyperdiversification in Mediterranean hotspots. *Proceedings of the National Academy of Sciences, USA*, **106**, 221–225.

Sauquet, H., P.H. Weston, N.P. Barker, *et al.* (2009b). Using fossils and molecular data to reveal the origins of the Cape proteas (subfamily Proteoideae). *Molecular Phylogenetics and Evolution*, **51**, 31–43.

Savolainen, V., M.-C. Anstett, C. Lexer, *et al.* (2006). Sympatric speciation in palms on an oceanic island. *Nature*, **441**, 210–213.

Schaefer, H., and S.S. Renner. (2011). Phylogenetic relationships in the order Cucurbitales and a new classification of the gourd family (Cucurbitaceae). *Taxon*, **60**, 122–138.

Schaefer, H., C. Heibl, and S.S. Renner. (2009). Gourds afloat: a dated phylogeny reveals an Asian origin of the gourd family (Cucurbitaceae) and numerous oversea dispersal events. *Proceedings of the Royal Society, London, B*, **276**, 843–851.

Schaefer, H., I.R.H. Telford, and S.S. Renner. (2008). *Austrobryonia* (Cucurbitaceae), a new Australian endemic genus, is the closest living relative to the Eurasian and Mediterranean *Bryonia* and *Ecballium*. *Systematic Botany*, **33**, 125–132.

Schatz, G.E., and P.J.M. Maas. (2010). Synoptic revision of *Stenanona* (Annonaceae). *Blumea*, **55**, 205–223.

Schellart, W.P. (2007). North-eastward subduction followed by slab detachment to explain ophiolite obduction and Early Miocene volcanism in Northland, New Zealand. *Terra Nova*, **19**, 211–218.

Schellart, W.P., B.L.N. Kennett, W. Spakman, and M. Amaru. (2009). Plate reconstructions and tomography reveal a fossil lower mantle slab below the Tasman Sea. *Earth and Planetary Science Letters*, **278**, 143–151.

Schellart, W.P., G.S. Lister, and V.G. Toy. (2006). A Late Cretaceous and Cenozoic reconstruction of the Southwest Pacific region: tectonics controlled by subduction and slab rollback processes. *Earth-Science Reviews*, **76**, 191–233.

Schmidt, O. (2005). Revision of *Scotocyma* Turner (Lepidoptera: Geometridae: Larentiinae). *Australian Journal of Entomology*, **44**, 257–278.

Schodde, R. (1982). Origin, adaptation, and evolution of birds in arid Australia. In *Evolution of the Flora and Fauna of Arid Australia*, eds. W.R. Barker and P.J.M. Greenslade. Frewville, SA: Peacock, pp. 191–224.

Schubart, C.D., and P.K.L. Ng. (2002). The sesarmid genus *Neosarmatium* (Decapoda: Brachyura): new distribution records and a new species from Sulawesi. *Crustacean Research*, **31**, 28–38.

Schuster, T.M., J.L. Reveal, and K.A. Kron. (2011a). Phylogeny of Polygoneae (Polygonaceae: Polygonoideae). *Taxon*, **60**, 1653–1666.

Schuster, T.M., K.L. Wilson, and K.A. Kron. (2011b). Phylogenetic relationships of *Muehlenbeckia*, *Fallopia*, and *Reynoutria* (Polygonaceae) investigated with chloroplast and nuclear sequence data. *International Journal of Plant Sciences*, **172**, 1053–1066.

Schweizer, M., M. Güntert, and S.T. Hertwig. (2012). Phylogeny and biogeography of the parrot genus *Prioniturus* (Aves: Psittaciformes). *Journal of Zoological Systematics and Evolutionary Research*, **50**, 145–156.

Schweizer, M., M. Güntert, and S.T. Hertwig. (2013). Out of the Bassian province: historical biogeography of the Australasian platycercine parrots (Aves, Psittaciformes). *Zoologica Scripta*, **42**, 13–27.

Schweizer, M., O. Seehausen, M. Güntert, and S.T. Hertwig. (2010). The evolutionary diversification of parrots supports a taxon pulse model with multiple trans-oceanic dispersal events and local radiations. *Molecular Phylogenetics and Evolution*, **54**, 984–994.

Schweizer, M., O. Seehausen, and S.T. Hertwig. (2011). Macroevolutionary patterns in the diversification of parrots: effects of climate change, geological events and key innovations. *Journal of Biogeography*, **38**, 2176–2194.

Sdrolias, M., R.D. Müller, and C. Gaina. (2003). Tectonic evolution of the southwest Pacific using constraints from backarc basins. *Geological Society of Australia Special Publications*, **22**, and *Geological Society of America Special Papers*, **372**, 343–359.

Selvi, F. (2007). Diversity, geographic variation and conservation of the serpentine flora of Tuscany. *Biodiversity and Conservation*, **16**, 1423–1439.

Sequeira, A.S., B.B. Normark, and B.D. Farrell. (2000). Evolutionary assembly of the conifer fauna: distinguishing ancient from recent associations in bark beetles. *Proceedings of the Royal Society, London, B*, **267**, 2359–2366.

Setoguchi, H., T.A. Osawa, J.-C. Pintaud, T. Jaffré, and J.-M. Veillon. (1998). Phylogenetic relationships within Araucariaceae based on *rbcL* gene sequences. *American Journal of Botany*, **85**, 1507–1516.

Sharma, P., and G. Giribet. (2009). A relict in New Caledonia: phylogenetic relationships of the family Troglosironidae (Opiliones: Cyphophthalmi). *Cladistics*, **25**, 279–294.

Sharma, P.P., and G. Giribet. (2012). Out of the Neotropics: Late Cretaceous colonization of Australasia by American arthropods. *Proceedings of the Royal Society B*, **279**, 3501–3509.

Sheldon, F.H., D.J. Lohman, H.C. Lim, *et al.* (2009). Phylogeography of the magpie–robin species complex (Aves: Turdidae: *Copsychus*) reveals a Philippine species, an interesting isolating barrier and unusual dispersal patterns in the Indian Ocean and Southeast Asia. *Journal of Biogeography*, **36**, 1070–1083.

Shepherd, K.A., M. Waycott, and A. Calladine. (2004). Radiation of the Australian Salicornioideae (Chenopodiaceae) – based on evidence from nuclear and chloroplast DNA sequences. *American Journal of Botany*, **91**, 1387–1397.

Shepherd, L.D., and D.M. Lambert. (2007). The relationships and origins of the New Zealand wattlebirds (Passeriformes, Callaeatidae) from DNA sequence analysis. *Molecular Phylogenetics and Evolution*, **43**, 480–492.

Shepherd, L.D., T.H. Worthy, A.J.D. Tennyson, *et al.* (2012). Ancient DNA analyses reveal contrasting phylogeographic patterns amongst kiwi (*Apteryx* spp.) and a recently extinct lineage of spotted kiwi. *PLoS One*, **7**(8), e42384, 1–9.

Sherwood, A.R., and R.G. Sheath. (1999). Biogeography and systematics of *Hildenbrandia* (Rhodophyta, Hildenbrandiales) in North America: inferences from morphometrics and *rbc*L and 18S rRNA gene sequence analyses. *European Journal of Phycology*, **34**, 523–532.

Sherwood, A.R., and R.G. Sheath. (2003). Systematics of the Hildenbrandiales (Rhodophyta): gene sequence and morphometric analyses of global collections. *Journal of Phycology*, **39**, 409–422.

Shoo, L.P., R. Rose, P. Doughty, J.J. Austin, and J. Melville. (2008). Diversification patterns of pebble-mimic dragons are consistent with historical disruption of important habitat corridors in arid Australia. *Molecular Phylogenetics and Evolution*, **48**, 528–542.

Siddoway, C.S. (2008). Tectonics of the West Antarctic Rift System: new light on the history and dynamics of distributed intracontinental extension. In *Antarctica: A Keystone in a Changing World*, eds. A.K. Cooper, P.J. Barrett, H. Stagg, *et al.* and the 10th ISAES editorial team. Washington, DC: The National Academies Press, pp. 91–114.

Siler, C.D., and R. M. Brown. (2010). Phylogeny-based species delimitation in Philippines slender skinks (Reptilia: Squamata: Scincidae: *Brachymeles*): taxonomic revision of pentadactyl species groups and description of three new species. *Herpetological Monographs*, **24**, 1–54.

Siler, C.D., A.C. Alcala, A.C. Diesmos, and R.M. Brown. (2009). A new species of limestone-forest frog, genus *Platymantis* (Amphibia: Anura: Ceratobatrachidae) from eastern Samar Island, Philippines. *Herpetologica*, **65**, 92–104.

Siler, C.D., A.C. Diesmos, A.C. Alcala, and R.M. Brown. (2011). Phylogeny of Philippine slender skinks (Scincidae: *Brachymeles*) reveals underestimated species diversity, complex biogeographical relationships, and cryptic patterns of lineage diversification. *Molecular Phylogenetics and Evolution*, **59**, 53–65.

Siler, C.D., A.C. Diesmos, C.W. Linkem, M. Diesmos, and R.M. Brown. (2010a). A new species of limestone-forest frog, genus *Platymantis* (Amphibia: Anura: Ceratobatrachidae) from central Luzon Island, Philippines. *Zootaxa*, **2482**, 49–63

Siler, C.D., J.R. Oaks, J.A. Esselstyn, A.C. Diesmos, and R.M. Brown. (2010b). Phylogeny and biogeography of Philippine bent-toed geckos (Gekkonidae: *Cyrtodactylus*) contradict a prevailing model of Pleistocene diversification. *Molecular Phylogenetics and Evolution*, **55**, 699–710.

Siler, C.D., J.R. Oaks, L.J. Welton, *et al.* (2012a). Did geckos ride the Palawan raft to the Philippines? *Journal of Biogeography*, **39**, 1217–1234.

Siler, C.D., C.H. Oliveros, A. Santanen, and R.M. Brown. (2013). Multilocus phylogeny reveals unexpected diversification patterns in Asian wolf snakes (genus *Lycodon*). *Zoologica Scripta*, **42**, 262–277.

Siler, C.D., J.C. Swab, C.H. Oliveros, *et al.* (2012b). Amphibians and reptiles, Romblon Island Group, central Philippines: comprehensive herpetofaunal inventory. *Check List*, **8**, 443–462.

Simonsen, T.J., E.V. Zakharov, M. Djernaes, *et al.* (2011). Phylogenetics and divergence times of Papilioninae (Lepidoptera) with special reference to the enigmatic genera *Teinopalpus* and *Meandrusa*. *Cladistics*, **27**, 113–137.

Sinclair, B.J. (2003). Taxonomy, phylogeny and zoogeography of the subfamily Ceratomerinae of Australia (Diptera: Empidoidea). *Records of the Australian Museum*, **55**, 1–44.

Sinclair, E.A., F.W. Fetzner Jr., and K.A. Crandall. (2004). Proposal to complete a phylogenetic taxonomy and systematic revision for freshwater crayfish (Astacidea). *Freshwater Crayfish*, **14**, 21–29.

Sipman, H.J.M. (1983). A monograph of the lichen family Megalosporaceae. *Bibliotheca Licheno-logica*, **18**, 1–241.

Slack, K.E., F. Delsuc, P.A. Mclenachan, U. Arnason, and D. Penny, D. (2007). Resolving the root of the avian mitogenomic tree by breaking up long branches. *Molecular Phylogenetics and Evolution*, **42**, 1–13.

Sleumer, H. (1967). Ericaceae (continued). *Flora Malesiana I*, **6**, 669–914.

Smedmark, J.E.E., and T. Eriksson. (2002). Phylogenetic relationships of *Geum* (Rosaceae) and relatives inferred from the nrITS and *trnL-trnF* regions. *Systematic Botany*, **27**, 303–317.

Smissen, R.D., I. Breitwieser, and J.M. Ward. (2007). Genetic characterization of hybridization between the New Zealand everlastings *Helichrysum lanceolatum* and *Anaphalioides bellidioides* (Asteraceae: Gnaphaliaeae). *Botanical Journal of the Linnean Society*, **154**, 89–98.

Smissen, R. D., G.K. Chambers, and P.J. Garnock-Jones. (2005). Dating nodes on molecular phylogenies: older or younger than the Earth itself? *Cladistics*, **21**, 403.

Smissen, R.D., P.J. Garnock-Jones, and G.K. Chambers. (2003). Phylogenetic analysis of ITS sequences suggests a Pliocene origin for the bipolar distribution of *Scleranthus* (Caryophyllaceae). *Australian Systematic Botany*, **16**, 301–313.

Smith, A.B. (2007). Marine diversity through the Phanerozoic: problems and prospects. *Journal of the Geological Society*, **164**, 731–745.

Smith, A.B., and K.J. Peterson. (2002). Dating the time of origin of major clades: molecular clocks and the fossil record. *Annual Review of Earth and Planetary Science*, **30**, 65–88.

Smith, A.C. (1979–96). *Flora Vitiensis Nova: A New Flora of Fiji (Spermatophytes Only)*. Lawai, HI: Pacific Tropical Botanical Garden.

Smith, A.D. (2007). A plate model for Jurassic to Recent intraplate volcanism in the Pacific Ocean basin. *Geological Society of America Special Papers*, **430**, 471–495.

Smith, D.R., L. Villafuerte, G. Otis, and M.R. Palmer. (2000). Biogeography of *Apis cerana* F. and *A. nigrocincta* Smith: insights from mtDNA studies. *Apidologie*, **31**, 265–279.

Smith, J.F., A.C. Stevens, E.J. Tepe, and C. Davidson. (2008). Placing the origin of two species-rich genera in the Late Cretaceous with later species divergence in the Tertiary: a phylogenetic, biogeographic and molecular dating analysis of *Piper* and *Peperomia* (Piperaceae). *Plant Systematics and Evolution*, **275**, 9–30.

Smith, J.V., E.L. Braun, and R.T. Kimball. (2013). Ratite non-monophyly: independent evidence from 40 novel loci. *Systematic Biology*, **62**, 35–49.

Smith, P.J., and K.J. Collier. (2001). Allozyme diversity and population genetic structure of the caddisfly *Orthopsyche fimbriata* and the mayfly *Acanthophlebia cruentata* in New Zealand streams. *Freshwater Biology*, **46**, 795–805.

Smith, S.A., J.M. Beaulieu, and M.J. Donoghue. (2010). An uncorrelated relaxed-clock analysis suggests an earlier origin for flowering plants. *Proceedings of the National Academy of Sciences USA*, **107**, 5897–5902.

Smith, W.L., and M.T. Craig. (2007). Casting the percomorph net widely: the importance of broad taxonomic sampling in the search for the placement of serranid and percid fishes. *Copeia*, **2007**, 35–55.

Smith-Ramírez, C. (2004). The Chilean Coastal Range: a vanishing center of biodiversity and endemism in southern temperate rain forests. *Biodiversity and Conservation*, **13**, 373–393.

Smith-Ramírez, C., J.J. Armesto, and C. Valdovinos. (2005). *Historia, Biodiversidad y Ecología de los Bosques Costeros de Chile*. Santiago: Editoriale Universitaria.

Soligo, C., O. Will, S. Tavaré, C.R. Marshall, and R.D. Martin. (2007). New light on the dates of primate origins and divergence. In *Primate Origins: Adaptations and Evolution*, eds. M.J. Ravosa and M. Dagosto. New York: Springer, pp. 29–49.

Soltis, D.E., S.A. Smith, N. Cellinese, *et al.* (2011). Angiosperm phylogeny: 17 genes, 640 taxa. *American Journal of Botany*, **98**, 704–730.

Sota, T., Y. Takami, G.B. Monteith, and B.P. Moore. (2005). Phylogeny and character evolution of endemic carabid beetles of the genus *Pamborus* based on mitochondrial and nuclear gene sequences. *Molecular Phylogenetics and Evolution*, **36**, 391–404.

Soutullo, A., S. Dodsworth, S.B. Heard, and A.Ø. Mooers. (2005). Distribution and correlates of carnivore phylogenetic diversity across the Americas. *Animal Conservation*, **8**, 249–258.

Spalik, K., M. Piwczyński, C.A. Danderson, *et al.* (2010). Amphitropic amphiantarctic disjunctions in Apiaceae subfamily Apioideae. *Journal of Biogeography*, **37**, 1977–1994.

Spandler, C., D. Rubatto, and J. Hermann. (2005a). Late Cretaceous–Tertiary tectonics of the Southwest Pacific: insights from U-Pb sensitive, high-resolution ion microprobe (SHRIMP) dating of eclogite facies rocks from New Caledonia. *Tectonics*, **24**, TC3003, doi:10.1029/2004TC001709.

Spandler, C., K. Worden, R. Arculus, and S. Eggins. (2005b). Igneous rocks of the Brook Street Terrane, New Zealand: implications for Permian tectonics of eastern Gondwana and magma genesis in modern intra-oceanic volcanic arcs. *New Zealand Journal of Geology and Geophysics*, **48**, 167–183.

Sparks, J.S. (2004). Molecular phylogeny and biogeography of the Malagasy and South Asian cichlids (Teleostei: Perciformes: Cichlidae). *Molecular Phylogenetics and Evolution*, **30**, 599–614.

Sparks, J.S., and W.L. Smith. (2004a). Phylogeny and biogeography of cichlid fishes (Teleostei: Perciformes: Cichlidae): a multilocus approach to recovering deep intrafamilial divergences and the cichlid sister-group. *Cladistics*, **20**, 1–17.

Sparks, J.S., and W.L. Smith. (2004b). Phylogeny and biogeography of the Malagasy and Australasian rainbowfishes (Teleostei: Melanotaenioidei): Gondwanan vicariance and evolution in freshwater. *Molecular Phylogenetics and Evolution*, **33**, 719–734.

Sparks, J.S., and W.L. Smith. (2005). Freshwater fishes, dispersal ability, and nonevidence: 'Gondwana life rafts' to the rescue. *Systematic Biology*, **54**, 158–165.

Springer, M.S., R.W. Meredith, J.E. Janecka, and W.J. Murphy. (2011). The historical biogeography of Mammalia. *Philosophical Transactions of the Royal Society B*, **366**, 2478–2502.

Stadelmann, B., L.-K. Lin, T.H. Kunz, and M. Ruedi. (2007). Molecular phylogeny of New World *Myotis* (Chiroptera, Vespertilionidae) inferred from mitochondrial and nuclear DNA genes. *Molecular Phylogenetics and Evolution*, **43**, 32–48.

Stech, M., M. Sim-Sim, M.G. Esquível, *et al.* (2008). Explaining the 'anomalous' distribution of Echinodium (Bryopsida: Echinodiaceae): independent evolution in Macaronesia and Australasia. *Organisms Diversity and Evolution*, **8**, 282–292.

Stefanović, S., M. Kuzmina, and M. Costea. (2007). Delimitation of major lineages within *Cuscuta* subgenus *Grammica* (Convolvulaceae) using plastid and nuclear DNA sequences. *American Journal of Botany*, **94**, 568–589.

Steffen, S., L. Mucina, and G. Kadereit. (2010). Revision of *Sarcocornia* (Chenopodiaceae) in South Africa, Namibia and Mozambique. *Systematic Botany*, **35**, 390–408.

Stegman, D.R., J. Freeman, W.P. Schellart, L. Moresi, and D. May. (2006). Influence of trench width on subduction hinge retreat rates in 3-D models of slab rollback. *Geochemistry, Geophysics, Geosystems*, **7**, Q03012.

Stelbrink, B., C. Albrecht, R. Hall, and T. von Rintelen. (2012). The biogeography of Sulawesi revisited: is there evidence for a vicariant origin of taxa on Wallace's "anomalous island"? *Evolution*, **66**, 2252–2271.

Steppan, S.J., B.L. Storz, and R.S. Hoffmann. (2004). Nuclear DNA phylogeny of the squirrels (Mammalia: Rodentia) and the evolution of arboreality from c-myc and RAG1. *Molecular Phylogenetics and Evolution*, **30**, 703–719.

Steppan, S.J., C. Zawadski, and L.R. Heaney. (2003). Molecular phylogeny of the endemic Philippine rodent *Apomys* (Muridae) and the dynamics of diversification in an oceanic archipelago. *Biological Journal of the Linnean Society*, **80**, 699–715.

Stevanović, V., K. Tan, and G. Iatrou. (2003). Distribution of the endemic Balkan flora on serpentine I. – Obligate serpentine endemics. *Plant Systematics and Evolution*, **242**, 149–170.

Stevens, P.F. (2012). *Angiosperm phylogeny website*. www.mobot.org/MOBOT/Research/APweb/ Accessed December 2012.

Stöckler, K., I.L. Daniel, and P.J. Lockhart. (2002). New Zealand kauri (*Agathis australis* (D.Don) Lindl., Araucariaceae) survives Oligocene drowning. *Systematic Biology*, **51**, 827–832.

Stoltzfus, A. (2006). Mutationism and the dual causation of evolutionary change. *Evolution and Development*, **8**, 304–317.

Stoltzfus, A., and L.Y. Yampolsky. (2009). Climbing Mount Probable: mutation as a cause of nonrandomness in evolution. *Journal of Heredity*, **100**, 637–647.

Strandberg, J., and K.A. Johanson. (2011). The historical biogeography of *Apsilochorema* (Trichoptera, Hydrobiosidae) revised, following molecular studies. *Journal of Zoological Systematics and Evolutionary Research*, **49**, 110–118.

Stuessy, T.F. (2006). Sympatric speciation in islands? *Nature*, **443**, E12.

Su, Y.-C., Y.-H. Chang, S.-C. Lee, and I-M. Tso. (2007). Phylogeography of the giant wood spider (*Nephila pilipes*, Araneae) from Asian–Australian regions. *Journal of Biogeography*, **34**, 177–191.

Su, Y.C.F., and R.M.K. Saunders. (2009). Evolutionary divergence times in the Annonaceae: evidence of a late Miocene origin of *Pseuduvaria* in Sundaland with subsequent diversification in New Guinea. *BMC Evolutionary Biology*, **9** (153), 1–19.

Sulloway, F.J. (1979). Geographic isolation in Darwin's thinking: the vicissitudes of a crucial idea. *Studies in the History of Biology*, **3**, 23–65.

Sun, G., D.L. Dilcher, H. Wang, and Z. Chen. (2011). A eudicot from the Early Cretaceous of China. *Nature*, **471**, 625–628.

Sutherland, J.E., S.C. Lindstrom, W.A. Nelson, *et al.* (2011). A new look at an ancient order: generic revision of the Bangiales (Rhodophyta). *Journal of Phycology*, **47**, 1131–1151.

Swenson, U. (1995). Systematics of *Abrotanella*, an amphipacific genus of Asteraceae (Senecioneae). *Plant Systematics and Evolution*, **197**, 149–193.

Swenson, U., Nylinder, S., and S. Wagstaff. (2012). Are Asteraceae 1.5 billion years old? A reply to Heads. *Systematic Biology*, **61**, 522–532.

Tangney, R. (2007). Biogeography of austral pleurocarpous mosses: distribution patterns in the Australasian region. In *Pleurocarpous Mosses: Systematics and Evolution*, eds. A.E. Newton and R.S. Tangney. Boca Raton, FL: CRC Press, pp. 393–407.

Tangney, R.S., S. Huttunen, M. Stech, and D. Quandt. (2010). A review of the systematic position of the pleurocarpous moss genus *Acrocladium* Mitten. *Tropical Bryology*, **31**, 164–170.

Tank, D.C., and M.J. Donoghue. (2010). Phylogeny and phylogenetic nomenclature of the Campanulidae based on an expanded sample of genes and taxa. *Systematic Botany*, **35**, 425–441.

Tassin, J., G. Derroire, and J.-N. Rivière. (2004). Gradient altitudinal de la richesse spécifique et de l'endémicité de la flore ligneuse indigéne á l'île de La Réunion (archipel des Mascareignes). *Acta Botanica Gallica*, **151**, 181–196.

Taylor, B. (1999). Background and regional setting. *Proceedings of the Ocean Drilling Program, Initial Reports*, **180**, 2–20.

Taylor, B. (2006). The single largest oceanic plateau: Ontong Java–Manihiki–Hikurangi. *Earth and Planetary Science Letters*, **241**, 372–380.

Tejada, M.L.G., J.J. Mahoney, P.R. Castillo, *et al.* (2004). Pin-pricking the elephant: evidence on the origin of the Ontong Java Plateau from Pb-Sr-Hf-Nd isotopic characteristics of ODP Leg 192 basalts. *Geological Society of London, Special Publications*, **229**, 33–150.

Templeton, A.R. (2008). The reality and importance of founder speciation in evolution. *Bioessays*, **30**, 470–479.

Tennyson, A.J.D., and P. Martinson. (2006). *Extinct Birds of New Zealand*. Wellington: Te Papa Press.

Tennyson, A.J.D., T.H. Worthy, C.M. Jones, R.P. Scofield, and S.J. Hand. (2010). Moa's Ark: Miocene fossils reveal the great antiquity of moa (Aves: Dinornithiformes) in Zealandia. *Records of the Australian Museum*, **62**, 105–114.

Teske, P.R., M.I. Cherry, and C.A. Matthee. (2004). The evolutionary history of seahorses (Syngnathidae: *Hippocampus*): molecular data suggest a West Pacific origin and two invasions of the Atlantic Ocean. *Molecular Phylogenetics and Evolution*, **30**, 273–286.

Thomas, D.C., M. Hughes, T. Phutthai, *et al.* (2012a). West to east dispersal and subsequent rapid diversification of the mega-diverse genus *Begonia* (Begoniaceae) in the Malesian archipelago. *Journal of Biogeography*, **39**, 98–113.

Thomas, D.C., S. Surveswaran, B. Xue, *et al.* (2012b). Molecular phylogenetics and historical biogeography of the *Meiogyne-Fitzalania* clade (Annonaceae): generic paraphyly and late Miocene-Pliocene diversification in Australasia and the Pacific. *Taxon*, **61**, 559–575.

Thompson, F.C. (2008). A conspectus of New Zealand flower flies (Diptera: Syrphidae) with the description of a new genus and species. *Zootaxa*, **1716**, 1–20.

Thordarson, T. (2004). Accretionary-lapilli-bearing pyroclastic rocks at ODP Leg 192 Site 1184: a record of subaerial phreatomagmatic eruptions on the Ontong Java Plateau. *Geological Society of London, Special Publications*, **229**, 275–306.

Thorne, R.F. (1965). Floristic relationships of New Caledonia. *University of Iowa Studies in Natural History*, **20** (7), 1–14.

Tian, B., R. Liu, L. Wang, Q. Qiu, and J. Liu. (2009). Phylogeographic analyses suggest that a deciduous species (*Ostryopsis davidiana* Decne., Betulaceae) survived in northern China during the Last Glacial Maximum. *Journal of Biogeography*, **36**, 2148–2155.

Timm, C., K. Hoernle, R. Werner, *et al.* (2011). Age and geochemistry of the oceanic Manihiki Plateau, SW Pacific: new evidence for a plume origin. *Earth and Planetary Science Letters*, **304**, 135–146.

Titus, S.J., S.M. Maes, B. Benford, E.C. Ferré, and B. Tikoff. (2011). Fabric development in the mantle section of a paleotransform fault and its effect on ophiolite obduction, New Caledonia. *Lithosphere*, **3**, 221–244.

Tokeshi, M. (1999). *Species Coexistence: Ecological and Evolutionary Perspectives*. Oxford: Blackwell.

Tokuoka, T. (2008). Molecular phylogenetic analysis of Violaceae (Malpighiales) based on plastid and nuclear DNA sequences. *Journal of Plant Research*, **121**, 253–260.

Toon, A., J.J. Austin, G. Dolman, L. Pedler, and L. Joseph. (2012). Evolution of arid zone birds in Australia: leapfrog distribution patterns and mesic-arid connections in quail-thrush (*Cinclosoma*, Cinclosomatidae). *Molecular Phylogenetics and Evolution*, **62**, 286–295.

Toon, A., M. Pérez-Losada, C.E. Schweitzer, *et al.* (2010). Gondwanan radiation of the southern hemisphere crayfishes (Decapoda: Parastacidae): evidence from fossils and molecules. *Journal of Biogeography*, **37**, 2275–2290.

Toonen, R.J., K.R. Andrews, I.B. Baums, *et al.* (2011). Defining boundaries for ecosystem-based management: a multispecies case study of marine connectivity across the Hawaiian Archipelago. *Journal of Marine Biology*, Article 460173, 1–13.

Toxopeus, L.J. (1950). The geological principles of species evolution in New Guinea. In *Proceedings of the 8th International Congress of Entomology, Stockholm*, 1948, ed. A.R. Elfstrums. Stockholm: International Congress of Entomology, pp. 508–522.

Trénel, P., M.H. Gustafsson, W.J. Baker, *et al.* (2007). Mid-Tertiary dispersal, not Gondwanan vicariance explains distribution patterns in the wax palm subfamily (Ceroxyloideae: Arecaceae). *Molecular Phylogenetics and Evolution*, **45**, 272–288.

Trewick, S.A. (2000). Molecular evidence for dispersal rather than vicariance as the origin of flightless insect species on the Chatham Islands, New Zealand. *Journal of Biogeography*, **27**, 1189–1200.

Trewick, S.A., and G.C. Gibb. (2010). Vicars, tramps and assembly of the New Zealand avifauna: a review of molecular phylogenetic evidence. *Ibis*, **152**, 226–253.

Trewick, S.A., and M. Morgan-Richards. (2005). After the deluge: mitochondrial DNA indicates Miocene radiation and Pliocene adaptation of tree and giant weta (Orthoptera: Anostostomatidae). *Journal of Biogeography*, **32**, 295–309.

Tronchet, F., G.M. Plunkett, J. Jérémie, and P.P. Lowry II. (2005). Monophyly and major clades of *Meryta* (Araliaceae). *Systematic Botany*, **30**, 657–670.

Tulloch, A., M. Beggs, J. Kula, T. Spell, and N. Mortimer. (2006). Cordillera Zealandia, the Sisters Shear Zone, and their influence on the early development of the Great South basin. Paper presented at the New Zealand Petroleum Conference 2006, Crown Minerals, Auckland, New Zealand. http://www.nzpam.govt.nz/cms/pdf-library/petroleum-conferences-1/2006/papers/Papers_59.pdf

Turner, S., L.B. Bean, M. Dettmann, *et al.* (2009). Australian Jurassic sedimentary and fossil successions: current work and future prospects for marine and non-marine correlation. *GFF – Journal of the Geological Society of Sweden*, **131**, 49–70.

Uhl, N.W., and J. Dransfield. (1987). *Genera Palmarum: a Classification of Palms Based on the Work of Harold E. Moore, Jr.* Lawrence, KS: Allen.

Uit de Weerd, D.R., W.H. Piel, and E. Gittenberger. (2004). Widespread polyphyly among Alopiinae snail genera: when phylogeny mirrors biogeography more closely than morphology. *Molecular Phylogenetics and Evolution*, **33**, 533–548.

Umhoefer, P.J. (2011). Why did the Southern Gulf of California rupture so rapidly? – Oblique divergence across hot, weak lithosphere along a tectonically active margin. *GSA Today*, **21**, 4–10.

Unmack, P.J., and T.E. Dowling. (2010). Biogeography of the genus *Craterocephalus* (Teleostei: Atherinidae) in Australia. *Molecular Phylogenetics and Evolution*, **55**, 968–984.

Unmack, P.J., G.R. Allen, and J.B. Johnson. (2013). Phylogeny and biogeography of rainbowfishes (Melanotaeniidae) from Australia and New Guinea. *Molecular Phylogenetics and Evolution*, **67**, 15–27.

Unmack, P.J., M.P. Hammer, M. Adams, and T.E. Dowling. (2011). Phylogenetic analysis of pygmy perches (Teleostei: Percichthyidae) with an assessment of the major historical influences on aquatic biogeography in southern Australia. *Systematic Biology*, **60**, 797–812.

Vallejo, B., Jr. (2005). Inferring the mode of speciation in Indo-West Pacific *Conus* (Gastropoda: Conidae). *Journal of Biogeography*, **32**, 1429–1439.

Vallejo, B., Jr. (2011). The Philippines in Wallacea. In *Biodiversity, Biogeography and Nature Conservation in Wallacea and New Guinea*, ed. D. Telnov. Riga: Entomological Society of Latvia, pp. 27–42

van Balgooy, M.M.J. (1984). Coronanthereae. *Pacific Plant Areas*, **4**, 186–187.

van Balgooy, M.M.J., and R. van der Meijden. (1993). Moutabeeae. *Pacific Plant Areas*, **5**, 166–167.

van Tol, J. (2005). Revision of the Platystictidae of the Philippines (Odonata), excluding the *Drepanosticta halterata* group, with descriptions of twenty-one new species. *Zoologische Mededelingen*, **79**, 195–282.

van Tol, J., B.T. Reijnen, and H.A. Thomassen. (2009). Phylogeny and biogeography of the Platystictidae (Odonata). In *Phylogeny and Biogeography of the Platystictidae (Odonata)*, ed. J. van Tol. PhD thesis: Leiden University.

van Tuinen, M., and E.A. Hadly. (2004). Error in estimation of rate and inferred from the early amniote fossil record and avian molecular clocks. *Journal of Molecular Evolution*, **59**, 267–276.

van Ufford, A.Q., and M. Cloos. (2005). Cenozoic tectonics of New Guinea. *American Association of Petroleum Geologists Bulletin*, **89**, 119–140.

van Welzen, P.C., J.A.N. Parnell, and J.W.F. Slik. (2011). Wallace's Line and plant distributions: two or three phytogeographical areas and where to group Java? *Biological Journal of the Linnean Society*, **103**, 531–545.

van Wijk, J.W., J.F. Lawrence, and N.W. Driscoll. (2008). Formation of the Transantarctic Mountains related to extension of the West Antarctic Rift. *Tectonophysics*, **458**, 117–126.

Vandenberg, A.H.M. (2010). Paleogene basalts prove early uplift of Victoria's eastern uplands. *Australian Journal of Earth Sciences*, **57**, 291–315.

Vanderpoorten, A., S.R. Gradstein, M.A. Carine, and N. Devos. (2010). The ghosts of Gondwana and Laurasia in modern liverwort distributions. *Biological Reviews*, **85**, 471–487.

Vane-Wright, R.I., C.J. Humphries, and P.H. Williams. (1991). What to protect? Systematics and the agony of choice. *Biological Conservation*, **55**, 235–254.

Vanormelingen, P., E. Verleyen, and W. Vyverman. (2007). The diversity and distribution of diatoms: from cosmopolitanism to narrow endemism. *Biodiversity and Conservation*, **3**, 393–405.

Vaughan, A.P.M., P.T. Leat, and R.J. Pankhurst. (2005). Terrane processes at the margin of Gondwana: introduction. *Geological Society of London Special Publications*, **245**, 1–21.

Veevers, J.J. (1988). *Phanerozoic Earth History of Australia*. New York: Oxford University Press.

Veevers, J.J. (2012). Reconstructions before rifting and drifting reveal the geological connections between Antarctica and its conjugates in Gondwanaland. *Earth-Science Reviews*, **111**, 249–318.

Vega, R., C. Fløjgaard, A. Lira-Noriega, *et al.* (2010). Northern glacial refugia for the pygmy shrew *Sorex minutus* in Europe revealed by phylogeographic analyses and species distribution modelling. *Ecography*, **33**, 260–271.

Vences, M. (2004). Origin of Madagascar's extant fauna: a perspective from amphibians, reptiles and other non-flying vertebrates. *Italian Journal of Zoology Supplement*, **2**, 217, 228.

Venning, J., and A. Prescott. (1984). Aizoaceae. *Flora of Australia*, **4**, 19–62.

Verboom, G.A., J.K. Archibald, F.T. Bakker, *et al.* (2009). Origin and diversification of the Greater Cape flora: ancient species repository, hot-bed of recent radiation, or both? *Molecular Phylogenetics and Evolution*, **51**, 44–53.

Vidal, N, J.-C. Rage, A. Couloux, and S.B. Hedges. (2009). Snakes (Serpentes). In *The Timetree of Life*, eds. S.B. Hedges and S. Kumar. New York: Oxford University Press, pp. 390–397.

Vidal-Russell, R., and D.L. Nickrent. (2008). Evolutionary relationships in the showy mistletoe family (Loranthaceae). *American Journal of Botany*, **95**, 1015–1029.

Vijverberg, C.A. (2001). *Adaptive Radiation of Australian and New Zealand* Microseris *(Asteraceae): a Case Study Based on Molecular and Morphological Markers*. PhD Thesis: University of Amsterdam.

Vijverberg, K., T.H.M. Mes, and K. Bachmann. (1999). Chloroplast DNA evidence for the evolution of *Microseris* (Asteraceae) in Australia and New Zealand after long-distance dispersal from western North America. *American Journal of Botany*, **86**, 1448–1463.

Villanueva, R.J.Y., and J.R.S. Gil. (2011). Odonata fauna of Catanduanes Island, Philippines. *International Dragonfly Fund – Report*, **39**, 1–38.

Villeneuve, M., R. Martini, H. Bellon, *et al.* (2010). Deciphering of six blocks of Gondwanan origin within Eastern Indonesia (South East Asia). *Gondwana Research*, **18**, 420–437.

Virot, R. (1956). La végétation canaque. *Mémoires du Muséum national d'Histoire naturelle, Paris, sér. B, Botanique*, **7**, 1–398.

Viseshakul, N., W. Charoennitikul, S. Kitamura, *et al.* (2011). A phylogeny of frugivorous hornbills linked to the evolution of Indian plants within Asian rainforests. *Journal of Evolutionary Biology*, **24**, 1533–1545.

Voelker, G., and J. Klicka. (2008). Systematics of *Zoothera* thrushes, and a synthesis of true thrush molecular systematic relationships. *Molecular Phylogenetics and Evolution*, **49**, 377–381.

Voelker, G., S. Rohwer, R. Bowie, and D.C. Outlaw. (2007). Molecular systematics of a speciose, cosmopolitan songbird genus: defining the limits of, and relationships among, the *Turdus* thrushes. *Molecular Phylogenetics and Evolution*, **42**, 422–434.

Voje, K.L., C. Hemp, O. Flagstad, G.-P. Sætre, and N.C. Stenseth. (2009). Climatic change as an engine for speciation in flightless Orthoptera species inhabiting African mountains. *Molecular Ecology*, **18**, 93–108.

Vorontsova, M.S., P. Hoffmann, H. Kathriarachchi, D.A. Kolterman, and M.W. Chase. (2007a). *Andrachne cuneifolia* (Phyllanthaceae; Euphorbiaceae s.l.) is a *Phyllanthus*. *Botanical Journal of the Linnean Society*, **155**, 519–525.

Vorontsova, M.S., P. Hoffmann, O. Maurin, and M.W. Chase. (2007b). Molecular phylogenetics of tribe Poranthereae (Phyllanthaceae; Euphorbiaceae sensu lato). *American Journal of Botany*, **94**, 2026–2040.

Vyverman, W., E. Verleyen, K. Sabbe, *et al.* (2007). Historical processes constrain patterns in global diatom diversity. *Ecology*, **88**, 1924–1931.

Waddell P.J., Y. Cao, M. Hasegawa, and D.P. Mindell. (1999). Assessing the Cretaceous super-ordinal divergence times within birds and placental mammals by using whole mitochondrial protein sequences and an extended statistical framework. *Systematic Biology*, **48**, 119–137.

Wade, C.M., P.B. Mordan, and F. Naggs. (2006). Evolutionary relationships among the pulmonate land snails and slugs (Pulmonata, Stylommatophora). *Biological Journal of the Linnean Society*, **87**, 593–610.

Wagner, M. (1873). *The Darwinian Theory and the Law of the Migration of Organisms*. London: Edward Stanford.

Wagstaff, S.J., and I. Breitwieser. (2004). Phylogeny and classification of *Brachyglottis* (Senecioneae, Asteraceae): an example of a rapid species radiation in New Zealand. *Systematic Botany*, **29**, 1003–1010.

Wagstaff, S.J., and B.R. Clarkson. (2012). Systematics and ecology of the Australasian genus *Empodisma* (Restionaceae) and description of a new species from peatlands in northern New Zealand. *PhytoKeys*, **2012** (13), 39–79.

Wagstaff, S.J., and M.I. Dawson. (2000). Classification, origin, and patterns of diversification of *Corynocarpus* (Corynocarpaceae) inferred from DNA sequences. *Systematic Botany*, **25**, 134–149.

Wagstaff, S.J., and F. Hennion. (2007). Evolution and biogeography of *Lyallia* and *Hectorella* (Portulacaceae), geographically isolated sisters from the southern hemisphere. *Antarctic Science*, **19**, 417–426.

Wagstaff, S.J., and J.A. Tate. (2011). Phylogeny and character evolution in the New Zealand endemic genus *Plagianthus* (Malveae, Malvaceae). *Systematic Botany*, **36**, 405–418.

Wagstaff, S.J., M.J. Bayly, P.J. Garnock-Jones, and D.C. Albach. (2002). Classification, origin, and diversification of the New Zealand hebes (Scrophulariaceae). *Annals of the Missouri Botanical Garden*, **89**, 38–63.

Wagstaff, S.J., I. Breitwieser, and U. Swenson. (2006). Origin and relationships of the austral genus *Abrotanella* (Asteraceae) inferred from DNA sequences. *Taxon*, **55**, 95–106.

Wagstaff, S.J., M.I. Dawson, S. Venter, *et al.* (2010a). Origin, diversification, and classification of the Australasian genus *Dracophyllum* (Richeeae, Ericaceae). *Annals of the Missouri Botanical Garden*, **97**, 235–2581

Wagstaff, S.J., K. Martinsson, and U. Swenson. (2000). Divergence estimates of *Tetrachondra hamiltonii* and *T. patagonica* (Tetrachondraceae) and their implications for austral biogeography. *New Zealand Journal of Botany*, **38**, 587–596.

Wagstaff, S.J., B.P.J. Molloy, and J.A. Tate. (2010b). Evolutionary significance of long-distance dispersal and hybridisation in the New Zealand endemic genus *Hoheria* (Malvaceae). *Australian Systematic Botany*, **23**, 112–130.

Wahlberg, N., and D. Rubinoff. (2011). Vagility across *Vanessa* (Lepidoptera: Nymphalidae): mobility in butterfly species does not inhibit the formation and persistence of isolated sister taxa. *Systematic Entomology*, **36**, 362–370.

Wall, S.P. (2005). Origin and rapid diversification of a tropical moss. *Evolution*, **59**, 1413–1424.

Wall, W.A., N.A. Douglas, Q.-Y. (J.) Xiang, *et al.* (2010). Evidence for range stasis during the latter Pleistocene for the Atlantic Coastal Plain endemic genus, *Pyxidanthera* Michaux. *Molecular Ecology*, **19**, 4302–4314.

Wallace, A.R. (1876). *The Geographical Distribution of Animals*. London: Macmillan.

Wallach, V., R.M. Brown, A.C. Diesmos, and G.V.A. Gee. (2007). An enigmatic new species of blind snake from Luzon Island, northern Philippines, with a synopsis of the genus *Acutotyphlops* (Serpentes: Typhlopidae). *Journal of Herpetology*, **41**, 690–702.

Wallis, G.P., and S.A. Trewick. (2001). Finding fault with vicariance: a critique of Heads (1998). *Systematic Biology*, **50**, 602–609.

Wallis, G.P., and S.A. Trewick. (2009). New Zealand phylogeography: evolution on a small continent. *Molecular Ecology*, **18**, 3548–3580.

Wandres, A.M., and J.D. Bradshaw. (2005). New Zealand tectonostratigraphy and implications from conglomeratic rocks for the configuration of the SW Pacific margin of Gondwana. *Geological Society of London Special Publications*, **246**, 179–216.

Wang, H., L. Qiong, K. Sun, *et al.* (2010). Phylogeographic structure of *Hippophae tibetana* (Elaeagnaceae) highlights the highest microrefugia and the rapid uplift of the Qinghai–Tibetan Plateau. *Molecular Ecology*, **19**, 2964–2979.

Wanntorp, L., V. Vajda, and J.I. Raine. (2011). Past diversity of Proteaceae on subantarctic Campbell Island, a remote outpost of Gondwana. *Cretaceous Research*, **32**, 357–367.

Ward, J.B., R.A.B. Leschen, B.J. Smith, and J.C. Dean. (2004). Phylogeny of the caddisfly (Trichoptera) family Hydrobiosidae using larval and adult morphology, with the description of a new genus and species from Fiordland, New Zealand. *Records of the Canterbury Museum*, **18**, 23–43.

Ware, J.L., J.P. Simaika, and M.J. Samways. (2009). Biogeography and divergence time estimation of the relict Cape dragonfly genus *Syncordulia*: global significance and implications for conservation. *Zootaxa*, **2216**, 22–36.

Warnock, R.C.M., Z. Yang, and P.C.J. Donoghue. (2012). Exploring uncertainty in the calibration of the molecular clock. *Biology Letters*, **8**, 156–159.

Wasson, R.J. (1982). Landform development in Australia. In *Evolution of the Flora and Fauna of Arid Australia*, eds. W.R. Barker and P.J.M. Greenslade. Frewville, SA: Peacock, pp. 23–34.

Waters, J.M., and D. Craw. (2006). Goodbye Gondwana? New Zealand biogeography, geology, and the problem of circularity. *Systematic Biology*, **55**, 351–356.

Waters, J.M., G.A. McCulloch, and J.A. Eason. (2007). Marine biogeographical structure in two highly dispersive gastropods: implications for trans-Tasman dispersal. *Journal of Biogeography*, **34**, 678–687.

Watson, C. (2009). A new species of *Clavisyllis* Knox 1957 (Polychaeta: Syllidae): a genus with the unusual distribution of New Zealand and the Great Barrier Reef, northern Queensland. *Beagle: Records of the Museums and Art Galleries of the Northern Territory*, **25**, 77–84.

Webb, C.O., and R. Ree. (2012). Historical biogeography inference in Malesia. In *Biotic Evolution and Environmental Change in Southeast Asia*, eds. D. Gower, K. Johnson, J. Richardson, *et al.* Cambridge: Cambridge University Press.

Weber, A. (2004a). Gesneriaceae and Scrophulariaceae: Robert Brown and now. *Telopea*, **10**, 543–571.

Weber, A. (2004b). Gesneriaceae. In *The Families and Genera of Vascular Plants. Vol. 7. Flowering Plants, Dicotyledons: Lamiales (except Acanthaceae including Avicenniaceae)*, ed. J.W. Kadereit. Berlin: Springer, pp. 63–158.

Weese, T.L., and L. Bohs, (2010). A three-gene phylogeny of the genus *Solanum* (Solanaceae). *Systematic Botany*, **32**, 445–463.

Weimarck, H. (1934). *Monograph of the Genus* Cliffortia. Lund: Kåkan Ohlsson.

Welton, L.J., C.D. Siler, D. Bennett, *et al.* (2010). A spectacular new Philippine monitor lizard reveals a hidden biogeographic boundary and a novel flagship species for conservation. *Biology Letters*, **6**, 654–658.

Westergaard, K.B., I.G. Alsos, M. Popp, *et al.* (2011). Glacial survival may matter after all: nunatak signatures in the rare European populations of two west-arctic species. *Molecular Ecology*, **20**, 376–393.

Westerman, M., B.P. Kear, K. Aplin, *et al.* (2012). Phylogenetic relationships of living and recently extinct bandicoots based on nuclear and mitochondrial DNA sequences. *Molecular Phylogenetics and Evolution*, **62**, 97–108.

Westerman, M., J. Young, and C. Krajewski. (2008). Molecular relationships of species of *Pseudantechinus, Parantechinus* and *Dasykaluta* (Marsupialia: Dasyuridae). *Australian Mammalogy*, **29**, 201–212.

Weston, P.H. (2007). Proteaceae. In *The Families and Genera of Vascular Plants. Vol. 9. Flowering Plants, Eudicots*, ed. K. Kubitzki. Berlin: Springer, pp. 364–404.

Whattam, S.A. (2009). Arc-continent collisional orogenesis in the SW Pacific and the nature, source and correlation of emplaced ophiolitic nappe components. *Lithos*, **113**, 88–114.

Whattam, S.A., J. Malpas, J.R. Ali, C.-H. Lo, and I.E.M. Smith. (2005). Formation and emplacement of the Northland ophiolite, northern New Zealand: SW Pacific tectonic implications. *Journal of the Geological Society, London*, **162**, 225–241.

Whattam, S.A., J.G. Malpas, J.R. Ali, I.E.M. Smith, and C.-H. Lo. (2004). Origin of the Northland Ophiolite, northern New Zealand: discussion of new data and reassessment of the model. *New Zealand Journal of Geology and Geophysics*, **47**, 383–389.

Whinnett, A., M. Zimmermann, K.R. Willmott, *et al.* (2005). Strikingly variable divergence times inferred across an Amazonian butterfly 'suture zone'. *Proceedings of the Royal Society, London, B*, **272**, 2525–2533.

White, G. [1789] (1977). *The Natural History of Selborne*. London: Penguin.

White, N.E., M.J. Phillips, M.T.P. Gilbert, *et al.* (2011). The evolutionary history of cockatoos (Aves: Psittaciformes: Cacatuidae). *Molecular Phylogenetics and Evolution*, **59**, 615–622.

Wiens, J.J., C.A Kuczynski, S.A. Smith, *et al.* (2008). Branch length, support, and congruence. Testing the phylogenomic approach with 20 nuclear loci in snakes. *Systematic Biology*, **57**, 420–431.

Wikström, N., M. Avino, S.G. Razafimandimbison, and B. Bremer. (2010). Historical biogeography of the coffee family (Rubiaceae, Gentianales) in Madagascar: case studies from the tribes Knoxieae, Naucleeae, Paederieae and Vanguerieae. *Journal of Biogeography*, **37**, 1094–1113.

Wikström, N., V. Savolainen, and M.W. Chase. (2001). Evolution of the angiosperms: calibrating the family tree. *Proceedings of the Royal Society, London, B*, **268**, 2211–2220.

Wiley, E.O. (1988). Vicariance biogeography. *Annual Review of Ecology and Systematics*, **19**, 513–542.

Wilkinson, R.D., M.E. Steiper, C. Soligo, *et al.* (2011). Dating primate divergences through an integrated analysis of palaeontological and molecular data. *Systematic Biology*, **60**, 16–31.

Willdenow, K. (1798). *Grundriss des Kräuterkunde*. Berlin: Haude und Spener.

Williams, D.J., M. O'Shea, R.L. Daguerre, *et al.* (2008). Origin of the eastern brownsnake, *Pseudonaja textilis* (Duméril, Bibron and Duméril) (Serpentes: Elapidae: Hydrophiinae) in

New Guinea: evidence of multiple dispersals from Australia, and comments on the status of *Pseudonaja textilis pughi* Hoser 2003. *Zootaxa*, **1703**, 47–61.

Wilson, A.J.G. (ed.). (1999). *Flora of Australia, vol. 17B. Proteaceae part 3, Hakea to Dryandra.* Melbourne: Australian Biological Resources Study.

Wilson, G.D.F. (2008). Gondwanan groundwater: subterranean connections of Australian phreatoicidean isopods (Crustacea) to India and New Zealand. *Invertebrate Systematics*, **22**, 301–310.

Wilson, G.D.F., and R.T. Johnson. (1999). Ancient endemism among freshwater isopods (Crustacea, Phreatoicidea). In *The Other 99%: The Conservation and Biodiversity of Invertebrates*, eds. W. Ponder and D. Lunney. Mosman: Royal Zoological Society of New South Wales, pp. 264–268.

Wilson, P.G. (1996). Myrtaceae in the Pacific, with special reference to Metrosideros. In *The Origin and Evolution of Pacific Island Biotas, New Guinea to Eastern Polynesia: Patterns and Processes*, eds. A. Keast and S.E. Miller. Amsterdam: SPB Academic Publishing, pp. 233–245.

Wilson, S., and G. Swan. (2003). *A Complete Guide to Reptiles of Australia.* Sydney: Reed New Holland.

Winkworth, R.C., and M.J. Donoghue. (2005). *Viburnum* phylogeny based on combined molecular data: implications for taxonomy and biogeography. *American Journal of Botany*, **92**, 653–666.

Winkworth, R.C., J. Grau, A.W. Robertson, and P.J. Lockhart. (2002a). The origins and evolution of the genus *Myosotis* L. (Boraginaceae). *Molecular Phylogenetics and Evolution*, **24**, 180–193.

Winkworth, R.C., J. Lundberg, and M.J. Donoghue. (2008). Toward a resolution of campanulid phylogeny, with special reference to the placement of Dipsacales. *Taxon*, **57**, 53–65.

Winkworth, R.C., S.J. Wagstaff, D. Glenny, and P.J. Lockhart. (2002b). Plant dispersal N.E.W.S. from New Zealand. *Trends in Ecology and Evolution*, **17**, 514–520.

Woo, V.L., M.M. Funke, J.F. Smith, P.J. Lockhart, and P.J. Garnock-Jones. (2011). New World origins of southwest Pacific Gesneriaceae: multiple movements across and within the South Pacific. *International Journal of Plant Science*, **172**, 434–457.

Woodruff, R.E. (2009). A new fossil species of stag beetle from Dominican Republic amber, with Australasian connections (Coleoptera: Lucanidae). *Insecta Mundi*, **98**, 1–10.

Worthington, T.J., R. Hekinian, P. Stoffers, T. Kuhn, and F. Hauff. (2006). Osbourn Trough: structure, geochemistry, and implications of a mid-Cretaceous paleospreading ridge in the South Pacific. *Earth and Planetary Science Letters*, **245**, 685–701.

Worthy, T.H. (1987). Osteology of *Leiopelma* (Amphibia, Leiopelmatidae) and descriptions of three new sub-fossil *Leiopelma* species. *Journal of the Royal Society of New Zealand*, **17**, 201–251.

Worthy, T.H., and G.M. Wragg. (2008). A new genus and species of pigeon (Aves: Columbidae) from Henderson Island, Pitcairn group. In *Islands of Inquiry: Colonisation, Seafaring and the Archaeology of Maritime Landscapes*, eds. G. Clark, F. Leach, and S. O'Connor. Canberra: Australian National University Press, pp. 499–510.

Worthy, T.H., S.J. Hand, J.M.T. Nguyen, *et al.* (2010). Biogeographical and phylogenetic implications of an Early Miocene wren (Aves: Passeriformes: Acanthisittidae) from New Zealand. *Journal of Vertebrate Paleontology*, **30**, 479–498.

Worthy, T.H., A.J.D. Tennyson, M. Archer, *et al.* (2006). Miocene mammal reveals a Mesozoic ghost lineage on insular New Zealand, southwest Pacific. *Proceedings of the National Academy of Sciences, USA*, **103**, 19419–19423.

Wright, T.F., E.E. Schirtzinger, T. Matsumoto, *et al.* (2008). A multilocus molecular phylogeny of the parrots (Psittaciformes): support for a Gondwanan origin during the Cretaceous. *Molecular Biology and Evolution*, **25**, 2141–2156.

Wurdack, K.J. (2009). The South American genera of Hemerocallidaceae (*Eccremis* and *Pasithea*): two introductions to the New World. *Taxon*, **58**, 1122–1134.

Wurdack, K.J., P. Hoffmann, and M.W. Chase. (2005). Molecular phylogenetic analysis of uniovulate Euphorbiaceae (Euphorbiaceae *sensu stricto*) using plastid *rbcL* and *trnL-F* DNA sequences. *American Journal of Botany*, **92**, 1397–1420.

Wüster, W., A.J. Dumbrell, C. Hay, *et al.* (2005). Snakes across the Strait: trans-Torresian phylogeographic relationships in three genera of Australasian snakes (Serpentes: Elapidae: *Acanthophis*, *Oxyuranus*, and *Pseudechis*). *Molecular Phylogenetics and Evolution*, **34**, 1–14.

Yan, J., H. Li, and K. Zhou. (2008). Evolution of the mitochondrial genome in snakes: gene rearrangements and phylogenetic relationships. *BMC Genomics*, **9** (569), 1–7.

Yang, Y., and P.E. Berry. (2007). Phylogenetics and evolution of *Euphorbia* subgenus *Chamaesyce* (Euphorbiaceae). *Botany and Plant Biology Joint Congress 2007*. http://2007.botanyconference.org/

Yin, A. (2010). Cenozoic tectonic evolution of Asia: a preliminary synthesis. *Tectonophysics*, **488**, 293–325.

Yokoyama, J., M. Suzuki, K. Iwatsuki, and M. Hasebe. (2000). Molecular phylogeny of *Coriaria*, with special emphasis on the disjunct distribution. *Molecular Phylogenetics and Evolution*, **14**, 1055–1073.

Yu, S.B., Y.-J. Hsu, T. Bacolcol, *et al.* (2013). Present-day crustal deformation along the Philippine Fault in Luzon, Philippines. *Journal of Asian Earth Sciences*, **65**, 64–74.

Yuan, Y.-M., S. Wohlhauser, M. Möller, *et al.* (2005). Phylogeny and biogeography of *Exacum* (Gentianaceae): a disjunctive distribution in the Indian Ocean Basin resulted from long distance dispersal and extensive radiation. *Systematic Biology*, **54**, 21–34.

Yumul, G.P., Jr., C.B. Dimalanta, V.B. Maglambayan, and E.J. Marquez. (2008). Tectonic setting of a composite terrane: a review of the Philippine island arc system. *Geosciences Journal*, **12**, 7–17.

Yumul, G.P., Jr., C.B. Dimalanta, K. Queaño, and E.J. Marquez. (2009a). Philippines, geology. In *Encyclopedia of Islands*, eds. R. Gillespie and D. Clague. Berkeley, CA: University of California Press, pp. 732–738.

Yumul, G.P., Jr., C.B. Dimalanta, E.J. Marquez, and K.L. Queaño. (2009b). Onland signatures of the Palawan microcontinental block and Philippine mobile belt collision and crustal growth process: a review. *Journal of Asian Earth Sciences*, **34**, 610–623.

Zamoras, L.R., M.G.A. Montes, K.L. *et al.* (2008). Buruanga Peninsula and Antique Range: two contrasting terranes in Northwest Panay, Philippines featuring an arc–continent collision zone. *Island Arc*, **17**, 443–457.

Zerega, N.J.C., W.L. Clement, S.L. Datwyler, and G.D. Weiblen. (2005). Biogeography and divergence times in the mulberry family (Moraceae). *Molecular Phylogenetics and Evolution*, **37**, 402–416.

Zhang, M. (2011). A cladistic scenario of Southern Pacific biogeographical history based on *Nothofagus* dispersal and vicariance analysis. *Journal of Arid Land*, **3**, 104–113.

Zhou, L., Y.C.F. Su, D.C. Thomas, and R.M.K. Saunders. (2012). 'Out-of-Africa' dispersal of tropical floras during the Miocene climatic optimum: evidence from *Uvaria* (Annonaceae). *Journal of Biogeography*, **39**, 322–335.

Zielske, S., M. Glaubrecht, and M. Haase. (2011). Origin and radiation of rissooidean gastropods (Caenogastropoda) in ancient lakes of Sulawesi. *Zoologica Scripta*, **40**, 221–237.

Zirakparvar, N.A., S.L. Baldwin, and J.D. Vervoort. (2013). The origin and geochemical evolution of the Woodlark Rift of Papua New Guinea. *Gondwana Research*, **23**, 931–943.

Zona, S., J. Francisco-Ortega, B. Jestrow, W.J. Baker, and C.E. Lewis. (2011). Molecular phylogenetics of the palm subtribe Ptychospermatinae (Arecaceae). *American Journal of Botany*, **98**, 1716–1726.

Zuccon, D., and P.G.P. Ericson. (2012). Molecular and morphological evidences place the extinct New Zealand endemic *Turnagra capensis* in the Oriolidae. *Molecular Phylogenetics and Evolution*, **62**, 414–426.

Index